Birkhäuser

Monographs in Mathematics

Volume 99

The foundations of the Monographs in Mathematics Birkhäuser book series were laid in 1944. Until the end of the 1970s, a total of 77 volumes appeared in this outstanding series, including works of distinguished mathematicians such as Carathéodory, Nevanlinna and Shafarevich, to name a few. The series came to its name and present appearance in the 1980s. In keeping its well-established tradition, only monographs of excellent quality are published in this collection. Comprehensive, in-depth treatments of areas of current interest are presented to a readership ranging from graduate students to professional mathematicians. Concrete examples and applications both within and beyond the immediate domain of mathematics illustrate the import and consequences of the theory under discussion.

Yoshikazu Giga • Qing Liu

Surface Evolution Equations

A Level Set Approach

Second Edition

Birkhäuser

Yoshikazu Giga
Graduate School of Mathematical Sciences
University of Tokyo
Tokyo, Japan

Qing Liu
Geometric Partial Differential Equations Unit
Okinawa Institute of Science and Technology
Okinawa, Japan

ISSN 1017-0480 ISSN 2296-4886 (electronic)
Monographs in Mathematics
ISBN 978-3-032-20302-1 ISBN 978-3-032-20303-8 (eBook)
https://doi.org/10.1007/978-3-032-20303-8

Mathematics Subject Classification: 53-02, 35K10, 74N22, 53E10, 49L25, 35K55

This book is published under the imprint Birkhäuser, www.birkhauser-science.com by the registered company Springer Nature Switzerland AG
The registered company address is: Gewerbestrasse 11, 6330 Cham, Switzerland

Dedicated to Yoshi's late parents, Kazuyo and Kenjiro, and to Qing's parents, Youling and Zhongming

Preface to the Second Edition

In this new edition, we have added two new chapters related to time-discrete differential game approaches. For the reader's convenience, an appendix has been included to provide a proof of a modified version of the maximum principle for semicontinuous functions. At the end of each chapter, we briefly highlight several developments that have appeared since the publication of the first edition; these notes are not intended to be exhaustive. In addition, a collection of exercises has been added after Chap. 8 to help understanding of the material. We also took the opportunity to correct numerous typos and a few minor flaws from the first edition.

We are grateful to Dr. Longjie Zhang for carefully checking the details of the first edition and identifying many typographical errors as well as a couple of minor issues. We also thank Professor Nao Hamamuki for permitting us to include, as an appendix, a note he wrote during his undergraduate studies at the University of Tokyo. In addition, we are indebted to Professor Keisuke Takasao for his valuable suggestions on Brakke's solutions.

Last but not least, we are grateful to Ms. Satoko Kimura for preparing the second edition especially for her careful checking of the bibliography of this book.

Tokyo, Japan — Yoshikazu Giga
Okinawa, Japan — Qing Liu
December 2025

Declaration of Competing Interests The authors declare that they have no competing interests.

Preface to the First Edition

This book is intended to be a self-contained introduction to analytic foundations of a level set method for various surface evolution equations including curvature flow equations. These equations are important in various fields including material sciences, image processing and differential geometry. The goal of this book is to introduce a generalized notion of solutions allowing singularities and solve the initial-value problem globally-in-time in a generalized sense. Various equivalent definitions of solutions are studied. Several new results on equivalence are also presented.

We present here a rather complete introduction to the theory of viscosity solutions which is a key tool for the level set method. Also a self-contained explanation is given for general surface evolution equations of the second order. Although most of the results in this book are more or less known, they are scattered in several references, sometimes without proof. This book presents these results in a synthetic way with full proofs. However, the references are not exhaustive at all.

The book is suitable for applied researchers who would like to know the detail of the theory as well as its flavour. No familiarity with differential geometry and the theory of viscosity solutions is required. The prerequisites are calculus, linear algebra and some familiarity with semicontinuous functions. This book is also suitable for upper level under graduate students who are interested in the field.

I am grateful to Professor Herbert Amann for inviting me to write this book which is based on my Lipschitz lectures in Bonn 1997. I am also grateful to its audience for their interest. The first version of the book was included in Series of Lipschitz Lecture Notes as volume 44 (2002). It was also included in Hokkaido University Technical Report Series in Mathematics as volume 71 (2002). However, since then the author has been fully occupied with the Center of Excellence Programme 'Mathematics of Nonlinear Structures via Singularities' (Hokkaido University) and with editing the 4-th version of the Encyclopedic Dictionary of Mathematics. Moreover, the author moved to Tokyo from Sapporo in the middle of 2004. So it has taken a rather long time to complete the present version of this book. After the first version appeared, the field continued to grow and many new articles have been published. They could not all be included without significant expansion of the text.

Although some effort was made to cite them, the reference list is not intended to be exhaustive.

I am grateful to Dr. Mi-Ho Giga, Professor Katsuyuki Ishii, Professor Masaki Ohnuma, Dr. Takeshi Ohtsuka, Professor Reiner Schätzle and Professor Kazuyuki Yamauchi for their critical remarks on an earlier version of this book. I am also grateful to Professor Naoyuki Ishimura who read the first version line by line and provided numerous useful suggestions. Without his careful reading it would have been almost impossible to prepare a final version. I am also grateful to anonymous referees for their constructive remarks on the first version. The major part of the book was written when the author was a faculty member of the Hokkaido University. I am grateful to my colleague in Hokkaido University for encouragement. The financial support of the Japan Society for the Promotion of Science (no. 10304010, 11894003, 12874024, 13894003, 14204011, 15634008, 17654037), the formation of COE 'Mathematics of Nonlinear Structures via Singularities' (Hokkaido University) is gratefully acknowledged. Finally I am grateful to Ms. Hisako Morita (neé Iwai) and Ms. Mika Marubishi for careful typing of respectively, the first and the final version of the manuscripts in latex style.

Tokyo, Japan
October 2005

Y. Giga

Contents

Chapter 1
Introduction

In various fields of science there often arise phenomena in which phases (of materials) can coexist without mixing. A surface bounding the two phases is called a *phase boundary*, an *interface*, or *front* depending upon the situation. In the process of phase transition a phase boundary moves by thermodynamical driving forces. Since evolution of a phase boundary is unknown and it should be determined as a part of the solution, the problem including such a phase boundary is called in general a free boundary problem. The motion of a phase boundary between ice and water is a typical example, and it has been well studied—the Stefan problem. For classical Stefan problems the reader is referred to the books of L. I. Rubenšteĭn [384] and of A. M. Meirmanov [347]. The reader is referred to the book of A. Visintin [424] for free boundary problems related to phase transition. In the Stefan problem, evolution of a phase boundary is affected by the physical situation of the exterior of the surface. However, there is a special but important class of problems where evolution of a phase boundary does not depend on the physical situation outside the phase boundary but only on its geometry. The equation that describes such motion of the phase boundary is called a *surface evolution equation* or *geometric evolution equations*. There are several examples in material sciences and the equation is also called an *interface controlled model*. Examples are not limited to material sciences. Some of them come from geometry, crystal growth problems, and image processing. An important subclass of surface evolution equations consists of equations that arise when the normal velocity of the surface depends locally on its normal and the second fundamental form as well as on position and time. In this book we describe an analytic foundation of the level set method which is useful to analyze such surface evolution equations including the mean curvature flow equation as a typical example. We intend to give a systematic and synthetic approach since the results are scattered in the literature, although there are several review articles, in particular a lecture note by L. Ambrosio and N. Dancer [12]. This book also includes several new results on barrier solutions (Chap. 6).

Y. Giga, Q. Liu, *Surface Evolution Equations*, Monographs in Mathematics 99,
https://doi.org/10.1007/978-3-032-20303-8_1

We consider a family $\{\Gamma_t\}_{t\geq 0}$ of hypersurfaces embedded in N-dimensional Euclidean space $\mathbf{R}^N$ parametrized by time t. We assume that Γ_t is a compact hypersurface so that Γ_t is given as a boundary of a bounded open set D_t in $\mathbf{R}^N$ by Jordan–Brouwer's decomposition theorem. Physically, we regard Γ_t as a phase boundary bounding D_t and $\mathbf{R}^N \backslash D_t$ each of which is occupied by different phases. To write down a surface evolution equation we assume that Γ_t is smooth and changes its shape smoothly in time. Let $\mathbf{n}$ be a unit normal vector field of Γ_t outward from D_t. Let $V = V(x, t)$ be the *normal velocity* in the direction of $\mathbf{n}$ at a point of Γ_t. If V depends locally on normal $\mathbf{n}$ and the second fundamental form $-\nabla\mathbf{n}$ of Γ_t, as well as on position x and time t, a general form of surface evolution equation is

$$V = f(x, t, \mathbf{n}, \nabla\mathbf{n}) \quad \text{on } \Gamma_t, \tag{1.1.1}$$

where f is a given function. We list several examples of (1.1.1).

(1) Mean curvature flow equation: $V = H$, where H is the sum of all principal curvatures in the direction of $\mathbf{n}$ and is called the mean curvature throughout in this book (although many authors since Gauss call the average of principal curvatures the mean curvature). The mean curvature is expressed as $H = -\mathrm{div}\ \mathbf{n}$, where div is the surface divergence on Γ_t. This equation was first proposed by W. W. Mullins [360] to describe motion of grain boundaries in annealing metals.
(2) Gaussian curvature flow equation: $V = K$, where K is the Gaussian curvature of Γ_t, that is, the product of all principal curvatures in the direction of $\mathbf{n}$. For this problem we take $\mathbf{n}$ inward so that a sphere shrinks to a point in a finite time if it evolves by $V = K$. This equation was proposed by W. J. Firey [175] to describe shapes of rocks on the seashore.
(3) General evolutions of isothermal interface:

$$\beta(\mathbf{n})V = -a\ \mathrm{div}\ \xi(\mathbf{n}) - c(x, t), \tag{1.1.2}$$

where β is a given positive function on a unit sphere S^{N-1} and a is nonnegative constant and c is a given function. The quantity ξ is the *Cahn–Hoffman vector* defined by the gradient of a given nonnegative positively homogeneous function γ of degree 1, i.e., $\xi = \nabla\gamma$ in $\mathbf{R}^N$. In problems of crystal growth we should often consider the anisotropic property of the surface structure of phase boundaries; in one direction the surface is easy to grow, but in the other direction it is difficult to grow. This kind of thing often happens. Equation (1.1.2) includes this effect and was derived by M. E. Gurtin [244, 245] and by S. Angenent and M. E. Gurtin [27] from the fundamental laws of thermodynamics and the balance of forces. Note that if $\gamma(p) = |p|$ and $\beta(p) \equiv 1$ with $c \equiv 0$, $a \equiv 1$, then (1.1.2) becomes $V = H$. If $a = 0$, Eq. (1.1.2) becomes simpler:

$$V = -c/\beta(\mathbf{n}) \quad \text{on } \Gamma_t. \tag{1.1.3}$$

This equation is a kind of Hamilton–Jacobi equation. If $\beta \equiv 1$ and $c < 0$ is a constant, this equation describes the wave front propagation based on Huygens' principle.

(4) Affine curvature flow equations: $V = K^{1/(N+1)}$ or $V = (tK)^{1/(N+1)}$ which were axiomatically derived by L. Alvarez et al. [11] for applications in image processing. The feature of these equations is that they are invariant by affine transform of coordinates. For this problem we take $\mathbf{n}$ inward as for the Gaussian curvature flow equation.

Examples of surface evolution equations are provided by the singular limit of reaction-diffusion equations as many authors have studied. See, for example, papers of Xu-Yan Chen [104] and Xinfu Chen [102].

As we summarized above, surface evolution equations are by now very popular among various branches of sciences, especially in image processing. The reader is referred to, for example, books of G. Sapiro [386], F. Guichard and J.-M. Morel [243], F. Cao [82], and R. Kimmel [307] for applications of equations in image processing.

A fundamental question of analysis is to construct a unique family $\{\Gamma_t\}_{t \geq 0}$ satisfying (1.1.1) for given initial hypersurface Γ_0 in $\mathbf{R}^N$. In other words it is the question whether there exists a unique solution $\{\Gamma_t\}_{t \geq 0}$ of the initial value problem for (1.1.1) with $\Gamma_t|_{t=0} = \Gamma_0$. This problem is classified as unique existence of a local solution or of a global solution depending on whether one can construct a solution of (1.1.1) in a short time interval or for infinite time. If Eq. (1.1.1) is strictly parabolic in a neighborhood of initial hypersurface Γ_0, then there exists a unique local smooth solution $\{\Gamma_t\}$ for given initial data provided that the dependence of variables in f is smooth. It applies to the mean curvature flow equation and its generalization (1.1.2) with $a > 0$ and smooth β and c for general initial data Γ_0 provided that the Frank diagram

$$\text{Frank}\, \gamma = \{p \in \mathbf{R}^N;\ \gamma(p) \leq 1\} \tag{1.1.4}$$

has a smooth, strictly convex boundary in the sense that all inward principal curvatures are positive. For the Gaussian curvature flow equations and the affine curvature flow equation the equation may not be parabolic for general initial data. It resembles solving the heat equation backward in time, so for general initial data it is not solvable. However, if we restrict ourselves to strictly convex initial surfaces, the problems are strictly parabolic around the initial surfaces and locally uniquely solvable. A standard method to construct a unique local solution is to analyze an equation of a "height" function, where the evolving surface is parametrized by the height (or distance) from the initial surface. See, for example, a paper by Xu-Yan Chen [104], where he discussed (1.1.2) with $\gamma(p) = |p|$, $\beta \equiv 1$, $a = 1$. The major machinery is the classical parabolic theory in a book of O. A. Ladyženskaja et al. [317] since the equation of a height function is a strictly parabolic equation of second order (around zero height), although it is nonlinear. For (1.1.3) the equation of a height function is of first order so a local smooth solution can be constructed by a method of characteristics. However, as we see later, such a local smooth solution may cease to be smooth in a finite time, and singularities may develop even for the mean curvature flow equation where a lot of regularizing effects exist. Since the phenomena may continue after a solution ceases to be smooth, it is natural to continue the solution by generalizing the notion of solution. The *level set method* is a powerful tool in constructing global solutions allowing singularities. It also provides a correct notion of solutions, if (1.1.1) is degenerate parabolic but not of first order so that a "solution" may lose smoothness even instantaneously and that a smooth local solution may not exist for smooth initial data. The analytic foundation of this theory was established independently by Y.-G. Chen et al. [105, 106] and by L. C. Evans and J. Spruck [166], where the latter work concentrated on the mean curvature flow equation while the former work handled a more general equation of form (1.1.1). These works were preceded by a numerical computation by S. Osher and J. A. Sethian [376] using a level set method. (We point out that there are two preceding works implicitly related to the level set method for first order problems. One is by G. Barles [44] and the other one is by L. C. Evans and P. E. Souganidis [165].) Since the analytic foundation was established, a huge number of articles on the level set method has been published. One purpose of this book is to explain the analytic foundation of the level set method in a systematic and synthetic way so that the reader can access the field without scratching for references. For the development of numerical aspects of the theory the reader is referred to a book of J. A. Sethian [393] and the book of S. Osher and R. Fedkiw [374]. Before explaining the level set method we discuss phenomena of formation of singularities for solutions of surface evolution equations.

Formation of Singularities We consider the mean curvature flow equation $V = H$. If the initial surface Γ_0 is a sphere of radius R_0, a direct calculation shows that the sphere with radius $R(t) = (R_0^2 - 2(N-1)t)^{1/2}$ is the exact solution. At time $t_* = R_0^2/(2(N-1))$ the radius of the sphere is zero so it is natural to interpret Γ_t as the empty set after t_*. If Γ_0 is a smooth, compact, and convex hypersurface, G. Huisken [259] showed that the solution Γ_t with initial data Γ_0 remains smooth, compact, and convex until it shrinks to a "round point" in a finite time; the asymptotic shape of Γ_t just before it disappears is a sphere. He proved this result for hypersurfaces of $\mathbf{R}^N$ with $N \geq 3$ but his method does not apply to the case $N = 2$. Later M. Gage and R. S. Hamilton [182] showed that it still holds when $N = 2$, for simple convex curves in the plane. The methods used by G. Huisken [259] and M. Gage and R. S. Hamilton [182] do not resemble each other. M. Gage and R. S. Hamilton [182] also observed that any smooth family of plane-immersed curves moved by its curvature remains embedded if it is initially embedded. M. A. Grayson [241] proved the remarkable fact that such a family must become convex before it becomes singular. Thus, in the plane $\mathbf{R}^2$ if the initial data Γ_0 is a smooth, compact, (and embedded) curve, then the solution Γ_t remains smooth (and embedded), becomes convex in a finite time, and remains convex until it shrinks to a "round point." The situation is quite different for higher dimensions. While it is still true that smooth immersed solutions remain embedded if their initial data is embedded, M. A. Grayson [242] also showed that there exist smooth solutions Γ_t that become singular before they shrink to a point. His example consisted of a barbell: two spherical surfaces connected by a sufficiently thin "neck" (Fig. 1.1). In this example the inward curvature of the neck is so large that it will force the neck to pinch before the two spherical ends can shrink appreciably. In this example the surface has genus zero but there also is an example of the surface with genus 1 that becomes singular in a finite time. For example, if the initial data Γ_0 is a fat doughnut (Fig. 1.2), then Γ_t become singular before they shrink to a point, since the outward curvature of the hole is so large that it will force the hole to pinch. Of course if a doughnut is thin and axisymmetric it is known that it shrinks to a ring, as shown by K. Ahara and N. Ishimura [3]. See also a paper of K. Smoczyk [396] for symmetric surfaces under rotation fixing $\mathbf{R}^m$ with $m \geq 2$.

We consider (1.1.3) with $\beta \equiv 1$ and $c \equiv -1$ so that (1.1.3) is $V = 1$. For this equation it is natural to think that the "solution" Γ_t with initial data Γ_0 equals the set of all points x with $\operatorname{dist}(x, \Gamma_0) = t$ that lies in the positive direction $\mathbf{n}$ from Γ_0. If initial Γ_0 has a "hollow", then Γ_t becomes singular in a finite time even if Γ_0 is smooth. The solution Γ_t given by the distance function provides a natural candidate for an extension of a smooth solution after singularities develop. For the second order equation including the mean curvature flow, it is not expected that Γ_t can be expressed by a simple explicit function like a distance function. So one introduces a generalized notion of solutions after singularities develop (Fig. 1.3).

Fig. 1.1 Barbell with thin neck

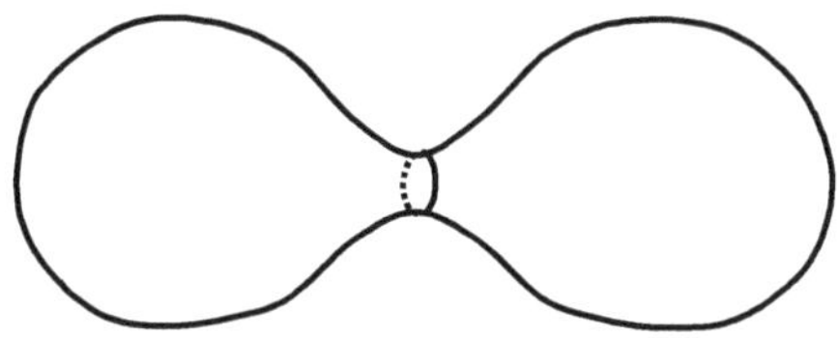

Fig. 1.2 Fat doughnut

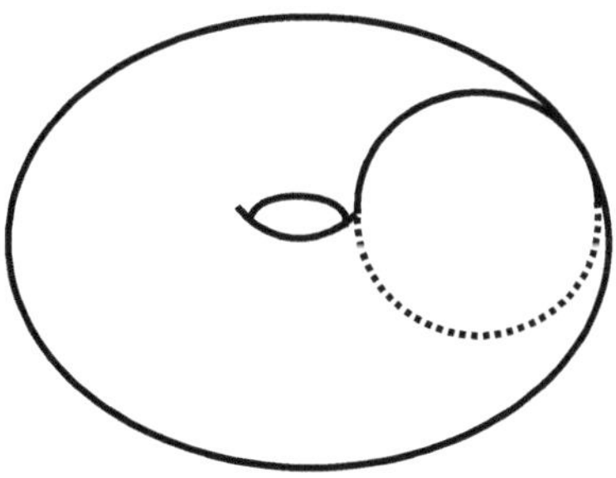

Fig. 1.3 Formation of singularity

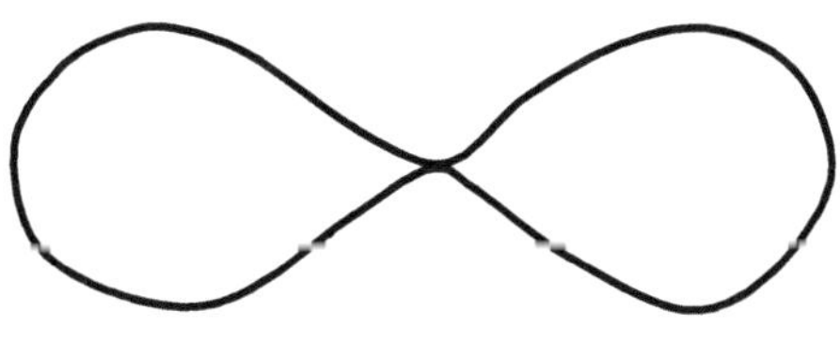

For the Gaussian curvature and the affine curvature flow equations as for the mean curvature flow equation if Γ_0 is a strictly convex smooth hypersurface, then Γ_t is a strictly convex smooth hypersurface and stays convex until it shrinks to a point. This was proved by K. Tso [420] and B. Andrews [17]. (For the affine curvature flow equation the shrinking pattern may not be a "round point" since the equation is affine invariant so that the solution is a shrinking ellipse.) If Γ_0 is convex but not strictly convex, these equations are degenerate parabolic, so one should be afraid that Γ_t may lose smoothness even instantaneously. For (1.1.2) if Frank γ is merely convex, then it actually happens that Γ_t ceases to be C^1 instantaneously even if initial data Γ_0 is smooth as shown in the work of Y. Giga [199]. We need a generalized notion of solution to track these phenomena.

Tracking Solutions After Formation of Singularities Suppose that singularities develop when a hypersurface evolves by a surface evolution equation. In what way should one construct the solution after that? From the geometric point of view, one method may be to classify all possible singularities which may appear and then construct the solution restarting with the shape having such singularities as initial data. This method has a drawback when it is difficult to classify singularities. In fact, even for the mean curvature flow equation, the classification of singularities is not completed. (We do not pursue this topic in the present version of the book. The reader is referred to the review by Y. Giga [200].) On the other hand, in analysis it is standard to introduce a suitable generalized notion of solutions so that it allows singularities. Such a solution is called a weak solution or generalized solution (Fig. 1.3).

The notion of generalized solution for the mean curvature flow equation was first introduced by K. A. Brakke [67]. He showed, how, using geometric measure theory, one can construct a theory of generalized solutions for variables in $\mathbf{R}^N$ of arbitrary dimension and co-dimension. His generalized solution is a kind of family of measure supported in Γ_t that satisfies transport inequality in a generalized sense. He proved the existence of at least one global solution for any initial data. In fact, he showed in an example that one initial datum may have many different solutions. This feature of his theory is related to the *fattening phenomena* in the level set method.

The second way to try to construct weak solutions is to study the singular limit of reaction-diffusion equations approximating the surface evolution equations. For the mean curvature flow equation, we consider the semilinear heat equation called the Allen–Cahn equation of the form

$$\frac{\partial u}{\partial t} - \Delta u + \frac{W'(u)}{\varepsilon^2} = 0 \quad \text{in } \mathbf{R}^N \times (0, \infty) \tag{1.1.5}$$

which was introduced by S. M. Allen and J. W. Cahn [8] to describe the motion of the anti-phase boundary in material sciences. Here W is a function that has only two equal minima; its typical form is $W(v) = \frac{1}{2}(v^2 - 1)^2$ and W' denotes the derivative. In the Allen–Cahn equation (1.1.5), ε is a positive parameter. For given initial data we solve the Allen–Cahn equation globally in time and denote its solution by u^ε. Equation (1.1.5) is semilinear parabolic and there is no growing effect for the nonlinear term so it is globally solvable. We guess that if ε tends to 0, u^ε tends to the minimal values of W, i.e., -1 or 1. In fact it is easy to check this if there is no Laplace term Δu (unless its value is unstable stationary point of W). The boundary between regions where the limit equals -1 and 1 is known to evolve by the mean curvature flow equation up to constant multiplication. This is well-known, at least formally, by asymptotic analysis. In fact, if the mean curvature flow equation has a smooth solution, the convergence was proved rigorously, in a time interval where the smooth solution exists, by several authors including L. Bronsard and R. V. Kohn [69], P. de Mottoni and M. Schatzman [140], and Xinfu Chen [102] in various settings. It is expected that a weak solution is obtained as a limit of u^ε (as $\varepsilon \to 0$) since u^ε itself exists globally in time. This idea turns out to produce Brakke's generalized solution as shown by T. Ilmanen [268] and improved by H. M. Soner [401]. Extending these ideas, a Brakke's type generalized solution was also constructed by H. M. Soner [400] and Xinfu Chen [103], respectively, for the Mullins–Sekerka problem with or without kinetic undercooling as a singular limit of the phase-field model and of the Cahn–Hilliard equation. We won't touch these problems in this book but we note that this method is based on variational structure of problems and does not appeal to the *comparison principle* which says that the solution Γ_t is always enclosed by another solution $\Gamma_t{}'$ if initially Γ_0 is enclosed by $\Gamma_0{}'$. This is why this method works for higher order nonlocal surface evolution equation like the Mullins–Sekerka problem.

A level set method, which is the main topic of this book, is another analytic method to construct weak solutions. It is based on comparison principle (or order-preserving structure of the solutions). It does not depend on variational surface evolution equation (1.1.1) even if other methods fail to apply. For the mean curvature flow equation the relations among the above three methods were clarified by T. Ilmanen [268] preceded by the work of L. C. Evans et al. [164]. We shall explain these works after we explain the idea of the level set method.

Level Set Method To describe a hypersurface, one can represent it as the zero set of a function, i.e., the zero level set. Compared with the method of representation by local coordinates, there is an advantage since the zero level set is allowed to have singularities even if the function is smooth. In other words, a hypersurface with singularities may be represented by a smooth function. The idea of the level set method is to regard a hypersurface Γ_t as the zero level set of some auxiliary function $u : \mathbf{R}^N \times [0, \infty) \to \mathbf{R}$ and to derive an equation which guarantees that its zero level set will evolve by a surface evolution equation (1.1.1). Many equations will have this property, but if one requires that not only the zero level set but *also all* level sets of the function u evolves by the same surface evolution equation (1.1.1), then, as in Y. Giga and S. Goto [207], there exists a unique partial differential equation of the form

$$\frac{\partial u}{\partial t} + F(x, t, \nabla u, \nabla^2 u) = 0; \tag{1.1.6}$$

such an equation is called the *level set equation* of (1.1.1). Here ∇u denotes the spatial gradient and $\nabla^2 u$ denotes the Hessian; F is a function determined by the values of $(x, t, \nabla u(x,t), \nabla^2 u(x,t))$.

For example for the surface evolution equation $V = 1$ its level set equation is

$$\frac{\partial u}{\partial t} - |\nabla u| = 0, \tag{1.1.7}$$

once we take the orientation $\mathbf{n} = -\nabla u/|\nabla u|$ of each level set so that $V = \partial u/\partial t/|\nabla u|$. For a mean curvature flow equation its level set equation is

$$\frac{\partial u}{\partial t} - |\nabla u| \operatorname{div}\left(\frac{\nabla u}{|\nabla u|}\right) = 0. \tag{1.1.8}$$

The idea to represent hypersurfaces as level sets is of course common in differential geometry. In the present context it goes back to T. Ohta et al. [370] who used the level set equation (1.1.8) to derive a scaling law for "dynamic structure functions" from a physical point of view. There were some previous articles on combustion theory implicitly including (1.1.8), but it seems that their paper was the first one to use (1.1.8) effectively. The idea to use (1.1.8) to study motion by mean curvature numerically was used by S. Osher and J. A. Sethian [376].

A level set method for the initial value problem of (1.1.1) is summarized as follows.

1° For a given initial hypersurface Γ_0 which is the boundary of a bounded open set D_0, we take an auxiliary function u_0 which is at least continuous such that

$$\Gamma_0 = \{x \in \mathbf{R}^N;\ u_0(x) = 0\},\ D_0 = \{x \in \mathbf{R}^N;\ u_0(x) > 0\}.$$

(For convenience we often arrange so that u_0 equals a negative constant α outside some big ball.)

2° We solve the initial value problem globally-in-time for the level set equation (1.1.6) with initial condition $u(x, 0) = u_0(x)$.

3° We then set

$$\begin{aligned} \Gamma_t &= \{x \in \mathbf{R}^N;\ u(x,t) = 0\}, \\ D_t &= \{x \in \mathbf{R}^N;\ u(x,t) > 0\} \end{aligned} \tag{1.1.9}$$

and expect that Γ_t is a kind of generalized solution (Fig. 1.4).

The first step is easy. The second step is not easy since the level set equation is not very nice from the point of view of analysis. If (1.1.1) is parabolic, it is a parabolic equation but it is very degenerate. There is no diffusion effect normal to its level set since each level set of u moves independently of the others. Thus classical techniques and results in the theory of parabolic equations cannot be expected to apply. We do not expect to have a global smooth solution for (1.1.6) even if initial

Fig. 1.4 Auxiliary function

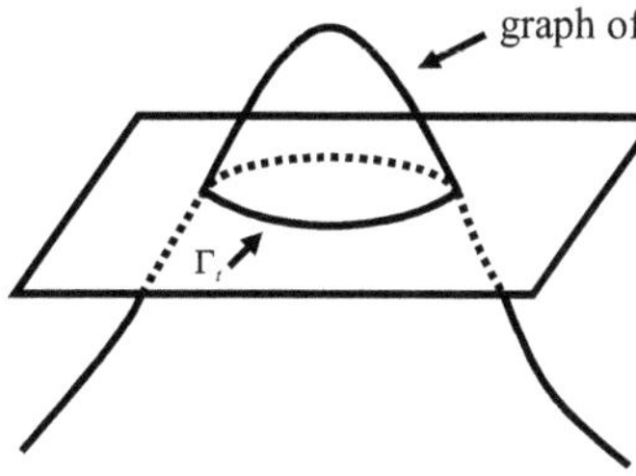

data is smooth. It is necessary to introduce the notion of weak solutions to (1.1.6). As known by Y.-G. Chen et al. [105, 106] for (1.1.6) and L. C. Evans and J. Spruck [166] for (1.1.8) the concept that fits this situation perfectly is a notion of viscosity solutions initiated by M. G. Crandall and P.-L. Lions [129, 130]. The reader is referred to the review article by M. G. Crandall et al. [128] for development of the theory of viscosity solutions. The theory of viscosity solutions applies nonlinear degenerate elliptic and parabolic single equations, including equations of first order where the comparison principle is expected. The key step of the theory is to establish a comparison principle for viscosity solutions. For (1.1.7) the theory for first order equations applies. For (1.1.8) the equation is singular at $\nabla u = 0$ which is a new aspect of problems in the theory of viscosity solutions. Since the mean curvature flow equation has a comparison principle or order-preserving properties for smooth solutions, the comparison principle for its level set equation is expected. It turns out that the extended theory of viscosity solutions yields a unique global continuous solution u of (1.1.6) with $u(x, 0) = u_0(x)$ (with the property $u(x, t) - \alpha$ compactly supported as a function of space variables for all $t \geq 0$) provided that (1.1.1) is degenerate parabolic and f in (1.1.1) does not grow superlinearly in $\nabla \mathbf{n}$. To apply this theory for the Gaussian curvature flow equation we need to extend the theory so that f is allowed to grow superlinearly in $\nabla \mathbf{n}$. This extension was done by S. Goto [239] and independently by H. Ishii and P. Souganidis [284]. We note that order-preserving structure of (1.1.1) is essential to get a global continuous solution to (1.1.6).

The method to construct Γ_t by 1°–3° is extrinsic. There is huge freedom to choose u_0 for given Γ_0. Although the solution u of (1.1.6) for given initial data u_0 is unique, we wonder whether Γ_t and D_t in (1.1.9) are determined by Γ_0 and D_0, respectively, independent of the choice of u_0. The problem is the uniqueness of the level set of the initial value problem for (1.1.6). Since F in (1.1.6) has a scaling property (called *geometricity*):

$$F(x, t, \lambda p, \lambda X + \sigma p \otimes p) = \lambda F(x, t, p, X)$$

for all $\lambda > 0, \sigma \in \mathbf{R}$, real symmetric matrix X, $p \in \mathbf{R}^N \backslash \{0\}$, $x \in \mathbf{R}^N$, $t \in [0, \infty)$, Eq. (1.1.6) has the invariance property: u solves (1.1.6) so does $\theta(u)$ for every nondecreasing continuous function θ in the viscosity sense. Using the invariance and the comparison principle, we get the uniqueness of level sets. In other words

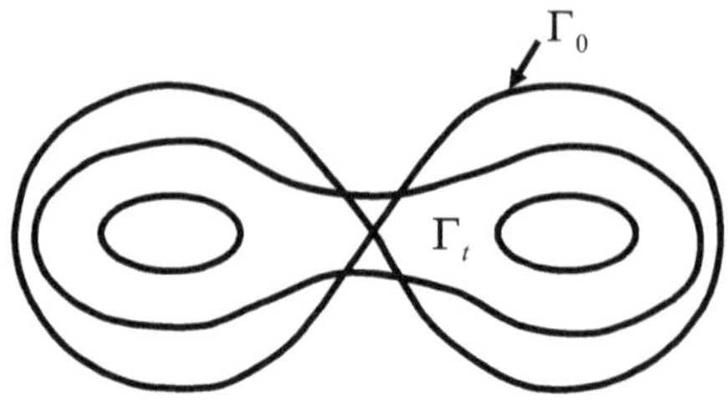

Fig. 1.5 Fattening

Γ_t and D_t in (1.1.9) are uniquely determined by Γ_0 and D_0, respectively. It is also possible to prove that Γ_t is an extended notion of a smooth solution.

Fattening One disturbing aspect of the solution Γ_t defined by (1.1.9) is that for $t > 0$ the solution Γ_t may have a nonempty interior even if the initial hypersurface is smooth, except for a few isolated singularities. An example is provided by L. C. Evans and J. Spruck [166] for the mean curvature flow equation, where it is argued that the solution in $\mathbf{R}^2$ whose initial shape is a *figure eight* has nonempty interior (Fig. 1.5). Such phenomena were studied by many authors in various settings and several sufficient conditions of nonfattening were provided. For the mean curvature flow problem it is observed that Γ_t may fatten (i.e., have no empty interior) even if the initial hypersurface is smooth, as numerically observed by S. Angenent et al. [29] for $N = 3$, and proved by S. B. Angenent et al. [30] for $4 \leq N \leq 8$. For $N = 3$ such an example is given by B. White [427] with a rigorous proof. We do not pursue this problem in this book. If we introduce the notion of set-theoretic solutions, the fattening phenomena can be interpreted as nonuniqueness of the set-theoretic solution. Such a notion was first introduced by H. M. Soner [399] for (1.1.1) when f is independent of x by using distance function from a set. It turns out to be more natural that a family of set Ω_t is a set-theoretic solution of (1.1.1) if the characteristic function χ_{Ω_t} is a solution of its level set solution in the viscosity sense. D_t and $D_t \cup \Gamma_t$ defined by (1.1.9) are typical examples of set-theoretic solutions. If fattening occurs, $\overline{D_t}$ and $D_t \cup \Gamma_t$ are essentially different and we have at least two solutions for given initial data D_0; here $\overline{D_t}$ denotes the cross section at t of the closure of the set where u is positive in $\mathbf{R}^N \times [0, \infty)$ and not the closure of D_t in $\mathbf{R}^N$. Intuitively, one can understand this fattening phenomena as follows. If we approximate Γ_0 from inside by a smooth hypersurface $\Gamma_i \to \Gamma_0$ $(i \to \infty)$, then taking the limit of corresponding solution $(\Gamma_i)_t$ as $i \to \infty$, we expect to have a solution with initial data Γ_0. In particular, one can approximate Γ_0 from the inside and obtain one solution, and one can approximate Γ_0 from the outside to obtain another solution. If these "inner" and "outer" solutions coincide, then it follows from the comparison principle that any sequence will have the same limiting solution. This corresponds to the case of nonfattening. If they are different, there is no preferred smooth solution, and Γ_t in (1.1.9) will consist of the entire region between the inner and outer solutions. Thus D_t is a minimal set-theoretic solution while $D_t \cup \Gamma_t$ is a maximal set theoretic solution.

There is another notion of set-theoretic solution called a barrier solution. Our solution D_t in (1.1.9) is a kind of barrier solution, since by the comparison principle any smooth evolving hypersurface Σ_t solving (1.1.1) remains contained in D_t or $\mathbf{R}^N \backslash \overline{D}_t$ for $t \geq t_0$ if Σ_{t_0} is contained in D_{t_0} or $\mathbf{R}^N \backslash \overline{D}_{t_0}$. In other words $(\partial D)_t$ avoids all smooth evolutions. It turns out that a barrier solution is an equivalent notion of set-theoretic solutions even without the comparison principle for (1.1.6). The notion of barrier solutions was first introduced by E. De Giorgi [138] and T. Ilmanen [267] (see also T. Ilmanen [266]) for the mean curvature flow equations, and many authors have developed the theory. However, the above equivalence has not been observed in the literature so the theory developed here (Chap. 6) is new. This characterization provides an alternative way to prove the comparison principle for the level set equation in a set theoretic way, at least for (1.1.1) with f independent of x, including the mean curvature flow equation. Actually, the idea of the proof is also useful to establish the comparison theorem for the level set equation for the crystalline curvature flow equation in the plane; the equation is formally written as (1.1.2) but the Frank diagram of γ in (1.1.4) is a convex polygon so that (1.1.2) is no longer a partial differential equation. See M.-H. Giga and Y. Giga [187]. For the background of motion by crystalline curvature the reader is referred to a book of M. E. Gurtin [246] or a review paper by J. E. Taylor [412]. We won't touch this problem in this book except in the end of Sect. 4.8. For the level set method for crystalline curvature flow equations the reader is referred to papers by M.-H. Giga and Y. Giga [187–191], a review paper by Y. Giga [202] and references cited there. The idea of barrier solution provides an alternate proof for the convergence of solutions of the Allen–Cahn equation to our generalized solution Γ_t and D_t which was originally proved by L. C. Evans et al. [164] by using distance functions. This was remarked by G. Barles and P. E. Souganidis [56]. The convergence results are global. For example, it reads: $u^\varepsilon(x, 0) = 2\chi_{D_0}(x) - 1$ then u^ε converges to 1 (as $\varepsilon \to 0$) on D_t and to -1 outside $\Gamma_t \cup D_t$, where Γ_t and D_t are our generalized solutions defined in (1.1.9). We do not know the behavior of u^ε on Γ_t if Γ_t fattens. The Brakke type solution is always a kind of set-theoretic solution as proved by T. Ilmanen [268] and L. Ambrosio and H. M. Soner [14] so it is contained in Γ_t. Moreover, T. Ilmanen [268] proved that as a limit Brakke solution is obtained. Thus his method also recovers the above convergence results on D_t and outside of $D_t \cup \Gamma_t$.

The idea of the level set method is fundamentally important to study behavior of solutions. For example the first rigorous proof of existence of type II singularity for the mean curvature flow is given by using the level set method. (See a paper by S. Altschuler et al. [10].) We do not mention such applications of the method in our book. The reader is referred to a review paper of Y. Giga [200].

This book is organized as follows. In Chap. 2 we formulate surface evolution equations rigorously by defining several relevant quantities. We pay attention to modify the Gaussian curvature flow equation and related equations so that the equation becomes parabolic. We also derive level set equations and study their structural properties. We conclude Chap. 2 by giving several explicit solutions for typical surface evolution equation having curvature effects. In Chap. 3 we prepare the theory of viscosity solution which is necessary to analyze level set equations. In Chap. 4 we mainly discuss stability and Perron's method. A comparison principle which is always fundamental in the theory of viscosity solutions is discussed in Chap. 4. In Chaps. 3 and 4 we do not use geometricity of the equation so that the theory applies to other equations including the p-Laplace diffusion equation. Some of the comparison theorems presented in Chap. 4 are standard but there are several versions for the equation depending on the space variables. In Chap. 4 we also establish convexity and Lipschitz preserving properties for spatially homogeneous equations. In Chap. 5 we apply the theory of viscosity solutions to get a generalized solution by a level set method. In Chap. 6 we consider the set-theoretic approach of the level set method. In particular we introduce the notions of set-theoretic solutions and barrier solutions. We give an alternate approach to establish comparison results for level set equations. Most of the contents in Chap. 6 is new at least in this generality. In this book the level set method is adjusted so that it applies to evolutions of noncompact hypersurfaces. Also the evolution with boundary conditions is discussed.

This book is written so that no knowledge of differential geometry is required, although such knowledge is helpful for complete understanding. No knowledge of the theory of viscosity solutions is required except the standard maximum principle for semicontinuous functions.

Finally, we note that since the level set method of the present version depends on the comparison principle, it is impossible to extend it directly to higher order surface evolution equations, for example the surface diffusion equation:

$$V = -\Delta H \quad \text{on } \Gamma_t,$$

where Δ denotes the Laplace–Beltrami operator of Γ_t. This equation was analyzed by C. M. Elliott and H. Garcke [155] and J. Escher et al. [157] where local existence and global existence near equilibrium are established. Even for curves in the plane, the solution behaves differently from that of the curvature flow equation. An embedded curve may lose embeddedness in a finite time as proved in Y. Giga and K. Ito [214].

In the new edition, we have added two chapters related to a game-theoretic approach to level set curvature flow equations. This approach has emerged as one of the most fruitful developments in the analysis of nonlinear partial differential equations over the past two decades. The idea of using deterministic two-player games to approximate mean curvature motion was introduced by R. V. Kohn and S. Serfaty [312], and it has since been extended to a wide class of fully nonlinear elliptic and parabolic equations. The guiding principle is that the *value function* of a suitably designed discrete-time game, in which two players make optimal moves according to prescribed rules, converges to the viscosity solution of the corresponding PDE as the step size tends to zero. Such an interpretation provides not only an approximation scheme, but also a conceptual bridge between discrete dynamics and continuous nonlinear analysis. Level set curvature flow equations provide a particularly appealing example of this approach. The game-theoretic formulation offers a discrete geometric mechanism that naturally reproduces curvature motion through the interplay of local strategic choices made by the players. It deepens our understanding of curvature-driven evolution and highlights the versatility of viscosity solution methods in connecting discrete and continuous viewpoints.

Chapter 7 of this book is devoted to introducing this game approximation for level set curvature flow equations. We begin by formulating the discrete two-player game and explaining its intuitive connection to curvature-driven motion. We then develop the analytical framework required to prove convergence of the value functions to viscosity solutions of the corresponding PDEs. We discuss the role of *dynamic programming principles*, comparison arguments, and boundary barrier constructions. The emphasis of this chapter is methodological: how games are constructed, how their strategies reflect curvature effects, and why the approximation is valid. In this sense, Chap. 7 serves as both a general introduction and a technical foundation of the subject.

Chapter 8 turns to the applications of the game-theoretic approximation. We explore how this framework can be used to address analytical and geometric problems associated with curvature flows. Examples include the study of fattening phenomena and the preservation of convexity properties in level set motion. We also discuss such properties for boundary value problems, including dynamic and Neumann boundary conditions. The flexibility of the game-theoretic approach makes it particularly well-suited to handling boundary interactions, where traditional PDE techniques may encounter significant difficulties.

Chapters 7 and 8 are intended to show that the game-theoretic approach is not merely an alternative approximation technique, but rather a new method that offers novel insights into nonlinear partial differential equations. This two-part presentation is intended to provide the reader with both the tools and the perspective necessary to appreciate the growing role of game-theoretic methods in the theory of viscosity solutions and geometric evolution equations.

We take this opportunity to mention a few recent development (not exhaustive at all) on variational weak solutions started by K. A. Brakke [67]. It is well known that original Brakke's (multi-phase) solution lacks uniqueness or continuity in time; the solution may instantaneously disappear at any time. To overcome the latter problem, L. Kim and Y. Tonegawa [306] constructed a special Brakke's solution so that enclosed volume changes continuously in time. Their construction ruled out sudden disappearance of the solution. The original work of K. A. Brakke [67] is very pioneering. It was even earlier than the establishment of the monotonicity formula by G. Huisken [260]. Thus, it is sometimes difficult to follow the proofs. Fortunately, Y. Tonegawa and his collaborators clarified Brakke's theory with several extensions; see Y. Tonegawa [415, 416]. Since a level set method is based on comparison principle, it is difficult to handle multi-grain (multi-phase) problem. A variational weak solution is useful to describe multi-grain problem after it develops "singularities".

In fact, S. Stuvard and Y. Tonegawa [408] constructed a Brakke's solution having "nice" properties to describe multi-grain problems and called it a canonical Brakke flow. There are other notions of variational weak solutions. They include a BV

solution obtained by S. Luckhaus and T. Sturzenhecker [338], L^2-flow obtained by L. Mugnai and M. Röger [358] and a varifold solution by S. Hensel and T. Laux [258] which is based on De Giorgi's energy dissipation inequality. S. Stuvard and Y. Tonegawa [408] proved that their canonical Brakke flow is also an L^2 flow and a (generalized) BV solution. Moreover, they showed that the time derivative of the volume of each grain equals its boundary integral of (measure-theoretic) mean curvature, which is physically quite reasonable.

Although uniqueness for general initial data is unexpected, weak-strong uniqueness has been established in several settings. By weak-strong uniqueness, we mean that a weak solution agrees with a strong (classical) solution as far as the latter exists. This is carried out by J. Fischer et al. [174] for (multi-phase) BV solution and canonical Brakke flow in the plane by introducing a relative entropy together with calibration of a vector field defined on grain boundaries. It is extended for 3-dimensional setting with three phases by S. Hensel and T. Laux [257, Theorem 6]. A variational solution based on De Giorgi's energy dissipation also enjoys the weak-strong uniqueness as shown by S. Hensel and T. Laux [258]. As already observed from nonuniqueness of the set-theoretic solution, unique solvability is not expected unless initial hypersurface is smooth and embedded. Recently, S. Stuvard and Y. Tonegawa [409] show that under Dirichlet condition there exists a time-dependent Brakke flow starting from immersed minimal hypersurface which itself is a stationary Brakke flow. This is an example of essential nonuniqueness of Brakke flow, which shows instability with respect to initial data.

As in the first edition, we do not mention the tremendously many applications of the level set method across various fields; see, for example, the excellent book by S. Osher and R. Fedkiw [374] for computational aspects of a wide range of problems. In computer vision, the method has been used to find the shortest path on a surface by R. Kimmel et al. [308] by solving certain eikonal equations. This approach was further adjusted by H. M. Yahia et al. [431] for surface matching. This is just one example among many. In contrast to numerous studies devoted to applications, the present book focuses on the rigorous mathematical foundations of the level set approach to curvature flow equations.

Chapter 2
Surface Evolution Equations

There are several interesting examples of equations governing motion of hypersurfaces bounding two phases of materials in various sciences. Such a hypersurface is called an interface or a phase boundary. When the motion depends only on geometry of the hypersurface as well as position and time, the governing equation is often called a *surface evolution equation* or a *geometric evolution equation*. In material sciences it is also called an interface controlled model. Although there are several types of surface evolution equations, we focus on equations of evolving hypersurfaces whose speed depend on their shape through their local geometric quantities such as normals and curvatures. In this chapter we formulate such equations. We derive various useful expressions of these quantities especially when the hypersurface is given as a level set of a function. The main goal of this chapter is to study structural properties of level set equations obtained by level set formulation of surface evolution equations. We introduce the notion of geometric equations which is fundamental in a level set method. We also give a few self-similar shrinking solutions as examples of exact solutions.

2.1 Representation of a Hypersurface

There are at least three ways of representing (locally) a hypersurface embedded in $\mathbf{R}^N$. These are representation by local coordinates, by zero level set of a function and by a graph of a function. To fix the idea a set Γ in $\mathbf{R}^N$ is called a C^m hypersurface around a point x_0 of Γ if there is a C^m ($m \geq 1$) function $u(x)$ defined in a neighborhood U of x_0 such that

$$\Gamma \cap U = \{x \in U; u(x) = 0\} \tag{2.1.1}$$

and that the gradient

Y. Giga, Q. Liu, *Surface Evolution Equations*, Monographs in Mathematics 99,
https://doi.org/10.1007/978-3-032-20303-8_2

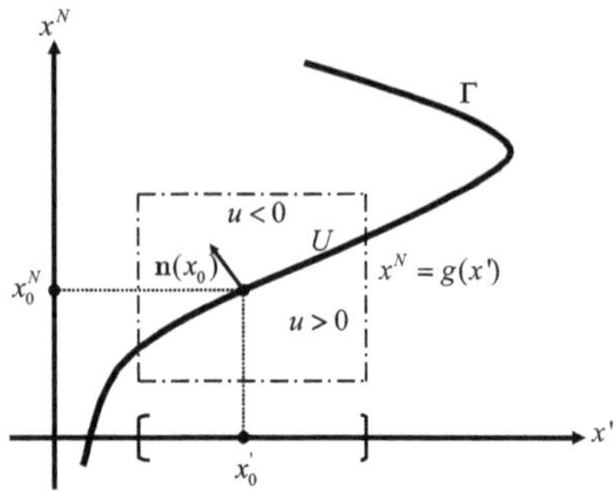

Fig. 2.1 Graph representation

$$\nabla u = \left(\frac{\partial u}{\partial x_1}, \ldots, \frac{\partial u}{\partial x_N}\right) = (u_{x_1}, \ldots, u_{x_N})$$

of u does not vanish on Γ. We call this representation a *level set representation*. If u can be taken C^∞ in U, i.e., u is C^m in U for all $m \geq 1$, Γ is called smooth around x_0. If Γ is a (C^m) hypersurface around every point x_0 of Γ, Γ is simply called a (C^m) hypersurface.

Of course, by the implicit function theorem, one may assume that Γ is locally represented by a graph of a function. By rotating coordinates and shrinking U if necessary, there is a C^m function of $N-1$ variables defined in a neighborhood U' of x_0' that satisfies

$$\Gamma \cap U = \{x_N = g(x'); x' \in U' \subset \mathbf{R}^{N-1}\}$$

with $g(x_0') = x_{0N}$, where $x = (x', x_N)$. This representation is called a *graph representation*. If Γ is represented by the graph of g, it is represented by the zero level of $u = -x_N + g(x')$ (Fig. 2.1).

Another representation is by local coordinates and it includes a graph representation as a special case. It represents a hypersurface Γ around x_0 by the image $\varphi(U')$ of some C^m mapping φ (of full rank) from some open set U' in $\mathbf{R}^{N-1}$ to $\mathbf{R}^N$. By full rank we mean that the Jacobi matrix $\nabla\varphi$ of φ has the maximal rank (i.e., in this case the rank of $\nabla\varphi$ equals $N-1$) at each point of U'. This representation is called a *parametric representation* and U' is called a space of parameters. The equivalence of a level set representation and a parametric representation is guaranteed by the implicit function theorem.

Tangents and Normals Let Γ be a hypersurface around x. A vector τ in $\mathbf{R}^N$ is called a *tangent vector* of Γ at x if there is a (C^1) curve ζ on Γ that satisfies $\zeta = x$, $d\zeta/dt = \tau$ at $t = 0$, where ζ is defined at least in a neighborhood of 0. The space of tangent vectors at x is called the *tangent space* of Γ at x and is denoted $T_x\Gamma$.

We shall calculate $T_x\Gamma$ when Γ is represented by a graph of a function. We may assume that $x = 0$ and that Γ is represented by the graph $x_N = g(x')$ around a point $x = 0$. A curve ζ on Γ through 0 is of the form

$$\zeta(t) = (\sigma(t), g(\sigma(t)))$$

with a curve $\sigma(t)$ in $\mathbf{R}^{N-1}$ through 0 of $\mathbf{R}^{N-1}$. Note that

$$\frac{d\zeta}{dt} = \left(\frac{d\sigma}{dt}, \left\langle \nabla' g(\sigma(t)), \frac{d\sigma}{dt} \right\rangle\right) \in \mathbf{R}^N,$$

where ∇' denotes the gradient in x' and $\langle\ ,\ \rangle$ denotes the standard inner product in the Euclidean space. For given $\tau' \in \mathbf{R}^{N-1}$ there is a curve σ that satisfies $d\sigma/dt = \tau'$ at $t = 0$ with $\sigma(0) = 0$ so τ is a tangent vector (of Γ at 0) if and only if

$$\tau = (\tau', \langle \nabla' g(0), \tau' \rangle).$$

In other words

$$T_x\Gamma = \{(\tau', \langle \nabla' g(x), \tau' \rangle), \tau' \in \mathbf{R}^{N-1}\},$$

which in particular implies that $T_x\Gamma$ is an $N-1$ dimensional vector subspace of $\mathbf{R}^N$.

A *unit normal vector* $\mathbf{n}(x)$ at x of Γ is a unit vector of $\mathbf{R}^N$ orthogonal to $T_x\Gamma$ with respect to the standard inner product $\langle\ ,\ \rangle$ of $\mathbf{R}^N$. It is unique up to multiplier ± 1 since $T_x\Gamma$ is an $N-1$ dimensional space.

Suppose that Γ is a hypersurface around $x_0 \in \Gamma$. If $\mathbf{n}(x)$ is a unit normal vector at x of Γ near x_0 and $\mathbf{n}$ depends on x at least continuously, we say that $\mathbf{n}$ is a *unit normal vector field* of *orientation* (around x_0) of Γ. Such a field $\mathbf{n}$ of course exists around x_0. To see this we use a level set representation (2.1.1) of Γ. For $x \in \Gamma \cap U$ let $\zeta(t)$ be a curve on Γ through x at $t = 0$, i.e., $\zeta(0) = x$. Differentiate $u(\zeta(t)) = 0$ in t and evaluate at $t = 0$ to get

$$\left\langle \nabla u(x), \frac{d\zeta}{dt}(0) \right\rangle = 0.$$

This implies that $\nabla u(x)$ is orthogonal to $T_x\Gamma$. Since u is C^m ($m \geq 1$), ∇u is C^{m-1} and at least continuous. Since ∇u does not vanish around x_0,

$$\mathbf{n}(x) = -\frac{\nabla u(x)}{|\nabla u(x)|}$$

is a unit normal vector field around x_0. Here $|p|$ denotes the Euclidean length of vector p, i.e., $|p| = \langle p, p \rangle^{1/2}$. There are exactly two unit normal vector fields around x_0. The other field is of course $-\mathbf{n}(x)$. We take $\mathbf{n}$ as above just to fix the idea. If Γ is given as the graph $x_N = g(x')$, setting $u(x) = -x_N + g(x')$ yields the upward unit normal:

$$\mathbf{n}(x') = \frac{(-\nabla' g(x'), 1)}{(1 + |\nabla' g(x)|^2)^{1/2}}. \tag{2.1.2}$$

In any case if $\mathbf{n}$ is a unit normal vector field around x_0, then

$$T_x\Gamma = \{\tau \in \mathbf{R}^N; \langle \tau, \mathbf{n}(x)\rangle = 0\}$$

for $x \in \Gamma$ around x_0. We shall often suppress the words 'vector field'. If a hypersurface Γ is a topological boundary ∂D of a domain D, the unit normal (vector field) pointing outward from D is called the *outward* unit normal (vector field).

Evolving Hypersurface Suppose that Γ_t is a set in $\mathbf{R}^N$ depending on the time variable t. We say that a family $\{\Gamma_t\}$ or simply Γ_t is a ($C^{2m,m}$) *evolving hypersurface* around (x_0, t_0) (with $x_0 \in \Gamma_{t_0}$) if there is a $C^{2m,m}$ ($m \geq 1$) function $u(x,t)$ defined for $t_0 - \delta < t < t_0 + \delta$, $x \in U$ for some $\delta > 0$ and some neighborhood U of x_0 in $\mathbf{R}^N$ such that

$$\Gamma_t \cap U = \{x \in U; u(x,t) = 0\} \tag{2.1.3}$$

and that the spatial gradient ∇u of u does not vanish on Γ_t. (This is a level set representation of Γ_t.) Here by a $C^{2m,m}$ function we mean that derivatives $\nabla^{(k)}\partial_t^h u$ are continuous for $k + 2h \leq 2m$, where ∂_t^h denotes the h-th differentiation in the time variable and $\nabla^{(k)}$ denotes the k-th differentiation in the space variables. If u can be taken C^∞, i.e., $C^{2m,m}$ for all $m \geq 1$, we just say that Γ_t is a smoothly evolving hypersurface around (x_0, t_0). If Γ_t is a $C^{2m,m}$ (resp. smoothly) evolving hypersurface around all (x,t) with $x \in \Gamma_t$ and t belonging to an interval I, we say that Γ_t is a $C^{2m,m}$ (resp. smoothly) evolving hypersurface on I. In this chapter we always assume that Γ_t is a smoothly evolving hypersurface in some time interval unless otherwise claimed.

2.2 Normal Velocity

Let $\mathbf{n}$ be a unit normal vector field of Γ_t so that it depends on time t smoothly. The reasonable quantity which describes the motion of Γ_t is a normal velocity, that is the speed in the direction of $\mathbf{n}$. Note that there is a chance that each point of Γ_t moves but the set Γ_t is independent of time like a rotating sphere.

Definition 2.2.1 Let x_0 be a point of Γ_{t_0}. Let $x(t)$ be a (C^1) curve defined on $(t_0 - \delta, t_0 + \delta)$ for some $\delta > 0$ such that $x(t)$ is a point on Γ_t and $x(t_0) = x_0$. The quantity

$$V = \left\langle \frac{dx}{dt}(t_0), \mathbf{n} \right\rangle$$

is called the *normal velocity* at x_0 of Γ_t at the time t_0 in the direction of $\mathbf{n}$.

As we will see later V is independent of the choice of curve $x(t)$. We shall give various expressions of V.

Level Set Representation Suppose that Γ_t is represented by (2.1.3). We shall calculate the normal velocity V at (x_0, t_0) in the direction of $\mathbf{n}$ defined by

$$\mathbf{n}(x_0, t_0) = -\frac{\nabla u(x_0, t_0)}{|\nabla u(x_0, t_0)|}.$$

Let $x(t)$ be a curve in $\mathbf{R}^N$ on Γ_t with $x(t_0) = x_0$. Since Γ_t is the zero level set of $u(\cdot, t)$, $u(x(t), t) = 0$ for t near t_0. Differentiate $u(x(t), t)$ in t and evaluate at (x_0, t_0) to get

$$u_t(x_0, t_0) + \left\langle \frac{dx}{dt}(t_0), \nabla u(x_0, t_0) \right\rangle = 0,$$

where $u_t = \partial_t u$. Recalling $\mathbf{n} = -\nabla u / |\nabla u|$, we obtain

$$V = \frac{u_t(x_0, t_0)}{|\nabla u(x_0, t_0)|}. \tag{2.2.1}$$

Clearly, this shows that V is independent of the choice of $x(t)$.

Graph We shall give a formula for V when Γ_t is represented by a graph of a function. By rotating coordinates we may assume that Γ_t is expressed as

$$\Gamma_t = \{x_N = g(x', t);\, x' \in \mathbf{R}^{N-1}\}$$

around (x_0, t_0), where $g(x_0') = x_{0N}$ and $x' = (x_1, \dots, x_{N-1})$. If $\mathbf{n}$ is taken upward, Γ_t is given as a zero level set of

$$u(x, t) = -x_N + g(x', t)$$

with $\mathbf{n} = -\nabla u / |\nabla u|$ as in (2.1.2). Then by (2.2.1) we see that

$$V = \frac{g_t(x_0', t_0)}{(1 + |\nabla' g(x_0', t_0)|^2)^{1/2}}. \tag{2.2.2}$$

Axisymmetric Surface Suppose now that Γ_t is obtained by rotating the graph of a function $\varphi(x_1, t)$ around the x_1-axis and that x_0 is not on the axis. In other words, around (x_0, t_0) Γ_t is of the form

$$\Gamma_t = \left\{ r = \varphi(x_1, t);\, r = \left(\sum_{j=2}^{N} x_j^2 \right)^{1/2} \right\}.$$

Since Γ_t is regarded as the zero level set of

$$u(x,t) = -r + \varphi(x_1, t),$$

the normal velocity V in the direction of

$$\mathbf{n} = \frac{\left(-\varphi_{x_1}(x_{01}, t_0), x_0''/|x_0''|\right)}{\left(1 + (\varphi_{x_1}(x_{01}, t_0))^2\right)^{1/2}} = -\frac{\nabla u(x_0, t_0)}{|\nabla u(x_0, t_0)|} \tag{2.2.3}$$

at (x_0, t_0) is of the form

$$V = \frac{\varphi_t(x_{01}, t_0)}{\left(1 + \left(\varphi_{x_1}(x_{01}, t_0)\right)^2\right)^{1/2}} \tag{2.2.4}$$

where $x_0'' = (x_{02}, \ldots, x_{0N})$ and $\varphi_{x_1} = \partial\varphi/\partial x_1$.

Rescaled Motion If we would like to study the behavior of Γ_t near $x_* \in \mathbf{R}^N$ as t tends to t_* with $t < t_*$, we often magnify Γ_t near (x_*, t_*) by rescaling. There are of course several ways to rescale but here we only give a typical example. Let (y, s) be defined by

$$y = (t_* - t)^{-1/2}(x - x_*), \ s = -\log(t_* - t) \tag{2.2.5}$$

for $t < t_*$. Let $\hat{V}$ denote the normal velocity of

$$\hat{\Gamma}_s = \{y \in \mathbf{R}^N; \ y = (t_* - t)^{-1/2}(x - x_*), \ x \in \Gamma_t\}$$

at (y_0, s_0), where s is regarded as the new time variable. Here the unit normal vector $\hat{\mathbf{n}}(y_0, s_0)$ of $\hat{\Gamma}_{s_0}$ is taken so that its direction is the same as the unit normal $\mathbf{n}(x_0, t_0)$ of Γ_{t_0} at x_0, where (x_0, t_0) is determined by (2.2.5) by setting $(y, s) = (y_0, s_0)$. Let V be the normal velocity of Γ_{t_0} at x_0 in the direction of $\mathbf{n}(x_0, t_0)$. Then

$$\hat{V} = e^{-s_0/2}\, V + \frac{1}{2}\,\langle\, y_0, \ \hat{\mathbf{n}}(y_0, s_0)\,\rangle. \tag{2.2.6}$$

Note that the behavior of Γ_t as t tends to t_* with $t < t_*$ corresponds to the large time behavior of $\hat{\Gamma}_s$ by (2.2.5).

To see this formula we use level set representation of Γ_t near (x_0, t_0). Suppose that Γ_t is represented by (2.1.3). We may assume $\mathbf{n}(x_0, t_0)$ is of form

$$\mathbf{n}(x_0, t_0) = -\frac{\nabla u(x_0, t_0)}{|\nabla u(x_0, t_0)|}.$$

We set

$$w(y, s) = u((t_* - t)^{1/2}\, y + x_*, \ t), \ s = -\log(t_* - t)$$

so that $\hat{\Gamma}_s$ is represented as a zero level set of w near (y_0, s_0). Since

$$\hat{\mathbf{n}}(y_0, s_0) = -\frac{\nabla w(y_0, s_0)}{|\nabla w(y_0, s_0)|},$$

the formula (2.2.1) yields

$$\hat{V} = \frac{w_s(y_0, s_0)}{|\nabla w(y_0, s_0)|}.$$

A direct calculation shows that

$$\left(w_s + \frac{1}{2}\langle\, y,\ \nabla w\,\rangle\right)(y_0, s_0) = e^{-s_0}\, u_t(x_0, t_0),\ \ \nabla w(y_0, s_0) = e^{-s_0/2}\,\nabla u(x_0, t_0).$$

These two identities together with representation of V, $\hat{\mathbf{n}}$, $\hat{V}$ yields (2.2.6).

If we consider a little bit more general rescaling than (2.2.5) of form

$$y = (t_* - t)^{-\alpha}\,(x - x_*),\ \ s = -\log(t_* - t) \quad \text{with} \quad \alpha > 0,$$

then

$$\hat{V} = e^{-(1-\alpha)s_0}\, V + \alpha\,\langle\, y_0,\ \hat{\mathbf{n}}(y_0,\ s_0)\,\rangle$$

instead of (2.2.6).

2.3 Curvatures

Let Γ be a hypersurface in $\mathbf{R}^N$. For a point x_0 let τ be a tangent vector of Γ at x_0. Let X be a (C^1) vector field on Γ around x_0, i.e., let X be a C^1 function from Γ to $\mathbf{R}^N$ around x_0. Here C^1 means that X can be extended to a C^1 function in a neighborhood of x_0 in $\mathbf{R}^N$. The vector field X need not be tangential to Γ. Let ζ be a curve on Γ that satisfies

$$\zeta(0) = x_0, \quad \frac{d\zeta}{dt}(0) = \tau.$$

A tangential derivative in the direction of τ is defined by

$$(D_\tau X)(x_0) = \frac{d}{dt}(X(\zeta(t))|_{t=0}.$$

If X is extended to a neighborhood of x_0 in $\mathbf{R}^N$, we observe that

$$(D_\tau X)(x_0) = (\tau \cdot \nabla)X = \sum_{j=1}^{N} \tau_j \frac{\partial}{\partial x_j} X, \qquad \tau = (\tau_1, \dots, \tau_N).$$

This shows that the operator D_τ is independent of the choice of the curve ζ. By the definition of D_τ the quantity $(\tau \cdot \nabla)X$ is independent of the extension of X outside Γ. Thus the operator D_τ is well defined for any vector field on Γ.

Second Fundamental Form Suppose now that Γ is a (C^2) hypersurface around x on Γ. Let $\mathbf{n}$ be a unit normal vector field around x. From the level set representation it is clear that $\mathbf{n}$ is C^1 on Γ. For each $\tau \in T_x\Gamma$ we set

$$\mathbf{A}\tau = -(D_\tau \mathbf{n})(x) \in \mathbf{R}^N.$$

Since $|\mathbf{n}| = 1$,

$$\langle \mathbf{A}\tau, \mathbf{n}(x) \rangle = -\frac{1}{2} D_\tau(|\mathbf{n}|^2) = 0$$

so that $\mathbf{A}\tau \in T_x\Gamma$. The linear operator $\mathbf{A} = \mathbf{A}_x$ from $T_x\Gamma$ into itself is called the *Weingarten map* (in the direction of $\mathbf{n}(x)$). The bilinear form on $T_x\Gamma \times T_x\Gamma$ associated with $\mathbf{A}$ defined by

$$B_x(\tau, \eta) = \langle \mathbf{A}\tau, \eta \rangle \tag{2.3.1}$$

is called the *second fundamental form* (in the direction of $\mathbf{n}(x)$) at $x \in \Gamma$.

To see the geometric meaning of B_x for $\tau, \eta \in T_x\Gamma$ let ϕ be a function from a neighborhood of the origin of $\mathbf{R}^2$ to $\Gamma \subset \mathbf{R}^N$ that satisfies

$$\frac{\partial \phi}{\partial x_1}(0,0) = \tau, \ \frac{\partial \phi}{\partial x_2}(0,0) = \eta, \ \phi(0,0) = x.$$

Since

$$\left\langle \mathbf{n}(\phi(x_1, x_2)), \frac{\partial \phi}{\partial x_1}(x_1, x_2) \right\rangle = 0 \tag{2.3.2}$$

near $(x_1, x_2) = (0, 0) \in \mathbf{R}^2$, differentiating in x_2 and evaluating at zero yields

$$\left\langle \left(\frac{\partial \phi}{\partial x_2}(0,0) \cdot \nabla \right) \mathbf{n}, \frac{\partial \phi}{\partial x_1}(0,0) \right\rangle + \left\langle \mathbf{n}, \frac{\partial^2 \phi}{\partial x_1 \partial x_2}(0,0) \right\rangle = 0,$$

or

$$\langle (\eta \cdot \nabla)\mathbf{n}, \tau \rangle + \left\langle \mathbf{n}, \frac{\partial^2 \phi}{\partial x_1 \partial x_2}(0,0) \right\rangle = 0.$$

Fig. 2.2 Normal curvature

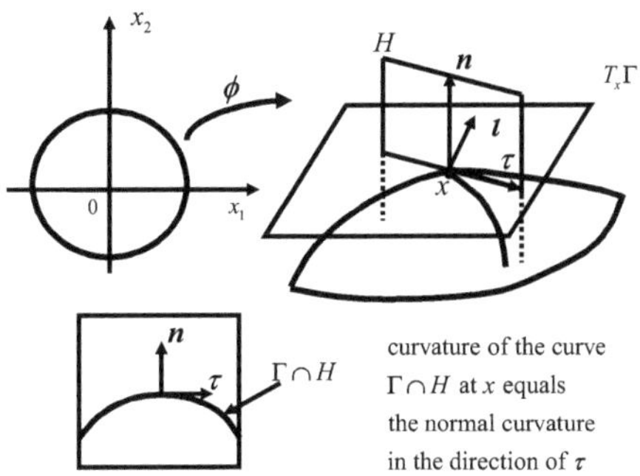

By definition this yields

$$B_x(\eta,\ \tau) = \left\langle \mathbf{n},\ \frac{\partial^2 \phi}{\partial x_1\, \partial x_2}(0,0) \right\rangle. \tag{2.3.3}$$

In particular B_x is a symmetric bilinear form and the Weingarten map $\mathbf{A}$ is a symmetric linear operator. Thus its eigenvalues are all real and called *principal curvatures* of Γ at x (in the direction of $\mathbf{n}(x)$). The principal curvatures are denoted $\kappa_1, \ldots, \kappa_{N-1}$. Differentiating (2.3.2) in x_1 and evaluating at zero, we get

$$B_x(\tau,\ \tau) = \left\langle \mathbf{n},\ \frac{\partial^2 \phi}{\partial x_1^2}(0,0) \right\rangle$$

instead of (2.3.3). If $|\tau| = 1$, this quantity is called the *normal curvature* of Γ in the direction of τ (Fig. 2.2).

Surfaces of Higher Codimension As well known the second fundamental form is defined even for any embedded manifold M in $\mathbf{R}^N$ whose *dimension* k is strictly less than $N-1$ (i.e., *codimension* $N-k$ is strictly greater than 1). We briefly review the definition since it is almost the same as (2.3.1). For the tangent space $T_x M$ of M at x let $N_x M$ denote its orthogonal complement in $\mathbf{R}^N$. The space $N_x M$ is called the *normal space* of M at x. Since M has dimension k, the dimensions of $T_x M$ and $N_x M$ equal k and $N-k$, respectively. Let $\mathbf{n}^1, \ldots, \mathbf{n}^{N-k}$ be (C^1) vector fields near $x \in M$ such that $\{\mathbf{n}^i(z);\ 1 \le i \le N-k\}$ is an orthonormal basis of $N_z M$ for every z near x. The *second fundamental form* of M at x is defined by

$$\mathbf{B}_x(\tau,\ \eta) = -\sum_{j=1}^{N-k} \langle\, (D_\tau \mathbf{n}^i)(x),\ \eta\,\rangle\, \mathbf{n}^i(x), \qquad \tau,\ \eta \in T_x M$$

as a mapping

$$\mathbf{B}_x\ :\ T_x M \times T_x M \to N_x M.$$

As in the same way we derived (2.3.3), we have

$$\mathbf{B}_x(\eta,\ \tau) = \pi\left(\frac{\partial^2\phi}{\partial x_1\,\partial x_2}(0,0)\right) \text{ with } \frac{\partial\phi}{\partial x_1}(0,0)$$
$$= \tau,\ \frac{\partial\phi}{\partial x_2}(0,0) = \eta,\ \ \phi(0,0) = x,$$

where π denotes the orthogonal projection from $\mathbf{R}^N$ onto $N_x M$. In particular $\mathbf{B}_x$ is symmetric and $\mathbf{B}_x$ is independent of the choice of $\mathbf{n}^1, \dots, \mathbf{n}^{N-k}$ forming an orthonormal basis of $N_z M$ for z near x. Note that the definition does not require $k < N-1$. If $k = N-1$, then, as expected

$$\langle\ \mathbf{B}_x(\tau,\eta),\ \mathbf{n}\ \rangle\ = B_x(\tau,\eta),$$

where B_x is the second fundamental form in the direction of $\mathbf{n}$. This shows that $\mathbf{B}_x$ is a natural generalization of B_x.

Surface Divergence Let X be a C^1 vector field on a hypersurface Γ. For $x \in \Gamma$ let $\{\tau^\ell;\ 1 < \ell \leq N-1\}$ be an orthonormal basis of $T_x\Gamma$. The *surface divergence* of X is denoted by $\operatorname{div}_\Gamma X$ and is defined by

$$(\operatorname{div}_\Gamma X)(x) = \sum_{\ell=1}^{N-1} \langle\ (D_{\tau^\ell} X)(x),\ \tau^\ell\ \rangle. \tag{2.3.4}$$

If we extend X around Γ so that ∇X is well defined as an $N \times N$ matrix, then (2.3.4) yields

$$(\operatorname{div}_\Gamma X)(x) = \operatorname{trace}\left(\sum_{\ell=1}^{N-1} \tau^\ell \otimes \tau^\ell\right)(\nabla X)(x) \tag{2.3.5}$$

since $D_\tau X = (\tau\ \cdot\ \nabla)X$, where $\otimes$ denotes the tensor product of N-vectors and trace denotes the trace of an $N \times N$ matrix. The definition of $\operatorname{div}_\Gamma X$ is independent of the extension of X, so the right-hand side of (2.3.5) is independent of the extension of X. Since

$$I_N - \sum_{\ell=1}^{N-1} \tau^\ell \otimes \tau^\ell = \mathbf{n}(x) \otimes \mathbf{n}(x)$$

as matrices, the identity (2.3.5) is rewritten as

$$(\operatorname{div}_\Gamma X)(x) = \operatorname{trace}\{(I - \mathbf{n}(x) \otimes\ \mathbf{n}(x))(\nabla X(x))\}, \tag{2.3.6}$$

where I denotes the $N \times N$ unit matrix. From (2.3.6) it follows that the surface divergence is defined independent of the choice of orthonormal basis of $T_x\Gamma$ and the orientation $\mathbf{n}$.

The surface divergence $\operatorname{div}_M X$ is also defined for a vector field X on an embedded manifold M of dimension k in $\mathbf{R}^N$ whose codimension $N-k>1$. It is defined as (2.3.4), i.e.,

$$(\operatorname{div}_M X)(x) = \sum_{\ell=1}^{k} \langle (D_{\tau^\ell} X)(x),\ \tau^\ell \rangle,$$

where $\{\tau^\ell;\ 1 \le \ell \le k\}$ is an orthonormal basis of $T_x M$. As for a hypersurface, $\operatorname{div}_M X$ is independent of the choice of orthonormal basis of $T_x M$ which can be proved directly from (2.3.5) with $N-1$ replaced by k.

Mean Curvature Let $\mathbf{n}$ be a unit normal vector field around x on a (C^2) hypersurface Γ. Let H be the sum of all principal curvatures $\kappa_1(x), \dots, \kappa_{N-1}(x)$ at x in the direction of $\mathbf{n}$, i.e.,

$$H = \kappa_1(x) + \cdots + \kappa_{N-1}(x).$$

We say that H is the *mean curvature* (of Γ) at x (in the direction of $\mathbf{n}$). We do *not* take the average of principal curvatures although many authors since the time of Gauss have taken the average to define the mean curvature. Since H is the trace of the Weingarten map,

$$H = \sum_{\ell=1}^{N-1} \mathbf{B}_x(\tau^\ell,\ \tau^\ell) = -\sum_{\ell=1}^{N-1} (D_{\tau^\ell}\,\mathbf{n}(x),\ \tau^\ell) = -(\operatorname{div}_\Gamma \mathbf{n})(x),$$

where τ^ℓ $(\ell = 1, \dots, m-1)$ is an orthonormal basis of $T_x\Gamma$. If $N=2$, H is simply called the *curvature* at $x \in \Gamma$ (in the direction of $\mathbf{n}$) and is denoted κ.

Even if a manifold M in $\mathbf{R}^N$ has higher codimension, or its dimension $k < N-1$, the mean curvature is defined as a vector. We say that

$$\mathbf{H} = \sum_{\ell=1}^{N-1} \mathbf{B}_x(\tau^\ell,\ \tau^\ell) \in N_x M$$

is the *mean curvature vector* at $x \in M$. By definition of the second fundamental form and the surface divergence

$$\mathbf{H} = -\sum_{i=1}^{N-k} (\operatorname{div}_M \mathbf{n}^i)\mathbf{n}^i,$$

where $\mathbf{n}^i$ $(1 \leq i \leq N-k)$ is the same as in the definition of $\mathbf{B}_x$. If $k = N-1$, then by definition

$$H = \langle \mathbf{H}, \mathbf{n} \rangle,$$

where H is the mean curvature in the direction of $\mathbf{n}$. Since $N_x\Gamma$ is one dimensional, the mean curvature H has all information of the mean curvature vector if we take the sign of H into account.

Symmetric Curvatures Let $\kappa_1(x), \ldots, \kappa_{N-1}(x)$ be principal curvatures at $x \in \Gamma$ (in the direction of $\mathbf{n}$), where Γ is a (C^2) hypersurface in $\mathbf{R}^N$. We shall consider elementary symmetric polynomials of principal curvatures. Let $e_m(1 \leq m \leq N-1)$ denote the m-th *elementary symmetric polynomial* of $\lambda_1, \ldots, \lambda_{N-1}$, i.e.,

$$e_m(\lambda_1, \ldots, \lambda_{N-1}) = \sum \lambda_{i_1} \cdots \lambda_{i_m}$$

where the sum is taken over all integers $i_1, \ldots, i_m$ that satisfy $1 \leq i_1 < i_2 < \cdots < i_m \leq N-1$. In particular

$$e_1(\lambda_1, \ldots, \lambda_{N-1}) = \lambda_1 + \cdots + \lambda_{N-1},$$
$$e_{N-1}(\lambda_1, \ldots, \lambda_{N-1}) = \lambda_1 \cdot \lambda_2 \cdots \lambda_{N-1}.$$

Clearly, the mean curvature is of the form

$$H = e_1(\kappa_1, \ldots, \kappa_{N-1}).$$

The quantity

$$K = e_{N-1}(\kappa_1, \ldots, \kappa_{N-1})$$

is called the *Gaussian curvature*. In general we say that

$$H_m = e_m(\kappa_1, \ldots, \kappa_{N-1})$$

is the m-th *symmetric curvature*. Sometimes we consider a little more complicated curvature defined by the ratio H_m/H_ℓ with $m > \ell$. The quantity H_{N-1}/H_{N-2} for $N \geq 3$ is called the *harmonic curvature* since

$$H_{N-1}/H_{N-2} = \left(\sum_{i=1}^{N-1} \frac{1}{\kappa_i} \right)^{-1}.$$

Anisotropic Curvatures It is well known that the mean curvature H is the change ratio of surface area of Γ per change of volume of D enclosed by Γ, where Γ is a (C^2) hypersurface in $\mathbf{R}^N$. If the hypersurface Γ has anisotropic structure depending on its normal direction, it is natural to consider surface energy instead of area. Let γ_0 be a positive function defined on the unit sphere

$$S^{N-1} = \left\{ p \in \mathbf{R}^N;\ |p| = 1 \right\}.$$

We extend γ_0 on $\mathbf{R}^n$ so that

$$\gamma(p) = \gamma_0(p/|p|)\,|p|. \tag{2.3.7}$$

Clearly, $\gamma(p)$ is *positively homogeneous of degree* 1, , i.e.,

$$\gamma(\lambda p) = \lambda\gamma(p) \quad \text{for all} \quad \lambda > 0,\ p \in \mathbf{R}^N. \tag{2.3.8}$$

Conversely, γ satisfying (2.3.8) with $\gamma|_{S^{N-1}} = \gamma_0$ is always expressed by (2.3.7). The *surface energy* of Γ with *surface energy density* γ_0 is defined by

$$\int_\Gamma \gamma_0(\mathbf{n})d\sigma$$

where $d\sigma$ denotes the surface element. Of course, this is nothing but a surface area if $\gamma_0 \equiv 1$. We shall define a quantity (called the anisotropic mean curvature) which describes the change ratio of surface energy per change of volume of D enclosed by Γ as a generalization of the mean curvature.

For a given surface energy density γ_0 let ξ be the gradient of the homogenization γ of γ_0 given by (2.3.7), i.e., $\xi = \nabla\gamma$. The vector ξ is called the *Cahn–Hoffman vector* of γ_0. We say

$$h(x) = -(\mathrm{div}_\Gamma\, \xi(\mathbf{n}))(x)$$

is the *anisotropic* (or *weighted*) *mean curvature* of Γ at x (in the direction of $\mathbf{n}$) with respect to surface energy density γ_0. To define h as a continuous function we need to assume that γ is C^2 outside the origin. Of course, if $\gamma_0 \equiv 1$, then $\xi(p) = p/|p|$ so that $\xi(\mathbf{n}) = \mathbf{n}$. Thus the anisotropic mean curvature agrees with usual mean curvature when $\gamma_0 \equiv 1$ as expected.

2.4 Expression of Curvature Tensors

Let $\Gamma \subset \mathbf{R}^N$ be a (C^2) hypersurface around $x_0 \in \Gamma$. Let $\mathbf{A} = \mathbf{A}_{x_0}$ denote the Weingarten map in the direction of $\mathbf{n}(x_0)$, where $\mathbf{n}$ is a unit normal vector field on Γ around x_0. We shall give various expressions of $\mathbf{A}$ and curvatures.

Level Set Representation Suppose that Γ is represented by (2.1.1) with

$$\mathbf{n}(x_0) = -\frac{\nabla u(x_0)}{|\nabla u(x_0)|}. \tag{2.4.1}$$

Then for $\tau \in T_{x_0}\Gamma$,

$$\begin{aligned}\mathbf{A}_{x_0}\tau &= -(D_\tau \mathbf{n})(x_0) = (\tau \cdot \nabla)\frac{\nabla u}{|\nabla u|} \\ &= \frac{1}{|\nabla u|}\left\{(\tau \cdot \nabla)\nabla u - \langle\, (\tau \cdot \nabla)\nabla u, \nabla u \,\rangle \frac{\nabla u}{|\nabla u|^2}\right\} \quad \text{at} \quad x = x_0.\end{aligned} \tag{2.4.2}$$

It is convenient to introduce the orthogonal projection Π_{x_0} from $T_{x_0}\mathbf{R}^n$ to $T_{x_0}\Gamma$. Its matrix expression is

$$R_{\mathbf{n}(x_0)} = I - \mathbf{n}(x_0) \otimes \mathbf{n}(x_0)$$

so that

$$\Pi_{x_0}\zeta = R_{\mathbf{n}(x_0)}\zeta, \quad \zeta \in T_{x_0}\mathbf{R}^N = \mathbf{R}^N,$$

where ζ is regarded as a column vector. Using the notation

$$R_p = I - p \otimes p/|p|^2 \quad \text{for} \quad p \in \mathbf{R}^N,\ p \neq 0, \tag{2.4.3}$$

we observe from (2.4.2) that

$$\mathbf{A}_{x_0}\tau = \frac{1}{|\nabla u|} R_p\, (\nabla^2 u)\tau \quad \text{at } x_0 \text{ with} \quad p = \nabla u(x_0),$$

where τ is regarded as a column vector. Since

$$R_p\tau = \tau,\ p = \nabla u(x_0)$$

by definition, we see

$$\mathbf{A}_{x_0}\tau = \frac{1}{|\nabla u(x_0)|} Q_p(\nabla^2 u(x_0))\tau \quad \text{at} \quad x_0$$

with

$$Q_p(X) = R_p X R_p, \tag{2.4.4}$$

where X is an $n \times n$ real symmetric matrix. Although $R_p \nabla^2 u$ may not be symmetric, $Q_p(\nabla^2 u)$ is now symmetric. It is not difficult to see that the symmetric operator $\tilde{\mathbf{A}}_{x_0}$ from $T_{x_0}\mathbf{R}^N$ to $T_{x_0}\mathbf{R}^N$ defined by

$$\tilde{\mathbf{A}}_{x_0}\zeta = \frac{1}{|\nabla u(x_0)|} Q_{\nabla u(x_0)}(\nabla^2 u(x_0))\zeta,\ \zeta \in T_{x_0}\mathbf{R}^N = \mathbf{R}^N \tag{2.4.5}$$

is a unique linear operator with the property that

$$\tilde{\mathbf{A}}_{x_0}\zeta = \mathbf{A}_{x_0}\Pi_{x_0}\zeta.$$

We often identify the Weingarten map $\mathbf{A}_{x_0}$ by $\tilde{\mathbf{A}}_{x_0}$. By definition $\tilde{\mathbf{A}}_{x_0}$ is given by a direct sum of operators

$$\tilde{\mathbf{A}}_{x_0} = \mathbf{A}_{x_0} \oplus 0$$

corresponding to the decomposition of the tangent space $\mathbf{R}^N = T_{x_0}\mathbf{R}^N$ of form

$$\mathbf{R}^N = T_{x_0}\Gamma \oplus N_{x_0}\Gamma,$$

where $N_{x_0}\Gamma$ denotes the normal vector space at x_0. Thus the eigenvalues of $\tilde{\mathbf{A}}_{x_0}$ consist of principal curvatures $\kappa_1, \dots, \kappa_{N-1}$ and 0.

We shall derive the level set representation of various curvatures from (2.4.5). The mean curvature H at $x_0 \in \Gamma$ in the direction of $\mathbf{n}(x_0)$ (defined by (2.4.1)) is

$$\begin{aligned} H &= \kappa_1 + \cdots + \kappa_{N-1} + 0 = \operatorname{trace} \tilde{\mathbf{A}}_{x_0} \\ &= \frac{1}{|p|} \operatorname{trace} Q_p(X) \quad \text{with} \quad p = \nabla u(x_0),\ X = \nabla^2 u(x_0). \end{aligned}$$

Since $R_p^2 = R_p$ by (2.4.3) and trace $(R_p X R_p) = \operatorname{trace}(R_p^2 X)$, H is of the form

$$\begin{aligned} H &= \frac{1}{|\nabla u(x_0)|} \operatorname{trace}\left(\left(I - \frac{\nabla u(x_0) \otimes \nabla u(x_0)}{|\nabla u(x_0)|^2}\right)\nabla^2 u(x_0)\right) \\ &= \frac{1}{|\nabla u|}\left(\Delta u - \sum_{1 \le 1,\ j \le N} \frac{u_{x_i} u_{x_j}}{|\nabla u|^2} u_{x_i x_j}\right) \quad \text{at} \quad x = x_0, \end{aligned} \tag{2.4.6}$$

where $u_{x_i} = \partial u/\partial x_i$, $u_{x_i x_j} = \partial^2 u/\partial x_i \partial x_j$ and $\Delta u = \sum_{j=1}^N u_{x_j x_j}$. For m-th symmetric curvature with $m \le N-1$ by the form of $\tilde{\mathbf{A}}_x$ we see

$$H_m = e_m(\kappa_1, \dots, \kappa_{N-1}, 0)$$

where $\kappa_1, \dots, \kappa_{N-1}, 0$ are eigenvalues of $Q_p(X)/|p|$ with $p = \nabla u(x_0)$, $X = \nabla^2 u(x_0)$, since

$$e_m(\kappa_1, \dots, \kappa_{N-1}, 0) = e_m(\kappa_1, \dots, \kappa_{N-1}).$$

If $m = N - 1$, then the Gaussian curvature K is of the form

$$K = e_{N-1}(\kappa_1, \ldots, \kappa_{N-1}, 0) = e_{N-1}(\kappa_1, \ldots, \kappa_{N-1}) = e_N(\kappa_1, \ldots, \kappa_{N-1}, 1).$$

This observation gives a simple representation of the Gaussian curvature

$$K = \det\left(\frac{Q_p(X)}{|p|} + \frac{p \otimes p}{|p|^2}\right), \quad p = \nabla u(x_0), \ X = \nabla^2 u(x_0) \tag{2.4.7}$$

since the eigenvalues of $Q_p(X)/|p| + p \otimes p/|p|^2$ are $\kappa_1, \ldots, \kappa_{N-1}, 1$. If we arrange $\kappa_1 \leq \cdots \leq \kappa_{N-1}$, κ_i $(1 \leq i \leq N-1)$ is written as

$$\kappa_i = k_i(p, X), \ p = \nabla u(x_0) \neq 0, \ X = \nabla^2 u(x_0)$$

where $k_i(p, X) \leq k_2(p, X) \leq \cdots \leq k_{N-1}(p, X)$ are the eigenvalues of the linear operator $Q_p(X)/|p|$ the orthogonal complement of the vector p. By this expression it is possible to express the m-th symmetric curvature by u.

There is another way to derive the level set representation of the mean curvature and anisotropic mean curvature without using the representation (2.4.5) but using surface divergence. We set $\mathbf{m} = -\nabla u/|\nabla u|$ and note that the vector field $\mathbf{m}$ is defined not only on Γ but also on some neighborhood of Γ near x_0. It is an extension of $\mathbf{n}$ in a neighborhood of Γ. By (2.3.6) we observe that

$$\begin{aligned} H &= -\mathrm{div}_\Gamma \mathbf{n} = -\mathrm{trace}((I - \mathbf{n}(x_0) \otimes \mathbf{n}(x_0))(\nabla \mathbf{m})(x_0)) \\ &= -\mathrm{trace}(\nabla \mathbf{m}(x_0)) + \mathrm{trace}(\mathbf{n}(x_0) \otimes \mathbf{n}(x_0)(\nabla \mathbf{m})(x_0)) \\ &= \mathrm{div}\left(\frac{\nabla u}{|\nabla u|}\right)(x_0) + \sum_{1 \leq i,\ j \leq N} n_i(x_0) n_j(x_0) (\frac{\partial}{\partial x_j} m_i)(x_0), \end{aligned} \tag{2.4.8}$$

where $\mathbf{n} = (n_1, \ldots, n_N)$, $\mathbf{m} = (m_1, \ldots, m_N)$. Here div denotes the divergence of a vector field on $\mathbf{R}^N$. Since $\mathbf{m}$ is a *unit* vector field near x_0,

$$\sum_{i=1}^N n_i(x_0)\left(\frac{\partial}{\partial x_j} m_i\right)(x_0) = \frac{1}{2}\left(\frac{\partial}{\partial x_j} \sum_{i=1}^N m_i^2\right)(x_0) = \frac{1}{2}\frac{\partial}{\partial x_j} 1 = 0.$$

Thus the last term of (2.4.8) disappears and we obtain

$$H = \left(\mathrm{div}\left(\frac{\nabla u}{|\nabla u|}\right)\right)(x_0) \tag{2.4.9}$$

which is the same as (2.4.6).

We shall give a level set representation of anisotropic mean curvature. We recall (2.3.6) to calculate

$$\begin{aligned} h &= -\mathrm{div}_\Gamma \xi(\mathbf{n}) = -\mathrm{trace}\{(I - \mathbf{n}(x_0) \otimes \mathbf{n}(x_0))(\nabla \xi(\mathbf{m}))(x_0)\} \\ &= -\mathrm{trace}(\nabla \xi(\mathbf{m}))(x_0) + \mathrm{trace}\{(\mathbf{n}(x_0) \otimes \mathbf{n}(x_0))(\nabla \xi(\mathbf{m}))(x_0)\}. \end{aligned} \tag{2.4.10}$$

The second term is of form

$$\sum_{1 \le i,k,\ell \le N} n_i(x_0) n_j(x_0) \frac{\partial \xi^i}{\partial p_\ell}(\mathbf{n}(x_0)) \frac{\partial m_\ell}{\partial x_j}(x_0)$$

with $\xi = (\xi^1, \dots, \xi^N)$. Since $\xi = \nabla \gamma$ and γ is positively homogeneous of degree 1 so that

$$\gamma(\lambda p) = \lambda \gamma(p),\ \lambda > 0,$$

differentiating in p_ℓ yields

$$\lambda \xi^\ell(\lambda p) = \lambda \xi^\ell(p) \quad \text{or} \quad \xi^\ell(\lambda p) = \xi^\ell(p)$$

In other words ξ^ℓ is positively homogeneous of degree zero. Differentiating in λ and setting $\lambda = 1$ yields the Euler equation

$$\sum_{i=1}^N p_i \frac{\partial \xi^\ell}{\partial p_i}(p) = 0,\ 1 \le \ell \le N. \tag{2.4.11}$$

Since $\partial \xi^\ell / \partial p_i = \partial \xi^i / \partial p_\ell$, the second term of (2.4.10) can be rewritten as

$$\begin{aligned} &\sum_{1 \le i,j,\ell \le N} n_i(x_0) n_j(x_0) \frac{\partial \xi^\ell}{\partial p_i}(\mathbf{n}(x_0)) \frac{\partial m_\ell}{\partial x_j}(x_0) \\ &= \sum_{1 \le i,j,\ell \le N} \left(\sum_{i=1}^N n_i(x_0) \frac{\partial \xi^\ell}{\partial p_i}(\mathbf{n}(x_0)) \right) \frac{\partial m_\ell}{\partial x_j}(x_0) n_j(x_0) = 0 \quad \text{by (2.4.11).} \end{aligned}$$

Thus we have

$$h = -(\mathrm{div}\, \xi(\mathbf{m}))(x_0); \tag{2.4.12}$$

note that this formula holds for any extension $\mathbf{m}$ of $\mathbf{n}$ in a tubular neighborhood of Γ near x_0 since we do not use the property $|\mathbf{m}| = 1$. Instead, ξ should be a Cahn–Hoffman vector. If $\gamma(p) = |p|$ so that $\xi = p/|p|$, we recover the formula (2.4.9) with $\mathbf{m} = -\nabla u / |\nabla u|$. Since ξ is homogeneous of degree zero, (2.4.12) yields

$$\begin{aligned} h &= -\mathrm{div}(\xi(-\nabla u))(x_0) \\ &= \sum_{1 \le i,j \le N} \frac{\partial^2 \gamma}{\partial p_i \partial p_j}(-\nabla u(x_0)) \frac{\partial^2 u}{\partial x_i \partial x_j}(x_0). \end{aligned} \tag{2.4.13}$$

(Of course if $\gamma(p) = |p|$, this again yields (2.4.9).) By (2.4.11) we see $R_p \nabla^2 \gamma(p) = \nabla^2 \gamma(p) = \nabla^2 \gamma(p) R_p$. From (2.4.13) it now follows that

$$h = \frac{1}{|\nabla u(x_0)|} \operatorname{trace}\left(\nabla^2 \gamma(\mathbf{n}(x_0)) Q_{\mathbf{n}(x_0)}(\nabla^2 u)(x_0)\right) \tag{2.4.14}$$

with $\mathbf{n}(x_0) = -\nabla u(x_0)/|\nabla u(x_0)|$ since $\partial^2 \gamma / \partial p_i \partial p_j$ is homogeneous of degree -1 so that

$$\frac{\partial^2 \gamma}{\partial p_i \partial p_j}(-\nabla u) = \frac{1}{|\nabla u|} \frac{\partial^2 \gamma}{\partial p_i \partial p_j}\left(-\frac{\nabla u}{|\nabla u|}\right), \quad 1 \leq i, \ j \leq N.$$

Using the second fundamental form in (2.4.14) we have

$$h = \operatorname{trace}(\nabla^2 \gamma(\mathbf{n}(x_0)) \tilde{\mathbf{A}}_{x_0}). \tag{2.4.15}$$

The formula (2.4.15) may explain a reason why h is sometimes called a weighted mean curvature.

Graph Representation It is easy to derive a formula for the second fundamental form from (2.4.5) when Γ is given as the graph of a function. By rotating coordinates we may assume that Γ is expressed as

$$\Gamma = \left\{x_N = g(x'), \ x' \in \mathbf{R}^{N-1}\right\}$$

around $x_0 \in \Gamma$, where $g(x_0') = x_{0N}$. If $\mathbf{n}$ is taken upward, then Γ is given as the zero level set of

$$u(x) = -x_N + g(x')$$

with $\mathbf{n} = -\nabla u / |\nabla u|$ which is the same as (2.1.2). Plugging in (2.4.5) we obtain a formula of $\tilde{\mathbf{A}}_{x_0}$ written by g. A general formula is complicated so we do not give it here. However if $\nabla' g(x_0') = 0$ so that $\mathbf{n}(x_0) = (0, 0, \ldots, 0, 1)$, then the expression of $\tilde{\mathbf{A}}_{x_0}$ is simple. Indeed, since $\nabla u(x_0) = -\mathbf{n}(x_0)$, we see

$$Q_{\nabla u(x_0)}(\nabla^2 u(x_0)) = \begin{pmatrix} \nabla'^2 g(x_0) & 0 \\ 0 & 0 \end{pmatrix}.$$

Thus we obtain

$$\tilde{\mathbf{A}}_{x_0} \zeta = \begin{pmatrix} \nabla'^2 g(x_0) \zeta' \\ 0 \end{pmatrix}, \ \zeta' \in \mathbf{R}^{N-1} \quad \text{with } \zeta = (\zeta', \zeta_n) \in \mathbf{R}^N. \tag{2.4.16}$$

We shall calculate the mean curvature H at x_0 in the direction of $\mathbf{n}$. We plug u in (2.4.9) to get

$$\begin{aligned} H &= \left(\operatorname{div}'\left(\frac{\nabla' g}{\sqrt{1+|\nabla' g|^2}}\right)\right)(x_0') + \frac{\partial}{\partial x_N}\left(\frac{-1}{\sqrt{1+|\nabla' g|^2}}(x_0')\right) \\ &= \left(\operatorname{div}'\left(\frac{\nabla' g}{\sqrt{1+|\nabla' g|^2}}\right)\right)(x_0'), \end{aligned} \tag{2.4.17}$$

where div' denotes the divergence in x' variables.

Axisymmetric Surface Suppose that Γ is obtained by rotating the graph of a function $\varphi(x_1)$ around the x_1-axis and that $x_0 \in \Gamma$ is not on the axis. Around $x_0 \in \Gamma$ the hypersurface Γ is of form

$$\Gamma = \left\{ r = \varphi(x_1);\ r = \left(\sum_{j=2}^{N} x_j^2\right)^{1/2} \right\}.$$

If $\mathbf{n}$ is taken outward from x_1-axis, i.e., $\mathbf{n}$ is given by (2.2.3), then Γ is given as the zero level set of

$$u(x_1, \dots, x_N) = -r + \varphi(x_1)$$

with $\mathbf{n} = -\nabla u/|\nabla u|$ (around x_0). We obtain a formula for $\tilde{\mathbf{A}}_{x_0}$ in terms of φ by plugging above u into (2.4.5). However, we do not give its explicit formula. Here we only calculate the mean curvature H at x_0 in the direction of $\mathbf{n}$. We plug u in (2.4.9) and use the formula (2.2.3) to get

$$\begin{aligned} H &= \frac{\partial}{\partial x_1}\left(\frac{\varphi_{x_1}}{(1+\varphi_{x_1}^2)^{1/2}}\right) - \frac{1}{(1+\varphi_{x_1}^2)^{1/2}} \sum_{j=2}^{N} \frac{\partial}{\partial x_j}\frac{x_j}{r} \\ &= \frac{\varphi_{x_1 x_1}}{(1+\varphi_{x_1}^2)^{3/2}} - \frac{1}{(1+\varphi_{x_1}^2)^{1/2}} \frac{N-2}{r} \quad \text{at } x = x_0 = (x_{01}, x_0'') \end{aligned} \tag{2.4.18}$$

where $r = |x''|$ with $x'' = (x_2, \dots, x_N)$.

Gradient of Normal Vector Fields Let $\mathbf{n}$ be a unit normal vector field of Γ around x_0. Let $\mathbf{m}$ be a (C^1) extension of $\mathbf{n}$ to a tubular neighborhood of Γ around x_0. Since

$$\mathbf{A}_{x_0}\,\tau = -((\tau \cdot \nabla)\mathbf{m})(x_0),$$

for $\tau \in T_{x_0}\Gamma$,

$$\begin{aligned}\tilde{\mathbf{A}}_{x_0}\zeta &= -((\Pi_{x_0}\zeta \cdot \nabla)\mathbf{m})(x_0)\\ &= -((\zeta \cdot \nabla)\mathbf{m})(x_0) + \langle\, \zeta, \mathbf{n}(x_0)\,\rangle(\mathbf{n}(x_0) \cdot \nabla)\mathbf{m}(x_0).\end{aligned}$$

This implies that the matrix expression of $\tilde{\mathbf{A}}_{x_0}$ (with respect to the standard Euclidean basis) equals

$$-\frac{\partial}{\partial x_i} m_j(x_0) + n_i(x)(\mathbf{n}(x_0) \cdot \nabla)m_j(x_0),\ 1 \le i,\ j \le N$$

(which should be symmetric since $\tilde{\mathbf{A}}_{x_0}$ is symmetric). If the extension $\mathbf{m}$ has the property that

$$((\mathbf{n}(x_0) \cdot \nabla)\mathbf{m})(x_0) = 0, \tag{2.4.19}$$

then one may identify $-\nabla\mathbf{m}(x_0)$ by $\tilde{\mathbf{A}}_{x_0}$. We shall use the notation $\nabla\mathbf{n}$ by extending $\mathbf{n}$ to a tubular neighborhood of Γ around x_0 such that (2.4.19) holds. By this interpretation $-\nabla\mathbf{n} = \tilde{\mathbf{A}}_{x_0}$, so we shall often identify second fundamental form $\mathbf{A}_{x_0}$ with $-\nabla\mathbf{n}$.

We conclude this section by studying the range of $\nabla\mathbf{n}$. Let $\mathbf{S}^N$ denote the space of all $N \times N$ real symmetric matrices.

Lemma 2.4.1 *Let x_0 be a point in $\mathbf{R}^N$. For each $p \in S^{N-1}$ and $X \in \mathbf{S}^N$ there is a smooth hypersurface Γ around x_0 with the property that $\mathbf{n}(x_0) = p$, $\nabla\mathbf{n}(x_0) = Q_p(X)$, where $\mathbf{n}$ is a unit normal vector field on Γ around x_0 and is extended to a tubular neighborhood of Γ satisfying $((\mathbf{n}(x_0) \cdot \nabla)\mathbf{n})(x_0) = 0$.*

Proof We may assume $p = (0, 0, \ldots, 1)$ by rotation of coordinates and $x_0 = 0$ by translation. The matrix $Q_p(X)$ is of the form

$$\begin{pmatrix} & & 0\\ & Y & \vdots\\ 0 & \cdots & 0\end{pmatrix}$$

with $Y \in \mathbf{S}^{N-1}$. We set

$$g(x') = -\frac{1}{2}\langle\, Y x', x'\,\rangle \quad \text{for } x' \in \mathbf{R}^{N-1}$$

to get $g(0) = 0$, $\nabla' g(0) = 0$, $\nabla'^2 g(0) = -Y$. Since $\nabla\mathbf{n}(x_0) = -\tilde{\mathbf{A}}_{x_0}$, the formula (2.4.16) yields

$$\nabla\mathbf{n}(x_0) = Q_p(X).$$

□

2.5 Examples of Surface Evolution Equations

We give general examples of equations of an evolving hypersurface whose normal velocity V is determined by its normals and second fundamental forms. In general such an evolution equation is of the form

$$V = f(x, t, \mathbf{n}, \nabla\mathbf{n}) \tag{2.5.1}$$

on an evolving hypersurface Γ_t, where f is a given function and $\mathbf{n}$ is a unit normal vector field of Γ_t. For consistency with the literature we take the negative of the second fundamental form $\mathbf{A}$ of Γ_t as an independent variable of f and denote it by $\nabla\mathbf{n}$. Here V is the velocity in the direction of $\mathbf{n}$.

2.5.1 *General Evolutions of Isothermal Interfaces*

For an evolving hypersurface Γ_t we consider

$$\beta(\mathbf{n})V = -a \operatorname{div}_{\Gamma_t}\xi(\mathbf{n}) + c(x, t). \tag{2.5.2}$$

Here ξ is the Cahn–Hoffman vector of a surface energy density $\gamma_0 : S^{N-1} \to (0, \infty)$ and β is a given positive function on S^{N-1}, a is a nonnegative constant and c is a given function. We always assume that $N \geq 2$ if $a \neq 0$ since the curvature term equals zero when $N = 1$. As we observed in (2.4.15), this equation is an example of (2.5.1). If c is independent of x and $a = 1$, (2.5.2) is often used to describe motion of isothermal interface; there c is regarded as bulk free energy difference such as temperature difference in both phases. The function $1/\beta$ is called the *mobility*. It may be again anisotropic in the sense that it depends on the direction of normals. The mobility is determined by microstructure of the hypersurface. Sometimes it is proportional to γ_0, i.e., $\beta\gamma_0$ is constant independent of $\mathbf{n}$ but in general $1/\beta$ is not necessarily proportional to γ_0. Equation (2.5.2) has an energy structure. Indeed, we set

$$G(\Gamma) = \int_\Gamma a\, \gamma_0(\mathbf{n})d\sigma + \int_D c\, dx \tag{2.5.3}$$

for a hypersurface Γ surrounding D, then (2.5.2) is of the form

$$\beta(\mathbf{n})V = -\delta G/\delta\Gamma_t$$

where $\delta G/\delta\Gamma_t$ denotes the change ratio of (free energy) G per change of volume of D in the direction of $\mathbf{n}$. Here $\mathbf{n}$ is taken outward from D. In other words (2.5.2) is a gradient flow of G.

Mean Curvature Flow Equation Equation (2.5.2) includes several interesting important examples as a special case. If the mobility and the surface energy density is isotropic with no driving force c and $a = 1$, then (2.5.2) becomes

$$V = H \tag{2.5.4}$$

by rescaling time if necessary (or taking $\beta \equiv \gamma_0 \equiv 1$, $a = 1$, $c = 0$). This equation is called the *mean curvature flow equation*. If $N = 2$ so that Γ_t is a curve, (2.5.4) is called the *curve shortening equation*. If κ denotes the curvature in the direction of $\mathbf{n}$, the curve shortening equation is of the form

$$V = \kappa. \tag{2.5.5}$$

For the mean curvature flow equation the energy G in (2.5.3) is the surface area of Γ. Thus (2.5.4) gives a deformation so that decrease ratio of area is steepest. This is why (2.5.5) is called the curve shortening equation. Note that Eq. (2.5.4) is invariant under the change of orientation $\mathbf{n}$. In other words the evolution law (2.5.4) is the same even if we replace $\mathbf{n}$ by $-\mathbf{n}$.

Hamilton–Jacobi Equations If $a = 0$, then (2.5.2) becomes

$$\beta(\mathbf{n})V = c(x, t). \tag{2.5.6}$$

This equation is regarded as a special form of the Hamilton–Jacobi equation. Indeed, if Γ_t is represented by the graph of a function, i.e., $x_N = g(x', t)$, then by (2.1.2) and (2.2.2) Eq. (2.5.6) is of the form

$$g_t + H(x',\ t,\ g,\ \nabla' g) = 0 \tag{2.5.7}$$

with

$$H(x',\ t,\ r,\ p') = -c(x',\ r,\ t)\mu(p')/\beta(-p'/\mu(p'),\ \mu(p')^{-1}),$$
$$\mu(p') = (1 + |p'|^2)^{1/2}.$$

Equation (2.5.7) is a first order equation and the curvature plays no role in (2.5.6).

2.5.2 Evolution by Principal Curvatures

In the mean curvature flow equation the normal velocity depends only on principal curvatures of hypersurfaces. There are several other examples of form

$$V = g(\kappa_1, \ldots, \kappa_{N-1};\ \mathbf{n}), \tag{2.5.8}$$

where g is a given function of principal curvatures of $\kappa_1, \dots, \kappa_{N-1}$ and $\mathbf{n}$. In the mean curvature flow equation, g is independent of $\mathbf{n}$ and is taken as the first elementary symmetric polynomial e_1.

If g is taken as e_{N-1} so that $e_{N-1}(\kappa_1, \dots, \kappa_{N-1})$ equals the Gaussian curvature K, (2.5.8) becomes the *Gaussian curvature flow equation*

$$V = K. \tag{2.5.9}$$

Of course if $N = 2$, then this equation becomes the curve shortening equation $V = \kappa$. More generally, for each $1 \le m \le N-1$,

$$V = H_m \tag{2.5.10}$$

is called the *m-th symmetric curvature flow equation*, where H_m is the m-th symmetric curvature $e_m(\kappa_1, \dots, \kappa_{N-1})$. The equation

$$V = H_{N-1}/H_{N-2} \quad (N \ge 3) \tag{2.5.11}$$

is called the *harmonic curvature flow equation*, which is a special case of

$$V = H_m/H_\ell. \tag{2.5.12}$$

All equations (2.5.9)–(2.5.12) are examples of (2.5.8). As we see later, we impose the restriction $\ell < m$ in (2.5.12) so that the equation is parabolic at least for convex surfaces. Note that in general the evolution law (2.5.9) or (2.5.10) (with even m) may depend on the choice of orientation $\mathbf{n}$. For a closed evolving hypersurface Γ_t we take the inward normal vector field as $\mathbf{n}$ so that the sphere shrinks as time develops. The same remark applies to (2.5.12). We use this convention when we consider the evolution by principle curvatures (2.5.10) with even m and (2.5.12) with even $m-\ell$.

2.5.3 Other Examples

In general the right-hand side of (2.5.2) is not proportional to the velocity. A natural generalization is

$$V = h(-a \operatorname{div}_{\Gamma_t} \xi(\mathbf{n}) + c, \ \mathbf{n}) \tag{2.5.13}$$

where h is nondecreasing in the first variable and $h(0, \mathbf{n}) = 0$. If $N = 2$ so that Γ_t is a curve in the plane and $h(\sigma, \mathbf{n}) = \sigma_+^{1/3}$ with $c = 0$, $a = 1$, $\gamma_0 \equiv 1$, then (2.5.13) becomes

$$V = (\kappa_+)^{1/3} \tag{2.5.14}$$

where $\sigma_+ = \max(\sigma, 0)$. Equation (2.5.14) is called the *affine curvature flow equation* since the equation is invariant under affine transformation.

2.5.4 Boundary Conditions

It often happens that a hypersurface Γ_t moves in a domain Ω in $\mathbf{R}^N$ and the geometric boundary of Γ_t intersects the boundary $\partial\Omega$ of Ω. In this case in addition to Eq. (2.5.1) in Ω we have to impose the boundary condition so that evolution is determined by equations. We give here typical examples of them. We assume that the boundary $\partial\Omega$ is at least a C^1 hypersurface in $\mathbf{R}^N$. Let ν be the unit normal vector field of $\partial\Omega$ outward from Ω. Let $\mathbf{n}$ be the unit normal vector field of a smooth hypersurface $\{\Gamma_t\}$ in $\overline{\Omega}$.

Neumann Boundary Condition This condition imposes

$$\langle \nu, \mathbf{n} \rangle = 0$$

on the intersection of Γ_t and $\partial\Omega$. Geometrically speaking, Γ_t intersects $\partial\Omega$ orthogonally, i.e., $\Gamma_t \perp \partial\Omega$.

Prescribed Contact Angle Boundary Condition Let z be a given real-valued continuous function on $\partial\Omega$ that satisfies $|z| < 1$ on $\partial\Omega$. The prescribed contact boundary condition imposes

$$\langle \nu, \mathbf{n} \rangle = z$$

on the intersection of Γ_t and $\partial\Omega$ (Fig. 2.3). Of course if $z = 0$, this condition is exactly the Neumann boundary condition. Although the orientation of Γ_t is irrelevant to describe the Neumann boundary condition, the prescribed angle condition depends on the orientation of Γ_t. In the literature the prescribed contact angle condition is often referred to as the Neumann boundary condition.

Dirichlet Boundary Condition Let S be a given codimension 2-closed surface in $\mathbf{R}^N$. The Dirichlet condition imposes that the geometric boundary of Γ_t always

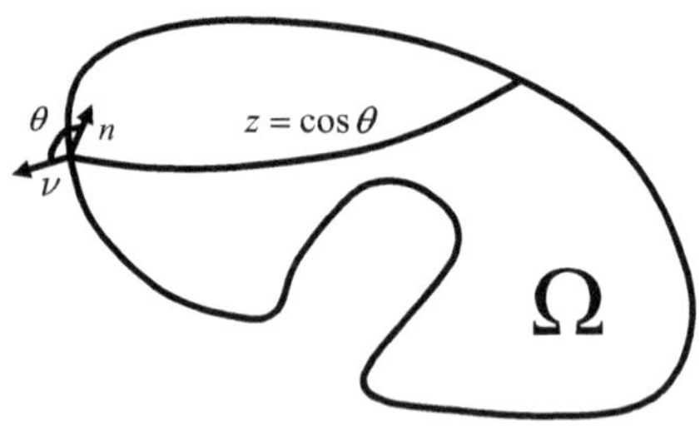

Fig. 2.3 Prescribed contact angle

equals S. This condition is so far not easy to treat in a level set method so we do not discuss this problem much in this book.

2.6 Level Set Equations

For a given surface evolution equation we shall introduce its level set equation. We shall study various properties of level set equations.

2.6.1 Examples

We consider a surface evolution equation

$$V = f(x,\ t,\ \mathbf{n},\ \nabla\mathbf{n}) \tag{2.6.1}$$

on an evolving hypersurface Γ_t in a domain Ω in $\mathbf{R}^N$. Here $f(x,\ t,\ \cdot,\ \cdot)$ for $(x,t) \in \Omega \times [0,T]$ is a given real-valued function defined in

$$\mathbf{E} = \{(p,\ Q_p(X));\ p \in S^{N-1},\ X \in \mathbf{S}^N\}.$$

By Lemma 2.4.1 the set $\mathbf{E}$ is a natural space so that $(\mathbf{n},\ \nabla\mathbf{n})$ lives. We say an equation

$$u_t(x,t) + F(x,\ t,\ \nabla u(x,t),\ \nabla^2 u(x,t)) = 0 \quad \text{for all } (x,t) \in \Omega \times (0,T) \tag{2.6.2}$$

is a *level set equation* of (2.6.1) if for *each* level set

$$\Gamma_t = \{x;\ u(x,t) = \text{constant}\}$$

of a $C^{2,1}$ solution u of (2.6.2) near $(x_0,t_0) \in \Omega$ satisfies (2.6.1) at (x_0,t_0) provided that $\nabla u(x_0,t_0) \neq 0$ and that the orientation $\mathbf{n}$ is chosen so that $\mathbf{n}(x_0,t_0) = -\nabla u(x_0,t_0)\ /|\nabla u(x_0,t_0)|$, where $x_0 \in \Gamma_{t_0}$. Here F is a real-valued function defined in $\Omega \times [0,T] \times (\mathbf{R}^N \backslash \{0\}) \times \mathbf{S}^N$.

Such an equation is uniquely determined by (2.6.1). Indeed, using the level set representation of V, $\mathbf{n}$ and $\nabla\mathbf{n}$, (2.6.1) is of form

$$\frac{u_t}{|\nabla u|} = f\left(x, t, -\frac{\nabla u}{|\nabla u|},\ -\frac{1}{|\nabla u|}\ Q_{\nabla u}(\nabla^2 u)\right)$$

on the evolving hypersurface. The representation of V, $\mathbf{n}$ and $\nabla\mathbf{n}$ is the same if the hypersurface is a c-level set instead of the zero level set of u. Thus, the equation

$$u_t - |\nabla u| f\left(x, t, -\frac{\nabla u}{|\nabla u|}, -\frac{1}{|\nabla u|} Q_{\nabla u}(\nabla^2 u)\right) = 0$$

is the unique level set equation. It may be rewritten as

$$u_t + F_f(x, t, \nabla u, \nabla^2 u) = 0 \tag{2.6.3}$$

with

$$F_f(x, t, p, X) = -|p| f\left(x, t, -\frac{p}{|p|}, -\frac{1}{|p|} Q_p(X)\right) \tag{2.6.4}$$

for $p \in \mathbf{R}^N \backslash \{0\}$, $X \in \mathbf{S}^N$, $(x, t) \in \Omega \times [0, T]$. Note that the function F_f is not defined for $p = 0$ in general as we will see in following examples.

Level Set Mean Curvature Flow Equation If (2.6.1) is the mean curvature flow equation (2.5.4) : $V = H$, then by (2.2.1) and (2.4.6), the level set equation is

$$u_t - \Delta u + \sum_{1 \le i,\, j \le N} \frac{u_{x_i} u_{x_j}}{|\nabla u|^2} u_{x_i x_j} = 0 \tag{2.6.5}$$

or (2.6.3) with

$$F_f(x, t, p, X) = F_f(p, X) = -\text{trace}\left(I - \frac{p \otimes p}{|p|^2}\right) X, \ p \neq 0, \ X \in \mathbf{S}^N. \tag{2.6.6}$$

Note that F is not defined for $p = 0$. Using (2.4.9) we often write the level set equation of (2.5.4) as

$$u_t - |\nabla u| \text{div}\left(\frac{\nabla u}{|\nabla u|}\right) = 0 \tag{2.6.7}$$

which is of course the same as (2.6.5). Equation (2.6.5) (and its equivalent form (2.6.7)) is called the *level set mean curvature flow equation.*

Level Set Hamilton–Jacobi Equation We consider the level set equation of the Hamilton–Jacobi equation (2.5.6)

$$\beta(\mathbf{n}) V = c(x, t).$$

By (2.2.1) and (2.4.1) its level set equation is of form

$$u_t - c(x, t)|\nabla u| \beta(-\nabla u / |\nabla u|) = 0 \tag{2.6.8}$$

or (2.6.3) with

$$F(x, t, p, X) = -|p|\beta(-p/|p|)c(x, t), \tag{2.6.9}$$

which is independent of X. Equation (2.6.8) is again the first order Hamilton–Jacobi equation

$$u_t + H(x, t, \nabla u) = 0$$

with the Hamiltonian $H(x, t, p) = F_f(x, t, p, X)$ which is not necessarily convex in p. We also note that $H(x, t, p)$ can be extended continuously to $p = 0$.

Anisotropic Version We consider the anisotropic version of the mean curvature flow given by (2.5.2). By (2.2.1), (2.4.1) and (2.4.13) its level set equation is

$$u_t - |\nabla u| \left(a \sum_{1\le i,\ j\le N} \frac{\partial^2 \gamma}{\partial p_i\, \partial p_j}(-\nabla u) \frac{\partial^2 u}{\partial x_i\, \partial x_j} + c \right) \frac{1}{\beta(-\nabla u/|\nabla u|)} = 0 \tag{2.6.10}$$

or (2.6.3) with

$$\begin{aligned} F_f(x, t, p, X) &= -\{a \operatorname{trace}(\nabla^2\gamma(-p)X) + c(x, t)\} \frac{|p|}{\beta(-p/|p|)} \\ &= -\{a \operatorname{trace}(\nabla^2\gamma(-p)R_p X R_p) + c(x, t)\} \frac{|p|}{\beta(-p/|p|)} \end{aligned} \tag{2.6.11}$$

by (2.4.14), where R_p is given by (2.4.3). This F_f depends on (x, t) through c and it is defined for all $p \in \mathbf{R}^N \backslash \{0\}$ and $X \in \mathbf{S}^N$.

Level Set Gaussian Curvature Flow Equation We consider the Gaussian curvature flow equation (2.5.9) : $V = K$. Its level set equation is of form

$$u_t - |\nabla u| \det \left(\left(I - \frac{\nabla u \otimes \nabla u}{|\nabla u|^2} \right) \frac{\nabla^2 u}{|\nabla u|} \left(I - \frac{\nabla u \otimes \nabla u}{|\nabla u|^2} \right) + \frac{\nabla u \otimes \nabla u}{|\nabla u|^2} \right) = 0 \tag{2.6.12}$$

since we have (2.4.7) for the Gaussian curvature. If we write it in the form (2.6.3),

$$F_f(x, t, p, X) = -|p| \det \left(R_p X R_p + \frac{\nabla u \otimes \nabla u}{|\nabla u|^2} \right). \tag{2.6.13}$$

In the above examples the level set equation is directly computable by (2.6.4) once the surface equation is given by (2.6.1) with explicit f. For example, the mean curvature flow equation is of the form

$$V = -\operatorname{trace} \nabla \mathbf{n}$$

so that $f(\mathbf{n},\ \nabla \mathbf{n}) = \operatorname{trace} \nabla \mathbf{n}$. Then by (2.6.4)

$$F_f(x,t,p,X) = -|p| \text{ trace } Q_p(X)/|p| = -\text{trace } Q_p(X),$$

which is the same as in (2.6.6).

Boundary Conditions If the boundary condition on $\partial\Omega$ is imposed for an evolving hypersurface Γ_t in a domain Ω, it should be included in the level set equation. In the level set representation the Neumann boundary condition

$$\langle\, \nu,\ \mathbf{n} \,\rangle = 0$$

is written as

$$\left\langle \nu,\ -\frac{\nabla u}{|\nabla u|} \right\rangle = 0$$

if we take $\mathbf{n}$ as in (2.4.1). More generally, the prescribed contact angle condition can be written as

$$\left\langle \nu,\ -\frac{\nabla u}{|\nabla u|} \right\rangle = z$$

or

$$\frac{\partial u}{\partial \nu} + z\,|\nabla u| = 0, \tag{2.6.14}$$

where ν is the outward unit normal vector field of $\partial\Omega$ and $|z| < 1$; $\partial u/\partial \nu$ denotes the directional derivative of u in the direction of ν, i.e., $\partial u/\partial \nu = (\nu \cdot \nabla)u$. If a boundary condition is imposed for (2.6.1) on Ω, its level set equation should include the boundary condition. For prescribed contact angle condition it is easy to include. However, for the Dirichlet problem, it is not clear in what way we include it. The level set equation of the boundary problem for (2.6.1) requires that each level set of solutions must satisfy the boundary condition. For example the level set equation of $V = H$ with the prescribed contact angle condition $\langle\, \nu,\ \mathbf{n} \,\rangle = z$ is

$$\begin{cases} u_t - \Delta u + \Sigma \dfrac{u_{x_i} u_{x_j}}{|\nabla u|^2}\, u_{x_i x_j} = 0, \text{ in } \Omega \times (0,T) \\ \qquad\qquad \dfrac{\partial u}{\partial \nu} + z\,|\nabla u| = 0 \text{ on } \partial\Omega \times (0,T). \end{cases}$$

2.6.2 General Scaling Invariance

The function F_f defined by (2.6.4) has special scaling properties in p and X. To see this we suppress the dependence in (x,t). Let f be a real-valued function defined in

$$\mathbf{E} = \{(p, Q_p(X));\ p \in S^{N-1},\ X \in \mathbf{S}^N\}. \tag{2.6.15}$$

We set

$$F_f(p, X) = -|p|\, f(-\frac{p}{|p|}, -\frac{1}{|p|}\, Q_p(X)),\ p \in \mathbf{R}^N\backslash\{0\},\ X \in \mathbf{S}^N. \tag{2.6.16}$$

Then F_f fulfills

(G1) $F_f(\lambda p,\ \lambda X) = \lambda F_f(p, X)$, for all $\lambda > 0$, $p \in \mathbf{R}^N\backslash\{0\}$, $X \in \mathbf{S}^N$,
(G2) $F_f(p, X+p\otimes y+y\otimes p) = F_f(p, X)$, for all $p \in \mathbf{R}^N\backslash\{0\}$, $y \in \mathbf{R}^N$, $X \in \mathbf{S}^N$.

Indeed, (G1) follows from definition of F_f since Q_p is a linear operator. To show (G2) we note the identity

$$(x \otimes p)(p \otimes y) = x \otimes y|p|^2.$$

Using this identity, we see that

$$R_p\ p \otimes y = \left(I - \frac{p \otimes p}{|p|^2}\right)\ p \otimes y = p \otimes y - p \otimes y = 0$$

and similarly $y \otimes p\ R_p = 0$, where R_p is given in (2.4.3). Thus

$$Q_p(p \otimes y + y \otimes p) = R_p(p \otimes y + y \otimes p)\ R_p = 0. \tag{2.6.17}$$

From this identity (G2) follows.

Definition 2.6.1 Let F be a real-valued function in $(\mathbf{R}^N\backslash\{0\}) \times \mathbf{S}^N$. We say that F is *strongly geometric* if F fulfills (G1) and (G2).

We have seen that F_f is always strongly geometric. We shall prove its converse: if F is strongly geometric, then there is (unique) f with $F = F_f$ (Theorem 2.6.4). To see this we study the structure of $\mathbf{E}$ as a bundle over a unit sphere S^{N-1}. The vector space $\mathbf{S}^N$ is equipped with an inner product

$$\langle\, A \mid B\, \rangle = \text{trace}\ AB,\ A,\ B \in \mathbf{S}^N.$$

For a given $p \in S^{N-1}$ we consider the linear operator from $\mathbf{S}^N$ into itself defined by

$$Q_p(X) = R_p X R_p,\ X \in \mathbf{S}^N$$

as in (2.4.3), (2.4.4). Let Q_p^* denote the adjoint operator of Q_p with respect to the inner product $\langle\, \cdot | \cdot\, \rangle$.

Lemma 2.6.2

(i) *The operator* Q_p *is an orthogonal projection on* $\mathbf{S}^N$, *i.e.,* $Q_p^2 = Q_p$ *and* $Q_p^* = Q_p$.
(ii) *The kernel of* Q_p *in* $\mathbf{S}^N$ *equals one-dimensional space*

$$L_p = \{p \otimes y + y \otimes p;\ y \in \mathbf{R}^N\}.$$

Proof

(i) The property $Q_p^2 = Q_p$ follows from $R_p^2 = R_p$. Since

$$\langle\, Q_p(X) \mid Y\,\rangle = \mathrm{trace}(R_p X R_p Y) = \mathrm{trace}(X R_p Y R_p) = \langle\, X \mid Q_p(Y)\,\rangle$$

for all $X, Y \in \mathbf{S}^N$, the operator Q_p is self-adjoint, i.e., $Q_p^* = Q_p$.
(ii) By (2.6.17) L_p is contained in the kernel of Q_p. It remains to prove that $Q_p(X) = 0$ for $X \in \mathbf{S}^N$ implies $X \in L_p$. By definition of Q_p we see

$$U^{-1} Q_p(X) U = Q_q(Y),\ q = pU,\ Y = U^{-1} X U,\ X \in \mathbf{S}^N$$

for any orthogonal matrix U, where p, q are regarded as row vectors. We take U so that $q = (1, 0, \ldots, 0)$ and observe that $Q_q(Y) = 0$ implies

$$Y = \begin{pmatrix} 2y_1 & y_2 & \ldots & y_N \\ y_2 & 0 & \ldots & 0 \\ \vdots & \vdots & & \vdots \\ y_N & 0 & \ldots & 0 \end{pmatrix},$$

since $X \in \mathbf{S}^N$ implies $Y \in \mathbf{S}^N$. Thus $Q_p(X) = 0$ implies $Y = q \otimes y + y \otimes q$ with $y = (y_1, \ldots, y_N)$, which is the same as $X \in L_p$.

□

The set $\mathbf{E}$ defined by (2.6.15) is regarded as a (smooth) vector subbundle of a trivial bundle $S^{N-1} \times \mathbf{S}^N$. The fibre dimension of $\mathbf{E}$ equals $N(N-1)/2$. Let Q be a bundle map

$$Q : S^{N-1} \times \mathbf{S}^N \to \mathbf{E}$$

defined by

$$Q(p, X) = (p, Q_p(X)).$$

Let $\mathbf{L}$ be a bundle over S^{N-1} of the form

$$\mathbf{L} = \{(p, X);\ p \in S^{N-1},\ X \in L_p\}.$$

Since Q is surjective to $\mathbf{E}$, Lemma 2.6.2 provides a direct sum decomposition of $S^{N-1} \times \mathbf{S}^N$.

Lemma 2.6.3 *The bundle $S^{N-1} \times \mathbf{S}^N$ is expressed as an orthogonal sum of the form $\mathbf{L} \oplus \mathbf{E}$ as bundles over S^{N-1}. The operator Q gives a projection to $\mathbf{E}$ on fibres.*

Let $\mathcal{G}$ be the set of all strongly geometric real-valued functions F defined in $(\mathbf{R}^N \backslash \{0\}) \times \mathbf{S}^N$. Let $\mathcal{I}$ be the set of all real-valued functions f defined in $\mathbf{E}$. Let $\mathbf{F}$ denote the mapping corresponding F_f to f, where F_f is defined by (2.6.16).

Theorem 2.6.4 *The mapping $\mathbf{F}$ is a bijection from $\mathcal{I}$ to $\mathcal{G}$.*

Proof Let $\mathcal{G}'$ be the set of all real-valued functions F' on $S^{N-1} \times \mathbf{S}^N$ satisfying (G2). Then the mapping $F' \mapsto F$ defined by

$$F(p, X) = |p|\, F'\left(\frac{p}{|p|}, \frac{X}{|p|}\right)$$

gives a bijection from $\mathcal{G}'$ to $\mathcal{G}$. By definition of $\mathbf{L}$ and L_p one may identify $F' \in \mathcal{G}'$ with a function on the quotient bundle $S^{N-1} \times \mathbf{S}^N/\mathbf{L}$. By Lemma 2.6.3, $S^{N-1} \times \mathbf{S}^N/\mathbf{L}$ is identified with $\mathbf{E}$ and the mapping $f \mapsto F'$ defined by

$$F'(p, X) = -f(-p, -Q_p(X)), \ p \in S^{N-1}, \ X \in \mathbf{S}^N/L_p$$

gives a bijection from $\mathcal{I}$ to $\mathcal{G}'$, since $Q_p = Q_{-p}$. Since $\mathbf{F} : f \mapsto F_f$ is a composition of $f \mapsto F'$ and $F' \mapsto F$, $\mathbf{F}$ is a bijection from $\mathcal{I}$ to $\mathcal{G}$. □

Remark 2.6.5 The bijective property of $\mathbf{F}$ is still valid for a function defined in a subset of $\mathbf{E}$ if $\mathcal{I}$ and $\mathcal{G}$ are appropriately modified. For a subset Σ of S^{N-1} let $\mathbf{E}_\Sigma$ be of form

$$\mathbf{E}_\Sigma = \{(p, Q_p(X));\ p \in \Sigma,\ X \in \mathbf{S}^N\}.$$

Let $\mathcal{I}$ be the set of all real-valued functions defined in $\mathbf{E}_\Sigma$. Let $\mathcal{G}$ be the set of all strongly geometric real-valued functions defined in $\tilde{\Sigma} \times \mathbf{S}^N$ with the cone $\tilde{\Sigma} = \{\lambda p;\ p \in \Sigma,\ \lambda > 0\}$; there we understand that (G1), (G2) holds for $p \in \tilde{\Sigma}$. Then $\mathbf{F}$ gives a bijection from $\mathcal{I}$ to $\mathcal{G}$ as before. The proof is the same.

2.6.3 *Ellipticity*

As we know the backward heat equation cannot be solvable for general smooth initial data even locally in time. We need some structural conditions for f to find a solution Γ_t with initial data Γ_0. We recall the notion of degenerate ellipticity and parabolicity for this purpose.

Definition 2.6.6 Let F be a real-valued function defined in $(\mathbf{R}^N \backslash \{0\}) \times \mathbf{S}^N$ (or in its subset $\tilde{\Sigma} \times \mathbf{S}^N$, where $\tilde{\Sigma} = \{\lambda p;\ p \in \Sigma \subset S^{N-1},\ \lambda > 0\}$). We say F is *degenerate elliptic* if

$$F(p, X) \leq F(p, Y), \quad p \in \mathbf{R}^N \backslash \{0\} \text{ (or } \tilde{\Sigma}) \tag{2.6.18}$$

for all $X, Y \in \mathbf{S}^N$ with $X \geq Y$. Here $X \geq Y$ means that $X - Y$ is a nonnegative matrix, i.e., $\langle\, (X - Y)\xi, \xi\, \rangle \geq 0$ for all $\xi \in \mathbf{R}^N$.

This condition is a kind of monotonicity of F in X. Fortunately, many examples F_f of level set equations fulfill this property as we see below.

Level Set Mean Curvature Flow Equation The function $F_f(p, X)$ defined by (2.6.6) is degenerate elliptic. Indeed, by definition

$$F_f(p, X) = -\text{trace}(R_p Y) - \text{trace}\left(R_p(X - Y)\right).$$

Since trace $AB \geq 0$ for $A \geq O,\ B \geq O$ and $R_p \geq O$, the last term is nonpositive if $X \geq Y$ so (2.6.18) follows. Here O denotes the $N \times N$ zero matrix.

Level Set Hamilton–Jacobi Equation The function $F_f(x, t, \cdot, \cdot)$ defined by (2.6.9) is independent of X so $F_f(x, t, \cdot, \cdot)$ is degenerate elliptic for all x, t. If a level set equation is of first order, F_f is always degenerate elliptic.

Anisotropic Version The function $F_f(x, t, \cdot, \cdot)$ defined by (2.6.11) is degenerate elliptic if $\nabla^2 \gamma \geq O$ as well as $a \geq 0$ and $\beta > 0$; as before we here assume that γ is C^2 outside the origin. The idea to prove (2.6.18) is the same as the proof for $-$ trace $(R_p X)$.

Surface Evolution Equation by Principal Curvatures The function F_f defined by (2.6.13) is no longer degenerate elliptic unless $N = 2$. We shall modify e_m so that F_f is degenerate elliptic. If we consider Eq. (2.5.8), F_f in the level set equation is of form

$$F_f(p, X) = -|p|\, g\left(k_1(p, X), \ldots, k_{N-1}(p, X),\ -\frac{p}{|p|}\right), \tag{2.6.19}$$

where the k_i's are eigenvalues of $Q_p(X)/|p|$ as defined in the paragraph on the Gaussian curvature in Sect. 2.4. There is a sufficient condition on g so that F_f is degenerate elliptic.

Proposition 2.6.7 *For each* $i = 1, \ldots, N-1,\ p \in S^{N-1}$ *and* $(\lambda_1, \ldots, \lambda_{i-1}, \lambda_{i+1}, \ldots, \lambda_{N-1}) \in \mathbf{R}^{N-2}$ *the function* $\lambda_i \mapsto g(\lambda_1, \ldots, \lambda_{N-1};\ p)$ *is nondecreasing in* $\mathbf{R}$. *Then* F_f *given by* (2.6.19) *is degenerate elliptic.*

Proof It suffices to prove that $X \geq Y$ implies $k_i(p, X) \geq k_i(p, Y)$ for $i = 1, 2, \ldots, N-1$. By rotation of p as in the proof of Lemma 2.6.2 we may assume that $p = (1, 0, \ldots, 0)$ and

$$Q_p(X) = \begin{pmatrix} 0 \dots 0 \\ \vdots \; X' \\ 0 \end{pmatrix}, \quad Q_p(Y) = \begin{pmatrix} 0 \dots 0 \\ \vdots \; Y' \\ 0 \end{pmatrix}$$

with $X', Y' \in \mathbf{S}^{N-1}$. Then $k_i(p, X)$ equals the i-th eigenvalue μ_i of $X'/|p|$ denoted $\mu_i(X'/|p|)$ where $\mu_1 \le \mu_2 \le \cdots \le \mu_{N-1}$. If $X \ge Y$, then evidently $X' \ge Y'$.

A minimax characterization (e.g. R. Courant and D. Hilbert [125]) of eigenvalues implies that $\mu_i(X'/|p|) \ge \mu_i(Y'/|p|)$, $1 \le i \le N-1$ for $X' \ge Y'$, so $k_i(p, X) \ge k_i(p, Y)$ for $i = 1, 2, \dots, N-1$. □

Remark 2.6.8 A minimax characterization of eigenvalues also implies that $k_i(p, X)$ is continuous on $(\mathbf{R}^N \setminus \{0\}) \times \mathbf{S}^N$; see for example a book of T. Kato [297]. By the observation we consider

$$g(\lambda_1, \dots, \lambda_{N-1};\; p) = \lambda_1^+ \cdots \lambda_{N-1}^+$$

instead of $\lambda_1 \cdots \lambda_{N-1}$ so that F_f in (2.6.19) is degenerate elliptic, where $\lambda_i^+ = \max(\lambda_i, 0)$. Moreover, F_f is continuous in $(\mathbf{R}^N \setminus \{0\}) \times \mathbf{S}^N$ since k_i's are continuous there.

The operation of the plus part depends upon the orientation $\mathbf{n}$. When we consider a convex surface, taking $\mathbf{n}$ inward is a way not to trivialize the problem.

More generally, when

$$g(\lambda_1, \dots, \lambda_{N-1};\; \mathbf{n}) = e_m(\lambda_1, \dots, \lambda_{N-1})$$

there is a way to modify g so that it satisfies the assumption on g in Proposition 2.6.7.

Lemma 2.6.9 *Let $N \ge 2$ and $1 \le m \le N-1$. There is a closed convex cone K_m in $\mathbf{R}^{N-1}$ with vertex at the origin such that*

(i) *$e_m(\lambda_1, \cdots, \lambda_{N-1})$ satisfies the monotonicity assumption of Proposition 2.6.7 as long as $(\lambda_1, \cdots, \lambda_{N-1}) \in K_m$.*
(ii) *$[0, \infty)^{N-1} \subset K_m$.*
(iii) *$e_m(\lambda_1, \cdots, \lambda_{N-1}) > 0$ for $(\lambda_1, \cdots, \lambda_{N-1})$ belonging to the interior int K_m of K_m.*
(iv) *$e_m(\lambda_1, \cdots, \lambda_{N-1}) = 0$ for $(\lambda_1, \cdots, \lambda_{N-1}) \in \partial K_m$.*
(v) *$K_m \setminus \{0\} \subset$ int K_ℓ for $\ell < m$.*

For the proof the reader is referred to a book of D. S. Mitrinović [354, p. 102, Theorem 1] and the article of N. S. Trudinger [418]. When $m = N-1$, it is easy to see that $K_m = [0, \infty)^{N-1}$ so that (i)–(iv) hold.

By this consideration we set

$$\hat{e}_m(\lambda_1, \ldots, \lambda_{N-1}) = \begin{cases} e_m(\lambda_1, \ldots, \lambda_{N-1}), & (\lambda_1, \ldots, \lambda_{N-1}) \in K_m, \\ 0 & \text{otherwise} \end{cases} \tag{2.6.20}$$

and observe by Lemma 2.6.9 that $\hat{e}_m$ fulfills the monotonicity condition of Proposition 2.6.7 as well as continuity on $\mathbf{R}^{N-1}$.

Theorem 2.6.10 *The function F_f defined by* (2.6.19) *with*

$$g(\lambda_1, \ldots, \lambda_{N-1};\ p) = \hat{e}_m(\lambda_1, \ldots, \lambda_{N-1}) \quad (2 \le m \le N-1)$$

is degenerate elliptic. Moreover, F_f is continuous in $(\mathbf{R}^N \backslash \{0\}) \times \mathbf{S}^N$.

When we consider a quotient e_m/e_ℓ, we note that $\hat{e}_m/e_\ell$ for $1 \le \ell < m \le N-1$ satisfies the monotonicity condition of Proposition 2.6.7 for

$$g(\lambda_1, \cdots, \lambda_{N-1};\ p) = (\hat{e}_m/e_\ell)(\lambda_1, \cdots, \lambda_{N-1});$$

see the book of D. S. Mintrinović [354, p. 102, Theorem 1] for the proof. Note that by Lemma 2.6.9 (v) $\hat{e}_m/e_\ell$ is a well-defined continuous function in $\mathbf{R}^{N-1}$ by assigning zero as the value at $(0, \ldots, 0)$. Thus the function F_f defined by (2.6.19) with this g is again degenerate elliptic and continuous in $(\mathbf{R}^N \backslash \{0\}) \times \mathbf{S}^N$.

Other Examples We finally consider the level set equations of (2.5.13) and (2.5.14). For (2.5.13) the function F_f is degenerate elliptic if h is nondecreasing and $\nabla^2 \gamma \ge O$ as well as $a \ge 0$. For (2.5.14) the function F_f is always degenerate elliptic.

The condition (2.6.18) for F_f defined by (2.6.16) is equivalent to saying

$$f(p, Q_p(X)) \le f(p, Q_p(Y)) \quad \text{whenever } Q_p(X) \ge Q_p(Y). \tag{2.6.21}$$

This is easy to prove since $Q_p(X) = Q_{-p}(X)$. We say that f defined in $\mathbf{E}$ (or $\mathbf{E}_\Sigma$) is *degenerate elliptic* if (2.6.21) is fulfilled for all $(p, Q_p, (X)),\ (p, Q_p(Y)) \in \mathbf{E}$ (or $\mathbf{E}_\Sigma$). We say that (2.6.1) is *degenerate parabolic* if $f(x, t, \cdot, \cdot)$ is degenerate elliptic for all $(x, t) \in \Omega \times [0, T]$. As we study in this section the mean curvature flow equation and the Hamilton-Jacobi equation (2.5.6) are of course degenerate parabolic. The anisotropic version (2.5.13) as well as (2.5.2) is degenerate parabolic when h is nondecreasing and $\nabla^2 \gamma \ge O$ and $a \ge 0$. The Gaussian curvature equation for $N \ge 3$ is not degenerate parabolic. More generally, m-th symmetric curvature flow equation (2.5.10) with $2 \le m \le N-1$ is not degenerate elliptic. We truncate these equations (2.5.10), (2.5.12) by replacing e_m by $\hat{e}_m$ as defined in (2.6.20). Then the truncated equations

$$V = \hat{e}_m(\kappa_1, \ldots, \kappa_{N-1}), \tag{2.6.22}$$

$$V = (\hat{e}_m/e_\ell)(\kappa_1, \ldots, \kappa_{N-1}), \quad 1 \le \ell < m \le N-1 \tag{2.6.23}$$

are degenerate parabolic.

If (2.6.18) is replaced by the strict monotonicity

$$F(p, X) < F(p, Y) \quad \text{for } X \geq Y \quad \text{with trace } (X - Y) > 0$$

is fulfilled, F is called *strictly elliptic*. A typical example of such F is

$$F(p, X) = -\text{trace } X$$

so that $F(\nabla u, \nabla^2 u) = -\Delta u$. We note that a strongly geometric F is not strictly elliptic because of condition (G2). Similarly, we say f is *strictly elliptic* if (2.6.21) is replaced by the strict monotonicity

$$f(p, Q_p(X)) < f(p, Q_p(Y))$$

for $Q_p(X) \geq Q_p(Y)$ with trace $Q_p(X - Y) > 0$. Equation (2.6.1) is called *strictly parabolic* if $f(x, t, \cdot, \cdot)$ is strictly elliptic. The mean curvature flow equation is strictly parabolic while the Hamilton–Jacobi equation is not strictly parabolic. The anisotropic version (2.5.13) as well as (2.5.2) is strictly parabolic when h is strictly increasing and $\nabla^2\gamma(p) + p \otimes p > O$ for $p \neq 0$. Here by $X > O$ we mean $X \geq O$ and det $X > 0$. Equation (2.6.23) with $1 \leq \ell < m \leq N - 1$ is not strictly parabolic because $K_m \neq \mathbf{R}^{N-1}$.

2.6.4 Geometric Equations

A familiar condition on scaling invariance of F_f defined by (2.6.16) appears to be a little bit weaker than (G1), (G2).

Definition 2.6.11 Let F be a real-valued function on $(\mathbf{R}^N \backslash \{0\}) \times \mathbf{S}^N$. We say that F is *geometric* if F fulfills (G1) and

(G2') $\quad F(p, X + \sigma\, p \otimes p) = F(p, X) \quad$ for all $p \in \mathbf{R}^N \backslash \{0\}$, $X \in \mathbf{S}^N$, $\sigma \in \mathbf{R}$.

Clearly (G2) implies (G2′). The condition (G2′) is certainly weaker than (G2). For example if we set

$$F_O(p, X) = ||R_p X||_2, \; R_p = I - \frac{p \otimes p}{|p|^2},$$

then F_O is geometric but not strongly geometric. Here $||Y||_2$ denotes the Hilbert–Schmidt norm of $N \times N$ matrix Y, i.e.,

$$||Y||_2 = \left(\sum_{1 \leq i,\, j \leq N} |Y_{ij}|^2 \right)^{1/2},$$

where Y_{ij} denotes the ij-component of the matrix Y. Since

$$\begin{aligned} F_O(\lambda p,\ \lambda X) &= ||R_p(\lambda X)||_2 = \lambda F_O(p, X), \\ F_O(p, X + \sigma\ p \otimes p) &= ||R_p X + R_p \sigma\ p \otimes p||_2 = ||R_p X||_2 = F_O(p, X), \end{aligned}$$

F_O fulfills (G1) and (G2′). However, F_O does not fulfill (G2). Indeed, if we take $p = (0, \dots 0, 1)$, then

$$F_O(p, X) = \left(\sum_{i=1}^{N-1} \sum_{j=1}^{N} |X_{ij}|^2 \right)^{1/2}.$$

We take $y = (\alpha, 0, \dots, 0)(\neq 0)$ and observe that

$$\begin{aligned} F_O(p, X + y \otimes p + p \otimes y)^2 - F_O(p, X)^2 &= F_O(p, X + y \otimes p)^2 - F_O(p, X)^2 \\ &= (X_{1N} + \alpha)^2 - X_{1N}^2 \neq 0 \end{aligned}$$

since X is symmetric. Thus (G2) is not fulfilled. However, if we assume that F is degenerate elliptic, then geometricity and strong geometricity are equivalent conditions.

Theorem 2.6.12 *For $p \in (\mathbf{R}^N \backslash \{0\})$ let $X \mapsto F(p, X)$ be a continuous function in $\mathbf{S}^N$. Assume that*

$$F(p, X + \sigma\ p \otimes p) = F(p, X)$$

for all $X \in \mathbf{S}^N$, $\sigma \in \mathbf{R}$ and that

$$F(p, X) \leq F(p, Y)$$

for all $X, Y \in \mathbf{S}^N$ with $X \geq Y$. Then

$$F(p, X + y \otimes p + p \otimes y) = F(p, X)$$

for all $X \in \mathbf{S}^N$, $y \in \mathbf{R}^N$. In particular, if a real-valued function $F = F(p, X)$ on $(\mathbf{R}^N \backslash \{0\}) \times \mathbf{S}^N$ is continuous in X and degenerate elliptic, then F is strongly geometric if and only if F is geometric. (The set $\mathbf{R}^N \backslash \{0\}$ may be replaced by a cone $\tilde{\Sigma}$ with vertex and $0 \in \tilde{\Sigma}$.)

Proof An elementary calculation shows that

$$\begin{pmatrix} c & 0 \\ 0 & d \end{pmatrix} \leq \begin{pmatrix} 0 & 1 \\ 1 & 0 \end{pmatrix} \leq \begin{pmatrix} a & 0 \\ 0 & b \end{pmatrix}$$

provided that $ab \geq 1,\ cd \geq 1,\ a > 0,\ c < 0$. This estimate yields

$$c\, p \otimes p + d\, y \otimes y \le p \otimes y + y \otimes p \le a\, p \otimes p + b\, y \otimes y \quad \text{in } \mathbf{S}^N.$$

By the invariance and monotonicity assumption on F we see

$$\begin{aligned} F(p, X + b\, y \otimes y) &= F(p, X + a\, p \otimes p + b\, y \otimes y) \\ &\le F(p, X + p \otimes y + y \otimes p) \\ &\le F(p, X + c\, p \otimes p + d\, y \otimes y) \\ &\le F(p, X + d\, y \otimes y). \end{aligned}$$

Keeping the relation $ab \ge 1,\ cd \ge 1,\ a > 0,\ c < 0$, we send b, d to zero to get

$$F(p, X) = F(p, X + p \otimes y + y \otimes p)$$

by continuity of F in X. □

In this book we consider degenerate parabolic equations so we do not need to distinguish strongly geometricity and geometricity. We shall mainly use the geometricity to describe the scaling property of level set equations.

It is sometimes convenient to extend the notion of the geometricity to general second order operators to handle boundary value problems.

Definition 2.6.13 Let E be a real-valued function defined in a dense subset W of $\mathbf{R}^d \times \mathbf{S}^d$. Assume that $(q, Y) \in W$ implies $(\lambda q, \lambda Y + \sigma\, q \otimes q) \in W$ for $\lambda > 0,\ \sigma \in \mathbf{R}$. We say that E is *geometric* on $E = 0$ if $E(q, Y) \le 0$ (resp. $E(q, Y) \ge 0$) implies $E(\lambda q, \lambda Y + \sigma\, q \otimes q) \le 0$ (resp. $E(\lambda q, \lambda Y + \sigma\, q \otimes q) \ge 0$) for all $\lambda > 0, \sigma \in \mathbf{R}$.

Let $E(z, \cdot, \cdot)$ be a real-valued function defined in W with $z \in \mathcal{O}$, where $\mathcal{O}$ is a locally compact subset of $\mathbf{R}^d$. We say an equation

$$E(z,\ Du,\ D^2u) = 0,\ z \in \mathcal{O}$$

is *geometric* in $\mathcal{O}$ if $E(z, \cdot, \cdot)$ is geometric on $E = 0$ for all $z \in \mathcal{O}$. Here $Du = (\partial u/\partial z_i)_{i=1}^d$, $D^2u = (\partial^2 u/\partial z_i \partial z_j)_{1 \le i,j \le d}$.

It is straightforward to see that

$$E(q, Y) = \tau + F(p, X),\ q = (\tau, p) \in \mathbf{R} \times (\mathbf{R}^N \backslash \{0\}),\ Y = \{0\} \oplus X,\ X \in \mathbf{S}^N$$

with $W = \mathbf{R} \times (\mathbf{R}^N \backslash \{0\}) \times \mathbf{S}^N$, $d = N + 1$ is geometric on $E = 0$ if and only if F is geometric. In particular a level set equation (2.6.2) is always a geometric equation in the sense of Definition 2.6.13. Thanks to Theorems 2.6.4 and 2.6.12, if an equation

$$u_t + F(x, t, \nabla u, \nabla^2 u) = 0,\ (x, t) \in \Omega \times (0, T)$$

is geometric in $\Omega \times (0, T)$, then it is a level set equation of some surface evolution equation (2.6.1) provided that $F(x, t, \cdot, \cdot)$ is degenerate elliptic for all $(x, t) \in \Omega \times (0, T)$, where Ω is a domain in $\mathbf{R}^N$. (We say that the preceding equation is *degenerate parabolic* if F is degenerate elliptic.)

Remark 2.6.14 We have introduced the notion of geometricity for E since it is convenient to handle boundary value problems. We consider

$$u_t + F(x, t, \nabla u, \nabla^2 u) = 0, \quad \text{in } \Omega \times (0, T)$$

with, for example, the boundary condition (2.6.14):

$$\frac{\partial u}{\partial \nu} + z|\nabla u| = 0 \quad \text{on } \partial\Omega \times (0, T),$$

where $|z| < 1$, $z \in \mathbf{R}$. As we see in Sect. 3.3, to study this problem it is natural to indroduce

$$E(x, \tau, p, X) = \begin{cases} \tau + F(p, X), & x \in \Omega, \\ (\tau + F(p, X)) \wedge (\langle \nu, p \rangle + z|p|), & x \in \partial\Omega \end{cases}$$

where $a \wedge b = \min(a, b)$ and regard the boundary value problem as

$$E(x, u_t, \nabla u, \nabla^2 u) = 0 \quad \text{in } \overline{\Omega} \times (0, T).$$

For this operator, it is easy to see that E is geometric at $E = 0$ so that the equation is geometric. We may replace (2.6.14) by a more general first order boundary condition

$$B(x, t, \nabla u) = 0$$

with $B(x, t, \lambda p) = \lambda B(x, t, p)$, $\lambda > 0$ so that E is geometric at $E = 0$; here E is defined by

$$E(x, t, \tau, p, X) = (\tau + F(p, X)) \wedge B(x, t, p).$$

In Chap. 5 we shall see how important the geometric property of equations are. The main observation is that if u solves a geometric equation so does $\theta \circ u = \theta(u)$ with nondecreasing θ, where $\theta \circ u$ denotes the composition of functions. This indicates that a geometric equation is invariant under coordinate change of dependent variables. Although we postpone a rigorous proof for general solutions, we here formally indicate how geometricity yields such a property. Assume that u solves a geometric equation

$$u_t + F(\nabla u, \nabla^2 u) = 0$$

and $\theta' \geq 0$. For $v = \theta \circ u$ we calculate

$$\begin{aligned}&v_t + F(\nabla v, \nabla^2 v)\\&= \theta'(u)u_t + F(\theta'(u)\nabla u,\ \theta'(u)\nabla^2 u + \theta''(u)\nabla u \otimes \nabla u)\\&= \theta'(u)(u_t + F(\nabla u,\ \nabla^2 u)) = 0\end{aligned}$$

by geometricity of F. This invariance property is natural if we recall that a geometric equation is the level set equation of some surface evolution equation so that motion of each level set of solutions is independent of other levels and the value of levels.

2.6.5 Singularities in Level Set Equations

Regularity of f in (2.6.1) of course reflects to F_f in (2.6.4). Here is a trivial observation.

Proposition 2.6.15 *Let f be a real-valued function defined in* $\mathbf{E}$ *given by* (2.6.15). *The associate function F_f defined by* (2.6.16) *is continuous in* $(\mathbf{R}^N \backslash \{0\}) \times \mathbf{S}^N$ *if and only if f is continuous in* $\mathbf{E}$.

Examples of Eqs. (2.6.1) with f continuous in its variables includes (2.5.2), (2.5.13), (2.5.14) with continuous h, $\beta > 0$, c with γ (provided that it is C^2 outside the origin) as well as the mean curvature flow equation. Examples also include (2.5.9)–(2.5.12) but these equations may not be degenerate parabolic so we rather consider modified equations (2.6.22), (2.6.23) instead of them. By Theorem 2.6.10 (and its following paragraph) these equations (2.6.22), (2.6.23) can be written in the form of (2.6.1) with continuous f. So for such an f the associate function F_f is continuous in $(\mathbf{R}^N \backslash \{0\}) \times \mathbf{S}^N$.

We next study the magnitude of singularity of $F_f(p, X)$ near $p = 0$. For this purpose we introduce the *upper semicontinuous envelope*

$$F^* : \mathbf{R}^N \times \mathbf{S}^N \to \mathbf{R} \cup \{\infty\}$$

of F defined on $(\mathbf{R}^N \backslash \{0\}) \times \mathbf{S}^N$ by setting

$$F^*(p, X) = \lim_{\varepsilon \downarrow 0} \sup\{F(q, Y);\ |q - p| \le \varepsilon,\ ||X - Y||_2 \le \varepsilon\}.$$

The *lower semicontinuous envelope* F_* is defined by $F_* = -(-F)^*$ with values in $\mathbf{R} \cup \{-\infty\}$. If F is continuous in $(\mathbf{R}^N \backslash \{0\}) \times \mathbf{S}^N$, then

$$F(p, X) = F^*(p, X) = F_*(p, X) \quad \text{for } (p, X) \in (\mathbf{R}^N \setminus \{0\}) \times \mathbf{S}^N.$$

If F is geometric and degenerate elliptic, so is F^* and F_*.

Lemma 2.6.16 *Assume that F is continuous $(\mathbf{R}^N \setminus \{0\}) \times \mathbf{S}^N$ and that F is geometric and degenerate elliptic. Let M and m denote*

$$M = \sup\{F(p, -I);\ |p| \le 1,\ p \ne 0\},$$
$$m = \inf\{F(p, I);\ |p| \le 1,\ p \ne 0\}.$$

Then the following three conditions are equivalent:

(a) $F^*(0, O) < \infty$ *(resp. $F_*(0, O) > -\infty$),*
(b) $M < \infty$ *(resp. $m > -\infty$),*
(c) $F^*(0, O) = 0$ *(resp. $F_*(0, O) = 0$).*

Proof Let $|X|$ denote the operator norm of X as a self-adjoint operator. In other words $|X|$ equals the largest modulus of eigenvalues of X. Since the Hilbert–Schmidt norm $||X||_2$ is equivalent to the operator norm $|X|$ for finite dimensional $\mathbf{S}^N$, we may replace $||X - Y||_2$ by $|X - Y|$ in the definition of F^*. We first note that $|X| \le \varepsilon$ implies

$$-\varepsilon I \le X \le \varepsilon I$$

for $\varepsilon > 0$. Since F is degenerate elliptic, we observe that

$$\sup_{|X| \le \varepsilon} F(p, X) \le F(p, -\varepsilon I),\ p \ne 0.$$

The converse inequality is trivial since $|-\varepsilon I| = \varepsilon$. We thus observe that

$$\sup_{\substack{|p| \le \varepsilon \\ p \ne 0}} \sup_{|X| \le \varepsilon} F(p, X) = \sup_{\substack{|p| \le \varepsilon \\ p \ne 0}} F(p, -\varepsilon I) = \varepsilon \sup_{\substack{|p| \le \varepsilon \\ p \ne 0}} F(p/\varepsilon, -I) = \varepsilon M$$

since F is geometric. Thus the equivalence of (a), (b), (c) is clear for F^*. The proof for F_* is symmetric. □

Remark 2.6.17 Even if F depends on $(x, t) \in \Omega \times (0, T)$ the same proof shows that

$$F^*(x, t, 0, O) = \lim_{\varepsilon \downarrow 0} (\varepsilon\, M^*(x, t)),$$
$$\text{with } M^*(x, t) = \sup\{F(x, t, p, -I);\ |p| \le 1,\ p \ne 0\},$$

provided that F is continuous in its variables and that $F(x, t, \cdot, \cdot)$ is geometric and degenerate elliptic for all (x, t). Here F^* denotes the upper semicontinuous envelope as a function of (x, t, p, Y) (see Sect. 3.1.1 for definition). Again if $M^*(x, t) < \infty$, then $F^*(x, t, 0, O) = 0$ and of course if $m_*(x, t) > -\infty$, then $F_*(x, t, 0, O) = 0$, where m is defined in the same way as M by replacing sup by inf.

Notice that the condition

$$-\infty < F_*(x, t, 0, O) = F^*(x, t, 0, O) < \infty$$

is equivalent to saying that F can be continuously extended to $(x, t, 0, O)$.

Proposition 2.6.18 *Let f be a real-valued continuous function defined in* $\mathbf{E}$ *by* (2.6.15). *Assume that f is degenerate elliptic. Then the associate function F_f defined by* (2.6.16) *can be continuously extended to* $(0, O)$ *with value zero if and only if*

$$\inf_{0<\rho\le 1} \rho \inf_{|p|=1} f(-p,\ R_p I/\rho) > -\infty, \quad \sup_{0<\rho\le 1} \rho \sup_{|p|=1} f(-p,\ -R_p I/\rho) < +\infty. \tag{2.6.24}$$

(The first (second) quantity equals $-M$ (resp. $-m$) defined in Lemma 2.6.16 with $F = F_f$.)

This follows from Lemma 2.6.16 since geometricity of F_f implies

$$\begin{aligned} M &= \sup\{F(p, -I);\ |p| \le 1,\ p \ne 0\} \\ &= \sup |p| F(p/|p|, -I/|p|);\ |p| \le 1,\ p \ne 0\} \\ &= -\inf_{0<\rho<1} \rho \inf_{|p|=1} f(-p,\ R_p I/\rho) \end{aligned}$$

and a similar expression is valid for m.

The condition (2.6.24) is a growth restriction of $f = f(\mathbf{n}, \nabla\mathbf{n})$ in $\nabla\mathbf{n}$. It roughly says that f grows either linearly or sublinearly in $\nabla\mathbf{n}$ as $|\nabla\mathbf{n}| \to \infty$. For example if f is positively homogeneously of degree 1 in the second variable, i.e., $f(p, \lambda Z) = \lambda f(p, Z)$ for $\lambda > 0$, $(p, Z) \in \mathbf{E}$ then (2.6.24) is fulfilled. If we write (2.5.2) in the form of (2.6.1), then evidently f satisfies (2.6.24) (with constant c) since f is linear in $\nabla\mathbf{n}$. In particular, f corresponding to the mean curvature flow equation fulfills (2.6.24). The condition (2.6.24) is also fulfilled for (2.5.13) and (2.5.14) provided that $\overline{\lim}_{z\to\infty} |h(z)|/|z| < \infty$; here, $\overline{\lim}$ denotes limsup.

Both estimates of (2.6.24) are violated for f corresponding to (2.6.22) (for $2 \le m \le N-1$) and (2.6.23) (for $1 \le \ell < m \le N-1,\ \ell - 1 < m$); if $m = \ell - 1$, the growth condition (2.6.24) is fulfilled for (2.6.23). Note that if both inequalities in (2.6.24) are, violated, then

$$F_*(0, O) = -\infty, \ F^*(0, O) = +\infty$$

by Lemma 2.6.16.

2.7 Exact Solutions

We give here some explicit solutions mainly for the mean curvature flow equation (2.5.4) and its anisotropic version (2.5.2).

2.7.1 *Mean Curvature Flow Equation*

Shrinking Sphere As expected there is a solution Γ_t of (2.5.4) that is a family of spheres of radius $R(t)$ centered at the origin. Equation (2.5.4) is now of form

$$-dR/dt = (N-1)/R \tag{2.7.1}$$

since the left-hand side is the inward velocity and the right hand side is the inward mean curvature; note that for the sphere of radius R all principal curvatures are $1/R$. Integrating (2.7.1), we see

$$R(t) = (R_0^2 - 2(N-1)t)^{1/2} \tag{2.7.2}$$

with $R(0) = R_0 > 0$. Thus an evolving sphere

$$\Gamma_t = \left\{x \in \mathbf{R}^N;\ |x| = R(t)\right\} \tag{2.7.3}$$

solves (2.5.4) if (and only if) $R(t)$ is of form (2.7.2). Note that if $t > t_* = R_0^2/(2(N-1))$, then $R(t)$ is not well defined as a real number. It is natural to interpret that Γ_t becomes empty after the time t_* when Γ_t shrinks to a point.

Similarly for the Gaussian curvature flow equation (2.5.9) there is a shrinking sphere solution Γ_t of form (2.7.3) provided that $R(t)$ solves

$$-dR/dt = 1/R^{N-1} \tag{2.7.4}$$

instead of (2.7.1). Integrating (2.7.4) yields

$$R(t) = (R_0^N - Nt)^{1/N}$$

instead of (2.7.2). For more general equation (2.5.12) it is still easy to find a shrinking sphere solution although we do not give its explicit form here.

Shrinking Cylinders We consider a little bit more general evolving hypersurface called a cylinder:

$$\Gamma_t = \{(x_1, \ldots, x_j, x_{j+1}, \ldots, x_N);\ (x_{j+1}^2 + \cdots + x_N^2)^{1/2} = R(t)\}; \tag{2.7.5}$$

of course Γ_t is a sphere if $j = 0$. Equation (2.5.4) is interpreted as

$$-\,dR/dt = (N-1-j)/R \tag{2.7.6}$$

which generalizes (2.7.1); note that $\kappa_1 = \cdots = \kappa_{N-1-j} = 1/R$, $\kappa_{N-j} = \kappa_{N-j+1} = \cdots = \kappa_{N-1} = 0$. Solving (2.7.6) to get

$$R(t) = (R_0^2 - 2(N-1-j)t)^{1/2}. \tag{2.7.7}$$

Thus an evolving cylinder Γ_t of (2.7.5) solves the mean curvature flow equation if and only if $R(t)$ is given by (2.7.7).

Since the Gaussian curvature of a cylinder of form (2.7.5) with $j \geq 1$ is always zero, any cylinder is a stationary solution of the Gaussian curvature flow equation (2.5.9).

In these exact solutions we notice that the shape of an evolving hypersurface is independent of time up to dilation. Such a solution is often called a *self-similar* solution. We here give a rigorous definition of self-similarity.

Definition 2.7.1 Let Γ_t be an evolving hypersurface $t \in I$ where I is a time interval. We say that Γ_t is *self-similar* if there is $x_0 \in \mathbf{R}^N$ and a hypersurface Γ (independent of time) in $\mathbf{R}^N$ such that for some $\lambda = \lambda(t)$, Γ_t is of form

$$\Gamma_t = \{\lambda(t)(x - x_0) + x_0;\ x - x_0 \in \Gamma\}.$$

For example the evolving cylinder Γ_t given (2.7.5) is self-similar. If (2.7.7) is fulfilled, then it is a self-similar solution of (2.5.4).

Level Set Approach It is sometimes convenient to find a self-similar solution by using level set equations. We seek a solution of the level set equation (2.6.7) of form

$$u(x,t) = -(t + \zeta(r)),\ r = |x|$$

with nondecreasing ζ. Then $\zeta(r(x))$ must solve

$$1 = |\nabla_x \zeta(r)| \operatorname{div}\left(\frac{\nabla\zeta(r)}{|\nabla\zeta(r)|}\right) = \zeta'(r)|\nabla_x r| \operatorname{div}\left(\frac{\nabla_r}{|\nabla r|}\right).$$

Since $\partial r/\partial x_i = x_i/r$ and $\partial/\partial x_j(x_i/r) = (\delta_{ij} - x_i x_j/r^2)r^{-1}$, this implies

$$1 = \zeta'(r)(N-1)/r.$$

By normalizing $\zeta(0) = 0$ we get $\zeta(r) = r^2/(2(N-1))$. Thus

$$u(x,t) = -(t + |x|^2/2(N-1))$$

solves (2.6.7) at least outside the place where $\nabla u = 0$. Each level set gives a shirinking sphere solution of (2.5.4). Similarly, we see

$$u(x,t) = -\left(t + \sum_{\ell=j+1}^{N} x_\ell^2/(2(N-1))\right)$$

solves (2.6.7) at least formally (see Remark 5.3.8). Each level set of u gives a shirinking cylindrical solution of (2.5.4). In these problems $-u$ also solves (2.6.7). This reflects the fact that the evolution law (2.5.4) is independent of the choice of the orientation $\mathbf{n}$.

2.7.2 Anisotropic Version

We shall consider a class of anisotropic curvature flow equations (2.5.2) with constant c which includes the mean curvature flow equation as a special case. We try to find a self-similar shrinking solution similar to a sphere. A typical property of a sphere is that its mean curvature is constant on the sphere. (If an (embedded) closed hypersurface has constant mean curvature, then it must be a sphere by a result of A. D. Aleksandrov [6, 7].) We shall seek a hypersurface whose anisotropic mean curvature is constant.

Wulff Shape Let γ_0 be a surface energy. We say that

$$\mathcal{W} = \bigcap_{|q|=1} \{x \in \mathbf{R}^N;\ \langle x, q\rangle \le \gamma_0(q)\}$$

is the *Wulff shape* associated with γ_0. If $\gamma_0 \equiv 1$, $\mathcal{W}$ is a unit sphere (Fig. 2.4).

According to Wulff's theorem, $\mathcal{W}$ minimizes the surface energy

$$\int_{\partial D} \gamma_0(\mathbf{n})d\sigma$$

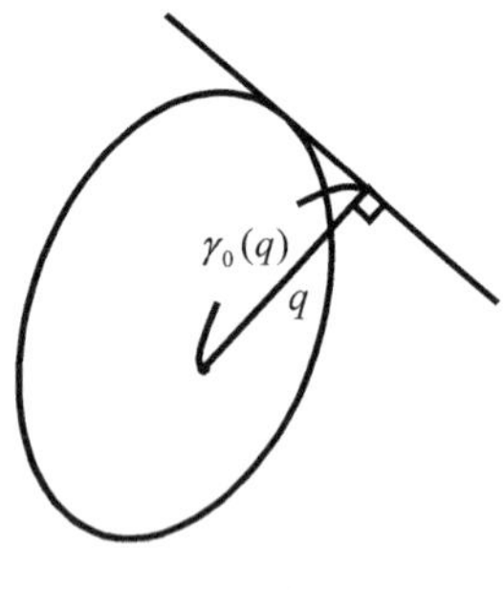

Fig. 2.4 Wulff shape

among all sets D with the same volume as $\mathcal{W}$. Such a minimizer is unique up to translation. In other words $\mathcal{W}$ is a unique solution of the anisotropic isoperimetric problem.

The Wulff shape is also characterized by the zero set of the conjugate convex function Γ of γ defined as

$$\gamma^{\natural}(x) = \sup\{\langle x, q\rangle - \gamma(q); q \in \mathbf{R}^N\},$$

where γ is given by (2.3.7). The function $\gamma^{\natural}$ is convex even if γ is not convex. Since γ is positively homogeneous of degree 1, we see that $\gamma^{\natural}$ is the *indicator function* of $\mathcal{W}$, i.e., $\gamma^{\natural} = 0$ on $\mathcal{W}$ and $\gamma^{\natural} = +\infty$ outside $\mathcal{W}$. Indeed, if $x \in \mathcal{W}$, then $\langle x, q\rangle - \gamma(q)$ attains zero so $\gamma^{\natural}(x) = 0$. If $x \notin \mathcal{W}$, then $c = \langle x, q\rangle - \gamma > 0$ for some q satisfying $|q| = 1$. Since γ is positively homogeneous of degree 1,

$$\langle x, \lambda q\rangle - \gamma(\lambda q) = \lambda c$$

for all $\lambda > 0$. This implies $\gamma^{\natural}(x) = \infty$.

Since $\gamma^{\natural}$ is convex and lower semicontinuous, $\mathcal{W}$ is convex and closed. Since the surface energy density γ_0 is always assumed to be positive, $\mathcal{W}$ contains the origin as an interior point. The next lemma shows that the boundary of a Wulff shape substitutes for the role of a sphere for anisotropic mean curvature. Let P be the *Minkowski function* of $\mathcal{W}$ defined by

$$P(x) = \inf\{\lambda \in (0, \infty);\ x/\lambda \in \mathcal{W}\}.$$

Clearly, P is positively homogeneous of degree 1 and convex. Since $\mathcal{W}$ contains the origin as an interior point, P is defined in all of $\mathbf{R}^N$. Clearly,

$$\mathcal{W} = \{x \in \mathbf{R}^N;\ P(x) \le 1\}.$$

Lemma 2.7.2 (Anisotropic Mean Curvature of the Wulff Shape) *Assume that surface energy density $\gamma_0 > 0$ is $C^m (m \ge 2)$ in the sense that γ is C^m outside the origin. Assume that γ satisfies a strict convexity assumption: $\nabla^2\gamma(p) + p \otimes p > O$ for $p \ne 0$ or equivalently,*

$$\sum_{i=1}^{N}\sum_{j=1}^{N} \frac{\partial^2\gamma}{\partial p_i \partial p_j}(p)\eta_i\eta_j > 0 \quad \textit{for all } \eta, p \in \mathbf{R}^N\backslash\{0\} \quad \textit{with } \langle \eta, p\rangle = 0,$$

where $\eta = (\eta_1, \dots, \eta_n)$. Then the boundary $\Gamma = \partial\mathcal{W}$ of the Wulff shape associated to $\mathcal{W}$ is C^m (so that P is C^m outside the origin). Moreover,

$$\gamma(\nabla P(x)) = 1 \quad \textit{in } \mathbf{R}^N\backslash\{0\}, \tag{2.7.8}$$

$$\xi(\nabla P(x)/|\nabla P(x)|) = \xi(\nabla P(x)) = x/P(x),\ x \in \mathbf{R}^N,\ x \ne 0, \tag{2.7.9}$$

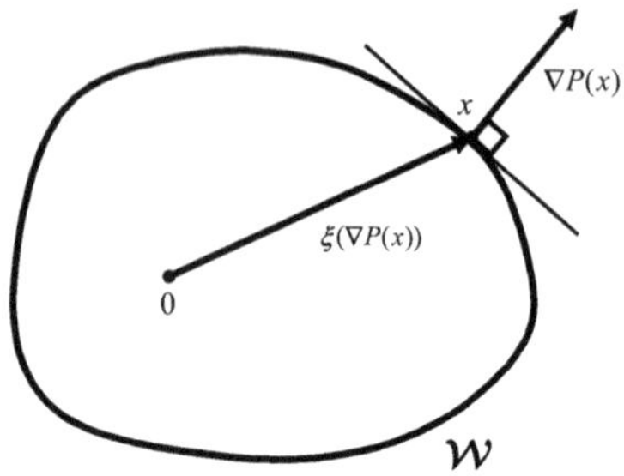

Fig. 2.5 Cahn-Hoffman vector

$$\left\langle \nabla \left(\frac{|x|}{P(x)} \right), \frac{x}{|x|} \right\rangle = 0, \ x \in R^N, \ x \neq 0, \tag{2.7.10}$$

where ξ is the Cahn–Hoffman vector of γ, i.e., $\xi = \nabla \gamma$. In particular $\xi(\mathbf{n}(x)) = x, x \in \Gamma$ so that the anisotropic mean curvature h in the direction of $\mathbf{n}$ equals $-(N-1)$, where $\mathbf{n}$ is the outward unit normal vector field of Γ (Fig. 2.5).

We postpone the proof in the next subsection. We seek a self-similar solution of (2.5.2), with $a \geq 0$ when c is a constant, by the level set approach as for (2.5.4). We set

$$u(x,t) = -(t + \zeta(P)) \tag{2.7.11}$$

with nondecreasing ζ defined on $[0, \infty)$, where P is the Minkowski function of $\mathcal{W}$. The level set equation of (2.5.2) is (2.6.10) or equivalently

$$u_t - |\nabla u|(-a \operatorname{div} \xi(-\nabla u) + c)/\beta(-\nabla u/|\nabla u|) = 0$$

by the identity (2.4.13). For the special form of u in (2.7.11), this equation is equivalent to

$$1 = \zeta'(P)|\nabla P|(a \operatorname{div} \xi(\nabla P) - c)/\beta(\nabla P/|\nabla P|). \tag{2.7.12}$$

Using (2.7.9) and (2.7.10), we have

$$\operatorname{div} \xi(\nabla P) = \operatorname{div}(x/P) = (|x|/P)\operatorname{div}(x/|x|) + \left\langle \nabla \frac{|x|}{P}, \frac{x}{|x|} \right\rangle = (N-1)/P + 0.$$

Thus if $\beta\gamma_0$ is a constant function with $\beta\gamma_0 = 1/\sigma_0$ on S^{N-1}, then, by (2.7.8), the identity (2.7.12) is equivalent to

$$1 = \zeta'(P)\sigma_0(a(N-1)/P - c). \tag{2.7.13}$$

Regarding P as an independent variable, (2.7.13) is an ordinary differential equation for ζ. It is easy to integrate (2.7.13). If $c \leq 0$, then a solution of (2.7.13) is a strictly increasing function of form

$$\zeta(\rho) = \int_0^\rho \frac{\tau}{\sigma_0(a(N-1) - c\tau)} d\tau \tag{2.7.14}$$

by normalizing $\zeta(0) = 0$. For example, if $c = 0$ with $a > 0$, then $\zeta(\rho) = \rho^2/(2\sigma_0 a(N-1))$. If $a = 0$ and $c < 0$, $\zeta(\rho) = -\rho/\sigma_0 c$. We thus observe that u of (2.7.11) with ζ of form (2.7.14) solves the level set equation of (2.5.2) at least where $\nabla u \neq 0$. Since each level set of u solves (2.5.2) at least where $\nabla u \neq 0$,

$$\Gamma_t = \{x \in \mathbf{R}^N;\ P(x) = R(t)\} \tag{2.7.15}$$

solves (2.5.2) provided that

$$R(t) = \zeta^{-1}(\zeta(R(0)) - t), \tag{2.7.16}$$

where ζ^{-1} denotes the inverse function of ζ. This generalizes (2.7.2). If $c = 0$, Γ_t disappears in a finite time; it disappears at $t = \zeta(R(0))$. If $c > 0$, then we cannot find an increasing function ζ solving (2.7.13). This corresponds to the phenomena that there is a growing solution of form (2.7.15) to (2.5.2) with $c > 0$ when $\mathbf{n}$ is taken outward.

We give another way to find a self-similar solution to see these properties depending on c. We argue as in the same way to derive (2.7.1) from (2.5.4). We note that

$$R(t) = |\nabla P(x)| \langle \mathbf{n},\ x \rangle,\ \ x \in \Gamma_t$$

since $R(t) = P(x) = \langle x, \nabla P(x) \rangle$ by the homogeneity of P. We also note that dR/dt may not equal the outward normal velocity V of Γ_t given by (2.7.15) but

$$dR/dt = V/\langle \mathbf{n}, x/R(t) \rangle.$$

By homogeneity of γ we see

$$\langle \xi(\mathbf{n}), \mathbf{n} \rangle = \gamma(\mathbf{n});$$

indeed differentiating $\gamma(\lambda p) = \lambda \gamma(p)$ in λ and setting $\lambda = 1$ yields the Euler equation

$$\langle \xi(p), p \rangle = \gamma(p). \tag{2.7.17}$$

By (2.7.9) this yields

$$\gamma(\mathbf{n}) = \left\langle \frac{x}{R(t)},\ \mathbf{n} \right\rangle$$

so we obtain

$$dR/dt = V/\gamma_0(\mathbf{n}).$$

If $\beta\gamma_0 = 1/\sigma_0$, Eq. (2.5.2) becomes

$$V = \sigma_0\gamma_0(\mathbf{n})(-a \operatorname{div}_{\Gamma_t}\xi(\mathbf{n}) + c). \tag{2.7.18}$$

This yields

$$dR/dt = \sigma_0(-a(N-1)/R + c) \tag{2.7.19}$$

since $\operatorname{div}_{\Gamma_t}\xi(\mathbf{n}) = (N-1)/R$ by Lemma 2.7.2. If $c \leq 0, a \geq 0$, $R(t)$ given by (2.7.16) is the unique solution of (2.7.18) with given $R(0)$. If $c > 0$, then if initial data $R(0) = R_0$ is small, say

$$a(N-1)/R_0 > c,$$

then the solution disappears in a finite time. If $a(N-1)/R_0 = c$, then $R(t) \equiv R_0$. If $a(N-1)/R_0 < c$, the solution $R(t)$ exists globally in time and its asymptotics as $t \to \infty$ is $\sigma_0 ct$. Thus in any case we have a self-similar solution since

$$\Gamma_t = \{R(t)x;\ x \in \Gamma\}.$$

Theorem 2.7.3 *Assume the same hypothesis of Lemma 2.7.2 concerning γ_0 and γ. Assume that $\beta\gamma_0$ is a constant function on S^{N-1} with value $1/\sigma_0 > 0$ and that c is a constant. Let R be a solution of* (2.7.19). *Then Γ_t given by* (2.7.15) *or*

$$\Gamma_t = \{R(t)x;\ x \in \Gamma = \partial\mathcal{W}\} \tag{2.7.20}$$

is a self-similar solution of (2.7.18) *(or* (2.5.2) *with $\beta\gamma_0 = 1/\sigma_0, a \geq 0,\ c \in \mathbf{R}$), where $\mathcal{W}$ denotes the Wulff shape of γ_0.*

For $a \geq 0, c \leq 0$ with $(c, a) \neq (0, 0)$ all solutions of form (2.7.20) *disappear in a finite time while for $c > 0$, a solution with $R(0) < a(N-1)/c$ disappears in a finite time while $R(0) > a(N-1)/c$ the solution exists globally in time and grows with $\lim_{t\to\infty} R(t)/t = \sigma_0 c$. If $R(0) = a(N-1)/c$, Γ_t with $R(t) = R(0)$ is a stationary solution.*

2.7.3 Anisotropic Mean Curvature of the Wulff Shape

We shall prove Lemma 2.7.2. For the Minkowski function P of the Wulff shape we set

$$P(x) = |x|/w(x)$$

so that w is positively homogeneous of degree zero. We use the convention $\hat{x} = x/|x|$ for $x \in \mathbf{R}^N$, $x \neq 0$.

Proposition 2.7.4 *Assume the same hypothesis of Lemma 2.7.2 concerning γ_0 and γ. Let ξ be the Cahn–Hoffman vector of γ, i.e., $\xi = \nabla\gamma$.*

(i) *For $x \in \mathbf{R}^N$ with $x \neq 0$,*

$$w(x) = \min\left\{\gamma(q)/\langle q, \hat{x}\rangle; q \in S^{N-1} \text{ and } \langle q, \hat{x}\rangle > 0\right\} > 0. \tag{2.7.21}$$

(ii) *For $x \neq 0$ let $\Theta(x) \subset S^{N-1}$ be the set of minimizers of the right-hand side of (2.7.21):*

$$\Theta(x) = \left\{q \in S^{N-1};\ w(x) = \gamma(q)/\langle q, \hat{x}\rangle \text{ and } \langle q, \hat{x}\rangle > 0\right\}. \tag{2.7.22}$$

Then $\Theta(x)$ is a singleton $\{q(x)\}$. The mapping $q : (\mathbf{R}^N\backslash\{0\}) \to S^{N-1}$ is C^{m-1}. Moreover,

$$\xi(q(x))\langle q(x), \hat{x}\ \rangle - \gamma(q(x))\hat{x} = 0. \tag{2.7.23}$$

(iii) *For $x \neq 0$,*

$$\gamma(q(x)) = w(x)\langle q(x), \hat{x}\rangle = \max\{w(y)\langle q(x), \hat{y}\rangle;\ y \neq 0\}. \tag{2.7.24}$$

In particular, differentiating $w(y)\langle q(x), \hat{y}\rangle$ in y yields

$$\nabla w(x)\langle q(x), \hat{x}\rangle + w(x)\frac{q(x)}{|x|} - w(x)\langle q(x), \hat{x}\rangle\frac{x}{|x|^2} = 0. \tag{2.7.25}$$

(iv) *For $x \neq 0$,*

$$\nabla P(x) = q(x)/\gamma(q(x)), \tag{2.7.26}$$

$$\xi(q(x)) = x/P(x). \tag{2.7.27}$$

Lemma 2.7.2 easily follows Proposition 2.7.4. Indeed, $\partial\mathcal{W}$ is C^1 by (2.7.26) and the implicit function theorem. The C^m regularity follows from C^m regularity of w which follows from C^{m-1} regularity of q and (2.7.26). The formula (2.7.26) yields (2.7.8) since γ is positively homogeneous of degree 1. The formula (2.7.27) yields (2.7.9) by (2.7.26) and the homogeneity of ξ. The formula (2.7.10) follows from (2.7.25) since $|x|/P = w$. Since $P(x) = 1$ on $\partial\mathcal{W}$, by (2.7.27) the anisotropic curvature h on $\partial\mathcal{W}$ equals

$$\begin{aligned} h &= -\mathrm{div}_\Gamma \xi(\mathbf{n}(x)) = -\mathrm{div}_\Gamma x \\ &= -\mathrm{trace}((I - \mathbf{n}(x) \otimes \mathbf{n}(x))\nabla x) = -(N-1) \end{aligned}$$

by (2.3.6).

Proof of Proposition 2.7.4

(i) Let w_0 denote the right-hand side of (2.7.21). Since $|x| \leq w_0(x)$ is equivalent to

$$\langle q, x \rangle - \gamma(q) \leq 0 \quad \text{for all } q \in S^{N-1},\ \langle q, x \rangle > 0,$$

we see, by interpreting $|x|/w_0(x) = 0$ for $x = 0$, that

$$\mathcal{W} = \{x \in \mathbf{R}^N;\ |x|/w_0(x) \leq 1\}$$

by definition of $\mathcal{W}$; apparent extra condition $\langle q, x \rangle > 0$ does not play a role at all. By definition of the Minkowski function w_0 must equal w.

(ii) We put

$$\begin{aligned} G(p, x) &= (G_1(p, x), \ldots, G_N(p, x)) \\ G_i(p, x) &= \left(\frac{\partial}{\partial p_i}\left(\frac{\gamma(p)}{\langle p, \hat{x} \rangle}\right)\right)\langle\, p, \hat{x}\, \rangle^2 \\ &= \langle p, \hat{x} \rangle \frac{\partial \gamma}{\partial p_i}(p) - \gamma(p)\hat{x}_i,\ 1 \leq i \leq N. \end{aligned}$$

Since $\gamma(p)/\langle p, \hat{x} \rangle$ is invariant under positive multiplication of p and $q \in \Theta$ is a minimizer, we see

$$G(q, x) = 0 \quad \text{for } q \in \Theta(x).$$

We differentiate G_i in p_j to get

$$\frac{\partial G_i}{\partial p_j}(p, x) = \langle p, \hat{x} \rangle \frac{\partial^2 \gamma}{\partial p_i \partial p_j}(p) + \hat{x}_j \frac{\partial \gamma}{\partial p_i}(p) - \hat{x}_i \frac{\partial \gamma}{\partial p_j}(p).$$

We then obtain

$$\sum_{1 \leq i, j \leq N} \frac{\partial G_i}{\partial p_j}(p, x)\eta_i \eta_j = \langle p, \hat{x} \rangle \sum_{1 \leq i, j \leq N} \frac{\partial^2 \gamma}{\partial p_i \partial p_j}(p)\eta_i \eta_j + 0$$

for $\eta = (\eta_1, \ldots, \eta_N) \in \mathbf{R}^N$. The strict convexity assumption of γ now yields

$$\sum_{1\le i,j\le N} \frac{\partial G_i}{\partial p_j}(p,x)\eta_i\eta_j > 0 \text{ for all } \eta,\, p \in \mathbf{R}^N\backslash\{0\} \text{ with } \langle p, \hat{x}\rangle > 0,\, \langle \eta, p\rangle = 0.$$

Thus for every x, there is a unique solution $p = q$ of $G(p, x) = 0$ so that $\Theta(x)$ is a singleton. The implicit function theorem implies that $x \mapsto q(x)$ is C^{m-1} since G is C^{m-1}. The identity (2.7.23) is the same as $G(q(x), x) = 0$.

(iii) By definition of $w(y)$ we see

$$w(y)\langle q(x), \hat{y}\rangle \le \gamma(q(x)) = w(x)\langle q(x), \hat{x}\rangle,\ \ y \ne 0,\ \ x \ne 0.$$

This yields (2.7.24).

(iv) For $x \ne 0$ we differentiate P to get

$$\nabla P(x) = \frac{x}{|x|w(x)} - \frac{|x|}{(w(x))^2}\nabla w(x).$$

Since

$$\nabla w(x) = w(x)\frac{x}{|x|^2} - w(x)\frac{q(x)}{|x|\langle q(x), \hat{x}\rangle},$$

by (2.7.25) we obtain

$$\nabla P(x) = q(x)/(w(x)\langle q(x), \hat{x}\rangle) = q(x)/\gamma(q(x)).$$

It remains to prove (2.7.27). Since

$$\gamma(q(x)) = \langle q(x), \xi(q(x))\rangle$$

by homogeneity (2.7.17), from (2.7.24) it follows that $\xi(q(x)) = \lambda\hat{x}$ with some $\lambda \in \mathbf{R}$. Plugging this into (2.7.23) yields

$$\lambda\hat{x}\langle q(x), \hat{x}\rangle = \gamma(q(x))\hat{x}.$$

Thus $\lambda = \gamma(q(x))/\langle q(x), \hat{x}\rangle = w(x)$ and $\xi(q(x)) = w(x)\hat{x} = x/P(x)$.

□

Remark 2.7.5 The strict convexity assumption

$$\nabla^2\gamma(p) + p \otimes p > O \quad \text{for} \quad p \ne 0$$

in Lemma 2.7.2 is equivalent to saying that

$$\nabla^2\gamma^2(p) > O \quad \text{for} \quad p \ne 0.$$

This condition is equivalent to saying that the *Frank diagram*

$$\mathcal{F} = \{p \in \mathbf{R}^N \ ; \ r(p) \le 1\}$$

is strictly convex in the sense that all inward principal curvatures of $\partial\mathcal{F}$ are positive on its boundary.

Indeed, assume that $\nabla^2\gamma^2(p) > O$. Since

$$\nabla^2\gamma^2 = 2\gamma\nabla^2\gamma + 2\nabla\gamma \otimes \nabla\gamma,$$

we see that

$$\text{trace } (\nabla^2\gamma(p)\eta \otimes \eta) > 0 \ \text{ for } \ \langle \eta, \nabla\gamma(p)\rangle = 0 \ \text{ and } \ \eta \ne 0. \qquad (*)$$

We know that p is a zero eigenvector of $\nabla^2\gamma$ by the homogeneity of γ. To prove strict positivity of $\nabla^2\gamma(p) + p \otimes p$ it suffices to show that the kernel of $\nabla^2\gamma(p)$ is the one dimensional space spanned by p. Suppose that the kernel were more than one dimensional. Then there would exist a zero eigenvector q orthogonal to p. By a dimension relation there would exist a zero eigenvector $\xi \ne 0$ which is a linear combination of p and q such that $\langle \nabla\gamma(p), \xi\rangle = 0$. By definition

$$\nabla^2\gamma(p)\xi = 0 \quad \text{so that } \ \text{trace}(\nabla^2\gamma(p)\xi \otimes \xi) = 0.$$

This contradicts $(*)$. So the kernel is one dimensional. We thus proved that $\nabla^2\gamma(p) + p \otimes p > O$.

Suppose now that $\nabla^2\gamma^2$ had a zero eigenvalue. Suppose that $q \ne 0$ were a zero eigenvector of $\nabla^2\gamma^2$. Then by the above relation between $\nabla^2\gamma^2$ and $\nabla^2\gamma$ we see that two nonnegative quantities

$$\text{trace } (\nabla^2\gamma \ q \otimes q) \quad \text{and} \quad \text{trace } ((\nabla\gamma \otimes \nabla\gamma)q \otimes q) = \langle \nabla\gamma(p), q\rangle^2$$

must be zero. If the first quantity equals zero, then q is proportional to p since $\nabla^2\gamma(p) + p \otimes p > O$. We may assume that $q = p$ so that $\langle \nabla\gamma(p), q\rangle^2 = \gamma(p)^2 > 0$. This is a contradiction. So there is no zero eigenvector of $\nabla^2\gamma^2$ so that $\nabla^2\gamma^2 > O$.

The curvature condition of the boundary of the Frank diagram is equivalent to saying that $\nabla^2\gamma(p)$ is strictly positive on the space orthogonal to $\nabla\gamma(p)$, i.e., the condition $(*)$. Its equivalence to $\nabla^2\gamma^2 > O$ is now clear by the above expression of $\nabla^2\gamma^2$.

As a corollary, we observe that

$$\text{trace } (\nabla^2\gamma(p)\eta \otimes \eta) > 0 \ \text{ for } \ \langle \eta, p\rangle = 0 \ \text{ and } \ \eta \ne 0$$

is equivalent to $\nabla^2\gamma^2(p) > O$.

2.7.4 Affine Curvature Flow Equation

We consider the affine curvature flow equation (2.5.14) and its generalization

$$V = k_+^\alpha \tag{2.7.28}$$

for $\alpha > 0$ in the plane. As expected, this equation admits a shrinking circle as a self-similar solution. However, for $\alpha = 1/3$ (corresponding to the affine curvature flow equation), it also admits a shrinking ellipse as a self-similar solution while for other α the only self-similar shrinking ellipse is a circle. We study this aspect below.

We interpret V and k in (2.5.14) as inward velocity and inward curvature, respectively when Γ_t is a closed curve. For level set representation we take $\nabla u/|\nabla u|$ as inward normal and obtain the level set equation of (2.7.28) of form

$$u_t = -|\nabla u| \left(-\mathrm{div}\, \frac{\nabla u}{|\nabla u|}\right)_+^\alpha. \tag{2.7.29}$$

As in Sect. 2.7.1 we set

$$u(x,t) = -(t + \zeta(S)),\ S(x) = \left(\frac{x_1}{a_1}\right)^2 + x_2^2,\ a_1 > 0,\ a_1 \neq 0 \tag{2.7.30}$$

with nondecreasing $\zeta : [0,\infty) \to [0,\infty)$. Plugging (2.7.30) in (2.7.29) yields

$$1 = \zeta'(S)|\nabla S|(\mathrm{div}\,(\nabla S/|\nabla S|))^\alpha. \tag{2.7.31}$$

We calculate

$$\nabla S = 2(x_1/a_1^2,\ x_2)$$

to get

$$|\nabla S| = 2\rho,\ \rho = ((x_1/a_1^2)^2 + x_2^2)^{1/2}.$$

We further calculate

$$\mathrm{div}\, \frac{\nabla S}{|\nabla S|} = \frac{x_2^2}{a_1^2\rho^3} + \frac{(x_1/a_1^2)^2}{\rho^3} = \frac{S}{a_1^2\rho^3}$$

and observe that (2.7.31) is equivalent to

$$1 = \zeta'(S)2\rho[S/(a_1^2\rho^3)]^\alpha.$$

If $a_1 = 1$, so that $S = \rho^2$, the equation for ζ is solvable for all $\alpha > 0$. However, if $a_1 \neq 1$, then the dependence of ρ must disappear so that the equation is solvable for ζ. The dependence of ρ disappears if and only if $\alpha = 1/3$. If $\alpha = 1/3$ we proceed to get

$$1 = \zeta'(S) 2 a_1^{-2/3} S^{1/3}.$$

Integrating the last equation to get

$$\zeta(S) = 3a_1^{2/3} S^{2/3}/4 \tag{2.7.32}$$

by normalizing $\zeta(0) = 0$. We now conclude

$$u(x, t) = -(t + \zeta(S))$$

with (2.7.32) solving (2.7.29) when $\alpha = 1/3$. To see its level set we conclude:

Theorem 2.7.6 *Assume that* $\alpha = 1/3$. *Then for each ellipse* Γ *there is a self-similar evolving curve* Γ_t *defined as in Definition 2.7.1 which solves* (2.7.28). *If*

$$\Gamma_t|_{t=1} = \left\{ (x_1, x_2);\ \frac{x_1^2}{a_1^2} + x_1^2 = s_0 \right\},$$

then it vanishes at time $\zeta(s_0)$. *If* $\alpha \neq 1/3$, *there is no self-similar solution of an ellipse unless the ellipse is a circle.*

2.8 Notes and Comments

In the first four sections we review several notions of geometric quantities such as curvature and give their various representation. Except for normal velocity and anisotropic curvature these notions are standard in differential geometry. For further background the reader is referred to classical books by S. Kobayashi and K. Nomizu [309, 310]. Explanation of curvature and the second fundamental form follows that of a book of L. Simon [395]. Surface energy has been popular in material sciences. The Cahn–Hoffman vector has been introduced by J. W. Cahn and D. W. Hoffman [77]. For further background the reader is referred to a nice review article on anisotropic curvature by J. E. Taylor [412].

The mean curvature flow equation was first introduced by W. W. Mullins [360] to model motion of grain boundaries in material sciences. The Eikonal equation $V = 1$ in geometric optics is a typical example of the Hamilton–Jacobi equation. Even for general β the equation is well studied in material sciences to describe growth of crystals. For the development of this topic the reader is referred to a review by

J. W. Cahn et al. [78]. The anisotropic version was derived by S. Angenent and M. E. Gurtin [27] from balance of forces and the second law of thermomechanics. However, even before it was used to describe a crystal growth phenomena by H. Müller-Krumbhaar et al. [359] (see also a book by A. A. Chernov [111]). There are nice review articles on anisotropic curvature flow equations by M. E. Gurtin [246] and J. E. Taylor et al. [413]. The Gaussian curvature flow was first introduced by W. J. Firey [175] to describe motion of surface of stones worn on the seashore. The affine curvature flow was axiomatically derived by L. Alvarez et al. [11] to propose a way of deformation of image in image processing.

The level set mean curvature flow equation was first effectively used to derive scaling law for "dynamic structure functions" of motion by mean curvature flow equations in T. Ohta et al. [370]. S. Osher and J. A. Sethian [376] used the level set equations to track the evolution numerically. Except for Sect. 2.6.1 most of the contents of Sect. 2.6 is taken from the paper by Y. Giga and S. Goto [207]. The definition of geometricity is due to Y.-G. Chen et al. [106]. The property in Lemma 2.6.9 is well-studied to solve the Dirichlet problem for Monge–Ampère type equations. It is known that $\partial e_m/\partial\lambda_j > 0$ $(1 \le j \le N-1, 1 \le m \le N-1)$ and $e_m^{1/m}$ is concave in K_m; see L. Caffarelli et al. [75]. Using the concavity properties, these authors proved a necessary and sufficient condition on domain $\Omega \subset \mathbf{R}^{N-1}$ such that the Dirichlet problem $\ell_m(\lambda_1, \dots, \lambda_{N-1}) = \psi$ in Ω with $u = \text{const.}$ on $\partial\Omega$ is solvable, where λ_j denotes the eigenvalues of the Hessian $\nabla^2 u$ of u in Ω. Later L. Caffarelli et al. [76] also studied the problem where λ_j is a principal curvature of the graph of u. N. S. Trudinger [418] also studied this problem by using viscosity solutions. For further generalization the reader is referred to the paper of N. M. Ivochkina et al. [291].

For surfaces of higher codimension it is possible to consider the mean curvature flow equation by assigning its velocity vector by its mean curvature vector defined in Sect. 2.3. A level set method is proposed for such an equation by L. Ambrosio and H. M. Soner [14].

Wulff's Theorem and Wulff Shape (Sect. 2.7.2) G. Wulff [429] formulated the generalized isoperimetric problem "Find a set minimizing the surface energy with fixed volume" and conjecture that the answer is a dilation of the Wulff shape $\mathcal{W}$. A. Dinghas [143] gave a formal proof. J. E. Taylor [410] gave a precise proof for very general surface energies and a very general class of sets for which the surface energy is defined by using geometric measure theory. B. Dacorogna and C.-E. Pfister [132] gave an analytic proof when $N = 2$. I. Fonseca [177] and I. Fonseca and S. Müller [179] gave a simpler proof for arbitrary dimensions. The minimizer of the generalized isoperimetric problem, or Wulff's problem is also unique up to translation and it is a dilation of $\mathcal{W}$. We do not discuss Wulff's problem further. See also the book of F. Morgan [355, Chapter 10] for an elementary proof when $N = 2$. For more information on convex bodies $\mathcal{W}$ see a book of R. Schneider [392].

The self-similar solution in Theorem 2.7.3 for (2.5.2) was constructed by H. M. Soner [399] by proving Lemma 2.7.2. Lemma 2.7.2 says that anisotropic mean curvature is constant if the surface is the boundary of the Wulff shape. The converse problem seems to be open unless γ_0 is a constant. The problem is of form: if an embedded compact hypersurface has a constant anisotropic mean curvature, is the hypersurface a boundary of the Wulff shape up to translation and dilation? This problem is affirmatively answered by M. Koiso and B. Palmer [316] when the Wulff shape is rotationally symmetric in $\mathbf{R}^3$ under the strict convexity and smoothness assumption on interfacial energy.

The fact that an ellipse gives a self-similar solution for the affine curvature flow equation is easy if we admit that the equation is affine invariant. The higher dimensional version of affine curvature flow equations is of form $V = K^{1/(N+1)}$.

Existence of a Self-similar Solution for the Anisotropic Curvature Flow Equation If $c = 0$ in (2.5.2) and $a > 0$, is there still a self-similar solution Γ_t of form (2.7.20) even if $\beta\gamma_0$ is not a constant? The answer is affirmative if $N = 2$. In fact if γ_0 is C^2 and $\nabla^2\gamma^2 > O$ for $p \neq 0$ and β is continuous, then there is $\tilde{\gamma}$ such that (2.5.2) can be rewritten as

$$V = \tilde{\gamma}\,\mathrm{div}_\Gamma \tilde{\xi}(\mathbf{n}), \tag{2.8.1}$$

where $\tilde{\xi} = \nabla\tilde{\gamma}$; $\tilde{\gamma}_0$ another surface energy which satisfies the same property as γ_0. The existence of a self-similar solution follows from that for (2.8.1). This is proved by M. E. Gage and Y. Li [183] when γ_0 and β is C^4. Later, a direct proof for general γ_0 and β is given by C. Dohmen et al. [147]. It is also proved that such self-similar solution is unique (up to translation in time and space) if γ and β are even, i.e., $\gamma(p) = \gamma(-p)$, $\beta(p) = \beta(-p)$. This is proved by M. E. Gage [181]; see also C. Dohmen and Y. Giga [146] and Y. Giga [202]. Note that if the symmetry assumption is removed, the uniqueness may not hold, as proved by H. Yagisita [430] and independently by B. Andrews [22]. There are several researches on existence of self-similar solutions of $V = \kappa^\alpha$ and its anisotropic version $N = 2$. We do not intend to explain the detail. The reader is referred to a review article by Y. Giga [202] and a book by K.-S. Chou and X.-P. Zhu [119] with references cited there on this topic as well as the article by B. Andrews [19]. For self-similar solutions moved by the power of the Gaussian curvature the reader is referred to a review article by J. Urbas [422] and references cited there. For the harmonic curvature flow see a paper by K. Anada [15].

Asymptotic Self-similarity Although self-similar solutions are special solutions, they are important since they often represent a typical asymptotic behavior of solutions. For example for the mean curvature flow equation (2.5.4) with $N \geq 3$ G. Huisken [259] proved that a convex hypersurface shrinks to a point in finite time and the way of shrinking is asymptotically equal to the sphere shrinking. For the curve shortening equation (2.5.5) the corresponding result has been proved by M. Gage and R. S. Hamilton [182]. For its anisotropic version (2.8.1) M. E. Gage [181] proved that a convex curve shrinks to a point and the way of shrinking is asymptotically like the shrinking Wulff shape provided that the equation is orientation free. For further extensions of this result the reader is referred to papers of M. E. Gage and Y. Li [183], K.-S. Chou and X.-P. Zhu [117], X.-P. Zhu [434] and a book by K.-S. Chou and X.-P. Zhu [119]. If one starts from a nonconvex curve, it becomes convex in finite time for (2.5.5) as M. A. Grayson [241] proved. Such convexity formation is also generalized by J. A. Oaks [366] for an anisotropic orientation free equation including (2.8.1). For more development of the theory the reader is referred to papers of K.-S. Chou and X.-P. Zhu [118], X.-P. Zhu [434], books by K.-S. Chou and X.-P. Zhu [119] and X.-P. Zhu [435].

There are several related results for the Gaussian curvature flow equation (2.5.9) and its modification $V = K^{\alpha}(\alpha > 0)$. K. Tso [420] proved that solution of (2.5.9) remains smooth and strictly convex and shrinks to a point if the initial hypersurface is strictly convex. For $V = K^{1/(n-1)}$ B. Chow [120] proved that a strict convex hypersurface shrinks to a point in finite time and the way of shrinking is asymptotically equal to the sphere shrinking which corresponds to the results of G. Huisken [259] for the mean curvature flow equation. B. Andrews [17] extended the theory so that it includes both (2.5.4) and $V = K^{1/(n-1)}$. Note that the homogeneous degree with respect to principal curvatures are the same both for $K^{1/(n-1)}$ and H so they can be treated simultaneously. For the affine curvature flow equation $V = K^{1/(n+1)}$ the way of shrinking is asymptotically equal to an ellipsoid shrinking. This is proved by G. Sapiro and A. Tannenbaum [387] for strict convex curves moved by (2.5.14) and later by B. Andrews [18] for strict convex hypersurfaces including curves. The situation for $V = K^{\alpha}$ with $\alpha > 1/(n+1)$ is similar to the case $\alpha = 1/(n-1)$ according to works of B. Andrews. In fact, B. Andrews [20] confirmed it for $\alpha \in (1/(n+1), 1/(n-1)]$. On the other hand if $\alpha < 1/(n+1)$, then there seems to be no general asymptotic shrinking shapes. This conjecture was verified for $N = 2$ by B. Andrews [21].

Self-similar Solutions for the Mean Curvature Flow Equation Classification of self-similar solutions is a rather difficult topic even for the mean curvature flow equation with $N \geq 3$. There exists a torus type self-similar solution as proved by S. B. Angenent [26]. The existence of a self-similar solution whose genus is more than 1 is conjectured by D. L. Chopp [115] (Fig. 2.6). If a self-similar solution is

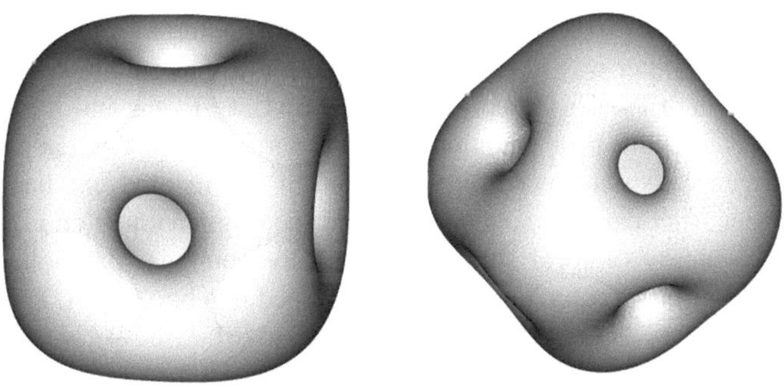

Fig. 2.6 Chopp's self-similar solution. Reprinted by permission of Informa UK Limited, trading Taylor & Francis Group from D. L. Chopp, Computation of Self-Similar Solutions for Mean Curvature Flow, Experimental Mathematics, 3 (1), pp. 1–15, 1994

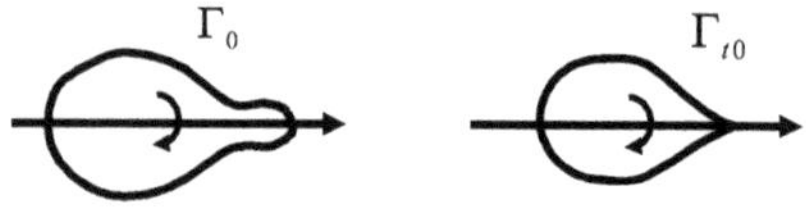

Fig. 2.7 Type II singularity

monotone shrinking and diffeomorphic to a sphere, it has been proved by G. Huisken [260] that it must be a shrinking sphere. However, without monotonicity it is not known whether there is another self-similar solution diffeomorphic to the sphere.

Singularities for the Mean Curvature Flow Equation The blow-up rate of curvatures near a singularity may be higher than the self-similar rate. Such a singularity is called type II otherwise it is called Type I. A shrinking sphere is of course type I. There exists a type II singularity as proved in S. Altschuler et al. [10], where they construct a smooth surface shrinking to a point without becoming convex (Fig. 2.7). They applied a level set method with topological argument; see also Y. Giga [200]. The existence of another type II singularity is constructed for a higher dimensional surface by J. J. L. Velázquez [423]. Later S. B. Angenent and J. J. L. Velázquez [31] give more explicit examples.

If the evolution is monotone in time, the asymptotic shape of a singularity is always convex (regardless of types of singularities). This statement has been proved by G. Huisken and C. Sinestrari [265] and independently by B. White [428] by a completely different method. For more development see a book of X.-P. Zhu [435].

Another Important Equation The equation $V = -1/H$ is used to prove the Riemannian Penrose inequality in cosmology by G. Huisken and T. Ilmanen [263].

Local Solvability In Chap. 5 we consider the initial value problem for (2.5.1). The first question would be whether there is a unique solution $\{\Gamma_t\}$ satisfying (2.5.1) with some time duration $(0, T)$ for a given initial data Γ_0. If (2.5.1) is strictly parabolic near initial data, the unique local existence of smooth solutions can be proved. One way is to analyze an equation of a height function, where the evolving surface is parametrized by the height from the initial surface. This idea has been carried out by Xu-Yan Chen [104] for a class of equations including the mean curvature flow equation. Another way is to solve the equation whose solution is the (signed) distance function of Γ_t. This idea is introduced by L. C. Evans and J. Spruck [167] for the mean curvature flow equation and is generalized by Y. Giga and S. Goto [208] for a general strictly parabolic equation. In this method one has to solve a fully nonlinear strict parabolic equation even if the original equation is quasilinear. However, the theory of local solvability for fully nonlinear parabolic equations (including higher order equations) is well developed in a book by A. Lunardi [341] improving the theory of O. A. Ladyženskaja et al. [317].

A general result of Y. Giga and S. Goto [208] yields the unique local existence of a smooth solution for (2.5.4) for arbitrary smooth initial data but for (2.5.9), (2.5.10) with $m > 1$, (2.5.12), (2.5.14), $V = K^\alpha$ for strict convex smooth initial data so that the equation is strictly parabolic. If (2.5.2) is strictly parabolic and γ and β are smooth enough, the same result for (2.5.4) holds. However, if there is degeneration of parabolicity in (2.5.2) with $a > 0$, we do not expect to have a smooth solution even locally-in-time for arbitrary smooth initial data, as observed by Y. Giga [199]. For the first order equation like (2.5.6) one can prove the unique local existence of a smooth solution for arbitrary smooth initial data by a method of characteristics.

There is a general theory of local solvability by A. Polden (See an article by G. Huisken and A. Polden [264]) for strictly parabolic equation (2.5.1) including higher order equations.

We conclude this section by pointing out that there is a nice review article by M. Struwe [407] on evolution equations related to geometry including the mean curvature flow equations as well as the harmonic map flow equations. This article explains singularities and weak solutions. It contains a short introduction to the level set method as well.

Further Comments for the Second Edition For the curve shortening equation, i.e., the mean curvature flow equation for $N = 2$, as mentioned in the introduction, a famous Grayson's theorem asserts that a solution curve stays embedded and smooth until it becomes convex. Since then several simplified proofs have been provided by many authors. The proofs of R. S. Hamilton [252] and G. Huisken [262] are based on isoperimetric estimates; note also that A. Magni and C. Mantegazza [342] gave a variational approach based on the work of R. S. Hamilton [252]. B. Andrews and P. Bryan [23] gave an estimate for extrinsic distance by intrinsic distance which directly yields the curvature bound. Although the both methods of G. Huisken [262] and B. Andrews and P. Bryan [23] yield Grayson's theorem, the latter is simpler because it does not need to appeal classification of possible singularities. Note that both methods are based on the maximum principle of two-point functions; see e.g. S. Brendle [68]. As noted in the paper of G. Huisken [262], if one considers the curve shortening equation of a non-closed curve with two fixed endpoints, one is able to prove that the solution curve stays embedded and smooth and converges to the straight line connecting two endpoints provided that the initial curve is in a convex domain and its endpoints are situated on the boundary of the convex domain. As anticipated, if one adjusted the method of B. Andrews and P. Bryan [23], the proof is simplified as shown by X. Bian et al. [65].

Let us discuss the case of surface evolution, i.e., $N \geq 3$. For behavior of a smooth solution up to first singularity is by now well studied for the mean curvature flow equations. See e.g. K. Ecker [151], C. Mantegazza [346], B. Andrews et al. [24].

A study of the asymptotic behavior of a mean curvature flow near singularities was initiated by G. Huisken [260] through his celebrated monotonicity formula; a similar monotonicity formula was found by Y. Giga and R. V. Kohn [215] to study asymptotic behavior of near blow-up point for a semilinear heat equation. G. Huisken [260] gave the asymptotic shape for neck pinching type for axisymmetric surfaces. For axisymmetric general hypersurfaces complete classification of singularities as well as its asymptotic shape was given by S. Altschuler et al. [10]. See also Chapter 3 of the book by M.-H. Giga et al. [198]. The asymptotic behavior of singularities of a mean curvature flow without assuming axisymmetry has been extensively studied; see a review paper by T. H. Colding et al. [124]. A major difficulty was to establish the uniqueness of the asymptotic limit. T. H. Colding and W. P. Minicozzi, II [121] overcame this difficulty by a smart adaptation of the Łojasiewicz-Simon inequality.

We take this opportunity to mention developments on constant anisotropic mean curvature surface that have appeared since the publication of the first edition. A simple form of the problem is: "If an embedded compact hypersurface has a constant anisotropic mean curvature, is this hypersurface a boundary of the Wulff shape (up to translation and dilation)?" This problem was posed in a preliminary version of this book by Y. Giga [204] and later by F. Morgan [356] with a proof for $N = 2$ in the case when the boundary of the Wulff shape is a smooth curve with positive curvature. If the anisotropy is weak, Y. Giga and J. Zhai [234] proved that any compact immersed constant anisotropic mean curvature surface in $\mathbf{R}^3$ must be a Wulff shape by using a global implicit function. For general smooth strictly convex anisotropy, i.e., $\nabla^2(\gamma^2) > 0$ and γ is smooth outside 0, Y. He et al. [256] proved that any compact embedded constant mean curvature hypersurface must be a Wulff shape by employing the Heintze-Karcher inequality. Additional early history of this problem can be found in an article by M.-H. Giga and Y. Giga [193]. More recently, an Aleksandrov-type problem has been extended to sets of finite perimeter by M. G. Delgadino and F. Maggi [139]. Its anisotropic version was established by A. De Rosa et al. [142] by extending the Heintze-Karcher inequality. They proved that finite unions of disjoint open Wulff shapes with equal radii are the only volume-constrained critical points of the anisotropic surface energy among all sets with finite perimeter and reduced boundary almost equal to its closure. The surface energy density γ could depend on space variables but totally it is assumed to be $C^{2,\alpha}$. It seems that in the case that γ_0 loses smoothness, like crystalline energy, the original problem remains open for $N \geq 3$. In the case $N = 2$, this problem was settled by P. B. Mucha and P. Rybka [357]; see also the review paper by M.-H. Giga and Y. Giga [193].

Chapter 3
Viscosity Solutions

The level set equations we would like to study are parabolic, but they are very degenerate. The classical techniques and results cannot be expected to apply. For example the initital value problem for degenerate parabolic equations may not admit a smooth solution for smooth initial data even locally in time. Moreover, level set equations may be singular at a place where the spatial gradient of solutions vanishes as noted for the level set mean curvature flow equations. Even for such singular degenerate equations it turns out that the theory of viscosity solutions provides a correct notion of generalized solutions. In this chapter we shall present the basic aspect of the theory especially for singular degenerate parabolic equations.

3.1 Definitions and Main Expected Properties

To motivate the definition of a viscosity solution we consider a C^2 function $u = u(z)$ which satisfies a differential inequality:

$$u_t(z) + F(z, u(z), \nabla u(z), \nabla^2 u(z)) \leq 0 \tag{3.1.1}$$

for all $z = (x, t)$ in $\Omega \times (0, T)$, where Ω is a domain in $\mathbf{R}^N$ and $T > 0$, and F is a function in $\Omega \times \mathbf{R} \times \mathbf{R}^N \times \mathbf{S}^N$. We assume that F is *(degenerate) elliptic* , i.e.,

$$F(z, r, p, X) \leq F(z, r, p, Y) \qquad \text{for } X \geq Y,$$

so that the equation

$$u_t + F(z, u, \nabla u, \nabla^2 u) = 0 \tag{3.1.2}$$

Y. Giga, Q. Liu, *Surface Evolution Equations*, Monographs in Mathematics 99,
https://doi.org/10.1007/978-3-032-20303-8_3

is a parabolic equation. Suppose that φ is also C^2 and $u-\varphi$ takes its local maximum at $\hat{z} = (\hat{x}, \hat{t})$ with $\hat{z} \in \Omega \times (0, T)$. Then the classical maximum principle which is just calculus implies that

$$u_t(\hat{z}) = \varphi_t(\hat{z}), \ \nabla u(\hat{z}) = \nabla\varphi(\hat{z}), \ \nabla^2 u(\hat{z}) \le \nabla^2\varphi(\hat{z}).$$

The degenerate ellipticity together with (3.1.1) implies that

$$\begin{aligned} &\varphi_t(\hat{z}) + F(\hat{z}, u(\hat{z}), \nabla\varphi(\hat{z}), \nabla^2\varphi(\hat{z})) \\ &\le u_t(\hat{z}) + F(\hat{z}, u(\hat{z}), \nabla u(\hat{z}), \nabla^2 u(\hat{z})) \le 0. \end{aligned}$$

The same inequality for φ is obtained even if $\Omega \times (0, T)$ is replaced by $\Omega \times (0, T]$ so that $\hat{t}$ may equal T, since we still have

$$u_t(\hat{z}) \le \varphi_t(\hat{z})$$

although $u_t(\hat{z})$ may not equal $\varphi_t(\hat{z})$. The inequality

$$\varphi_t(\hat{z}) + F(\hat{z}, u(\hat{z}), \nabla\varphi(\hat{z}), \nabla^2\varphi(\hat{z})) \le 0 \tag{3.1.3}$$

does not include derivatives of u so we are tempted to define an arbitrary function u to be a generalized subsolution of (3.1.2) in $\Omega \times (0, T)$ if for each $\hat{z} \in \Omega \times (0, T)$ an upper test function φ of u around $\hat{z}$ always solves (3.1.3). There are a couple of degrees of freedom to define viscosity solutions when F is not continuous. One should take a good definition of solutions which enjoy major expected properties including comparison principle, stability principle and Perron's construction of solutions. We shall give such definitions for a typical class of equations. For technical convenience we define a notion of solutions for arbitrary discontinuous functions.

3.1.1 Definition for Arbitrary Functions

To handle arbitrary functions we recall the upper and lower semicontinuous envelope defined in Sect. 2.6.5. Let h be a function defined on a set L of a metric space X with values in $\mathbf{R} \cup \{\pm\infty\}$. The *upper semicontinuous envelope* h^* and the *lower semicontinuous envelope* h_* of h are defined by

$$\begin{aligned} h^*(z) &= \lim_{r\downarrow 0} \sup\{h(\zeta); \zeta \in B_r(z) \cap L\} \\ h_*(z) &= \lim_{r\downarrow 0} \inf\{h(\zeta); \zeta \in B_r(z) \cap L\} \end{aligned} \qquad z \in \overline{L} \tag{3.1.4}$$

respectively, as functions defined on the closure $\overline{L}$ of L with values in $\mathbf{R}^{\cup}\{\pm\infty\}$, where $B_r(z)$ is a closed ball of radius r centered at z in X. In other words h_* is

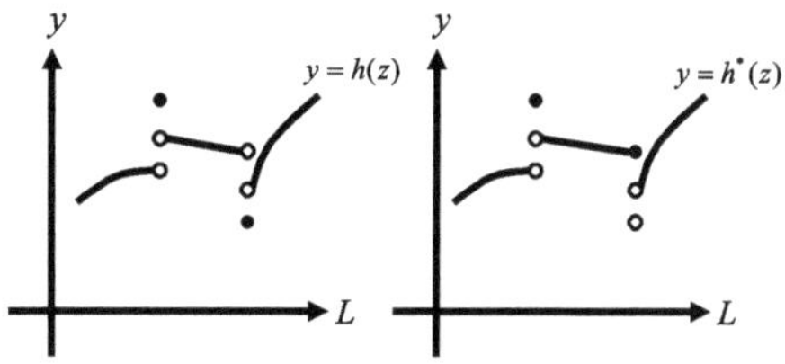

Fig. 3.1 Upper semicontinuous envelope

the greatest lower semicontinuous function on $\overline{L}$ which is smaller than h on L and similarly, h^* is the smallest upper semicontinuous function on $\overline{L}$ which is greater than h on L (Fig. 3.1). Clearly $h_* = h$ if h is lower semicontinuous in L and $h^* = h$ if h is upper semicontinuous in L. (We recall here that a function f on $Z \subset X$ is said to be *lower semicontinuous* if

$$\liminf_{x \to z} f(x) \geq f(z) \quad \text{for} \quad \text{all} \quad z \in Z.$$

This condition is equivalent to saying that any super level set

$$\{z \in Z; \, f(z) > \alpha\} \ (\alpha \in \mathbf{R})$$

is open. If $-f$ is lower semicontinuous, f is said to be *upper semicontinuous* .) For continuous φ on $\overline{L}$ it is clear that $(h - \varphi)_* = h_* - \varphi$. Also by definition $h_* = -(-h)^*$. For h^* the maximum over compact set K is always attained provided that $h^*(z) < \infty$ for $z \in K$. This property is often used in the sequel.

We first give a rigorous definition of a solution of (3.1.2) when F has no singularities.

Definition 3.1.1 Let Ω be an open set in $\mathbf{R}^N$ and $T > 0$. Let F be a continuous function on $\overline{\Omega} \times [0, T] \times \mathbf{R} \times \mathbf{R}^N \times \mathbf{S}^N$ with values in $\mathbf{R}$. Let $\mathcal{O}$ be an open set in $\Omega \times (0, T)$.

(i) A function $u : \mathcal{O} \to \mathbf{R} \cup \{-\infty\}$ is a (viscosity) *subsolution* of

$$u_t + F(z, u, \nabla u, \nabla^2 u) = 0 \tag{3.1.5}$$

in $\mathcal{O}$ (or equivalently, solution of $u_t + F(z, u, \nabla u, \nabla^2 u) \leq 0$) if

(a) $u^*(z) < \infty$ for $z \in \mathcal{O}$.
(b) If $(\varphi, \hat{z}) \in C^2(\mathcal{O}) \times \mathcal{O}$ satisfies

$$\max_{\mathcal{O}}(u^* - \varphi) = (u^* - \varphi)(\hat{z}), \tag{3.1.6}$$

then

$$\varphi_t(\hat{z}) + F(\hat{z}, u^*(\hat{z}), \nabla\varphi(\hat{z}), \nabla^2\varphi(\hat{z})) \leq 0. \tag{3.1.7}$$

Fig. 3.2 Upper test function

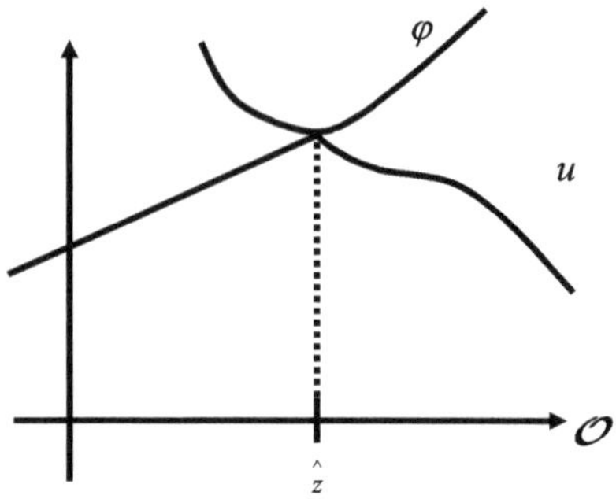

(ii) A function $u : \mathcal{O} \to \mathbf{R} \cup \{+\infty\}$ is a (viscosity) *supersolution* of (3.1.5) in $\mathcal{O}$ (or equivalently, solution of $u_t + F(z, u, \nabla u, \nabla^2 u) \geq 0$) if

(a) $u_*(z) > -\infty$ for $z \in \mathcal{O}$.
(b) If $(\varphi, \hat{z}) \in C^2(\mathcal{O}) \times \mathcal{O}$ satisfies

$$\min_{\mathcal{O}}(u_* - \varphi) = (u_* - \varphi)(\hat{z}), \tag{3.1.8}$$

then

$$\varphi_t(\hat{z}) + F(\hat{z}, u_*(\hat{z}), \nabla\varphi(\hat{z}), \nabla^2\varphi(\hat{z})) \geq 0. \tag{3.1.9}$$

By definition this is equivalent to saying that $v = -u$ is a solution of

$$v_t - F(z, -v, -\nabla v, -\nabla^2 v) \leq 0.$$

We shall often suppress the word "viscosity". In every circumstance when sub- and supersolutions are defined, we say a function is a (viscosity) *solution* of the equation in a given set if it is both a sub- and supersolution of the equation in the set. A function φ satisfying (3.1.6) is called an *upper test function* of u at $\hat{z}$ over $\mathcal{O}$ and φ satisfying (3.1.8) is called a *lower test function* of u at $\hat{z}$ over $\mathcal{O}$ (Fig. 3.2). As observed from (3.1.3) the notion of viscosity solutions is a natural extension of the notion of solutions if F is degenerate elliptic.

Proposition 3.1.2 *Assume that $u \in C^2(\mathcal{O})$ (i.e., u is C^2 on $\mathcal{O}$). Assume that F is degenerate elliptic. Then u is a subsolution (resp. supersolution) of* (3.1.5) *in $\mathcal{O}$ if and only if u satisfies*

$$u_t + F(z, u, \nabla u, \nabla^2 u) \leq 0 \quad (\textit{resp.} \geq 0)$$

in $\mathcal{O}$.

Remark 3.1.3 It is sometimes convenient to define solutions in an open set $\mathcal{O}$ in $\Omega \times (0, T]$ instead of $\Omega \times (0, T)$. It is defined in the same way as in Definition 3.1.1. As remarked before (3.1.3), the statement of Proposition 3.1.2 is still valid even if $\mathcal{O}$ intersects $t = T$.

The notation of viscosity solutions can be localized.

Localization Property

(i) *If u is a subsolution of an equation in a neighborhood of each point of $\Omega \times (0, T)$, then it is a subsolution (of the same equation) in $\Omega \times (0, T)$.*
(ii) *If u is a subsolution (of an equation) in $\Omega \times (0, T)$, then it is a subsolution (of the same equation) in every open set $\mathcal{O}$ in $\Omega \times (0, T)$.*
(iii) *The properties* (i), (ii) *are still valid if a supersolution replaces a subsolution.*

This is easily proved for Eq. (3.1.5) by restricting and extending test functions.

Unfortunately, if (3.1.5) is a level set equation, the value of F at $\nabla u = 0$ may not be defined. We have to extend the definition of viscosity solutions for (3.1.5) with discontinuous F to handle such equations.

Definition 3.1.4

(i) Assume that F is lower semicontinuous in $W = \overline{\Omega} \times [0, T] \times \mathbf{R} \times \mathbf{R}^N \times \mathbf{S}^N$ with values in $\mathbf{R} \cup \{-\infty\}$. A subsolution of (3.1.5) is defined as in Definition 3.1.1 (i).
(ii) Assume that F is upper semicontinuous in W with values in $\mathbf{R} \cup \{+\infty\}$. A supersolution of (3.1.5) is defined as in Definition 3.1.1 (ii). Here we use the convention that $a + \infty = \infty > 0$ and $a - \infty = -\infty < 0$ for any real number a.
(iii) Assume that F is defined only in a dense subset of W and that $F_* < \infty$, $F^* > -\infty$ in W. If u is a subsolution of

$$u_t + F_*(z, u, \nabla u, \nabla^2 u) = 0$$

in $\mathcal{O}$, then u is called a subsolution of (3.1.5) in $\mathcal{O}$. If u is a supersolution of

$$u_t + F^*(z, u, \nabla u, \nabla^2 u) = 0$$

in $\mathcal{O}$, then u is called a supersolution of (3.1.5). (In applications the case when F is continuous in $\overline{\Omega} \times [0, T] \times \mathbf{R} \times (\mathbf{R}^N \setminus \{0\}) \times \mathbf{S}^N$ has a particular importance.)

As stated for continuous F, the statement of Proposition 3.1.2 and the localization property are still valid for such an F.

3.1.2 Expected Properties of Solutions

We shall explain in an informal way three major properties which viscosity solutions are expected to satisfy.

Comparison Principle *If u is a subsolution of* (3.1.5) *in* $Q = \Omega \times [0, T)$ *and* v *is a supersolution of* (3.1.5) *in* Q, *then* $u^* \leq v_*$*in* Q *if* $u^* \leq v_*$ *on the parabolic boundary* $\partial_p Q$ *of* Q, *i.e.*,

$$\partial_p Q = \Omega \times \{0\} \cup \partial\Omega \times [0, T).$$

This is a comparison principle for semicontinuous functions so it is sometimes called a strong comparison principle to distinguish the comparison principle for continuous functions.

The major structural assumptions to guarantee the comparison principle are degenerate ellipticity of F and monotonicity of $F = F(z, r, p, X)$ in r in the sense that

$$r \mapsto F(z, r, p, X)$$

is nondecreasing. The last property can be weakened to the nondecreasing property of

$$r \mapsto F(z, r, p, X) + c_0 r$$

for some c_0 independent of (z, p, X). (Indeed, by the change of variable u by $e^{c_0 t} v$, this condition yields the stronger version of the monotonicity condition for the equation for v.) We need other regularity assumptions for F. A simplest set of assumptions for continuous F is that F is independent of z, r and is degenerate elliptic. In this case as we will see later the comparison principle holds at least for bounded Ω. (For unbounded Ω one should modify the interpretation of the inequality $u^* \leq v_*$.)

If $F = F(z, r, p, X)$ has a singularity (discontinuity) at $p = 0$ (but is continuous elsewhere), we assume that

$$-\infty < F_*(z, r, 0, O) = F^*(z, r, 0, O) < \infty \tag{3.1.10}$$

to get the comparison principle. (In Chap. 2 we have characterized a class of operators $F_f(p, X)$ such that this condition is fulfilled. See Lemma 2.6.16–Proposition 2.6.18. Many geometric F in level set equations including mean curvature flow equations (2.5.4) fulfill this condition.) This assumption cannot be removed completely.

Counterexample We set $F(p, X) = -X/|p|^\alpha$ with $N = 1, \alpha > 0$. Then both $u \equiv 0$ and

$$v(x,t) = \begin{cases} 1 & t \geq T, \\ 0 & t < T \end{cases}$$

are solutions of (3.1.5) in $\mathbf{R} \times (0, \infty)$. Of course F is degenerate elliptic but it violates (3.1.10); indeed

$$F_*(0, O) = -\infty, \ F^*(0, O) = +\infty.$$

Since both u and v are initially zero, the comparison principle is violated. One complaint lies in the fact that $\mathbf{R}$ is unbounded. But it still gives a counterexample for the comparison principle which is expected to hold for an unbounded domain. We do not pursue a counterexample for a bounded domain.

Stability Principle *We begin with its naive form. Assume that F_ε converges to F locally uniformly (as $\varepsilon \to 0$ with $\varepsilon > 0$) in the domain of definition of F. Assume that u_ε solves*

$$u_{\varepsilon t} + F_\varepsilon(z, u_\varepsilon, \nabla u_\varepsilon, \nabla^2 u_\varepsilon) \leq 0 \quad (\textit{resp.} \geq 0) \quad \textit{in} \quad \mathcal{O}$$

and that u_ε converges to u locally uniformly (as $\varepsilon \to 0$) in $\mathcal{O}$. Then u solves

$$u_t + F(z, u, \nabla u, \nabla^2 u) \leq 0 \quad (\textit{resp.} \geq 0) \quad \textit{in} \quad \mathcal{O}.$$

This property is useful when we want to construct a solution u of (3.1.5) by approximating the equation. There are several options to weaken the assumption of convergences $F_\varepsilon \to F$, $u_\varepsilon \to u$. We may replace "convergence" by "semi-convergence". Let h_ε be a function defined on a set L of a metric space X with values in $\mathbf{R} \cup \{\pm\infty\}$, where $\varepsilon > 0$. The *upper relaxed limit* $\overline{h} = \limsup^* h_\varepsilon$ and the *lower relaxed limit* $\underline{h} = \liminf_* h_\varepsilon$ are defined by

$$\begin{aligned} \overline{h}(z) = (\limsup_{\varepsilon \to 0}{}^* h_\varepsilon)(z) &= \limsup_{\substack{\zeta \to z \\ \varepsilon \to 0}} h_\varepsilon(\zeta) \\ &= \lim_{\varepsilon \to 0} \sup\{h_\delta(\zeta); \zeta \in L \cap B_\varepsilon(z), \ 0 < \delta < \varepsilon\}, \\ \underline{h}(z) = (\liminf_{\varepsilon \to 0}{}_* h_\varepsilon)(z) &= \liminf_{\substack{\zeta \to z \\ \varepsilon \to 0}} h_\varepsilon(\zeta) \\ &= \lim_{\varepsilon \to 0} \inf\{h_\delta(\zeta); \zeta \in L \cap B_\varepsilon(z), \ 0 < \delta < \varepsilon\}, \end{aligned}$$

respectively. These limits are a kind of 'lim sup' and 'lim inf' but take such an operation in *both* ε and ζ. The functions $\limsup^* h_\varepsilon$ and $\liminf_* h_\varepsilon$ are defined for all z in $\overline{L}$ allowing the values $\pm\infty$. Evidently, if $h_\varepsilon = h$, then

$$h^* = \limsup{}^* h_\varepsilon \quad \text{and} \quad h_* = \liminf{}_* h_\varepsilon.$$

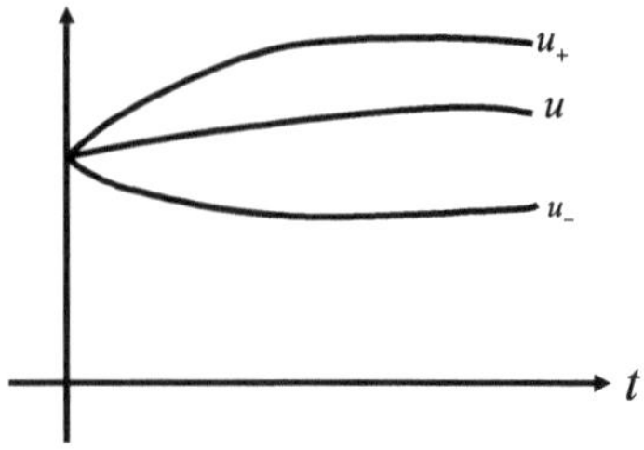

Fig. 3.3 Perron's method

By definition $\overline{h} = \limsup^*(h_\varepsilon^*)$ and $\underline{h} = \liminf_*(h_\varepsilon)_*$. If we set $k(z, \varepsilon) = h_\varepsilon(z)$ as a function of $z \in L$ and $\varepsilon > 0$, then by definition

$$\overline{h}(z) = k^*(z, 0),\ \underline{h}(z) = k_*(z, 0) \quad \text{for} \quad z \in \overline{L}.$$

If $\overline{h} = \underline{h}$ on $\overline{L}$, then k^* and k_* are continuous at $\varepsilon = 0$ as functions in $\overline{L} \times (0, 1)$. This implies that h_ε converges to $\overline{h} = \underline{h}$ locally uniformly in $\overline{L}$ as $\varepsilon \to 0$.

The convergence $u_\varepsilon \to u$ may be replaced by semi-convergence $u = \overline{u}$ if we consider sequences of subsolutions. For sequences of supersolutions we may replace $u_\varepsilon \to u$ by $u = \underline{u}$. The convergence $F_\varepsilon \to F$ is replaced by $F \leq \underline{F}$ and $F \geq \overline{F}$ for sub and supersolutions respectively. If convergence is replaced by semi-convergence in the stability principle, it is sometimes called the strong stability principle. The major advantage lies in the fact that $\overline{u}$ and $\underline{u}$ always exist allowing the values $\pm\infty$. So to apply the strong stability principle we need not estimate $\{u_\varepsilon\}$ to get locally uniform convergence of u_ε. Once we prove $(-\infty <)\ \overline{u} \leq \underline{u}\ (< \infty)$ for example by the comparison principle, we have $\overline{u} = \underline{u}$. This implies *a posteriori* locally uniform convergence $u_\varepsilon \to \overline{u} = \underline{u} = u$ and the limit u is continuous.

Perron's Method (Construction of Solutions) *If* u_+ *and* u_- *are super- and subsolution of* (3.1.5) *in* $\Omega \times [0, T)$, *respectively, then there is a solution* u *of* (3.1.5) *in* $\Omega \times [0, T)$ *that satisfies* $u_- \leq u \leq u_+$ on $\Omega \times (0, T)$ *provided that* $u_- \leq u_+$ *on* $\Omega \times (0, T)$ (Fig. 3.3).

This method reduces the business of construction of solutions to finding appropriate sub- and supersolutions satisfying the initial and boundary conditions. For example consider the initial value problem for (3.1.5) with continuous initial data u_0, where $\Omega = \mathbf{R}^N$. If there are sub- and supersolutions $u^\pm$ of (3.1.5) in $\mathbf{R}^N \times (0, T)$ that satisfy

$$u^{+*}(x, 0) = u_0(x) = u^-_*(x, 0),$$

then by Perron's method we have a solution u. Moreover, if the comparison principle applies, then $u^* \leq u_*$ in $\mathbf{R}^N \times [0, T)$. This implies that u is continuous on $\mathbf{R}^N \times [0, T)$ with $u(0, x) = u_0(x)$. The uniqueness of such a solution is clear from the comparison principle. Since $\mathbf{R}^N$ is unbounded, the comparison principle is not exactly as stated above, this discussion is a little bit informal but it explains the basic idea to get a solution.

Apparently, Definition 3.1.4 is a natural extension for discontinuous F. Indeed the definition for continuous F is included. Moreover, the stability principle and Perron's method is available. If (3.1.10) is fulfilled, as we see later, the comparison principle holds under reasonable regularity assumptions on F at least when $F(z, r, p, X)$ is continuous outside $p = 0$. This class of equations includes many important examples of level set equations satisfying (2.6.24) so for example it applies to the level set mean curvature flow equation (2.6.5). However, for the level set equation of the Gaussian curvature flow (2.6.12), condition (3.1.10) is not fulfilled (see Sect. 2.6.5). If F fails to satisfy (3.1.10), the comparison principle may not hold. One should modify the notion of viscosity solutions for such problems as discussed in the next subsection.

In this chapter we only discuss the stability principle and Perron's method. We discuss the comparison principle in the next chapter.

3.1.3 Very Singular Equations

Let F be continuous on $W_0 = \overline{\Omega}\times[0,T]\times\mathbf{R}\times(\mathbf{R}^N\setminus\{0\})\times\mathbf{S}^N$ with values in $\mathbf{R}$. We restrict the test functions. Let $\mathcal{F}_\Omega = \mathcal{F}_\Omega(F)$ be the set of functions $f \in C^2[0,\infty)$ such that

$$f(0) = f'(0) = f''(0) = 0 \quad \text{and} \quad f''(r) > 0 \quad \text{for} \quad r > 0$$

and that for $s = \pm 1$,

$$\lim_{p\to 0} \sup_{z\in\overline{\mathcal{O}}} \sup_{r\in\mathbf{R}} |F(z, r, \nabla_p(f(|p|)), s\nabla_p^2(f(|p|)))| = 0. \tag{3.1.11}$$

The formula (3.1.11) is often written as

$$\lim_{p\to 0} \sup_{z\in\overline{\mathcal{O}}} \sup_{r\in\mathbf{R}} |F(z, r, \nabla(f(|p|)), \pm\nabla^2(f(|p|))| = 0.$$

We suppress the Ω-dependence of $\mathcal{F}$ in this chapter since we fix Ω. We say that $\varphi \in C^2(\mathcal{O})$ is *compatible* with F if for any $\hat{z} = (\hat{x}, \hat{t})$ in $\mathcal{O}$ with $\nabla\varphi(\hat{z}) = 0$, there is a constant $\delta > 0$ and functions $f \in \mathcal{F}$ and $\omega_1 \in C[0,\infty)$ with $\lim_{\sigma\to 0} \omega_1(\sigma)/\sigma = 0$ that satisfies

$$|\varphi(x,t) - \varphi(\hat{z}) - \varphi_t(\hat{z})(t - \hat{t})| \le f(|x - \hat{x}|) + \omega_1(|t - \hat{t}|) \tag{3.1.12}$$

for all $(x,t) \in \mathcal{O}$ with $|x - \hat{x}| \le \delta$, $|t - \hat{t}| \le \delta$. The set of $\varphi \in C^2(\mathcal{O})$ compatible with F is denoted by $C_F^2(\mathcal{O})$.

Note that the set $\mathcal{F}$ can be empty. Indeed, for $N = 1$ let

$$F = F(p, X) = -X/|p|^\alpha. \tag{3.1.13}$$

If $\alpha \geq 1$, then, by the Gronwall inequality for f', a function $f \in C^2[0,\infty)$ that satisfies

$$\lim_{x\to 0} \frac{f''}{f'^\alpha} = 0 \quad \text{with} \quad f' \geq 0,\ f'(0) = 0,$$

is a constant near $x = 0$. This violates $f'(\sigma) > 0$ for $\sigma > 0$ so the set $\mathcal{F}$ is empty for $\alpha \geq 1$. If $0 < \alpha < 1$, $f(\sigma) = \sigma^\gamma$ belongs to $\mathcal{F}$ provided that $\gamma > (2-\alpha)/(1-\alpha)$. The case $\alpha \geq 1$, the initial value problem for (3.1.5) with (3.1.13), cannot be expected to be solvable except for monotone initial data. The case $0 < \alpha < 1$, Eq. (3.1.5) with (3.1.13), is the q-Laplace diffusion equation (with $1 < q < 2$)

$$u_t - c\,\mathrm{div}(|\nabla u|^{q-2}\nabla u) = 0, \tag{3.1.14}$$

where $q = 2-\alpha$, $c = 1/(q-1)$, $N = 1$. If $F = F(p, X)$ is geometric, degenerate elliptic and continuous outside $p = 0$, then it turns out that $\mathcal{F}$ is not empty. We shall discuss this point in the next chapter. It is easy to see that $\mathcal{F}$ is invariant under positive multiplication, i.e., $af \in \mathcal{F}$ if $f \in \mathcal{F}$ and $a > 0$ for (3.1.14) with $q > 1$. This important property also holds for geometric F.

Definition 3.1.5 Assume that F is continuous on $W_0 = \overline{\Omega} \times [0, T] \times \mathbf{R} \times (\mathbf{R}^N\backslash\{0\}) \times \mathbf{S}^N$ with values in $\mathbf{R}$. Assume that $\mathcal{F}$ is nonempty. An *$\mathcal{F}$-subsolution* of (3.1.5) with F is defined as in Definition 3.1.1 by replacing (i)(b) by the condition:

$$\begin{cases} \varphi_t(\hat{z}) + F(\hat{z},\ u^*(\hat{z}),\ \nabla\varphi(\hat{z}),\ \nabla^2\varphi(\hat{z})) \leq 0 & \text{if} \quad \nabla\varphi(\hat{z}) \neq 0, \\ \varphi_t(\hat{z}) \leq 0 & \text{otherwise} \end{cases}$$

for all $(\varphi, \hat{z}) \in C^2(\mathcal{O}) \times \mathcal{O}$ with $\varphi \in C^2_{\mathcal{F}}(\mathcal{O}')$ satisfying (3.1.6), where $\mathcal{O}' \subset \mathcal{O}$ is some neighborhood of $\hat{z}$. (The restriction of φ on $\mathcal{O}'$ is still denoted by φ.)

An *$\mathcal{F}$-supersolution* of (3.1.5) with F is defined as in Definition 3.1.1 by replacing (ii)(b) by the condition:

$$\begin{cases} \varphi_t(\hat{z}) + F(\hat{z},\ u_*(\hat{z}),\ \nabla\varphi(\hat{z}),\ \nabla^2\varphi(\hat{z})) \geq 0 & \text{if} \quad \nabla\varphi(\hat{z}) \neq 0, \\ \varphi_t(\hat{z}) \geq 0 & \text{otherwise} \end{cases}$$

for all $(\varphi, \hat{z}) \in C^2(\mathcal{O}) \times \mathcal{O}$ with $\varphi \in C^2_{\mathcal{F}}(\mathcal{O}')$ satisfying (3.1.8), where $\mathcal{O}'$ is some neighborhood of $\hat{z}$.

One may define $\mathcal{F}$-subsolutions and $\mathcal{F}$-supersolutions in a set $\mathcal{O}$ relatively open in $\Omega \times (0, T]$ as in Remark 3.1.3.

Remark 3.1.6 If

$$\lim_{\substack{p\to 0\\ X\to O}} \sup_{z\in\overline{\mathcal{O}}} \sup_{r\in\mathbf{R}} |F(z,r,p,X)| = 0, \tag{3.1.15}$$

(which in particular implies (3.1.10)), then clearly

$$\mathcal{F}=\{f \in C^2[0,\infty);\ f(0)=f'(0)=f''(0)=0 \quad \text{and} \quad f''(r)>0 \quad \text{for} \quad r>0\}.$$

It turns out that $C^2_F(\mathcal{O})$ equals

$$\mathcal{A}_0 = \{\varphi \in C^2(\mathcal{O});\ \nabla\varphi(\hat{z}) = 0 \quad \text{implies} \quad \nabla^2\varphi(\hat{z}) = O\}.$$

We postpone its proof. By this observation a subsolution is an $\mathcal{F}$-subsolution since $F_*(z,r,0,O) = 0$ by (3.1.15). It is curious whether or not an $\mathcal{F}$-subsolution agrees with a subsolution in Definition 3.1.4. It turns out that the converse is true if (3.1.15) is fulfilled and F is degenerate elliptic as proved in Proposition 3.2.8. Here we just give a simple equivalent definition of $\mathcal{F}$-subsolutions.

Proposition 3.1.7 *Assume that* (3.1.15) *holds for* F. *A function* $u : \mathcal{O} \to \mathbf{R}\cup\{-\infty\}$ *is an* $\mathcal{F}$*-subsolution of* (3.1.5) *if and only if* u *is a solution of*

$$u_t + F_\#(z,u,\nabla u,\nabla^2 u) \le 0 \quad \textit{in} \quad \mathcal{O}$$

with

$$F_\#(z,r,p,X) = \begin{cases} F(z,r,p,X) & \textit{if } p \ne 0, \\ 0 & \textit{if } (p,X) = (0,O), \\ -\infty & \textit{otherwise.} \end{cases}$$

A similar assertion holds for an $\mathcal{F}$*-supersolution by replacing the above inequality by*

$$u_t + F^\#(z,u,\nabla u,\nabla^2 u) \ge 0 \quad \textit{in} \quad \mathcal{O}$$

with

$$F^\#(z,r,p,X) = \begin{cases} F(z,r,p,X) & \textit{if } p \ne 0, \\ 0 & \textit{if } (p,X) = (0,O), \\ +\infty & \textit{otherwise.} \end{cases}$$

This follows from the definition of an $\mathcal{F}$-solution if we admit $\mathcal{A}_0 = C^2_F(\mathcal{O})$, which is proved in Proposition 3.1.8. Note that $F_\#(\le F_*)$ and $F^\#(\ge F^*)$ are, respectively, still lower and upper semicontinuous functions on $\overline{\Omega} \times [0,T] \times \mathbf{R} \times \mathbf{R}^N \times \mathbf{S}^N$ by (3.1.15). Here is a trivial remark. One may weaken (3.1.15) by

$$\lim_{\substack{p\to 0\\ X\to O}} \sup_{z\in\overline{\mathcal{O}}} \sup_{r\in\mathbf{R}} |F(z,r,p,X)-a_0| = 0$$

for a fixed constant a_0 so that $F_*(z,r,0,O) = a_0 = F^*(z,r,0,O)$. To define an $\mathcal{F}$-subsolution in this situation we should replace (3.1.11) by

$$\lim_{p\to 0} \sup_{z\in\overline{\mathcal{O}}} \sup_{r\in\mathbf{R}} |F(z,r,\nabla(f(|p|)),\pm\nabla^2(f(|p|)))-a_0| = 0$$

and replace $\varphi_t(\hat{z}) \leq 0$ by $\varphi_t(\hat{z}) + a_0 \leq 0$ and $\varphi_t(\hat{z}) \geq 0$ by $\varphi_t(\hat{z}) + a_0 \geq 0$ in Definition 3.1.5. The statement of Proposition 3.1.7 is still valid by replacing 0 by a_0 in the definition of $F_\#$ and $F^\#$. One may even replace a_0 by a continuous function $a_0(z,r)$ to handle the equation having an external force term.

For the level set equation (2.6.12) of the Gaussian curvature flow equation it is not difficult to see that $F_*(0,X) = -\infty$ and $F^*(0,X) = +\infty$ (see Sect. 2.6.5) so a subsolution in Definition 3.1.4 may not be an $\mathcal{F}$-subsolution, while an $\mathcal{F}$-subsolution is always a subsolution for (2.6.12) different from the case when (3.1.15) is fulfilled. In this case Definition 3.1.5 is more restrictive than Definition 3.1.4 which is indeed important to get the comparison principle.

Proposition 3.1.8 *Assume that* (3.1.15) *holds for* F. *Then* $C^2_{\mathcal{F}}(\mathcal{O}) = \mathcal{A}_0$.

Proof It is easy to see that $\mathcal{A}_0$ includes $C^2_{\mathcal{F}}(\mathcal{O})$. Indeed, the condition (3.1.12) implies that $\nabla^2\varphi(\hat{z}) = O$ for $\hat{z}$ with $\nabla\varphi(\hat{z}) = 0$.

It remains to prove that $C^2_{\mathcal{F}}(\mathcal{O})$ includes $\mathcal{A}_0$. If $\nabla\varphi(\hat{z}) = 0$ and $\nabla^2\varphi(\hat{z}) = O$, then

$$|\varphi(z)-\varphi(\hat{z})-\varphi_t(\hat{z})(t-\hat{t})| \leq \omega_1(|t-\hat{t}|)+\omega_2(|x-\hat{x}|)$$

for all (x,t) with $|x-\hat{x}| \leq \delta$, $|t-\hat{t}| \leq \delta$ for some $\delta > 0$ and $\omega_1, \omega_2 \geq 0$ with

$$\lim_{\sigma\to 0} \omega_k(\sigma)/\sigma^k = 0 \quad (k=1,2). \tag{3.1.16}$$

We would like to find $f_0 \in C^2[0,\infty)$ with $f_0(0) = f_0'(0) = f_0''(0) = 0$ and $f_0'' \geq 0$ that satisfies

$$\omega_2(\sigma) \leq f_0(\sigma) \quad \text{for} \quad 0 \leq \sigma \leq \delta. \tag{3.1.17}$$

If such an f_0 exists, $f(\sigma) = f_0(\sigma) + \sigma^4$ belongs to $\mathcal{F}$ since (3.1.11) is always fulfilled for $f \in C^2[0,\infty)$ with $f'(0) = f''(0) = 0$ under the condition (3.1.15). We shall prove the existence of f_0. If we set

$$\omega_0(\sigma) = \sup\{\omega_2(\eta)/\eta^2;\ 0 \leq \eta \leq \sigma,\ \eta \leq \delta\},$$

then ω_0 is a nondecreasing function on $[0, \infty)$ with $\omega_0(0) = 0$ which is continuous at zero by (3.1.16). Since

$$\omega_2(\sigma) \leq \omega_0(\sigma)\sigma^2 \quad \text{for} \quad 0 \leq \sigma \leq \delta,$$

the existence of f_0 that satisfies (3.1.17) follows from the next elementary lemma by taking $f_0 = \theta$ with $k = 2$. □

Lemma 3.1.9

(i) *Let ω_0 be a nondecreasing function on $[0, \infty)$ with $\omega_0(0) = 0$. Assume that ω_0 is continuous at zero. Then there is a modulus ω (i.e., ω is a nondecreasing continuous function on $[0, \infty)$ with $\omega(0) = 0$) with $\omega \in C^\infty(0, \infty)$ that satisfies $\omega_0 \leq \omega$ on $[0, \infty)$.*

(ii) *Let ω be a modulus and let k be a positive integer. Then there is $\theta \in C^k[0, \infty)$ such that $\theta^{(j)}(0) = 0$ and $\theta^{(j)}(\sigma) \geq 0$ for $\sigma \geq 0$ with $0 \leq j \leq k$ and $\omega(\sigma)\sigma^k \leq \theta(\sigma)$ for all $\sigma \geq 0$. Here $\theta^{(j)}$ denotes the j-th derivative of θ.*

Proof of Lemma 3.1.9

(i) We set $\omega(m) = \omega_0(m + 1)$, $m = 1, 2, \dots$ and $\omega(1/2^j) = \omega_0(1/2^{j-1})$, $j = 1, 2, \dots$. Interpolating the value of $\omega(\sigma)$ for $m \leq \sigma \leq m + 1$ and $1/2^j \leq \sigma \leq 1/2^{j-1}$ by an affine function, we get a continuous function ω on $(0, \infty)$. It is possible to mollify ω without changing values at $1/2^j$, $j = 1, 2, \dots$, so that $\omega \in C^\infty(0, \infty)$ and $\omega_0 \leq \omega$. Since ω_0 is nondecreasing so is ω and $\omega_0 \leq \omega$ on $(0, \infty)$. Since ω_0 is continuous at zero, we have

$$\lim_{\sigma \to 0} \omega(\sigma) = \lim_{j \to \infty} \omega_0(1/2^{j-1}) = 0.$$

Thus ω extends a continuous function on $[0, \infty)$ (still denoted ω) by setting $\omega(0) = 0$ (Fig. 3.4).

(ii) We set

$$\theta_j(\sigma) = \int_\sigma^{2\sigma} \theta_{j-1}(s)ds, \ j \geq 1, \ \theta_0 = \omega \quad \text{for} \quad \sigma \geq 0,$$

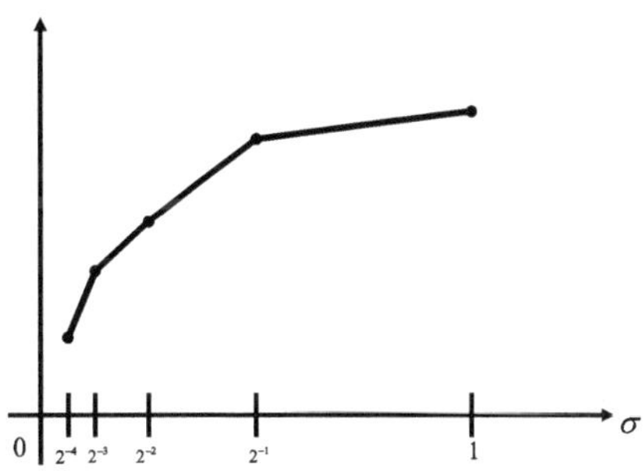

Fig. 3.4 Construction of ω

so that $\theta_j \in C^j[0,\infty)$ with $\theta_j^{(i)}(0) = 0$ for $0 \le i \le j$. Since θ_j is nondecreasing, we have

$$\theta_j(\sigma) \ge \sigma\theta_{j-1}(\sigma) \quad \text{for} \quad \sigma \ge 0$$

so that $\theta_j(\sigma) \ge \sigma^j\omega(\sigma)$. We thus observe that $\theta = \theta_k$ has all desired properties.

□

Remark 3.1.10 For $\mathcal{F}$-subsolutions the statement of Proposition 3.1.2 should be altered. Assume that $u \in C^2_{\mathcal{F}}(\mathcal{O})$ satisfies

$$u_t(\hat{z}) + F(\hat{z}, u(\hat{z}), \nabla u(\hat{z}), \nabla^2 u(\hat{z})) \le 0 \quad (\text{resp.} \ge 0)$$

for $\hat{z} \in \mathcal{O}$ with $\nabla u(\hat{z}) \ne 0$ and that

$$u_t(\hat{z}) \le 0 \quad (\text{resp.} \ge 0) \tag{3.1.18}$$

for $\hat{z} \in \mathcal{O}$ with $\nabla u(\hat{z}) = 0$. Then u is an $\mathcal{F}$-subsolution (resp. $\mathcal{F}$-supersolution) of (3.1.5) in $\mathcal{O}$ provided that F is degenerate elliptic. For $\hat{z}$ with $\nabla u(\hat{z}) = 0$, (3.1.18) is unnecessary if there are no $f \in \mathcal{F}$ and ω_1 (with $\omega_1(\sigma)/\sigma \to 0$ as $\sigma \to 0$) that satisfies

$$u(x,t) - u(\hat{z}) - u_t(\hat{z})(t-\hat{t}) \le (\text{resp.} \ \ge) f(|x-\hat{x}|) + \omega_1(t-\hat{t})$$

for all $z = (x,t)$ sufficiently close to $\hat{z} = (\hat{x},\hat{t})$. These statements immediately follow from the definition. We also note that the localization property is still valid for $\mathcal{F}$-solutions. The proof for (i) (in the localization property) is immediate. The proof for (ii) is not difficult. We set $Q = \Omega \times (0,T)$ and let $\mathcal{O}$ be an open subset in Q. Suppose that

$$\max_{\mathcal{O}}(u^* - \varphi) = (u^* - \varphi)(\hat{z})$$

with $\varphi \in C^2(\mathcal{O})$ satisfying $\varphi \in C^2_{\mathcal{F}}(\mathcal{O}')$, where $\mathcal{O}'$ is some open neighborhood of $\hat{z}$ in $\mathcal{O}$. We may assume that $\bar{\mathcal{O}}' \subset \mathcal{O}$ and $\bar{\mathcal{O}}'$ is compact. It is easy to find $\psi_0 \in C^1(Q)$ such that $\psi_0 = \varphi$ on $\mathcal{O}'$ and $\psi_0 \ge \varphi$ on $\mathcal{O}$. We set $\psi = \psi_0 + \eta$ with a nonnegative function $\eta \in C^2(Q)$ supported in $Q \backslash \bar{\mathcal{O}}'$. It is possible to take η large such that

$$\sup_{Q\backslash\mathcal{O}}(u^* - \psi) < (u^* - \varphi)(\hat{z}).$$

This yields

$$\max_{Q}(u^* - \psi) = (u^* - \psi)(\hat{z}).$$

If u is a subsolution in $\Omega \times (0, T)$, then we have

$$\begin{cases} \psi_t(\hat{z}) + F\left(\hat{z}, u^*(\hat{z}), \nabla\psi(\hat{z}), \nabla^2\psi(\hat{z})\right) \le 0 & \text{if} \quad \nabla\psi(\hat{z}) \neq 0 \\ \psi_t(\hat{z}) \le 0 & \text{otherwise.} \end{cases}$$

Since $\psi = \varphi$ in $\mathcal{O}'$, we may replace ψ by φ in the inequalities above. We thus conclude that u is a subsolution in $\mathcal{O}$. The property (iii) is obtained by a symmetric argument.

3.2 Stability Results

We shall give typical results on the stability principle in its strong form.

Theorem 3.2.1 *Let Ω be a domain in $\mathbf{R}^N$ and $T > 0$. Let $\mathcal{O}$ be an open set in $\Omega \times (0, T)$.*

(i) *Assume that F_ε and F are lower (resp. upper) semicontinuous in $W = \overline{\Omega} \times [0, T] \times \mathbf{R} \times \mathbf{R}^N \times \mathbf{S}^N$ with values in $\mathbf{R} \cup \{-\infty\}$ (resp. $\mathbf{R} \cup \{+\infty\}$) for $\varepsilon > 0$. Assume that*

$$F \le \liminf_{\varepsilon \to 0}{}_* F_\varepsilon \text{ in } W \quad (\text{resp.} F \ge \limsup_{\varepsilon \to 0}{}^* F_\varepsilon \text{ in } W). \tag{3.2.1}$$

Assume that u_ε is a subsolution (resp. supersolution) of

$$u_t + F_\varepsilon(z, u, \nabla u, \nabla^2 u) = 0 \tag{3.2.2}$$

in $\mathcal{O}$. Then $\overline{u} = \limsup_{\varepsilon \to 0}{}^* u_\varepsilon$ (resp. $\liminf_{\varepsilon \to 0}{}_* u_\varepsilon$) is a subsolution of

$$u_t + F(z, u, \nabla u, \nabla^2 u) = 0 \tag{3.2.3}$$

in $\mathcal{O}$ provided that $\overline{u}(z) < \infty$ (resp. $\underline{u}(z) > -\infty$) for each z in $\mathcal{O}$.

(ii) *Assume that F_ε and F are continuous in $W_0 = \overline{\Omega} \times [0, T] \times \mathbf{R} \times (\mathbf{R}^N \backslash \{0\}) \times \mathbf{S}^N$ with values in $\mathbf{R}$. Assume that $\mathcal{F}(F)$ is included in $\mathcal{F}_\varepsilon = \mathcal{F}(F_\varepsilon)$ for all (sufficiently small) ε and that F_ε converges to F locally uniformly in W_0 as $\varepsilon \to 0$. Assume that $\mathcal{F}$ is invariant under positive multiplication, i.e., $f \in \mathcal{F} = \mathcal{F}(F)$ implies $af \in \mathcal{F}$ for all $a > 0$. Assume that for any $f \in \mathcal{F}(F)$,*

$$\begin{aligned} \liminf\ & F_\varepsilon(\zeta, \rho, \nabla(f(|p|)), \nabla^2(f(|p|))) \ge 0, \\ (\text{resp.} \limsup\ & F_\varepsilon(\zeta, \rho, \nabla(f(|p|)), \nabla^2(f(|p|))) \le 0) \end{aligned} \tag{3.2.4}$$

as $\varepsilon \to 0$, $\zeta \to z$, $\rho \to r$, $p \to 0$ *for all* $z \in \mathcal{O}, r \in \mathbf{R}$. *Assume that* u_ε *is an* $\mathcal{F}_\varepsilon$*-subsolution (resp. supersolution) of* (3.2.2) *in* $\mathcal{O}$. *Then* $\overline{u}$ *is an* $\mathcal{F}$*-subsolution of* (3.2.3) *in* $\mathcal{O}$ *provided that* $\overline{u}(z) < \infty$ *(resp.* $\underline{u}(z) > -\infty$*) for each* $z \in \mathcal{O}$.

Clearly, the condition (3.2.1) is fulfilled if F_ε converges to F locally uniformly in W as $\varepsilon \to 0$. However, the uniform convergence in W is not expected if F has singularities. This is a reason why we assume (3.2.1) instead of uniform convergence in W. If F and $F_\varepsilon(\varepsilon > 0)$ satisfy (3.1.15), then $\mathcal{F} = \mathcal{F}_\varepsilon = \mathcal{A}_0$ by Proposition 3.1.8. In this case if $F_\varepsilon \to F_\#$ locally uniform in W_0 and

$$F_\# \leq \liminf_{\varepsilon \to 0}{}_* F_\varepsilon \quad \text{in} \quad W \tag{3.2.5}$$

(which is a special form of the left inequality in (3.2.1)), then it is easy to see that the first inequality of (3.2.4) holds, since $F_\#(z, r, 0, O) = 0$ by Proposition 3.1.8. However, the first inequality of (3.2.4) does not seem to imply (3.2.5) under the uniform convergence assumption in W_0 even if $F = F(z, r, p, X)$ does not depend on z and r, although both conditions are closely related.

The proof of Theorem 3.2.1 is not difficult if we are familiar with various equivalent definitions of viscosity solutions and convergence of maximum points explained in the next few subsections.

3.2.1 Remarks on a Class of Test Functions

We give several observations on test functions which are practically important to prove expected properties for viscosity solutions.

Proposition 3.2.2 *In* (3.1.6) *of Definitions 3.1.1, 3.1.4, and 3.1.5 one may replace* $\mathcal{O}$ *by some neighborhood of* $\hat{z}$. *The maximum may be replaced by a strict maximum in the sense that*

$$(u^* - \varphi)(z) < (u^* - \varphi)(\hat{z}), \ z \neq \hat{z}, \ z \in \mathcal{O} \tag{3.2.6}$$

or even by a local strict maximum in the sense that (3.2.6) *holds with* $\mathcal{O}$ *replaced by some neighborhood of* $\hat{z}$ *provided that* $\mathcal{F}$ *is invariant under positive multiplication. Similarly, in* (3.1.8) *the minimum may be replaced by a strict minimum or by a local strict minimum.*

Proof It is easy to see that a global maximum may be replaced by a local maximum in (3.1.6) and (3.2.6) in definitions similar to the proof of the localization property in Sect. 3.1.1. For $\mathcal{F}$-subsolution see also Remark 3.1.10.

If $\varphi \in C^2(\mathcal{O})$ satisfies (3.1.6), then $\psi(z) = \varphi(z) + |z - \hat{z}|^4$ satisfies (3.2.6). If (3.1.7) holds for ψ, so does φ. This shows that global maximum in (3.1.6) may be replaced by global strict maximum in Definitions 3.1.1 and 3.1.5. For $\mathcal{F}$-

subsolutions one should be a little bit careful. If $\varphi \in C_F^2(\mathcal{O})$ satisfies (3.1.6), then $\psi(z) = \varphi(z) + f(|x - \hat{x}|) + (t - \hat{t})^2$ with $f \in \mathcal{F}$ satisfying (3.2.6). However, ψ may not be in $C_F^2(\mathcal{O})$. So we choose ψ in another way. We may assume that $\nabla\varphi(\hat{z}) = 0$ and recall (3.1.12). Since φ is C^2, we may assume that

$$\omega_1(\sigma) = \varphi_{tt}(\hat{t})\sigma^2/2 + \omega(\sigma)\sigma^2$$

where ω is a modulus. By Lemma 3.1.9 (ii) there is $\theta_1 \in C^2[0, \infty)$ with $\theta_1(0) = \theta_1'(0) = 0$ and $\theta_2''(0) = \varphi_{tt}(\hat{t})$ that satisfies

$$\omega_1(|t - \hat{t}|) \leq \theta_1(|t - \hat{t}|).$$

If we set

$$\begin{aligned} \psi(x, t) = \varphi(\hat{z}) + \varphi_t(\hat{z})(t - \hat{t}) + f(|x - \hat{x}|) + \theta_1(|t - \hat{t}|) \\ + f(|x - \hat{x}|) + (t - \hat{t})^2, \end{aligned} \tag{3.2.7}$$

then $\varphi(\hat{z}) = \psi(\hat{z})$ and

$$\varphi(z) < \psi(z), \quad \text{for} \quad z \neq \hat{z},\ z = (x, t) \in \mathcal{O}'$$

with

$$\mathcal{O}' = \{(x, t); |x - \hat{x}| < \delta,\ |t - \hat{t}| < \delta\}.$$

The last two terms $f(|x - \hat{x}|)$ and $|t - \hat{t}|^2$ are added to get a strict inequality. Since $2f \in \mathcal{F}$, $\psi \in C_F^2(\mathcal{O}')$ and (3.2.6) holds with $\mathcal{O}$ replaced by $\mathcal{O}'$. Here the property that $\mathcal{F}$ is invariant under positive multiplication is invoked. If the second differential inequality in Definition 3.1.5 holds for ψ, i.e., $\psi_t(\hat{z}) \leq 0$, so does $\varphi_t(\hat{z}) \leq 0$. This explain the reason why we may replace maximum by strict maximum even for $\mathcal{F}$-subsolutions. The proof for supersolutions parallels that for subsolutions. □

Proposition 3.2.3

(i) *In Definitions 3.1.1 and 3.1.4 the class of test functions may be replaced by* $C^k(\mathcal{O}), C^\infty(\mathcal{O})$ *or*

$$A^k(\mathcal{O}) = \{\varphi(x, t) = b(x) + g(t) \in C^k(\mathcal{O})\}$$

where $\infty \geq k \geq 2$. *In Definition 3.1.5* $C_F^2(\mathcal{O})$ *may be replaced by*

$$A_F^k(\mathcal{O}) = \{\varphi(x, t) = b(x) + g(t) \in C_F^2(\mathcal{O}),\ g \in C^k(\mathbf{R})\}$$

where $\infty \geq k \geq 1$ *provided that* $\mathcal{F}$ *is invariant under positive multiplication.*

(ii) *The classes of test functions $C^2(\mathcal{O})$ and $C^2_{\mathcal{F}}(\mathcal{O})$ may be replaced also by $C^{2,1}(\mathcal{O})$ and*

$$C^{2,1}_{\mathcal{F}}(\mathcal{O}) = \text{ the set of } \varphi \in C^{2,1}(\mathcal{O}) \text{ that satisfies } (3.1.12),$$

where $C^{2,1}(\mathcal{O})$ is the space of φ whose derivatives φ_t, $\nabla^2\varphi$, $\nabla\varphi$ are continuous in $\mathcal{O}$. Of course for $C^{2,1}_{\mathcal{F}}$ the class $\mathcal{F}$ should be invariant under positive multiplication.

Remark 3.2.4 It is sometimes convenient to extend a class of test functions other than C^2 for continuous F in Definition 3.1.1. For example one would like to consider Sobolev spaces $W^{2,1}_p(\mathcal{O})$ as a class of test functions. Here $W^{2,1}_p(\mathcal{O})$ denotes the space of $\varphi \in L^p(\mathcal{O})$ that satisfies $\nabla^2\varphi \in L^p(\mathcal{O})$ and $\varphi_t \in L^p(\mathcal{O})$. By the Sobolev embedding for large p, say $p > N + 1$, $W^{2,1}_p(\mathcal{O}) \subset C(\mathcal{O})$ so that the value of φ at each point of $\mathcal{O}$ is meaningful. However, $\nabla^2\varphi$ may not be continuous, it is merely a p-th integrable measurable function and the value of $\nabla^2\varphi$ at each point of $\mathcal{O}$ is only determined up to a measure zero set in $\mathcal{O}$. The condition (i)(b) in Definition 3.1.1 should be interpreted as follows. If $(\varphi, \hat{z}) \in W^{2,1}_p(\mathcal{O}) \times \mathcal{O}$ satisfies

$$\varphi_t(z) + F(z, u^*(z), \nabla\varphi(z), \nabla^2\varphi(z)) \geq \varepsilon > 0$$

for some $\varepsilon > 0$ in some neighborhood of $\hat{z}$, then $u^* - \varphi$ does not attain its maximum over $\mathcal{O}$ at $\hat{z}$. It is immediate that for $\varphi \in C^2(\mathcal{O})$ this is equivalent to (i)(b). Such a type of definition is important to study regularity theory for fully nonlinear equations when z-dependence of F is just measurable.

To prove (i) we approximate a test function φ by a smoother function φ_ε. However, the maximum point z_ε of $u^* - \varphi_\varepsilon$ may be different from the maximum point $\hat{z}$ of $u^* - \varphi$ so we need to study the behavior of z_ε as $\varepsilon \to 0$. The next general lemma is useful not only to prove Proposition 3.2.3 but also to prove stability results.

For a sequence $\{S_j\}$ of subsets of a metric space the *upper relaxed limit* $\operatorname{Limsup}_{j\to\infty} S_j$ is defined by $\operatorname{Limsup}_{j\to\infty} S_j = \left\{x \in X;\ \limsup^*_{j\to\infty} \chi_{S_j}(x) = 1\right\}$, where χ_A is the characteristic function of a set A.

3.2.2 Convergence of Maximum Points

Lemma 3.2.5 *Let X be a metric space. For $\varepsilon > 0$ let U_ε be an upper semicontinuous function on X with values in $\mathbf{R} \cup \{-\infty\}$. Let $\overline{U}$ be a function on X defined by $\overline{U} = \limsup^*_{\varepsilon\to 0} U_\varepsilon$. Let B and S be compact sets in X. Assume that S is included in the interior of B, i.e., $S \subset \operatorname{int} B$ in the topology of X. Assume that $\overline{U}$ equals a constant M on S and that $\overline{U}(z) < M$ for $z \in B\backslash S$. (In other words, $\overline{U}$ takes a 'strict' maximum over B modulo points of S.) Let S_ε be the set of maximum points of U_ε on B, i.e.,*

$$S_\varepsilon = \{z \in B;\ U_\varepsilon(z) = \max_B U_\varepsilon\}.$$

(The set S_ε is nonempty and $U_\varepsilon(z_\varepsilon) < \infty$ for $z_\varepsilon \in S_\varepsilon$ since U_ε is upper semicontinuous and B is compact.) Then there exists subsequence $\varepsilon(j) \to \infty$ (as $j \to \infty$) such that

$$\mathit{Limsup}_{j\to\infty} S_{\varepsilon(j)} \subset S.$$

This is equivalent to saying that for each $r > 0$ there exists j_1 satisfying

$$S_{\varepsilon(j)} \subset B_r(S) = \{z \in B; d(z, S) \le r\} \quad \text{for all} \quad j \ge j_1,$$

where $d(z, S)$ denotes the distance from a point z to the set S. (In particular, $S_{\varepsilon(j)} \subset \mathrm{int}B$ for sufficiently large j since $S \subset \mathrm{int}B$.) Moreover, for any $z_{\varepsilon(j)} \in S_{\varepsilon(j)}$,

$$\lim_{j\to\infty} U_{\varepsilon(j)}(z_{\varepsilon(j)}) = M.$$

Proof

1. We may assume that $U_\varepsilon \not\equiv -\infty$ on B by taking a subsequence if necessary. By definition of $\overline{U}$ for each $\zeta \in S$ there exist a subsequence $\{U_{\varepsilon(j)}\}_{j=1}^\infty$ and a sequence $\zeta_j \in X$ converging to ζ as $j \to \infty$ that satisfies

$$M = \overline{U}(\zeta) = \lim_{j\to\infty} U_{\varepsilon(j)}(\zeta_j).$$

2. For a sequence $\{z_{\varepsilon(j)}\}(z_{\varepsilon(j)} \in S_{\varepsilon(j)})$ let A denote the set of its accumulation points, i.e.,

$$A = \{z \in B; \text{ there is a subsequence } \{z_{\varepsilon(j(k))}\} \text{ of } \{z_{\varepsilon(j)}\} \text{ that satisfies } z_{\varepsilon(j(k))} \to z \text{ as } k \to \infty\}.$$

The set A is a nonempty compact set since B is compact.

If $z_{\varepsilon(j(k))} \to z$, then by definition of $\overline{U}$, $\limsup_{k\to\infty} U_{\varepsilon(j(k))}(z_{\varepsilon(j(k))}) \le \overline{U}(z) \le \sup_A \overline{U}$; the last inequality is trivial since $z \in A$ by definition of A. Since any subsequence limit of $\{z_{\varepsilon(j)}\}$ belongs to A, we have

$$\limsup_{j\to\infty} U_{\varepsilon(j)}(z_{\varepsilon(j)}) \le \sup_A \overline{U}.$$

3. Since ζ_j converges to a maximum point $\zeta \in S$ of $\overline{U}$ over B and S is contained in int B, we observe $\zeta_j \in B$ for sufficiently large j, say $j \ge j_0$. Since

$$U_{\varepsilon(j)}(\zeta_j) \le U_{\varepsilon(j)}(z_{\varepsilon(j)})$$

for $j \geq j_0$, Steps 1 and 2 yield

$$M = \lim_{j\to\infty} U_{\varepsilon(j)}(\zeta_j) \leq \liminf_{j\to\infty} U_{\varepsilon(j)}(z_{\varepsilon(j)})$$
$$\leq \limsup_{j\to\infty} U_{\varepsilon(j)}(z_{\varepsilon(j)}) \leq \sup_A \overline{U}.$$

Since $M \geq \sup_A \overline{U}$ is trivial, we get

$$M = \sup_A \overline{U} = \lim_{j\to\infty} U_{\varepsilon(j)}(z_{\varepsilon(j)}).$$

Since S is the set of all maximum points of $\overline{U}$ in B, S includes A. This inclusion implies that

$$\operatorname{Limsup}_{j\to\infty} S_{\varepsilon(j)} \subset S.$$

□

Remark 3.2.6

(i) If U_ε converges to $\overline{U}$ locally uniformly in a neighborhood of S, then we need not take a subsequence $\varepsilon(j)$. If, moreover, S consists of only one point $\overline{z}$, so that $\overline{U}$ attains a strict local maximum at $\overline{z}$, then z_ε converges to $\overline{z}$ and $U_\varepsilon(z_\varepsilon) \to U(\overline{z})$ *without* taking subsequences.

Indeed, if U_ε converges to $\overline{U}$ locally uniformly in $B_r(\overline{z})$, for any sequence $\zeta_\varepsilon \to \overline{z}$ the formula

$$M = \overline{U}(\overline{z}) = \lim_{\varepsilon\to 0} U_\varepsilon(\zeta_\varepsilon)$$

is valid without taking any subsequence of U_ε. This is stronger than Step 1. We argue as in Step 2, 3 but z_ε, U_ε replaces $z_{\varepsilon(j)}$, $U_{\varepsilon(j)}$ respectively. Since A is contained in $S = \{\overline{z}\}$, z_ε converges to $\overline{z}$ without taking a subsequence.

This simple version is also important to develop the theory of viscosity solutions, for example, in proving Proposition 3.2.3.

(ii) Lemma 3.2.5 is a trivial modification of G. Barles [46, Lemma 4.2] where S is assumed to be a singleton. An essentially same proof is presented there. However, it seems that [p. 90, line 6] needs a further explanation, which is included in our proof. The inequality in [p. 90, line 6]

$$\overline{v}(y) \leq \limsup v_{\varepsilon'}(x_{\varepsilon'}) \text{ for } y \in \overline{\Omega} \cap B_r(x)$$

may not be true even for $y = x$ unless one first chooses a subsequence of v_ε such that $\overline{v}(x) = \lim_{j\to\infty} v_{\varepsilon(j)}(z_j)$ for some $z_j \to x$ and then takes further subsequences $v_{\varepsilon'}$ and $x_{\varepsilon'}$. In our terminology, we should first take subsequence $U_{\varepsilon(j)}$ as in Step 1 of our proof.

(iii) Similar statements for lower semicontinuous U_ε (with values in $\mathbf{R}\cup\{+\infty\}$) are obtained by replacing $\overline{U}$ by $\underline{U} = \liminf_* U_\varepsilon$, $\overline{U}(z) < M$ by $\underline{U}(z) > M$ and a maximum point by a minimum point. This assertion follows by Lemma 3.2.5 by replacing U_ε by $-U_\varepsilon$. Also, we may replace a maximum point by a minimum point in Remark 3.2.6 (i).

3.2.3 Applications

Proof of Proposition 3.2.3

(i) As in Proposition 3.2.2 we may interpret the maximum in (3.1.6) as a local maximum for each class of test functions. To show the first part it suffices to prove (3.1.7) (with lower semicontinuous F) for $\varphi \in C^2(\mathcal{O})$ satisfying (3.1.6) by assuming that (3.1.7) holds for all $z \in \mathcal{O}$ and all $\psi \in \cap_{k=2}^{\infty} A^k(\mathcal{O})$ such that $u^* - \psi$ takes a local maximum at z. Since φ is C^2 there is a modulus ω_0 that satisfies

$$\begin{aligned}\Phi &= \varphi(x,t) - \varphi(\hat{x},\hat{t}) - \varphi_t(\hat{x},\hat{t})(t-\hat{t}) - \langle \nabla\varphi(\hat{x},\hat{t}), x-\hat{x}\rangle \\ &\quad -\frac{1}{2}\langle \nabla^2\varphi(\hat{x},\hat{t})(x-\hat{x}), x-\hat{x}\rangle \\ &\le \omega_0(|x-\hat{x}|^2 + |t-\hat{t}|)(|x-\hat{x}|^2 + |t-\hat{t}|)\end{aligned}$$

for (x,t) sufficiently close to $(\hat{x},\hat{t})$. The right-hand side is dominated from above by

$$\begin{aligned}&2\omega_0(2|x-\hat{x}|^2)|x-\hat{x}|^2 \text{ for } |x-\hat{x}|^2 \ge |t-\hat{t}|,\\ &2\omega_0(2|t-\hat{t}|)|t-\hat{t}| \qquad \text{for } |x-\hat{x}|^2 \le |t-\hat{t}|.\end{aligned}$$

The left-hand side Φ is now dominated by

$$\Phi \le \overline{\omega_0}(|x-x|^2)|x-\hat{x}|^2 + \overline{\omega_0}(|t-\hat{t}|)|t-\hat{t}|$$

with $\overline{\omega_0}(\sigma) = 2\omega_0(2\sigma)$. Applying Lemma 3.1.9 (ii) yields the existence of $\theta_1 \in C^1[0,\infty)$ and $\theta_2 \in C^2[0,\infty)$ with $\theta_k^{(j)}(0) = 0$ for $0 \le j \le k$ that satisfies

$$\Phi \le \theta_2(|x-\hat{x}|) + \theta_1(|t-\hat{t}|)$$

for (x,t) sufficiently close to $(\hat{x},\hat{t})$. If we set

$$\psi_0(x,t) = \varphi - \Phi + \theta_2(|x-\hat{x}|) + \theta_1(|t-\hat{t}|) + |x-\hat{x}|^4 + |t-\hat{t}|^2,$$

then ψ_0 is a separable function

$$\begin{aligned}\psi_0(x,t) &= b(x) + g(t),\\ b(x) &= \langle \nabla\varphi(\hat{x},\hat{t}),\ x-\hat{x}\rangle + \frac{1}{2}\langle \nabla^2\varphi(\hat{x},\hat{t})(x-\hat{x}),\ x-\hat{x}\rangle\\ &\quad +\theta_2(|x-\hat{x}|) + |x-\hat{x}|^4,\\ g(t) &= \varphi_t(\hat{x},\hat{t})(t-\hat{t}) + \theta_1(|t-\hat{t}|) + |t-\hat{t}|^2,\end{aligned}$$

and $b \in C^2(\mathbf{R}^N)$, $g \in C^1(\mathbf{R})$. Since $\varphi < \psi_0$ near $(\hat{x},\hat{t})$ (except $(\hat{x},\hat{t})$) and $\varphi(\hat{x},\hat{t}) = \psi_0(\hat{x},\hat{t})$, $u^* - \psi_0$ takes its local strict maximum at $\hat{z} = (\hat{x},\hat{t})$ with $\nabla\varphi(\hat{x},\hat{t}) = \nabla b(\hat{x})$, $\nabla^2\varphi(\hat{x},\hat{t}) = \nabla^2 b(\hat{x})$, $\varphi_t(\hat{x},\hat{t}) = g'(\hat{t})$.

We approximate b by a smooth function b_ε so that b_ε converges to b locally uniformly together with its derivatives up to second order (as $\varepsilon \to 0$). We approximate g by a smooth function so that $g_\varepsilon \to g$ locally uniformly with its derivative. We now apply Lemma 3.2.5 or its simpler version Remark 3.2.6 (i) to $u^* - b_\varepsilon - g_\varepsilon$ with $X = \mathcal{O}$, $S = \{\hat{z}\}$ to get that there is a sequence $z_\varepsilon = (x_\varepsilon, t_\varepsilon)$ converging to $\hat{z}$ such that

$$u^* - b_\varepsilon - g_\varepsilon \le (u^* - b_\varepsilon - g_\varepsilon)(z_\varepsilon) \quad \text{near} \quad z_\varepsilon, \tag{3.2.8}$$

$$\lim_{\varepsilon\to 0}(u^* - b_\varepsilon - g_\varepsilon)(z_\varepsilon) = (u^* - b - g)(\hat{z}). \tag{3.2.9}$$

Since $b_\varepsilon \to b$, $g_\varepsilon \to g$ uniformly (together with its derivatives), we get

$$\begin{cases} b_\varepsilon(x_\varepsilon) \to b(\hat{x}),\ g_\varepsilon(t_\varepsilon) \to g(\hat{t}),\\ \nabla b_\varepsilon(x_\varepsilon) \to \nabla b(\hat{x}),\ g'_\varepsilon(t_\varepsilon) \to g'(\hat{t}),\\ \nabla^2 b_\varepsilon(x_\varepsilon) \to \nabla^2 b(\hat{x}) \text{ with } \quad z_\varepsilon = (x_\varepsilon, t_\varepsilon).\end{cases} \tag{3.2.10}$$

By (3.2.9) this implies

$$u^*(z_\varepsilon) \to u^*(\hat{z}). \tag{3.2.11}$$

Since (3.2.8) holds, we apply (3.1.7) at z_ε to get

$$g'_\varepsilon(t_\varepsilon) + F(z_\varepsilon,\ u^*(z_\varepsilon),\ \nabla b_\varepsilon(z_\varepsilon),\ \nabla^2 b_\varepsilon(z_\varepsilon)) \le 0. \tag{3.2.12}$$

Since F is lower semicontinuous, sending ε to 0 yields, by (3.2.10) and (3.2.11),

$$g'(\hat{t}) + F(\hat{z},\ u^*(\hat{z}),\ \nabla b(\hat{x}),\ \nabla^2 b(\hat{x})) \le 0,$$

which is

$$\varphi_t(\hat{t}) + F(\hat{z},\ u^*(\hat{z}),\ \nabla\varphi(\hat{z}),\ \nabla^2\varphi(\hat{z})) \le 0.$$

This is what we want to prove. The proof for supersolutions is similar so is omitted.

It remains to study whether test functions in Definition 3.1.5 may be replaced by $A_F^k(\mathcal{O})$. It suffices to prove $\varphi_t(\hat{z}) \leq 0$ for $\varphi \in C_F^2(\mathcal{O})$ satisfying (3.1.6) when $\nabla\varphi(\hat{z}) = 0$ by assuming that differential inequalities in Definition 3.1.5 hold for all $z \in \mathcal{O}$ and all $\psi \in \cap_{k=1}^{\infty} A_F^k(\mathcal{O})$ such that $u^* - \psi$ takes a local maximum at z. The basic idea of the proof is the same as above. We take $\psi_0 = \psi$ as defined in (3.2.7) and observe that $u^* - \psi_0$ takes a strict local maximum at $\hat{z}$ and that ψ_0 is of form

$$\psi_0(x,t) = f(|x - \hat{x}|) + g(t)$$

with some $f \in \mathcal{F}$ and $g \in C^1(\mathbf{R})$ (Note that $af \in \mathcal{F}, a > 0$). We approximate g by $g_\varepsilon \in C^\infty(\mathbf{R})$ as before. (Approximation of $f \in \mathcal{F}$ by $f_\varepsilon \in C^\infty(0,\infty) \cap \mathcal{F}$ may be impossible in this generality so we only approximate g. If F is geometric and continuous outside $\{p = 0\}$, f can be approximated by a smooth function provided that Ω is bounded (cf. Lemma 4.1.3.) As before we get (3.2.8)–(3.2.11), where we set $f_\varepsilon(x) = f(|x - \hat{x}|) = f(x)$ by abuse of notation. By differential inequalities in Definition 3.1.5 at z_ε we get (3.2.12) if $z_\varepsilon \neq \hat{z}$, and $g_\varepsilon'(t_\varepsilon) \leq 0$ if $z_\varepsilon = \hat{z}$. Sending m to ∞ we obtain $g'(\hat{t}) = \varphi_t(\hat{t},\hat{x}) \leq 0$ by using (3.1.11) and convergence results (3.2.10), (3.2.11). This is what we wanted to prove.

(ii) It suffices to prove that $\varphi \in C^{2,1}(\mathcal{O})$ (resp. $\varphi \in C_F^{2,1}(\mathcal{O})$) satisfy (3.1.7) (resp. differential inequalities in Definition 3.1.5) if φ is a test function at $\hat{z}$ of a (resp. $\mathcal{F}$-) subsolution u of (3.1.5). Let u be a subsolution in Definitions 3.1.1 and 3.1.4. As in (i) we may assume that max in (3.1.6) is a local strict maximum and that φ is a separable type : $\varphi(x,t) = b(x) + g(t)$ with $b \in C^2(\mathbf{R}^N)$ and $g \in C^1(\mathbf{R})$. We approximate g by C^2 functions g_ε so that $g_\varepsilon \to g$ locally uniformly with its first derivative. The rest of the proof is the same as in (i). For an $\mathcal{F}$-subsolution we may assume $\nabla\varphi(\hat{z}) = 0$ for $\varphi \in C_F^{2,1}(\mathcal{O})$. By (3.2.7) we may assume that φ is a separable type,

$$\varphi(x,t) = f(|x - \hat{x}|) + g(t)$$

with $f \in \mathcal{F}$, $g \in C^1(\mathbf{R})$. We do not need to approximate f; we only approximate g by C^2 function. The rest of the proof is the same as in (i) so it is safely left to the reader.

□

Remark 3.2.7 From the proof of Proposition 3.2.3 we obtain the following equivalent definition of an $\mathcal{F}$-subsolution of (3.1.5) if $\mathcal{F}$ is invariant under positive multiplication. Let $u : \mathcal{O} \to \mathbf{R} \cup \{-\infty\}$ fulfill $u^*(z) < \infty$ for all $z \in \mathcal{O}$. Then u is an $\mathcal{F}$-subsolution of (3.1.5) in $\mathcal{O}$ if and only if the next two conditions are fulfilled.

(A) $g'(\hat{t}) + F(\hat{z}, u^*(\hat{z}), \nabla b(\hat{x}), \nabla^2 b(\hat{x})) \le 0$ holds for all $\hat{z} = (\hat{x}, \hat{t}) \in \mathcal{O}$ and for all $\varphi = \varphi(x,t) := b(x) + g(t)$ which is $C^{2,1}$ near $\hat{z}$ provided that $u^* - \varphi$ attains its local strict maximum at $\hat{z}$ and that $\nabla b(\hat{x}) \neq 0$.

(B) $g'(\hat{t}) \le 0$ holds for all $\hat{z} = (\hat{x}, \hat{t}) \in \mathcal{O}$ and for all

$$\varphi = \varphi(x,t) := f(|x - \hat{x}|) + g(t)$$

with $f \in \mathcal{F}$ and with g which is C^1 near $\hat{t}$ provided that $u^* - \varphi$ attains its local strict maximum at $\hat{z}$.

Proof of Theorem 3.2.1 We only prove the subsolution case since the case of supersolution follows exactly in the same way.

(i) Since $\overline{u}$ is upper semicontinuous and $\overline{u}(z) < \infty$ for $z \in \mathcal{O}$, our goal is to prove

$$\varphi_t(\hat{z}) + F(\hat{z},\ \overline{u}(\hat{z}),\ \nabla\varphi(\hat{z}),\ \nabla^2\varphi(\hat{z})) \le 0 \tag{3.2.13}$$

by assuming that $\varphi \in C^2(\mathcal{O})$ and $\hat{z} \in \mathcal{O}$ fulfill

$$\max_{\mathcal{O}}(\overline{u} - \varphi) = (\overline{u} - \varphi)(\hat{z}).$$

By Proposition 3.2.2 we may assume that $\overline{u} - \varphi$ has a strict local maximum at $\hat{z}$, i.e.,

$$(\overline{u} - \varphi)(z) < (u - \varphi)(\hat{z}),\ \ z \in B\backslash\{\hat{z}\}$$

for some compact neighborhood of $\hat{z}$.

We set $U_\varepsilon = u_\varepsilon^* - \varphi$ and observe that

$$\overline{U} = \limsup_{\varepsilon\to 0}{}^*\ U_\varepsilon = \overline{u} - \varphi.$$

We now apply Lemma 3.2.5 with $X = \mathcal{O}$ and $S = \{\hat{z}\}$ to get a subsequence of $\{U_\varepsilon\}$ (still denoted $\{U_\varepsilon\}$) and a sequence $\{z_\varepsilon\}$ in $\mathcal{O}$ satisfies

$$\begin{aligned} &U_\varepsilon(z) \le U_\varepsilon(z_\varepsilon) \quad \text{for } z \text{ close to } z_\varepsilon, \\ &z_\varepsilon \to \hat{z} \quad \text{and} \quad U_\varepsilon(z_\varepsilon) \to \overline{U}(\hat{z}). \end{aligned} \tag{3.2.14}$$

Since u_ε is a subsolution of (3.2.2) and φ is a test function of u_ε at z_ε, we obtain, by Proposition 3.2.2, that

$$\varphi_t(z_\varepsilon) + F_\varepsilon(z_\varepsilon\ u_\varepsilon^*(z_\varepsilon),\ \nabla\varphi(z_\varepsilon),\ \nabla^2\varphi(z_\varepsilon)) \le 0. \tag{3.2.15}$$

Sending ε to 0 yields (3.2.13), since (3.2.1) holds and $z_\varepsilon \to \hat{z},\ u_\varepsilon^*(z_\varepsilon) \to \overline{u}(\hat{z})$.

(ii) By Proposition 3.2.2 our goal is to prove differential inequalities in Definition 3.1.5 by assuming that $\overline{u}-\varphi$ has a strict local maximum at $\hat{z}$ for $\varphi \in C^2_F(\mathcal{O})$. By (i) we may assume $\nabla\varphi(\hat{z}) = 0$. We may take ψ in (3.2.7) instead of φ.

Arguing as in (i), we observe that there is a subsequence of u_ε (still denoted u_ε) and $z_\varepsilon \in \mathcal{O}$ satisfying (3.2.14). Since u_ε is an $\mathcal{F}_\varepsilon$-subsolution of (3.2.2) we have (3.2.15) if $\nabla\psi(z_\varepsilon) \neq 0$ (i.e., $z_\varepsilon \neq \hat{z}$) and $\psi_t(z_\varepsilon) \leq 0$ if $\nabla\psi(z_\varepsilon) = 0$. We send ε to 0 and use (3.2.4) with $z_\varepsilon \to \hat{z}$, $u^*_\varepsilon(z_\varepsilon) \to \overline{u}(\hat{z})$ to get $\varphi_t(\hat{z}) = \psi_t(\hat{z}) \leq 0$.

□

We conclude this section by proving the equivalence of $\mathcal{F}$-subsolutions and subsolutions when (3.1.15) is fulfilled.

Proposition 3.2.8 *Assume that* (3.1.15) *holds for F and that F is degenerate elliptic. A function $u : \mathcal{O} \to \mathbf{R} \cup \{-\infty\}$ is an $\mathcal{F}$-subsolution of* (3.1.5) *(in $\mathcal{O}$) if and only if u is a subsolution of* (3.1.5).

Proof Since a subsolution is always an $\mathcal{F}$-subsolution under (3.1.15) it suffices to prove that an $\mathcal{F}$-subsolution is a subsolution. Assume that $(\varphi, \hat{z}) \in C^2(\mathcal{O}) \times \mathcal{O}$ satisfies

$$\max_{\mathcal{O}}(u^* - \varphi) = (u^* - \varphi)(\hat{z}).$$

We may assume that $u^* - \varphi$ takes its strict maximum at $\hat{z} = (\hat{x}, \hat{t})$ over $\mathcal{O}$ and $\nabla\varphi(\hat{z}) = 0$. For $\varepsilon > 0$ we set

$$\Phi_\varepsilon(x, y, t) = u^*(x, t) - \psi_\varepsilon(x, y, t),$$
$$\psi_\varepsilon(x, y, t) = |x - y|^4/4\varepsilon + \varphi(y, t)$$

and observe that

$$\overline{\Phi}(x, y, t) = (\limsup_{\varepsilon\to 0}{}^* \Phi_\varepsilon)(x, y, t) = \begin{cases} u^*(x, t) - \varphi(x, t) & \text{if } x = y, \\ -\infty & \text{if } x \neq y \end{cases}$$

for $(x, t), (y, t) \in \mathcal{O}$. Since $\overline{\Phi}$ takes its strict maximum at $(\hat{x}, \hat{x}, \hat{t})$, by convergence of maximum points (Lemma 3.2.5 and Remark 3.2.6 (i)) a maximizer $(x_\varepsilon, y_\varepsilon, t_\varepsilon)$ of Φ_ε converges to $(\hat{x}, \hat{x}, \hat{t})$ as $\varepsilon \to 0$. Moreover, $u^*(x_\varepsilon, t_\varepsilon) \to u^*(\hat{x}, \hat{t})$. Since $\Phi_\varepsilon(x_\varepsilon, y, t_\varepsilon)$ takes its maximum at $y = y_\varepsilon$, we see

$$\nabla_y \psi_\varepsilon(x_\varepsilon, y_\varepsilon, t_\varepsilon) = 0, \ \nabla^2_y \psi(x_\varepsilon, y_\varepsilon, t_\varepsilon) \geq O,$$

where ∇_y denotes the partial gradient in y. Since $\nabla|x|^4 = 4|x|^2 x$, $\nabla^2|x|^4 = 4|x|^2 I_N + 8x \otimes x$, this is equivalent to saying

$$-|x_\varepsilon - y_\varepsilon|^2(x_\varepsilon - y_\varepsilon)/\varepsilon + \nabla\varphi(y_\varepsilon, t_\varepsilon) = 0, \tag{3.2.16}$$

$$|x_\varepsilon - y_\varepsilon|^2\, I_N + 2(x_\varepsilon - y_\varepsilon) \otimes (x_\varepsilon - y_\varepsilon) + \varepsilon \nabla^2 \varphi(y_\varepsilon,\ t_\varepsilon) \geq O \quad (3.2.17)$$

where I_N denotes the $N \times N$ unit matrix. We note that $\psi_\varepsilon(\cdot,\ y_\varepsilon,\ \cdot)$ belongs to $\mathcal{A}_0$ as a function of (x, t) (Remark 3.1.6). This is equivalent to saying that $\psi_\varepsilon(\cdot,\ y_\varepsilon,\ \cdot) \in C^2_F(\mathcal{O})$ by Proposition 3.1.8.

Case A $\nabla\varphi(y_\varepsilon,\ t_\varepsilon) \neq 0$ for a sequence $\varepsilon = \varepsilon_j \to 0$.
Since $(x_\varepsilon,\ t_\varepsilon)$ is a maximum point of a function

$$(x, t) \mapsto u^*(x, t) - |x_\varepsilon - y_\varepsilon|^4/4\varepsilon - \varphi(x - (x_\varepsilon - y_\varepsilon), t)$$

for $\varepsilon = \varepsilon_j$ and since u is an $\mathcal{F}$-subsolution, we have

$$\varphi_t(y_\varepsilon,\ t_\varepsilon) + F(x_\varepsilon,\ t_\varepsilon, u_\varepsilon^*(x_\varepsilon, t_\varepsilon)\nabla\varphi(y_\varepsilon,\ t_\varepsilon), \nabla^2\varphi(y_\varepsilon,\ t_\varepsilon)) \leq 0$$

for $\varepsilon = \varepsilon_j$. Sending ε_j to zero yields

$$\varphi_t(\hat{x}, \hat{t}) + F_*(\hat{x}, \hat{t}, u^*(\hat{x}, \hat{t}), \nabla\varphi(\hat{x}, \hat{t}), \nabla^2\varphi(\hat{x}, \hat{t})) \leq 0.$$

Case B $\nabla\varphi(y_\varepsilon, t_\varepsilon) = 0$ for sufficiently small $\varepsilon > 0$.
This condition implies $x_\varepsilon = y_\varepsilon$ so that $\nabla_x \psi_\varepsilon(x_\varepsilon,\ y_\varepsilon,\ t_\varepsilon) = 0,\ \nabla_x^2\, \psi_\varepsilon(x_\varepsilon,\ y_\varepsilon,\ t_\varepsilon) = O$. Since $\Phi_\varepsilon(x,\ y_\varepsilon,\ t)$ takes its maximum at $(x_\varepsilon,\ t_\varepsilon)$ as a function of (x, t), definition of an $\mathcal{F}$-subsolution (Proposition 3.1.7) yields

$$\partial_t \psi_\varepsilon(x_\varepsilon,\ y_\varepsilon,\ t_\varepsilon) \leq 0.$$

Sending ε to zero yields

$$\varphi_t(\hat{x}, \hat{t}) \leq 0.$$

Since $x_\varepsilon = y_\varepsilon$, (3.2.16) and (3.2.17) yields

$$\nabla\varphi(y_\varepsilon,\ t_\varepsilon) = 0,\ \nabla^2\varphi(y_\varepsilon,\ t_\varepsilon) \geq O.$$

Sending ε to zero again we see

$$\nabla\varphi(\hat{x}, \hat{t}) = 0,\ \nabla^2\varphi(\hat{x}, \hat{t}) \geq O.$$

Since F is degenerate elliptic,

$$\begin{aligned}
&\varphi_t(\hat{x}, \hat{t}) + F_*(\hat{x}, \hat{t}, u^*(\hat{x}, \hat{t}), \nabla\varphi(\hat{x}, \hat{t}), \nabla^2\varphi(\hat{x}, \hat{t})) \\
&\leq 0 + F_*(\hat{x}, \hat{t}, u^*(\hat{x}, \hat{t}), 0, \nabla^2\varphi(\hat{x}, \hat{t})) \\
&\leq 0 + F_*(\hat{x}, \hat{t}, u^*(\hat{x}, \hat{t}), 0, O) = 0 \quad \text{by (3.1.15).}
\end{aligned}$$

□

3.3 Boundary Value Problems

So far we have discussed viscosity solutions in an open set. We seek a natural definition of solutions for the boundary value problem. It is desirable that solutions satisfy the stability principle. Suppose that $F_\varepsilon \in C^0(W)$ converges to F uniformly on every compact set in W, where $W = \overline{\Omega} \times [0, T] \times \mathbf{R} \times \mathbf{R}^N \times \mathbf{S}^N$. Suppose that $B_\varepsilon \in C^0(J)$ converges to B uniformly on every compact set with $J = \partial\Omega \times [0, T] \times \mathbf{R} \times \mathbf{R}^N \times \mathbf{S}^N$. Suppose that u_ε solves

$$\begin{aligned} u_t + F_\varepsilon(z, u, \nabla u, \nabla^2 u) &\le 0 \quad \text{in} \quad \Omega \times (0, T), \\ B_\varepsilon(z, u, \nabla u, \nabla^2 u) &\le 0 \quad \text{on} \quad \partial\Omega \times (0, T) \end{aligned} \tag{$*$}$$

in a classical sense. What problem does $u = \limsup^*_{\varepsilon\to 0} u_\varepsilon$ solve (when $u(z) < \infty$ for all $z \in \overline{\Omega} \times (0, T)$)?

We consider a function $\varphi \in C^2(\overline{\Omega} \times (0, T))$ such that $u - \varphi$ attains a strict maximum over $\overline{\Omega} \times (0, T)$ at $(\hat{x}, \hat{t})$ with $\hat{x} \in \partial\Omega$, $0 < \hat{t} < T$. By Remark 3.2.6 (i) there is a sequence $z_\varepsilon = (x_\varepsilon, t_\varepsilon) \to (\hat{x}, \hat{t})$ that satisfies

$$\max_{\overline{\Omega}\times(0,T)} (u_\varepsilon - \varphi) = (u_\varepsilon - \varphi)(x_\varepsilon, t_\varepsilon).$$

Since u_ε solves $(*)$, if $x_\varepsilon \in \partial\Omega$ then we expect

$$B_\varepsilon(z_\varepsilon,\ u_\varepsilon(z_\varepsilon),\ \nabla\varphi(z_\varepsilon),\ \nabla^2\varphi(z_\varepsilon)) \le 0$$

under a suitable monotonicity condition of B_ε (Proposition 3.3.3). If $x_\varepsilon \in \Omega$ then

$$\varphi_t(z_\varepsilon) + F_\varepsilon(z_\varepsilon,\ u_\varepsilon(z_\varepsilon),\ \nabla\varphi(z_\varepsilon),\ \nabla^2\varphi(z_\varepsilon)) \le 0$$

(under the ellipticity condition of F_ε). Unfortunately, for sufficiently small ε, x_ε may be an interior point of Ω (Example 3.3.6), so as $\varepsilon \to 0$ the best we expect is *either*

$$\varphi_t(\hat{z}) + F(\hat{z},\ u(\hat{z}),\ \nabla\varphi(\hat{z}),\ \nabla^2\varphi(\hat{z})) \le 0$$

or

$$B(\hat{z},\ u(\hat{z}),\ \nabla\varphi(\hat{z}),\ \nabla^2\varphi(\hat{z})) \le 0.$$

In other words

$$E(\hat{z},\ u(\hat{z}),\ \varphi_t(\hat{z}),\ \nabla\varphi(\hat{z}),\ \nabla^2\varphi(\hat{z})) \le 0$$

for $E(z, r, \tau, p, X) = (\tau + F(z, r, p, X)) \wedge B(z, r, p, X)$. This observation shows that the next general definition for viscosity solution is useful.

Definition 3.3.1

(i) Let $\mathcal{O}$ be a locally compact subset of $\mathbf{R}^d$ (so that every neighborhood of a point $\hat{z}$ of $\mathcal{O}$ includes some compact neighborhood of $\hat{z}$ in $\mathcal{O}$). Let E be a lower (resp. upper) semicontinuous function on $\overline{\mathcal{O}} \times \mathbf{R} \times \mathbf{R}^d \times \mathbf{S}^d$ with values in $\mathbf{R} \cup \{-\infty\}$ (resp. $\mathbf{R} \cup \{+\infty\}$). A function $u : \mathcal{O} \to \mathbf{R} \cup \{-\infty\}$ (resp. $\mathbf{R} \cup \{+\infty\}$) is a (viscosity) *subsolution* (resp. *supersolution*) of

$$E(z, u, Du, D^2u) = 0 \tag{3.3.1}$$

in $\mathcal{O}$ if

(a) $u^*(z) < \infty$ (resp. $u_*(z) > -\infty$) for all $z \in \mathcal{O}$.
(b) If $(\varphi, \hat{z}) \in C^2(\mathcal{O}) \times \mathcal{O}$ satisfies

$$\begin{aligned} &\max_{\mathcal{O}}(u^* - \varphi) = (u^* - \varphi)(\hat{z}) \\ &(\text{resp. } \min_{\mathcal{O}}(u_* - \varphi) = (u_* - \varphi)(\hat{z})), \end{aligned} \tag{3.3.2}$$

then

$$\begin{aligned} &E(\hat{z}, u^*(\hat{z}), D\varphi(\hat{z}), D^2\varphi(\hat{z})) \le 0 \\ &(\text{resp. } E(\hat{z}, u_*(\hat{z}), D\varphi(\hat{z}), D^2\varphi(\hat{z})) \ge 0), \end{aligned} \tag{3.3.3}$$

where $D\varphi = (\partial\varphi/\partial x_i)_{i=1}^d$, $D^2\varphi = (\partial^2\varphi/\partial x_i \partial x_j)_{1 \le i,j \le d}$. Here $C^k(\mathcal{O})$ denotes the set of C^k functions in some neighborhood of $\mathcal{O}$ in $\mathbf{R}^d$.

If E is defined in a dense subset of $\overline{\mathcal{O}} \times \mathbf{R} \times \mathbf{R}^d \times \mathbf{S}^d$ (with $E_* < +\infty$, $E^* > -\infty$ on $\overline{\mathcal{O}} \times \mathbf{R} \times \mathbf{R}^d \times \mathbf{S}^d$) then we say that u is a subsolution (resp. supersolution) of (3.3.1) in $\mathcal{O}$ if u is a subsolution (resp. supersolution) of $E_*(z, u, Du, D^2u) = 0$ (resp. $E^*(z, u, Du, D^2u) = 0$) in $\mathcal{O}$.

(ii) Let Ω be a domain in $\mathbf{R}^N$ and $T > 0$. Let F be a lower semicontinuous function in $W = \overline{\Omega} \times [0, T] \times \mathbf{R} \times \mathbf{R}^N \times \mathbf{S}^N$ with values in $\mathbf{R} \cup \{-\infty\}$. Let B be lower semicontinuous in $J = \partial\Omega \times [0, T] \times \mathbf{R} \times \mathbf{R}^N \times \mathbf{S}^N$ with values in $\mathbf{R} \cup \{-\infty\}$. Let $\mathcal{O}$ be an open set in $\overline{\Omega} \times (0, T)$. (For example we may take $\mathcal{O} = \overline{\Omega} \times (0, T)$.) A function $u : \mathcal{O} \to \mathbf{R} \cup \{-\infty\}$ (resp. $\mathbf{R} \cup \{+\infty\}$) is a *subsolution* (resp. supersolution)

$$\begin{aligned} u_t + F(z, u, \nabla u, \nabla^2 u) &= 0 \quad \text{in} \quad \mathcal{O} \cap \Omega \times (0, T), \\ B(z, u, \nabla u, \nabla^2 u) &= 0 \quad \text{on} \quad \mathcal{O} \cap \partial\Omega \times (0, T), \end{aligned} \tag{3.3.4}$$

if u is a subsolution (resp. supersolution) of

$$E(z, u, u_t, \nabla u, \nabla^2 u) = 0 \quad \text{in} \quad \mathcal{O}$$

in the sense of (i) with

$$E(z,r,\tau,p,X)=\begin{cases}\tau+F(z,r,p,X), & x\in\Omega\\ (\tau+F(z,r,p,X))\wedge B(z,r,p,X), & x\in\partial\Omega \text{ for } z=(x,t)\end{cases}$$

$$\left(\text{resp. } E(z,r,\tau,p,X)=\begin{cases}\tau+F(z,r,p,X), & x\in\Omega\\ (\tau+F(z,r,p,X))\vee B(z,r,p,X), & x\in\partial\Omega.\end{cases}\right)$$

Note that E (for subsolutions) is regarded as a lower semicontinuous function in $\overline{\mathcal{O}}\times\mathbf{R}\times\mathbf{R}^d\times\mathbf{S}^{d-1}$ with $d=N+1$. Here $a\vee b=\max(a,b)$.

Remark 3.3.2 By definition, u is a subsolution of (3.3.4) if and only if

(a) $u^*(z)<\infty$ for $z\in\mathcal{O}$.
(b) If $(\varphi,\hat{z})\in C^2(\mathcal{O})\times\mathcal{O}$ satisfies (3.1.6), then (3.1.7) holds for $\hat{z}\in\Omega\times(0,T)$, and for $\hat{z}\in\partial\Omega\times(0,T)$ *either*

$$B(\hat{z},u^*(\hat{z}),\ \nabla\varphi(\hat{z}),\ \nabla^2\varphi(\hat{z}))\le 0$$

or (3.1.7) is valid.

The similar equivalent definition is available for supersolutions.

We warn the reader that even for degenerate elliptic F, a function $u\in C^2(\mathcal{O})$ satisfying

$$\begin{aligned} u_t+F(z,u,\nabla u,\nabla^2 u)\le 0 \quad &\text{in} \quad \mathcal{O}\cap(\Omega\times(0,T)),\\ B(z,u,\nabla u,\nabla^2 u)\le 0 \quad &\text{on} \quad \mathcal{O}\cap(\partial\Omega\times(0,T)) \end{aligned} \tag{3.3.5}$$

pointwise may not be a viscosity subsolution in $\mathcal{O}$ unless we impose an extra condition on B. The main reason is that the condition

$$\max_{\mathcal{O}}(u-\varphi)=(u-\varphi)(\hat{x},\hat{t}) \tag{3.3.6}$$

does not deduce $\nabla u(\hat{x},\hat{t})=\nabla\varphi(\hat{x},\hat{t})$ if $\hat{x}\in\partial\Omega$ (Fig. 3.5). In fact, if $\partial\Omega$ is C^2 near $\hat{x}$ the condition (3.3.6) implies that either (i) or (ii) holds:

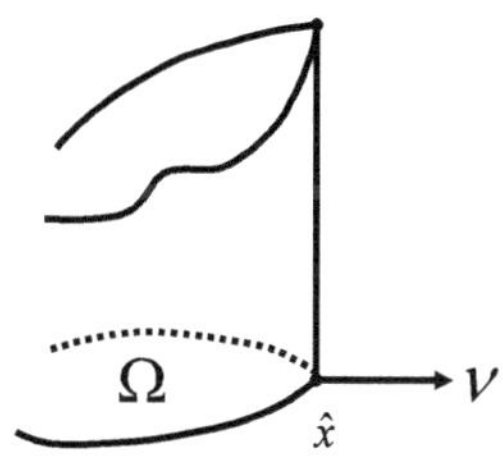

Fig. 3.5 Maximum at the boundary

(i) $\nabla u(\hat{x}, \hat{t}) = \nabla \varphi(\hat{x}, \hat{t})$ and $\nabla^2 u(\hat{x}, \hat{t}) \le \nabla^2 \varphi(\hat{x}, \hat{t})$,
(ii) $\nabla u(\hat{x}, \hat{t}) = \nabla \varphi(\hat{x}, \hat{t}) + \lambda \nu u(\hat{x})$ and

$$R_\nu \nabla^2 u(\hat{x}, \hat{t}) R_\nu \le R_\nu \nabla^2 \varphi(\hat{x}, \hat{t}) R_\nu - \lambda A_\nu \quad \text{for some} \quad \lambda > 0.$$

Here $\nu(\hat{x})$ is the outward unit normal and $R_\nu = I - \nu(\hat{x}) \otimes \nu(\hat{x})$; A_ν is the $N \times N$ symmetric matrix whose restriction on the tangent space $T_{\hat{x}} \partial\Omega$ is the Weingarten map of $\partial\Omega$ in the direction of $\nu(\hat{x})$ and $A_\nu \nu = 0$. By setting $v(x) = u(x, \hat{t}) - \varphi(x, \hat{t})$ this follows from the boundary version of the classical maximum principle:

Maximum Principle at the Boundary *Let Ω be a domain in $\mathbf{R}^N$. Assume that $\partial\Omega$ is C^2 near $\hat{x} \in \partial\Omega$. Let v be in $C^2(B_r(\hat{x}) \cap \overline{\Omega})$ with some $r > 0$. Assume that*

$$v(x) \le v(\hat{x}) \quad \text{for} \quad x \in B_r(\hat{x}) \cap \Omega.$$

Then either

(i) $\nabla v(\hat{x}) = 0$ *and* $\nabla^2 v(\hat{x}) \le O$, *or*
(ii) $\nabla v(\hat{x}) = \lambda \nu(\hat{x})$ *and* $R_\nu \nabla^2 v(\hat{x}) R_\nu + \lambda A_\nu \le 0$ *for some* $\lambda > 0$.

Proof We may assume $v(\hat{x}) = 0$. By translation and rotation of coordinates we may assume that $\hat{x} = 0$, $\nu(x) = (1, 0, \dots, 0)$, $R_\nu = \text{diag}\ (0, 1, \dots, 1)$, $A_\nu = \text{diag}\ (0, \lambda_2, \dots, \lambda_N)$. Since v has a local maximum at 0 over $\overline{\Omega} \cap B_r(0)$,

$$\frac{\partial v}{\partial x_1}(0) = \lambda \ge 0, \ \frac{\partial v}{\partial x_2}(0) = \dots = \frac{\partial v}{\partial x_N}(0) = 0.$$

By expression of A_ν for graphs, Ω is expressed as $x_1 < \psi(x_2, \dots, x_N)$ near 0 with some $\psi \in C^2$ satisfying $\nabla'^2 \psi(0) = A_\lambda$ with $\nabla' = (\partial_{x_2}, \dots, \partial_{x_N})$. Since

$$v(x_1, \dots, x_N) = \lambda x_1 + \frac{1}{2} \sum_{i,j=1}^{N} \frac{\partial^2 v}{\partial x_i \partial x_j}(0) x_i x_j + o(|x|^2) \quad \text{as} \quad |x| \to 0$$

and v takes maximum 0 over $\overline{\Omega} \cap B_r(0)$ at 0, we see that for $\nabla^2 v(0) \le O$ if $\lambda = 0$ and that

$$\lambda \psi(x_2, \dots, x_N) + \frac{1}{2} \langle \nabla^2 v(0) x, x \rangle + o(|x|^2) \le 0$$

for $\lambda > 0$. Since $\nabla' \psi(0) = 0$ and $\psi(0) = 0$, expanding ψ around zero (as a function of $x_2, \dots, x_N$) now yields

$$\lambda \nabla'^2 \psi(0) + \nabla'^2 v(0) \le O.$$

This yields (ii). □

We give a sufficient condition for B so that a classical solution of (3.3.5) is a viscosity solution by restricting B for the first order operators.

Proposition 3.3.3 *Let $u \in C^2(\mathcal{O})$ satisfy* (3.3.5). *Assume that F is degenerate elliptic and that B depends only on $(z, r, p) \in \mathcal{O} \times \mathbf{R} \times \mathbf{R}^N$. Assume that*

$$\lambda \mapsto B(z, r, p - \lambda\nu(x)) \text{ is nonincreasing in } \lambda \geq 0. \tag{3.3.7}$$

Then u is a (viscosity) subsolution of (3.3.4).

Indeed, if φ satisfies (3.3.6) for $\hat{x} \in \partial\Omega$ so that (i) or (ii) holds, then by the monotonicity assumption on B we have

$$\begin{aligned} B(\hat{z}, u(\hat{z}), \nabla\varphi(\hat{z})) &\leq B(\hat{z}, u(\hat{z}), \nabla\varphi(\hat{z}) + \lambda\nu(\hat{x})) \\ &= B(\hat{z}, u(\hat{z}), \nabla u(\hat{z})) \leq 0. \end{aligned}$$

Remark 3.3.4

(i) Under the same monotonicity assumption on B a classical supersolution of (3.3.4) is also a viscosity supersolution.
(ii) As in Proposition 3.2.2 we may replace the maximum by a local maximum, a strict maximum or a local strict maximum in Definition 3.3.1 and the same remark applies to the minimum. The proof is identical with that of Proposition 3.2.2.
(iii) As in Proposition 3.2.3 we may replace class of test functions by $C^k(\mathcal{O})$ or $C^\infty(\mathcal{O})$ for $k \geq 2$ in Definition 3.3.1 (i) and even by $A^k(\mathcal{O})(k \geq 2)$ or $C^{2,1}(\mathcal{O})$ in Definition 3.3.1 (ii). The proof is essentially the same and it is based on Lemma 3.2.5 or Remark 3.2.6 (i).
(iv) As in Theorem 3.2.1 we have a stability result for the boundary value problem. Since the proof is identical with that of Theorem 3.2.1 (i), we only state the results (for subsolutions) which extends Theorem 3.2.1 (i).
(v) By definition, the localization property in Sect. 3.1.1 is still valid for (3.3.1).

Theorem 3.3.5 *Let $\mathcal{O}$ be a locally compact subset of $\mathbf{R}^d$.*

(i) *Assume that E_ε and E are lower semicontinuous on $\overline{\mathcal{O}} \times \mathbf{R} \times \mathbf{R}^d \times \mathbf{S}^d$ with values in $\mathbf{R} \cup \{-\infty\}$ for $\varepsilon > 0$. Assume that*

$$E \leq \liminf_{\varepsilon \to 0}{}_* E_\varepsilon \quad \text{on} \quad \overline{\mathcal{O}} \times \mathbf{R} \times \mathbf{R}^d \times \mathbf{S}^d.$$

Assume that u_ε is a subsolution of

$$E_\varepsilon(z, u, Du, D^2u) \leq 0 \quad \text{in} \quad \mathcal{O}.$$

Then $\overline{u} = \limsup^*_{\varepsilon \to 0} u_\varepsilon$ is a subsolution of (3.3.1) *in $\mathcal{O}$ provided that $\overline{u}(z) < \infty$ for every $z \in \mathcal{O}$.*

(ii) *Under the notation of Definition 3.3.1 (ii) let F_ε and B_ε be lower semicontinuous on W and J with values in $\mathbf{R} \cup \{-\infty\}$, respectively for $\varepsilon > 0$. Assume that*

$$F \leq \liminf_{\varepsilon \to 0}{}_* F_\varepsilon \quad in\ W\ and\ B \leq \liminf_{\varepsilon \to 0}{}_* B_\varepsilon \quad in\ J.$$

Let u_ε be a subsolution of

$$\begin{aligned} u_t + F_\varepsilon(z, u, \nabla u, \nabla^2 u) &= 0 \quad in \quad \mathcal{O} \cap (\Omega \times (0, T)), \\ B_\varepsilon(z, u, \nabla u, \nabla^2 u) &= 0 \quad on \quad \mathcal{O} \cap (\partial\Omega \times (0, T)). \end{aligned}$$

Then $\overline{u} = \limsup_{\varepsilon \to 0}{}^* u_\varepsilon$ is a subsolution of (3.3.4) if $\overline{u}(z) < \infty$ for every $z \in \mathcal{O} \cap (\overline{\Omega} \times (0, T))$.

The part (ii) is a trivial corollary of part (i).

Example 3.3.6 We give a simple example that a maximum point z_ε of $u_\varepsilon^* - \varphi$ is not on $\partial\Omega \times (0, T)$ for all sufficiently small ε even if z_ε converges to $\hat{z} \in \partial\Omega \times (0, T)$ which is a maximum point of $\overline{u} - \varphi$ where we use the notation in Theorem 3.3.5 (ii). We set

$$u_\varepsilon(x, t) = \int_0^\infty \{g(t\varepsilon, x - y) + g(t\varepsilon, x + y)\}u_0(y)dy$$

with the Gauss kernel $g(t, x) = (4\pi t)^{-1/2} \exp(-|x|^2/4t)$ for $u_0 \in C^\infty[0, \infty)$ with $u_0'(0) < 0$. Clearly, $u_\varepsilon \in C^\infty(\mathcal{O})$ is a classical solution of

$$\begin{aligned} u_t - \varepsilon u_{xx} &= 0 \quad \text{in} \quad \Omega \times (0, \infty), \\ -u_x &= 0 \quad \text{on} \quad \partial\Omega \times (0, \infty) \end{aligned}$$

with $\Omega = (0, \infty)$ and $\mathcal{O} = \overline{\Omega} \times (0, \infty)$. (By Proposition 3.3.3, u_ε is also a viscosity solution.) We here wrote $-u_x$ instead of u_x above to emphasize that the outward unit normal to the region $\Omega = (0, \infty)$ at the boundary point 0 is -1.

By the representation of u_ε the limit $\overline{u} = \limsup_{\varepsilon \to 0}^* u_\varepsilon$ equals u_0. By the stability results (Theorem 3.3.5) u_0 is a viscosity subsolution of

$$\begin{aligned} u_t &= 0 \quad \text{in} \quad \Omega \times (0, \infty), \\ -u_x &= 0 \quad \text{on} \quad \partial\Omega \times (0, \infty). \end{aligned}$$

However, evidently u_0 fails to fulfill the subsolution property at the boundary $x = 0$ in the strong sense since it is assumed that $u_0'(0) < 0$.

We take $\varphi(x, t) = u_0(x) + x^4 + (t - \hat{t})^2$ for $\hat{t} > 0$ so that $u_0 - \varphi$ takes a strict maximum at $(0, \hat{t}) = \hat{z}$ over $\mathcal{O}$. Let z_ε be a maximum point of $u_\varepsilon - \varphi$ over $\mathcal{O}$. It converges to $\hat{z}$ as $\varepsilon \to 0$ by Lemma 3.2.5 and its remark. Suppose that there were a subsequence z_{ε_i} that belongs to $\partial\Omega \times (0, \infty)$. Since $u_{\varepsilon x}(0, t) = 0$, we would obtain $\varphi_x(z_{\varepsilon_i}) \geq 0$. Sending $\varepsilon_i \to 0$ we would obtain $\varphi_x(0, \hat{t}) = u_0'(0) \geq 0$, which is a contradiction. We conclude that $z_\varepsilon \in \Omega \times (0, T)$ for sufficiently small ε.

Definition 3.3.7 It is also possible to extend the notion of $\mathcal{F}$-solution for the boundary value problem. Let F be continuous in $W_0 = \overline{\Omega} \times [0, T) \times \mathbf{R} \times (\mathbf{R}^N \setminus \{0\}) \times \mathbf{S}^N$ with values in $\mathbf{R}$. Assume that F fulfills (3.1.11) and $B \in C(J)$. We say $u : \mathcal{O} \to \mathbf{R} \cup \{-\infty\}$ is an $\mathcal{F}$-*subsolution* of (3.3.4) if:

(a) $u^*(z) < \infty$ for $z \in \mathcal{O}$.
(b) If $(\varphi, \hat{z}) \in C_F^2(\mathcal{O}) \times \mathcal{O}$ satisfies (3.1.6), then

$$\begin{aligned} &\varphi_t(\hat{z}) + F(\hat{z}, u^*(\hat{z}), \nabla\varphi(\hat{z}), \nabla^2\varphi(\hat{z})) \leq 0 \text{ if } \nabla\varphi(\hat{z}) \neq 0, \\ &\varphi_t(\hat{z}) \leq 0 \qquad\qquad\qquad\qquad\qquad\qquad\qquad \text{otherwise} \end{aligned} \tag{3.3.8}$$

holds if $\hat{z} \in \Omega \times (0, T)$ and either (3.3.8) or

$$B(\hat{z}, u^*(\hat{z}), \nabla\varphi(\hat{z}), \nabla^2\varphi(\hat{z})) \leq 0$$

if $\hat{z} \in \partial\Omega \times (0, T)$.

The definition of $\mathcal{F}$-supersolution is similar so is omitted. Here $C_F^2(\mathcal{O})$ is the set of $\varphi \in C^2(\mathcal{O})$ compatible with F, i.e., φ satisfies (3.1.12). The stability result Theorem 3.2.1 (ii) extends to the boundary value problem with a trivial modification.

Example 3.3.8 We conclude this section by giving examples of boundary conditions that satisfy (3.3.7) other than the Dirichlet condition $B \equiv 0$. The operator $B = \langle p, \nu \rangle$ evidently fulfills (3.3.7). In this case the boundary condition $B = 0$ is called the homogeneous Neumann condition. The operator $B = \langle p, \nu \rangle - k(z)|p|$ fulfills (3.3.7) provided that $|k(z)| < 1$. The condition $B = 0$ is an oblique type boundary condition.

3.4 Perron's Method

We give typical results on Perron's method for constructing solutions for a general equation (3.3.1) rather than for (3.1.5).

Lemma 3.4.1 (Closedness Under Supremum) *Let $\mathcal{O}$ be a locally compact subset of $\mathbf{R}^d$. Let E be a lower semicontinuous function on $\overline{\mathcal{O}} \times \mathbf{R} \times \mathbf{R}^d \times \mathbf{S}^d$ with values in $\mathbf{R} \cup \{-\infty\}$. Let S be a set of subsolutions of*

$$E(z, u, Du, D^2u) = 0 \tag{3.4.1}$$

in $\mathcal{O}$. Let u be a function in $\mathcal{O}$ defined by

$$u(z) = \sup\{v(z); v \in S\} \quad \textit{for} \quad z \in \mathcal{O}. \tag{3.4.2}$$

If $u^*(z) < \infty$ for all $z \in \mathcal{O}$, then u is a subsolution of (3.4.1) *in $\mathcal{O}$.*

Lemma 3.4.2 (Maximal Subsolution) *Let $\mathcal{O}$ be as in Lemma 3.4.1. Let E be defined in a dense set of $\overline{\mathcal{O}} \times \mathbf{R} \times \mathbf{R}^d \times \mathbf{S}^d$ with $E_* < \infty$, $E^* > -\infty$ everywhere. Assume that Eq.* (3.4.1) *is degenerate elliptic (for subsolution) in the sense that all $u \in C^2(\mathcal{O}')$ satisfying*

$$E_*(z, u(z), Du(z), D^2u(z)) \le 0 \quad \textit{for} \quad z \in \mathcal{O}'$$

is a (viscosity) subsolution of (3.4.1) *in $\mathcal{O}'$ where $\mathcal{O}'$ is an open subset of $\mathcal{O}$. Let h be a supersolution of* (3.4.1)*. Let S be the collection of all subsolutions v of* (3.4.1) *in $\mathcal{O}$ that satisfies $v \le h$ in $\mathcal{O}$. If $\tilde{v} \in S$ is not a supersolution of* (3.4.1) *in $\mathcal{O}$ with $\tilde{v}_* > -\infty$ in $\mathcal{O}$, then there are a function $w \in S$ and a point $z_0 \in \mathcal{O}$ that satisfies $\tilde{v}(z_0) < w(z_0)$.*

Theorem 3.4.3 (Existence) *Let $\mathcal{O}$ be a locally compact subset of $\mathbf{R}^d$. Let E be a densely defined function on $Z = \mathcal{O} \times \mathbf{R} \times \mathbf{R}^d \times \mathbf{S}^d$ with $E_* < \infty$, $E^* > -\infty$ on Z. Assume that Eq.* (3.4.1) *is degenerate elliptic (for subsolutions). Let h_- and h_+ be a sub- and supersolution of* (3.4.1)*, respectively with $h_+^* < \infty$, $h_{-*} > -\infty$ in $\mathcal{O}$. Suppose that $h_- \le h_+$ in $\mathcal{O}$. Then there exists a solution u of* (3.4.1) *that satisfies $h_- \le u \le h_+$ in $\mathcal{O}$.*

If we admit Lemmas 3.4.1 and 3.4.2, it is easy to prove Theorem 3.4.3. Indeed, we take

$$S = \{v; v \text{ is a subsolution of (3.4.1) in } \mathcal{O} \text{ with } v \le h_+ \text{ in } \mathcal{O}\}.$$

Since $h_- \le h_+$, the set S is not empty. Since $h_{-*} > -\infty$, the function u defined by (3.4.2) fulfills $u_* > -\infty$ in $\mathcal{O}$. Since $h_+^* < \infty$, $u^* < \infty$ in $\mathcal{O}$. By Lemma 3.4.1 u is a subsolution of (3.4.1). If u were not a supersolution, then by Lemma 3.4.2 there would exist a function $w \in S$ and a point $z_0 \in \mathcal{O}$ with $u(z_0) < w(z_0)$. This contradicts the definition (3.4.2) of u. Thus u is a solution of (3.4.1) in $\mathcal{O}$. The inequality $h_- \le u \le h_+$ follows by definition of u. Note that the comparison principle is unnecessary to construct a solution.

Remark 3.4.4 By Proposition 3.3.3 and the localization property (Sect. 3.1.1), the results in Lemmas 3.4.1, 3.4.2 and Theorem 3.4.3 apply to the boundary value problem on $\mathcal{O} = \overline{\Omega} \times (0, T)$:

$$\begin{aligned} u_t + F(z, u, \nabla u, \nabla^2 u) &= 0 \text{ in } \quad \Omega \times (0, T), \\ B(z, u, \nabla u) &= 0 \quad \text{on} \quad \partial\Omega \times (0, T) \end{aligned}$$

provided that F is degenerate elliptic and B satisfies the monotonicity assumption (3.3.7), where Ω is a domain in $\mathbf{R}^N$. The same remark applies to (3.1.5) for degenerate elliptic F by Proposition 3.1.2. In both examples F need not be continuous as in Definition 3.1.4 (iii).

3.4.1 Closedness Under Supremum

We shall prove Lemma 3.4.1. Let $(\varphi, \hat{z}) \in C^2(\mathcal{O}) \times \mathcal{O}$ satisfy (3.3.2), i.e.,

$$\max_{\mathcal{O}}(u^* - \varphi) = (u^* - \varphi)(\hat{z}).$$

We must prove

$$E_*(\hat{z}, u^*(\hat{z}), D\varphi(\hat{z}), D^2\varphi(\hat{z})) \le 0. \tag{3.4.3}$$

We may assume that $(u^* - \varphi)(\hat{z}) = 0$ by replacing φ by $\varphi + (u^* - \varphi)(\hat{z})$. We set $\psi(z) = \varphi(z) + |z - \hat{z}|^4$ and observe that

$$(u^* - \psi)(z) \le -|z - \hat{z}|^4 \quad \text{for} \quad z \in \mathcal{O} \tag{3.4.4}$$

since

$$(u^* - \psi)(z) + |z - \hat{z}|^4 = (u^* - \varphi)(z) \le (u^* - \varphi)(\hat{z}) = 0.$$

By definition of an upper semicontinuous envelope there is a sequence $\{z_k\}$ in $\mathcal{O}$ converging to $\hat{z}$ as $k \to \infty$ such that

$$a_k := (u - \psi)(z_k) \to (u^* - \psi)(\hat{z}) = 0.$$

By definition (3.4.2) of u there exists a sequence $\{v_k\}$ in S with $v_k(z_k) > u(z_k) - 1/k$. This implies

$$(v_k^* - \psi)(z_k) \ge (v_k - \psi)(z_k) > a_k - 1/k. \tag{3.4.5}$$

Since $v_k \le u$ in $\mathcal{O}$, (3.4.4) implies

$$(v_k^* - \psi)(z) \le -|z - \hat{z}|^4 \quad \text{for} \quad z \in \mathcal{O}. \tag{3.4.6}$$

Since $\mathcal{O}$ is locally compact, there is a compact neighborhood of $\hat{z}$ denoted B. Since $v_k^* - \psi$ is upper semicontinuous and has an upper bound, $v_k^* - \psi$ attains its maximum over B at some point $y_k \in B$. From (3.4.5) and (3.4.6) it now follows that

$$a_k - 1/k < (v_k^* - \psi)(z_k) \le (v_k^* - \psi)(y_k) \le -|y_k - \hat{z}|^4$$

for sufficiently large k (so that $z_k \in B$). Since $a_k \to 0$ this implies that $y_k \to \hat{z}$ as $k \to \infty$ and that

$$\lim_{k\to\infty} (v_k^* - \psi)(y_k) = 0.$$

Hence

$$\lim_{k\to\infty} v_k^*(y_k) = \lim_{k\to\infty} \psi(y_k) = \psi(\hat{z}) = u^*(\hat{z}).$$

Since v_k is a subsolution, we see

$$E_*(y_k, v_k^*(y_k), D\psi(y_k), D^2\psi(y_k)) \le 0.$$

Since E_* is lower semicontinuous and $y_k \to \hat{z}$, $v_k^*(y_k) \to u^*(\hat{z})$, sending $k \to \infty$ yields

$$E_*(\hat{z}, u^*(\hat{z}), D\psi(\hat{z}), D^2\psi(\hat{z})) \le 0.$$

We now obtain (3.4.3) since $D\psi(\hat{z}) = D\varphi(\hat{z})$, $D^2\psi(\hat{z}) = D^2\varphi(\hat{z})$.

We have proved that a class of subsolutions is essentially closed under the operation of supremum. The symmetric argument yields the closedness of supersolutions under the operation of infimum.

Lemma 3.4.5 (Closedness Under Infimum) *Let $\mathcal{O}$ be a locally compact subset of $\mathbf{R}^d$. Let E be an upper semicontinuous function on $\overline{\mathcal{O}} \times \mathbf{R} \times \mathbf{R}^d \times \mathbf{S}^d$ with values in $\mathbf{R} \cup \{+\infty\}$. Let S be a set of supersolutions of* (3.4.1) *in $\mathcal{O}$. Let u be a function on $\mathcal{O}$ defined by*

$$u(z) = \inf\{v(z); v \in S\} \quad \textit{for} \quad z \in \mathcal{O}.$$

If $u_*(z) > -\infty$ for all $z \in \mathcal{O}$, then u is a supersolution of (3.4.1) *in $\mathcal{O}$.*

3.4.2 *Maximal Subsolution*

We shall prove Lemma 3.4.2. Since $\tilde{v}$ is not a supersolution of (3.4.1), there exist $\varphi \in C^2(\mathcal{O})$ and $\hat{z} \in \mathcal{O}$ that satisfies

$$\begin{aligned} &\min_{\mathcal{O}}(\tilde{v}_* - \varphi) = (\tilde{v}_* - \varphi)(\hat{z}) = 0, \\ &E^*(\hat{z}, \tilde{v}_*(\hat{z}), D\varphi(\hat{z}), D^2\varphi(\hat{z})) = E^*(\hat{z}, \varphi(\hat{z}), D\varphi(\hat{z}), D^2\varphi(\hat{z})) < 0. \end{aligned} \tag{3.4.7}$$

Modifying φ by $\varphi + |z - \hat{z}|^4$ as in the proof of Lemma 3.4.1, we may assume

$$(\tilde{v}_* - \varphi)(z) \geq |z - \hat{z}|^4 \quad \text{for} \quad z \in \mathcal{O}. \tag{3.4.8}$$

Evidently, $\tilde{v}_* \leq h_*$ in $\mathcal{O}$ so we see $\tilde{v}_*(\hat{z}) = \varphi(\hat{z}) < h_*(\hat{z})$. Indeed, if not, $\tilde{v}_*(\hat{z}) = h_*(\hat{z}) = \varphi(\hat{z})$ so that φ would be a lower test function of h_*. Since h is a supersolution, the inequality (3.4.7) is contradictory.

Since E^* is upper semicontinuous and $\varphi \in C^2(\mathcal{O})$, for sufficiently small $\delta > 0$ we have

$$E^*(z, \varphi(z) + \delta^4/2,\ D\varphi(z),\ D^2\varphi(z)) \leq 0 \quad \text{for} \quad z \in B_{2\delta} \tag{3.4.9}$$

with $B_{2\delta} = B \cap B_{2\delta}(\hat{z})$, where B is a compact neighborhood of $\hat{z}$ and $B_r(z)$ is the closed ball of radius r centered at z. (Such B exists since $\mathcal{O}$ is locally compact.) Since $\tilde{v}_*(\hat{z}) < h_*(\hat{z})$, we may assume

$$\varphi(z) + \delta^4/2 \leq h_*(z) \quad \text{for} \quad z \in B_{2\delta} \tag{3.4.10}$$

by taking δ smaller if necessary. Since (3.4.1) is degenerate elliptic, (3.4.9) indicates that $\varphi + \delta^4/2$ is a subsolution in the interior $\operatorname{int} B_{2\delta}$ of $B_{2\delta}$. From (3.4.8) it follows that

$$\tilde{v}(z) \geq \tilde{v}_*(z) - \delta^4/2 \geq \varphi(z) + \delta^4/2 \quad \text{for} \quad z \in B_{2\delta} \backslash B_\delta. \tag{3.4.11}$$

We now define w by

$$w(z) = \begin{cases} \max\{\varphi(z) + \delta^4/2, \tilde{v}(z)\} & z \in B_\delta, \\ \tilde{v}(z) & z \in \mathcal{O} \backslash B_\delta. \end{cases}$$

To see w is a subsolution one should be careful around a neighborhood of ∂B_δ. It follows from (3.4.11) that

$$w(z) = \max\{\varphi(z) + \delta^4/2, \tilde{v}(z)\} \quad \text{for} \quad z \in B_{2\delta}$$

not only for $z \in B_\delta$. Since $\tilde{v}$ is a subsolution in $\operatorname{int} B_{2\delta}$ by localization property (Sect. 3.1.1) and since $\varphi + \delta^4/2$ is a subsolution in $\operatorname{int} B_{2\delta}$, Lemma 3.4.1 implies that w is a subsolution in $B_{2\delta}$. By the localization property w is now a subsolution of (3.4.1) in $\mathcal{O}$. From (3.4.10) it follows that $w \in S$.

Since

$$0 = (\tilde{v}_* - \varphi)(\hat{z}) = \liminf_{r \downarrow 0} \{(\tilde{v} - \varphi)(z); z \in \mathcal{O} \quad \text{and} \quad |z - \hat{z}| \leq r\},$$

there is a point $z_0 \in B_\delta$ that satisfies

$$\tilde{v}(z_0) - \varphi(z_0) < \delta^4/2$$

which yields $v(z_0) < w(z_0)$. We have thus constructed $w \in S$ that satisfies $\tilde{v}(z_0) < w(z_0)$. The proof is now complete. □

Remark 3.4.6 By an argument symmetric to that of Lemma 3.4.2 for a class of supersolutions greater than or equal to a given subsolution, we observe that the infimum of the class is a solution if local boundedness condition is satisfied. We do not write a precise version of Lemma 3.4.2 for a class of supersolutions. This property of minimal supersolution yields also a solution in Theorem 3.4.3. Indeed, if the equation is degenerate elliptic (for supersolutions), i.e., $u \in C^2(\mathcal{O}')$ solving

$$E^*(z, u(z), Du(z), D^2u(z)) \geq 0 \quad \text{for} \quad z \in \mathcal{O}'$$

is a (viscosity) supersolution of (3.4.1) in $\mathcal{O}'$ for each open set $\mathcal{O}'$ in $\mathcal{O}$, then

$$u_-(z) = \inf\{v(z);\, v \text{ is a supersolution of (3.4.1) in } \mathcal{O} \text{ with } h_- \leq v \text{ in } \mathcal{O}\}$$

is a solution of (3.4.1) with $h_- \leq u_- \leq h_+$ in $\mathcal{O}$. As we have seen in the beginning of Chap. 3,

$$u_+(z) = \sup\{v(z);\, v \text{ is a subsolution of (3.4.1) in } \mathcal{O} \text{ with } v \leq h_+ \text{ in } \mathcal{O}\}$$

is a solution of (3.4.1) with $h_- \leq u_+ \leq h_+$ in $\mathcal{O}$. By definition u_- and u_+ are minimal and maximal solutions satisfying $h_- \leq u_- \leq u_+ \leq h_+$. If $h_{-*} = h_+^*$ on some portion Σ of the boundary of $\mathcal{O}$, we see $u_-^* \leq u_+^* \leq h_+^* = h_{-*} \leq u_{-*} \leq u_{+*}$ on Σ. If the comparison principle holds, then

$$u_+^* \leq u_{-*},\ u_+^* \leq u_{+*},\ u_-^* \leq u_{-*} \quad \text{in} \quad \mathcal{O} \cup \Sigma$$

so that $u_\pm \in C^0(\mathcal{O} \cup \Sigma)$ (i.e., $u_\pm$ is continuous in $\mathcal{O} \cup \Sigma$) and $u_+ = u_-$. In other words the minimal and maximal solutions agree and they are continuous.

3.4.3 Adaptation for Very Singular Equations

Perron's method applies $\mathcal{F}$-solutions for very singular equations in Definition 3.1.5. Although it applied the boundary problems in Definition 3.3.7 if the problem is degenerate elliptic, we restrict ourselves to the case without the boundary condition. We give rigorous statements and indication of the proof.

Lemma 3.4.7 (Closedness Under Supremum) *Let F and $\mathcal{F} \neq \emptyset$ be as in Definition 3.1.5. Assume that $\mathcal{F}$ is closed under positive multiplication. Let S be a set of $\mathcal{F}$-subsolutions of* (3.1.5)*:*

$$u_t + F(z, u, \nabla u, \nabla^2 u) = 0$$

in $\mathcal{O} = \Omega \times (0, T)$. *Let* u *be a function on* $\mathcal{O}$ *defined by* (3.4.2). *If* $u^*(z) < \infty$ *for all* $z \in \mathcal{O}$, *then* u *is a subsolution of* (3.1.5) *in* $\mathcal{O}$.

The proof essentially parallels that of Lemma 3.4.1 so we only point out the place to be altered. For $(\varphi, \hat{z}) \in C_F^2(\mathcal{O}) \times (\mathcal{O})$ satisfying (3.3.2) we must prove

$$\begin{cases} \varphi_t(\hat{z}) + F(\hat{z}, u^*(\hat{z}), \nabla\varphi(\hat{z}), \nabla^2\varphi(\hat{z})) \le 0 \text{ if } \nabla\varphi(\hat{z}) \ne 0, \\ \varphi_t(\hat{z}) \le 0 \quad \text{otherwise.} \end{cases}$$

The proof of the first case is the same as that of Lemma 3.4.1 so we may assume that φ is as in Remark 3.2.7 (B) with $\varphi(\hat{z}) = 0$ so that $\nabla\varphi(\hat{z}) = 0$. We set

$$\psi(z) = \varphi(z) + f(|x - \hat{x}|) + (t - \hat{t})^2.$$

Arguing as in Sect. 3.4.1 we end up with the existence of sequence $(x_k, t_k) \to (\hat{x}, \hat{t})(k \to \infty)$ with

$$u_k^*(x_k, t_k) \to \psi(\hat{z})$$

such that $v_k^* - \psi$ attains its local maximum at $y_k = (x_k, t_k)$. Since v_k is an $\mathcal{F}$-subsolution, we have

$$\begin{cases} \psi_t(y_k) + F(y_k, v_k^*(y_k), \nabla\psi(y_k), \nabla^2\psi(y_k)) \le 0 \text{ if } \nabla\psi(y_k) \ne 0, \\ \psi_t(y_k) \le 0 \qquad \text{otherwise.} \end{cases} \tag{3.4.12}$$

By definition $\nabla\psi(y_k) = 2\nabla(f(|x - x_k|)$,

$$\nabla^2\psi(y_k) = 2\nabla^2(f(|x - x_k|))$$

for sufficiently large k. Since $f \in \mathcal{F}$, $y_k \to \hat{z}$, $v_k^*(y_k) \to \psi(\hat{z})$ we observe that

$$\lim_{k\to\infty} F(y_k, v_k^*(y_k), \nabla\psi(y_k), \nabla^2\psi(y_k)) = 0. \tag{3.4.13}$$

Thus sending $k \to \infty$ in (3.4.13) we end up with $\psi_t(\hat{z}) \le 0$.

Lemma 3.4.8 (Maximal Subsolution) *Assume the same hypotheses of Lemma 3.4.7 concerning* F *and* $\mathcal{F}$. *Assume that* F *is degenerate elliptic. Let* h *be an* $\mathcal{F}$-*supersolution of* (3.1.5) *in* $\mathcal{O} = \Omega \times (0, T)$. *Let* S *be the collection of all* $\mathcal{F}$-*subsolutions* v *of* (3.4.1) *in* $\mathcal{O}$ *that satisfies* $v \le h$ *in* $\mathcal{O}$. *If* $\tilde{v} \in S$ *is not an* $\mathcal{F}$-*supersolution of* (3.1.5) *in* $\mathcal{O}$ *with* $\tilde{v}_* > -\infty$ *in* $\mathcal{O}$, *then there are a function* $w \in S$ *and a point* $z_0 \in \mathcal{O}$ *that satisfies* $\tilde{v}(z_0) < w(z_0)$.

The proof essentially parallels that of Lemma 3.4.2 so we give the idea of the way of modification. Since $\tilde{v}$ is not an $\mathcal{F}$-supersolution, there exist $\varphi \in C^2_{\mathcal{F}}(\mathcal{O})$ and a point $\hat{z}$ that satisfies

$$\min_{\mathcal{O}}(\tilde{v}_* - \varphi) = (\tilde{v}_* - \varphi)(\hat{z}) = 0$$

but does not fulfill the desired inequality in Definition 3.1.5. We may assume $\nabla\varphi(\hat{z}) = 0$ so that $\varphi_t(\hat{z}) < 0$, otherwise, the desired w and z_0 is constructed by the proof of Lemma 3.4.2; note, however one should take δ smaller so that $\nabla\varphi(z) \neq 0$ on $B_{2\delta}$. By Remark 3.2.7 we may assume φ is a separable function:

$$\varphi(x,t) = f(|x-\hat{x}|) + g(t), \quad f \in \mathcal{F}$$

and obtain, instead of (3.4.8), that

$$(\tilde{v}_* - \varphi)(z) \geq f(|x-\hat{x}|) + (t-\hat{t})^2 \quad \text{near} \quad (\hat{x}, \hat{t})$$

by adding $f(|x-\hat{x}|) + (t-\hat{t})^2$ to φ.

Since $\nabla\varphi(\hat{z}) = 0$ and $\varphi_t(\hat{z}) < 0$, for sufficiently small $\delta > 0$ we have

$$\varphi_t(z) + F(z, \varphi(z) + f(\delta)/2, \nabla\varphi(z), \nabla^2\varphi(z)) \leq 0, \quad z \neq \hat{z},$$
$$\varphi(z) + f(\delta)/2 \leq h_*(z) \quad \text{for} \quad z \in B_{2\delta}$$

by definition of $\mathcal{F}$. The first inequality implies that $\varphi + f(\delta)/2$ is an $\mathcal{F}$-subsolution of (3.1.5) in $B_{2\delta}$. We set

$$w(z) = \begin{cases} \max\{\varphi(z) + f(\delta)/2, \tilde{v}(z)\}, & z \in B_\delta, \\ \tilde{v}(z), & z \in \mathcal{O} \backslash B_\delta \end{cases}$$

and conclude that w is an $\mathcal{F}$-subsolution satisfying $w \leq h$ in $\mathcal{O}$ and $v(z_0) < w(z_0)$ with some point $z_0 \in B_\delta$ as in the proof of Lemma 3.4.2. Lemmas 3.4.7 and 3.4.8 yield:

Theorem 3.4.9 (Existence) *Assume that F is continuous in $W_0 = \overline{\Omega} \times [0,T] \times \mathbf{R} \times (\mathbf{R}^N \backslash \{0\}) \times \mathbf{S}^N$ with values in $\mathbf{R}$. Assume that $\mathcal{F} \neq \emptyset$ and $\mathcal{F}$ is invariant under positive multiplication. Assume that F is degenerate elliptic. Let h_- and h_+ be an $\mathcal{F}$-sub- and supersolution of* (3.1.5) *in $\mathcal{O} = \Omega \times (0,T)$, respectively with $h_+^* < \infty, h_{-*} > -\infty$ in $\mathcal{O}$. If $h_- \leq h_+$ in $\mathcal{O}$, then there exists an $\mathcal{F}$-solution u of* (3.4.1) *that satisfies $h_- \leq u \leq h_+$ in $\mathcal{O}$.*

3.4.4 Applicability

The theory we developed so far applies to a wide class of level set equations (2.6.3) of surface evolution equations (2.6.1). For example, Perron's method applies to (2.6.1) provided that f is continuous in its variables and that (2.6.1) is degenerate elliptic. As we shall see later (Sect. 4.1.3) we need the theory of $\mathcal{F}$-solutions if the growth of f with respect to $\nabla\mathbf{n}$ is superlinear. A typical example is the Gaussian curvature flow equation (2.5.9) ($N \geq 3$) provided that e_m is interpreted as $\hat{e}_m$ like (2.6.22). Generally, the theory developed so far needs fewer assumptions on f of (2.6.1) than those for the comparison principle in Chap. 4. It also applies to a level set equation supplemented by oblique type boundary conditions (2.6.14).

3.5 Notes and Comments

The notion of a viscosity solution was first introduced by M. G. Crandall and P.-L. Lions [129, 130] in a different way for first order Hamilton–Jacobi equations with non-convex Hamiltonian. For a survey of the early stage of the theory the reader is referred to a book of P.-L. Lions [320]. A few years later the theory extends to second order degenerate elliptic and parabolic equations. The User's Guide by M. G. Crandall et al. [128] gives a nice review for the development of the theory. There is a nice but more concise review by H. Ishii [274]. A book by G. Barles [46] covers related developments of the theory mainly for the Hamilton–Jacobi equations. See also a book by S. Koike [315] for an introduction to the theory of viscosity solutions.

A book of M. Bardi and I. Capuzzo-Dolcetta [40] discusses applications of the theory of viscosity solutions to control theory, differential games. There is a nice lecture note by M. Bardi et al. [41] where various applications of viscosity solutions including a level set method are presented. We do not intend to exhaust references and rather to suggest the readers to consult these books.

Subjects in Sects. 3.1.1 and 3.1.2 are standard. The definition for general discontinuous functions is essentially due to H. Ishii [269, 270]. There he constructed a solution by adjusting Perron's method for elliptic equations to viscosity solutions of the Hamilton–Jacobi equations. The advantage of the use of a semicontinuous envelope is to get continuity of solution by the comparison principle for semicontinuous functions as explained in Sect. 3.1.2 (Perron's method). The stronger version of the stability principle is due to G. Barles and B. Perthame [52, 53] and also H. Ishii [272]. We postpone discussion of comparison theorems to the next chapter. The definition of discontinuous viscosity solutions given here is not enough to characterize an entropy solution of scalar conservation laws. The reader is referred to a paper by Y. Giga [203] for the theory of viscosity solutions for such equations.

The degenerate equations (3.1.5) with singularity at $\nabla u = 0$ was first studied by Y.-G. Chen et al. [105, 106], to investigate level set equations of surface evolution equations, and independently by L. C. Evans and J. Spruck [166] for the level set equation of the mean curvature flow equations. The continuity condition (3.1.10) for comparison principle is explicitly stated in Y.-G. Chen et al. [105, 106] and this property is also used in L. C. Evans and J. Spruck [166]. The conterexample for the comparison principle for a very singular equation is due to Y.-G. Chen et al. [107].

A suitable notion of viscosity solutions for very singular equations was first introduced by S. Goto [239]. The definition of $\mathcal{F}$-subsolution in Sect. 3.1.3 is apparently different from his and it is essentially due to H. Ishii and P. Souganidis [284], where they only treated level set equations. It is not difficult to generalize this notion to other equations including p-Laplace diffusion equations (3.1.14) as in M. Ohnuma and K. Sato [368]. In Sect. 3.1.3 we compare an $\mathcal{F}$-subsolution with a usual subsolution. Although it is elementary, Propositions 3.1.7 and 3.1.8 are not found in the literature. The first part of Lemma 3.1.9 is standard. The second part of Lemma 3.1.9 is essentially found in M. G. Crandall et al. [126] where they showed equivalence of several definitions of viscosity solutions.

A convergence of maximum points and its various applications including strong stability principle are explained well in G. Barles [46] for first order equations. Results in Sect. 3.2 are a straightforward extension to singular and very singular equations. Since the equation is an evolution type, we note that separable type functions play the role of a class of test functions as explained in Proposition 3.2.3. The stability results for very singular equations is essentially found in H. Ishii and P. Souganidis [284] for level set equations and in M. Ohnuma and K. Sato [368] for general equations. An alternate definition of viscosity subsolution in Remark 3.2.4 is due to L. A. Caffarelli [71] and L. Wang [425]. This notion is useful when z-dependence of F is discontinuous and only measurable. Equivalence of $\mathcal{F}$-subsolution and subsolution is essentially due to G. Barles and C. Georgelin [50] where they proved Proposition 3.2.8 from Proposition 3.1.7 for the level set equation of the mean curvature flow equation.

The definition of solution for the boundary value problem goes back to P.-L. Lions [324]. Materials in Sect. 3.3 is essentially taken from the User's Guide with modification and adjustment for evolution problems. We do not consider Dirichlet problems in this weak sense in this book. There are also interesting other boundary conditions like the state constraint problem as studied in H. M. Soner [397, 398]. We do not touch these problems in this book. Those who are interested in these topics are encouraged to consult the User's Guide and a book by W. H. Fleming and H. M. Soner [176].

Perron's method presented in Sect. 3.4 is essentially due to H. Ishii [270], where first order equations are treated. Its extension to various other equations is usually not difficult so in many cases it is stated without proof. For the reader's convenience we give a full proof for general second order equations and also for $\mathcal{F}$-solutions. For $\mathcal{F}$-solutions Perron's method is stated in H. Ishii and P. Souganidis [284] without proof.

We won't mention any regularity theory for viscosity solutions. The reader is referred to nice books by L. A. Caffarelli and X. Cabré [72] and by Q. Han and F. Lin [253] for such topics as well as a paper by L. A. Caffarelli et al. [73].

Further Comments for the Second Edition The notion of viscosity solution has been adjusted for many other problems where comparison principle is formally expected. There are two directions for extension. The first direction is an extension of equations, for example, including fractional derivatives. For example, the notion of viscosity solutions has been adjusted to equations involving fractional time derivatives; see the work of Y. Giga and T. Namba [223], T. Namba [364] and E. Topp and M. Yangari [417].

The second direction is to extend the base space from an Euclidean domain to more general geometric structures, such as an infinite dimensional Hilbert space or a network. In general metric spaces, the gradient ∇u cannot be defined but $|\nabla u|$ can still be meaningfully interpreted. This allows for a viscosity solution framework that applies to first order nonlinear equations in metric spaces including the eikonal equation like $|\nabla u| = 1$ or $u_t - |\nabla u| = 0$. This is carried out by Y. Giga, N. Hamamuki and A. Nakayasu [213] and A. Nakayasu [361], who essentially reduced the problem to a one dimensional setting by composing the unknown with rectifiable curves in the metric space. A more general class of Hamilton-Jacobi equations in length spaces were studied by L. Ambrosio and J. Feng [13] and by W. Gangbo and A. Święch [184, 185], who introduced a class of viscosity test functions in general length spaces based on the squared distance function. This approach can be used to various properties of Hamilton-Jacobi equations in metric spaces. See A. Nakayasu and T. Namba [362] and S. Makida and A. Nakayasu [343] for stability results and large time behavior. A convexity preserving property was discussed by Q. Liu and A. Nakayasu [332] with this approach.

It is natural to ask whether the aforementioned two approaches are equivalent. The equivalence was established for the eikonal equation in complete length spaces by Q. Liu et al. [334] via a third alternative notion termed Monge solutions, which originates from the Euclidean setting studied by R. T. Newcomb, II and J. Su [365]. See also Q. Liu and M. B. P. Wiranata [336] for related discussions on the time-dependent case and Q. Liu et al. [335] for further developments concerning discontinuous eikonal equations in metric measure spaces.

In this chapter, the theory of viscosity solutions for second order equations is primarily employed to address singularities arising from vanishing gradients. There are, however, situations where the viscosity solution framework can be adapted to treat parabolic equations with discontinuities in the second-order derivatives. An example closely related to the theme of this book concerns the vanishing and large exponent behavior of power-type curvature flow equations, which play an important role in mathematical models of image denoising; see e.g. a book by F. Cao [82]. Consider the power-type curve shortening equation $V = \kappa^\alpha$ with $\alpha > 0$. For graph-like solutions, the equation becomes

$$u_t - (1 + u_x^2)^{\frac{1-3\alpha}{2}} |u_{xx}|^{\alpha-1} u_{xx} = 0.$$

Taking the formal limit as $\alpha \to 0$, we obtain

$$u_t - \sqrt{1 + u_x^2}\, \operatorname{sgn}(u_{xx}) = 0,$$

where $\operatorname{sgn}(a)$ denotes the sign of $a \in \mathbf{R}$. The standard theory of viscosity solutions does not apply directly to this limit equation due to its strong discontinuity caused by the sign function. Addressing such equations requires a suitable modification of the definition of viscosity solutions. We refer to a work by R. M. Chen and Q. Liu [101] for results on such parabolic equations in one dimensional case, and to a paper by Q. Liu [329] for further study of the associated level set equations in general dimensions.

Letting $\alpha \to \infty$ in the power-type curvature flow equation $V = \kappa^\alpha$ yields a nonlinear parabolic equation that exhibits a different type of discontinuity with respect to the second derivatives. In this case, we also need to suitably adapt the definition of viscosity solutions to ensure uniqueness and existence of solutions. See papers by Q. Liu and N. Yamada [337] and Q. Liu [330] for more details.

Chapter 4
Comparison Principle

The comparison principle is a key step in the theory of viscosity solutions. We present various versions of the comparison principle for singular degenerate equations which apply to level set equations as well as other equations.

4.1 Typical Statements

We consider an equation

$$u_t + F(z, u, \nabla u, \nabla^2 u) = 0 \tag{4.1.1}$$

in $Q = \Omega \times (0, T)$ for an open set Ω in $\mathbf{R}^N$ and $T > 0$. We list a typical set of assumptions which was fulfilled for degenerate parabolic level set equations in Chap. 2.

(F1) (Continuity) $F : W_0 = \overline{\Omega} \times [0, T] \times \mathbf{R} \times (\mathbf{R}^N \setminus \{0\}) \times \mathbf{S}^N \to \mathbf{R}$ is continuous.
(F2) (Degenerate ellipticity)

$$F(z, r, p, X) \leq F(z, r, p, Y) \quad \text{for } X \geq Y,\ X, Y \in \mathbf{S}^N$$

and $z \in \overline{\Omega} \times [0, T]$, $r \in \mathbf{R}$, $p \in \mathbf{R}^N \setminus \{0\}$.
For the class of level set equations (2.6.3) with (2.6.4) satisfying (2.6.24), which includes the level set mean curvature flow equations (2.6.5), the singularity of F at $p = 0$ is rather mild. To treat such equations it is reasonable to assume
(F3) $-\infty < F_*(z, r, 0, O) = F^*(z, r, 0, O) < \infty$.
In general (F3) is not fulfilled, for example for the level set Gaussian curvature flow equation (2.6.12) (see Sect. 2.6.5), a weaker assumption is necessary to treat such a problem.

Y. Giga, Q. Liu, *Surface Evolution Equations*, Monographs in Mathematics 99,
https://doi.org/10.1007/978-3-032-20303-8_4

(F3′) $\mathcal{F}(F) \neq \emptyset$.
Here $\mathcal{F}$ is defined in Sect. 3.1.3 and may depend on Ω and T if F depends on z. In this case we consider $\mathcal{F}_\Omega$-viscosity solutions instead of usual viscosity solutions. We often drop the prefix $\mathcal{F}$ unless confusion occurs. For level set equations F is independent of r but in general we need to assume monotonicity in r.

(F4) (Monotonicity). For some constant c_0,

$$r \mapsto F(z, r, p, X) + c_0 r$$

is a nondecreasing function.

4.1.1 Bounded Domains

When Ω is bounded, the comparison principle takes its naive form as in Chap. 3.

(BCP) Let u and v be sub- and supersolutions of (4.1.1) in $Q = \Omega \times (0, T)$, where Ω is bounded. If $-\infty < u^* \leq v_* < \infty$ on $\partial_p Q$, then $u^* \leq v_*$ in Q.

If $F(x, t, r, p, X)$ is independent of $x \in \Omega$, it is easy to state the conditions so that the comparison principle (BCP) holds.

Theorem 4.1.1 *Assume that Ω is bounded and that $F = F(x, t, r, p, X)$ is independent of x. Assume that* (F1)–(F3) *and* (F4) *hold. Then the comparison principle* (BCP) *is valid. If we assume* (F3′) *instead of* (F3), *then* (BCP) *is still valid (by replacing solutions by $\mathcal{F}_\Omega$-solutions) provided that $\mathcal{F}_\Omega$ is invariant under positive multiplication.*

Corollary 4.1.2 *Assume that Ω is bounded and that $F = F(x, t, r, p, X)$ is independent of x and r. Assume that $F : [0, T] \times (\mathbf{R}^N \setminus \{0\}) \times \mathbf{S}^N \to \mathbf{R}$ is continuous and degenerate elliptic. Then* (BCP) *holds if F is geometric.*

It is easy to see that Corollary 4.1.2 follows from Theorem 4.1.1 once we note:

Lemma 4.1.3 *Assume that $F : \overline{\Omega} \times [0, T] \times (\mathbf{R}^N \setminus \{0\}) \times \mathbf{S}^N \to \mathbf{R}$ is continuous and geometric, then $\mathcal{F}_\Omega = \mathcal{F}_\Omega(F)$ is nonempty provided that there exist $r_0 > 0$ and a continuous function $c \in C(0, r_0]$ such that*

$$|F(x, t, p, \pm I)| \leq c(|p|) \quad \text{on} \quad \overline{\Omega} \times [0, T] \times (B_{r_0}(0) \setminus \{0\}).$$

In particular, $\mathcal{F}_\Omega \neq \emptyset$ if Ω is bounded or F is independent of x. Moreover, $f \in \mathcal{F}_\Omega$ implies $af \in \mathcal{F}_\Omega$ for $a > 0$.

Proof of Lemma 4.1.3 We shall construct $f \in C^2[0,\infty)$ satisfying (3.1.11) with $f(0) = f'(0) = f''(0) = 0,\ f''(r) > 0$ for $r > 0$. We note that

$$\nabla_p(f(\rho)) = f'(\rho)p/\rho,\ \nabla_p^2(f(\rho)) = f''(\rho)\frac{p \otimes p}{\rho^2} + f'(\rho)\frac{1}{\rho}\left(I - \frac{p \otimes p}{\rho^2}\right)$$

with $\rho = |p| > 0$. Since F is geometric, we see

$$\begin{aligned} F(x,t,\nabla(f(\rho)),\ \pm\nabla^2(f(\rho))) &= F(x,t,f'(\rho)\frac{p}{\rho}, \pm f'(\rho)\frac{I}{\rho}) \\ &= \frac{f'(\rho)}{\rho}F(x,t,p,\pm I),\ \ \rho = |p|. \quad (4.1.2) \end{aligned}$$

By assumption we have

$$|F(x,t,p,\pm I)| \le c(|p|) \quad \text{on } \overline{\Omega} \times [0,T] \times (B_{r_0}(0) \setminus \{0\}).$$

We may assume that $c \in C^1(0, r_0]$ with $c > 0$ and that

$$(1/c)' > 0 \quad \text{in } (0, r_0],\ \lim_{r \downarrow 0} c(r) = \infty, \quad \text{and } \lim_{r \downarrow 0}(1/c)'(r) = 0.$$

Then $f : [0, r_0] \to \mathbf{R}$ defined by

$$f(r) = \begin{cases} \int_0^r \frac{s^2}{c(s)} ds & 0 < r \le r_0, \\ 0 & r = 0 \end{cases}$$

satisfies $f(0) = f'(0) = f''(0) = 0$ with $f''(r) > 0$ for $r > 0$. By definition of c we now observe that

$$\begin{aligned} |F(x,t,\nabla_p(f(\rho)),\ \pm\nabla_p^2(f(\rho)))| &= \frac{f'(\rho)}{\rho}|F(x,t,p,\pm I)| \quad (4.1.3) \\ &\le \frac{f'(\rho)}{\rho}c(\rho) = \rho \to 0 \quad \text{as } \rho \to 0. \end{aligned}$$

We extend f to $(0,\infty)$ in an appropriate way to get $f \in \mathcal{F}_\Omega(F)$.

If F is independent of x or Ω is bounded,

$$\begin{aligned} C(\rho) = \max\ \{&\sup\{F(x,t,p,I);\ x \in \overline{\Omega}, t \in [0,T], |p| = \rho\}, \\ &\sup\{F(x,t,p,-I);\ x \in \overline{\Omega}, t \in [0,T], |p| = \rho\}\} \end{aligned}$$

is continuous in $(0,\infty)$ so the above argument implies that $\mathcal{F}_\Omega(F)$ is nonempty.

The property that $f \in \mathcal{F}_\Omega$ implies $af \in \mathcal{F}_\Omega$ is clear since

$$F(x, t, \lambda p, \lambda X) = \lambda F(x, t, p, X)$$

for $\lambda > 0$, $p \in \mathbf{R}^N \setminus \{0\}$, $X \in \mathbf{S}^N$, $x \in \Omega$, $t \in [0, T]$. □

The proof of Theorem 4.1.1 is postponed to Sect. 4.4.

4.1.2 General Domains

For a general domain Ω we replace the comparison principle for a bounded domain (BCP) by (CP) stated below.

(CP) Let u and v be sub- and supersolutions of (4.1.1) in $Q = \Omega \times (0, T)$. Assume that u and $-v$ are bounded from above on Q. Assume that

$$\begin{aligned} \lim_{\delta \to 0} \sup\{u^*(x,t) - v_*(y,s);\ |x-y| \le \delta,\ |t-s| \le \delta, \\ \mathrm{dist}((x,t), \partial_p Q) \le \delta, \mathrm{dist}((y,s), \partial_p Q) \le \delta, \\ (x,t), (y,s) \in \overline{\Omega} \times [0, T']] \le 0 \end{aligned} \tag{4.1.4}$$

for each $T' \in (0, T)$ and that $u^* > -\infty$, $v_* < \infty$ on $\partial_p Q$. Then

$$\begin{aligned} \lim_{\delta \to 0} \sup\{u^*(x,t) - v_*(y,s);\ |x-y| \\ \le \delta,\ |t-s| \le \delta, (x,t), (y,s) \in \overline{\Omega} \times [0, T']\} \le 0 \end{aligned} \tag{4.1.5}$$

for each $T' \in (0, T)$.

If Ω is bounded, it is easy to check that $u^* \le v_*$ on $\partial_p Q$ is equivalent to (4.1.4) and $u^* \le v_*$ in $\overline{Q}$ is equivalent to (4.1.5). Thus (CP) and (BCP) is the same when Ω is bounded. Intuitively, (4.1.5) asserts the uniformity of $u^* \le v_*$.

Theorem 4.1.4 *Assume that $F(x, t, r, p, X)$ is independent of x. Assume that* (F1)–(F3) *and* (F4) *hold. Then the comparison principle* (CP) *is valid. If we assume* (F3′) *instead of* (F3), *then* (CP) *is still valid (by replacing solutions by $\mathcal{F}_\Omega$-solutions) provided that $\mathcal{F}_\Omega$ is invariant under positive multiplications.*

Corollary 4.1.5 *Assume that $F = F(x, t, r, p, X)$ is independent of x and r. Assume that $F : [0, T] \times (\mathbf{R}^N \setminus \{0\}) \times \mathbf{S}^N \to \mathbf{R}$ is continuous and degenerate elliptic. Then* (CP) *holds if F is geometric.*

Again Corollary 4.1.5 follows from Theorem 4.1.4 together with Lemma 4.1.3. We shall give a proof of Theorem 4.1.4 in Sect. 4.4.2.

4.1.3 Applicability

In Sect. 2.6 we have studied various properties of level set equations of surface evolution equations (2.6.1). Corollaries 4.1.2 and 4.1.5 apply to the level set equation (2.6.3) of (2.6.1) provided that:

(i) f in (2.6.1) is continuous in its variables; more precisely f is continuous on $\Omega \times [0, T] \times \mathbf{E}$.
(ii) f is independent of x.
(iii) (2.6.1) is degenerate parabolic.

Indeed, (F1) for F_f (defined by (2.6.16)) follows from (i) as in Proposition 2.6.15. The ellipticity of F_f follows from the definition of (iii); see (2.6.21). Geometricity follows from Sect. 2.6.2. The condition (F3) holds if the growth of f with respect to $\nabla \mathbf{n}$ is at most linear (Proposition 2.6.18).

We list examples of (2.6.1) satisfying (i), (ii), (iii) for the reader's convenience.

Equation (2.5.2) fulfills (i)–(iii) provided that $\beta > 0$ is continuous and γ in (2.3.7) is convex and C^2 except at the origin and that $a \geq 0$ and c is independent of x and is continuous on $[0, T]$. In particular, the mean curvature flow equation (2.5.4) and a Hamilton–Jacobi equation (2.5.6) evidently fulfill (i)–(iii) provided that c in (2.5.6) is independent of x and is continuous on $[0, T]$. The level set equations of these equations also fulfill (F3). Under the same assumption on γ, a, c as above, Eq. (2.5.13) fulfills (i)–(iii) provided that h is continuous and nondecreasing. Its level set equation fulfills (F3) if and only if growth of h is at most linear.

Equation (2.5.8) fulfills (i)–(iii) provided that g is continuous and that (2.5.8) is degenerate parabolic. The Gaussian curvature flow equation (2.5.9) and other related equations (2.5.10), (2.5.11), (2.5.12) fulfills (i)–(iii) provided that e_m is interpreted as $\hat{e}_m$ (Theorem 2.6.10). In other words the modified form (2.6.22), (2.6.23) fulfills (i)–(iii). Here (F3) is in general not expected to hold for the level set equation.

Another typical example, which is not geometric but Theorems 4.1.1 and 4.1.4 still apply, is the p-Laplace equation of parabolic type:

$$u_t - \operatorname{div}(|\nabla u|^{p-2} \nabla u) = 0$$

for $1 < p < 2$ as proved by M. Ohnuma and K. Sato [368].

4.2 Alternate Definition of Viscosity Solutions

4.2.1 Definition Involving Semijets

We recall the notion of the second-order semijets of a function which plays a role of derivatives up to the second-order in usual calculus. Semijets are infinitesimal quantities. We shall give an equivalent definition of viscosity solutions by using semijets. Such an infinitesimal interpretation of viscosity solutions is useful to prove comparison principles.

Definition 4.2.1 (Semijets) Let $\mathcal{O}$ be a locally compact subset of $\mathbf{R}^d$. Let $u : \mathcal{O} \to \mathbf{R} \cup \{-\infty\}$ be upper semicontinuous. Let $\hat{z}$ be a point in $\mathcal{O}$. An element $(q, Z) \in \mathbf{R}^d \times \mathbf{S}^d$ is called (the second-order) *superjet* of u at $\hat{z}$ in $\mathcal{O}$ if $u(\hat{z})$ is finite and

$$u(z) - u(\hat{z}) \le \langle q, z - \hat{z}\rangle + \frac{1}{2}\langle Z(z - \hat{z}), z - \hat{z}\rangle + o(|z - \hat{z}|^2) \tag{4.2.1}$$

for $z \in \mathcal{O}$ as $z \to \hat{z}$; here $o(h)$ denotes a function such that $o(h)/h \to 0$ as $h \to 0$. The set of all superjets of u at $\hat{z}$ in $\mathcal{O}$ is denoted by $J^{2,+}_{\mathcal{O}} u(\hat{z}) (\subset \mathbf{R}^d \times \mathbf{S}^d)$. For a lower semicontinuous function $u : \mathcal{O} \to \mathbf{R} \cup \{+\infty\}$ an element $(q, Z) \in \mathbf{R}^d \times \mathbf{S}$ is called a *subjet* of u at $\hat{z}$ in $\mathcal{O}$ if $(-q, -Z) \in J^{2,+}_{\mathcal{O}}(-u)(\hat{z})$. The set of all subjets of u at $\hat{z}$ in $\mathcal{O}$ is denoted by $J^{2,-}_{\mathcal{O}} u(\hat{z})$, so that $J^{2,-}_{\mathcal{O}} u(\hat{z}) = -J^{2,+}_{\mathcal{O}}(-u)(\hat{z})$. By a semijet we mean either a superjet or a subjet. Here $\langle , \rangle$ denotes the inner product in $\mathbf{R}^d$.

The set $J^{2,\pm}_{\mathcal{O}} u(\hat{z})$ does not vary even if $\mathcal{O}$ is replaced by a neighborhood $\mathcal{O}'$ of $\hat{z}$ in $\mathcal{O}$. In particular, $J^{2,\pm}_{\mathcal{O}} u(\hat{z})$ is independent of $\mathcal{O}$ if $\hat{z}$ is an interior point of $\mathcal{O}$, i.e., $\hat{z} \in \operatorname{int} \mathcal{O}$. In this case we often suppress the subscript $\mathcal{O}$ of $J^{2,\pm}_{\mathcal{O}} u(\hat{z})$ and simply write it by $J^{2,\pm} u(\hat{z})$. In general $J^{2,\pm}_{\mathcal{O}} u(\hat{z})$ may depend on $\mathcal{O}$. Indeed, if we consider $u \equiv 0$ in $\mathbf{R}$ we see

$$\begin{aligned} J^{2,+}_{\mathbf{R}} u(0) &= \{(0, Z);\ Z \ge 0\}, \\ J^{2,+}_{[0,\infty)} u(0) &= \{(q, Z);\ Z \in \mathbf{R},\ q > 0\} \cup \{(0, Z);\ Z \ge 0\}. \end{aligned}$$

By definition a function u has Taylor's expansion up to second order at $z \in \operatorname{int} \mathcal{O}$ if and only if

$$(J^{2,+} u)(\hat{z}) \cap (J^{2,-} u)(\hat{z}) \ne \emptyset.$$

(Actually, the intersection must be singleton if it is nonempty.) In particular, if u is C^2 at $\hat{z} \in \operatorname{int} \mathcal{O}$, then

$$(J^{2,+} u)(\hat{z}) \cap (J^{2,-} u)(\hat{z}) = \{(Du(\hat{z}),\ D^2 u(\hat{z}))\},$$

which is the set of the second-order jets of u at $\hat{z}$. Of course, the set $(J^{2,\pm}_{\mathcal{O}} u)(\hat{z})$ could be empty even if $u(\hat{z})$ is finite but this defines a mapping $J^{2,\pm}_{\mathcal{O}} u(\hat{z})$ from $\mathcal{O}$ to the set of all subsets of $\mathbf{R}^d \times \mathbf{S}^d$. Although the set $J^{2,\pm} u(\hat{z})$ could be empty for some $\hat{z}$, for generic $\hat{z}$, $J^{2,\pm}_{\mathcal{O}} u(\hat{z})$ is nonempty.

Lemma 4.2.2 *The set of z at which $J^{2,+}_{\mathcal{O}} u(\hat{z})$ is nonempty is dense in $\mathcal{O}' = \mathcal{O} \backslash \{z \in \mathcal{O};\ u(z) = -\infty\}$. The assertion is still valid if $J^{2,+}_{\mathcal{O}} u$ is replaced by $J^{2,-}_{\mathcal{O}} u$.*

Proof For a point $\hat{z} \in \mathcal{O}'$ we take a closed ball $B_r(\hat{z})$ of radius r centered at $\hat{z}$ in $\mathbf{R}^d$ such that $\mathcal{O} \cap B_r(\hat{z})$ is compact. We take an upper semicontinuous function φ on $\mathcal{O}$ which is C^2 in $\mathrm{int}B_r(\hat{z})$ and observe that

$$
\begin{aligned}
&J_{\mathcal{O}}^{2,\pm}(u+\varphi)(z) = J_{\mathcal{O}}^{2,\pm}u(z) + \{(D\varphi(z), D^2\varphi(z))\} \\
&:= \{(q + D\varphi(z), Z + D^2\varphi(z)); (q, Z) \in J_{\mathcal{O}}^{2,\pm}u(z)\}, \quad \text{for all } z \in \mathcal{O} \cap \mathrm{int}B_r(\hat{z}).
\end{aligned}
$$

We arrange $\varphi \equiv -\infty$ outside $B_r(\hat{z})$ and $-\varphi$ large near $\partial B_r(\hat{z})$ so that $u + \varphi$ attains its maximum over $\mathcal{O}$ at some point $z_r \in \mathcal{O} \cap \mathrm{int}B_r(\hat{z})$. By definition

$$
\{(0, Z); Z \geq 0\} \subset J_{\mathcal{O}}^{2,+}(u+\varphi)(z_r)
$$

so $J_{\mathcal{O}}^{2,+}u(z_r)$ is not empty. We may take r as small as we like. This implies that $\{z \in \mathcal{O}; J_{\mathcal{O}}^{2,+}(z) \neq \emptyset\}$ is dense in $\mathcal{O}'$. The same proof works for $J_{\mathcal{O}}^{2,-}u$ if we replace $+$ by $-$. □

For evolution equations (4.1.1) it is convenient to consider special components of semijets.

Definition 4.2.3 (Parabolic Semijets) Let u be an upper semicontinuous function from a locally compact subset $\mathcal{O}$ in $\mathbf{R}^N \times \mathbf{R}$ to $\mathbf{R} \cup \{-\infty\}$. Let $\hat{z} = (\hat{x}, \hat{t}) \in \mathbf{R}^N \times (0, \infty)$ be a point in $\mathcal{O}$. An element $(\tau, p, X) \in \mathbf{R} \times \mathbf{R}^N \times \mathbf{S}^N$ is called a *parabolic superjet* of u at $\hat{z}$ in $\mathcal{O}$ if

$$
\begin{aligned}
u(z) - u(\hat{z}) \leq \tau(t - \hat{t}) &+ \langle p, x - \hat{x}\rangle + \frac{1}{2}\langle X(x - \hat{x}), x - \hat{x}\rangle \\
&+ o(|x - \hat{x}|^2 + |t - \hat{t}|), \; z = (x, t) \in \mathcal{O}
\end{aligned}
\tag{4.2.2}
$$

as $z \to \hat{z}$, where $\langle , \rangle$ denotes the inner product in $\mathbf{R}^N$. The totality of parabolic superjets of u at $\hat{z}$ in $\mathcal{O}$ is denoted $\mathcal{P}_{\mathcal{O}}^{2,+}u(\hat{z})$. For a lower semicontinuous function $u : \mathcal{O} \to \mathbf{R} \cup \{+\infty\}$, the $-\mathcal{P}_{\mathcal{O}}^{2,+}(-u)(\hat{z})$ is denoted $\mathcal{P}_{\mathcal{O}}^{2,-}u(\hat{z})$ and its element is called a *parabolic subjet* of u at $\hat{z}$ in $\mathcal{O}$.

Proposition 4.2.4 *For $(p, \tau) \in \mathbf{R}^N \times \mathbf{R}$, $a \in \mathbf{R}$, $\ell \in \mathbf{R}^N$, $X \subset \mathbf{S}^N$ if*

$$
\left(\begin{pmatrix} p \\ \tau \end{pmatrix}, \begin{pmatrix} X & \ell \\ {}^t\ell & a \end{pmatrix} \right) \in J_{\mathcal{O}}^{2,+}u(\hat{x}, \hat{t}),
$$

then $(\tau, p, X) \in \mathcal{P}_{\mathcal{O}}^{2,+}u(\hat{x}, \hat{t})$, where $\mathcal{O}$ is a locally compact subset of $\mathbf{R}^N \times \mathbf{R}$ and ℓ denotes the transpose of a matrix ℓ. Here p and ℓ are column vectors.

Proof By definition

$$\begin{aligned}
&u(x,t)-u(\hat{x},\hat{t})\\
&\le \tau(t-\hat{t})+\langle p,x-\hat{x}\rangle+\tfrac{1}{2}a(t-\hat{t})^2\\
&\quad +(t-\hat{t})\langle \ell,x-\hat{x}\rangle+\tfrac{1}{2}\langle X(x-\hat{x}),x-\hat{x}\rangle+o(|t-\hat{t}|^2+|x-\hat{x}|^2)
\end{aligned}$$

as $(x,t)\to(\hat{x},\hat{t})$, $(x,t)\in\mathcal{O}$. We estimate the mixed term $(t-\hat{t})\langle \ell,x-\hat{x}\rangle$ by Young's inequality to get

$$\begin{aligned}
(t-\hat{t})\langle \ell,x-\hat{x}\rangle &\le |\ell|\,|t-\hat{t}|\,|x-\hat{x}|\le |\ell|(\frac{2}{3}|t-\hat{t}|^{3/2}+\frac{1}{3}|x-\hat{x}|^3)\\
&=o(|t-\hat{t}|+|x-\hat{x}|^2).
\end{aligned}$$

We thus observe that $(\tau,p,X)\in\mathcal{P}^{2,+}_{\mathcal{O}}u(\hat{x},\hat{t})$. □

We next recall the definition of the 'closures' of the set-valued mappings which is important to study the comparison principle. We set

$$\begin{aligned}
&\overline{J}^{2,+}_{\mathcal{O}}u(\hat{z})=\{(q,Z)\in\mathbf{R}^d\times\mathbf{S}^d;\quad \text{there is a sequence}\\
&(z_j,q_j,Z_j)\in\mathcal{O}\times\mathbf{R}^d\times\mathbf{S}^d\quad (j=1,2,\ldots)\quad \text{such that}\\
&(q_j,Z_j)\in J^{2,+}_{\mathcal{O}}u(z_j)\text{ and }(z_j,u(z_j),q_j,Z_j)\to(\hat{z},u(\hat{z}),q,Z)\text{ as }j\to\infty\},\\
&\overline{J}^{2,-}_{\mathcal{O}}u(\hat{z}):=-(\overline{J}^{2,+}_{\mathcal{O}}(-u))(\hat{z}).
\end{aligned}$$

These definitions are a little bit different from the standard closure of the set-valued mappings defined by the closure of the graph of the mapping $z\mapsto J^{2,+}_{\mathcal{O}}u(z)$ since there is the extra condition of convergence $u(z_j)\to u(\hat{z})$ $(j\to\infty)$. (In fact,

$$z\mapsto\left\{\left(u(z),q,Z\right);\,(q,Z)\in\overline{J}^{2,+}_{\mathcal{O}}u(z)\right\}$$

is the closure of the mapping

$$z\mapsto\left\{\left(u(z),q,Z\right)\,;\,\,(q,Z)\in J^{2,+}_{\mathcal{O}}u(z)\right\}.)$$

Remark 4.2.5 Clearly, the statement of Proposition 4.2.4 is still valid even if we replace $J^{2,+}_{\mathcal{O}}$ by $\overline{J}^{2,+}_{\mathcal{O}}$ and $\mathcal{P}^{2,+}_{\mathcal{O}}$ by $\overline{\mathcal{P}}^{2,-}_{\mathcal{O}}$ which is defined in Remark 4.2.9.

Proposition 4.2.6 (Infinitesimal Version of Definitions of Viscosity Solutions)

(i) *Let $\mathcal{O}$ be a locally compact subset of $\mathbf{R}^d$. Let E be defined in a dense subset of $\overline{\mathcal{O}}\times\mathbf{R}\times\mathbf{R}^d\times\mathbf{S}^d$ with the property $E_*<+\infty$ on $\overline{\mathcal{O}}\times\mathbf{R}\times\mathbf{R}^d\times\mathbf{S}^d$. A function $u:\mathcal{O}\to\mathbf{R}\cup\{-\infty\}$ (satisfying $u^*(z)<\infty$ for all $z\in\mathcal{O}$) is a subsolution of (3.3.1) in $\mathcal{O}$ if and only if*

$$E_*(\hat{z}, u^*(\hat{z}), q, Z) \le 0 \tag{4.2.3}$$

for all $\hat{z} \in \mathcal{O}$ *and* $(q, Z) \in J^{2,+}_{\mathcal{O}} u^*(\hat{z})$.

(ii) *Assume that* F *satisfies* (F1) *and* (F3′). *A function* $u : \mathcal{O} \to \mathbf{R} \cup \{-\infty\}$ *(satisfying* $u^*(z) < \infty$ *for all* $z \in \mathcal{O}$ *) is an* $\mathcal{F}$*-subsolution of* (4.1.1) *in* $\mathcal{O}$ *if and only if the following two conditions are fulfilled:*

(a)

$$\tau + F(\hat{x}, \hat{t}, u^*(z), p, X) \le 0 \tag{4.2.4}$$

for all $(\tau, p, X) \in \mathcal{P}^{2,+}_{\mathcal{O}} u(\hat{x}, \hat{t})$ *unless* $p = 0$.

(b) $\varphi_t(\hat{z}) \le 0$ *for all* $(\varphi, \hat{z}) \in C^2_F(Q') \times Q'$ *satisfying* (3.1.6), *for some neighborhood* Q' *of* $\hat{z}$ *provided that* $\mathcal{F}_\Omega$ *is invariant under positive multiplication.*

Here $\mathcal{O}$ *is an open set in* $\Omega \times (0, T)$, *where* Ω *is an open set in* $\mathbf{R}^N$.

These statements are easily verified by the following characterization of semijets; for (ii) we also invoke a localization property (Remark 3.1.10) and Proposition 3.2.3 (ii).

Lemma 4.2.7

(i) *For* $\hat{z} \in \mathcal{O}$,

$$J^{2,+}_{\mathcal{O}} u^*(\hat{z}) = \{(D\varphi(\hat{z}), D^2\varphi(\hat{z}));\ \ \varphi \in C^2(\mathcal{O}) \text{ that satisfies } \max_{\mathcal{O}}(u^* - \varphi) = (u^* - \varphi)(\hat{z})\}.$$

(ii) *For* $(\hat{x}, \hat{t}) \in Q$,

$$\mathcal{P}^{2,+}_{\mathcal{O}} u^*(\hat{x}, \hat{t}) = \{(\varphi_t(\hat{x}, \hat{t}), \nabla\varphi(\hat{x}, \hat{t}), \nabla^2\varphi(\hat{x}, \hat{t})); \varphi \in C^{2,1}(Q) \text{ that satisfies } \max_Q(u^* - \varphi) = (u^* - \varphi)(\hat{x}, \hat{t})\}.$$

(iii) *In* (i) *and* (ii) $\mathcal{O}$ *and* Q *may be replaced by an (open) neighborhood of* $\mathcal{O}'$ *of* $\hat{z}$ *in* $\mathcal{O}$ *and* Q' *of* $(\hat{x}, \hat{t})$ *in* Q, *respectively. In particular, for* $(\tau, p, X) \in \mathcal{P}^{2,+}_{\mathcal{O}} u^*(\hat{x}, \hat{t})$, $p \ne 0$ *there is always a neighborhood* Q' *of* $(\hat{x}, \hat{t})$ *in* Q *and* $\varphi \in C^2(\mathcal{O}')$ *that satisfies*

$$(\tau, p, X) = (\varphi_t(\hat{x}, \hat{t}), \nabla\varphi(\hat{x}, \hat{t}), \nabla^2\varphi(\hat{x}, \hat{t}))$$

and $\nabla\varphi(z) \ne 0$ *for all* $z \in Q'$.

Proof

(i) Let J denote the right-hand side of the equality. We may assume $\hat{z} = 0$ by translation. If $u^* - \varphi$ takes its maximum at $\hat{z} = 0$, then

$$u^*(z) - u^*(0) \le \varphi(z) - \varphi(0) = \langle D\varphi(0), z\rangle + \frac{1}{2}\langle D^2\varphi(0)z, z\rangle$$
$$+o(|z|^2) \quad \text{as } z \to 0$$

by Taylor's expansion. Thus $J_{\mathcal{O}}^{2,+}u^*(0) \supset J$.

The other side inclusion is less trivial. Assume now that $(q, Z) \in J_{\mathcal{O}}^{2,+}u^*(0)$, i.e.,

$$u^*(z) - u^*(0) \le \langle q, z\rangle + \frac{1}{2}\langle Zz, z\rangle + \varepsilon(z), \ z \in \mathcal{O}$$

where $\varepsilon(z)/|z|^2 \to 0$ as $z \to 0$. The problem is that $\varepsilon(z)$ itself is not C^2 at all. We set

$$\omega_0(\sigma) = \sup\{|\varepsilon(z)|/|z|^2;\ |z| \le \sigma, z \in \mathcal{O}\},$$

so that ω_0 is a nondecreasing function from $[0, \infty)$ to $[0, \infty)$ with the property that ω_0 is continuous at $\sigma = 0$ and $\omega_0(0) = 0$. By Lemma 3.1.9 there is $\theta \in C^2[0, \infty)$ that satisfies

$$\omega_0(\sigma)|\sigma|^2 \le \theta(\sigma)$$
$$\text{for } \sigma \ge 0, \theta(0) = \theta'(0) = \theta''(0) = 0, \theta'''(\sigma) \ge 0 \quad \text{for } \sigma \ge 0.$$

Since $|\varepsilon(z)| \le \omega_0(|z|)|z|^2$, we set a C^2 function by

$$\varphi(z) = \langle q, z\rangle + \frac{1}{2}\langle Zz, z\rangle + \theta(|z|)$$

to observe that $u^* - \varphi$ takes its maximum $u^*(0) - \varphi(0)$ over $\mathcal{O}$ at $\hat{z} = 0$ and that $q = D\varphi(0)$, $Z = D^2\varphi(0)$. Thus $J_{\mathcal{O}}^{2,+}u^*(0) \subset J$.

(ii) Let $\mathcal{P}$ denote the right-hand side of the equality. By Proposition 4.2.4 and $J \subset J_{\mathcal{O}}^{2,+}u^*(\hat{z})$ the inclusion $\mathcal{P} \subset \mathcal{P}_{\mathcal{O}}^{2,+}u^*(\hat{x}, \hat{t})$ is trivial.

To see the other inclusion we may assume $(\hat{x}, \hat{t}) = (0, 0)$. Assume that $(\tau, p, X) \in \mathcal{P}_{\mathcal{O}}^{2,+}u^*(0)$, i.e.,

$$u^*(x, t) - u^*(0) \le \langle \tau, t\rangle + \langle p, x\rangle + \frac{1}{2}\langle Xx, x\rangle + \varepsilon(x, t), (x, t) \in Q$$

where $\varepsilon(x, t)/(|x|^2 + |t|) \to 0$ as $|x| \to 0$, $|t| \to 0$. As for (i) we set

$$\omega_0(\sigma) = \sup\{|\varepsilon(x, t)/(|x|^2 + |t|)|;\ |x|^2 + |t| \le \sigma,\ (x, t) \in Q\}$$

and observe by Lemma 3.1.9 there is $\theta \in C^2[0, \infty)$ satisfying the same property as in (i). Since

$$|\varepsilon(x, t)| \le \omega_0((|x|^2 + |t|)^{1/2})(|x|^2 + |t|),$$

we set

$$\varphi(x, t) = \langle \tau, t \rangle + \langle p, x \rangle + \frac{1}{2}\langle Xx, x \rangle + \theta((|x|^2 + |t|)^{1/2})$$

(which is actually $C^{2,1}$ everywhere) to observe that $u^* - \varphi$ takes its maximum over Q at $(\hat{x}, \hat{t}) = 0$ and that $\tau = \varphi_t(0)$, $p = \nabla\varphi(0)$, $X = \nabla^2\varphi(0)$. Thus $\mathcal{P}_Q^{2,+} u^*(0) \subset \mathcal{P}$.

(iii) This follows from the proof of parts (i) and (ii).

□

Remark 4.2.8 Corresponding to Proposition 4.2.6 we have an infinitesimal version of a definition of viscosity supersolutions in a symmetric way, since we have a characterization of $J^{2,-}$ and $\mathcal{P}^{2,-}$ corresponding to Lemma 4.2.7 by replacing max by min. Proposition 4.2.6 (ii) applies boundary value problems by taking $\mathcal{O}$ by $\overline{\Omega} \times (0, T)$. It is easy to extend Proposition 4.2.6 (ii) to boundary value problems for $\mathcal{F}$-solutions.

Remark 4.2.9 (Closures of Semijets) If u is a subsolution of (3.3.1) so that (4.2.3) holds, then (4.2.3) still holds for all

$$(q, Z) \in \overline{J}_{\mathcal{O}}^{2,+} u^*(\hat{z}), \ \hat{z} \in \mathcal{O}$$

since E_* is lower semicontinuous. The same remark applies (ii) of Proposition 4.2.6. If (a) holds, then (4.2.4) still holds for $(\tau, p, X) \in \overline{\mathcal{P}}_{\mathcal{O}}^{2,+} u^*(\hat{x}, \hat{t})$ if $(\hat{x}, \hat{t}) \in Q$ since $F(x, t, r, p, X)$ is continuous outside the set where $p = 0$. The set $\overline{\mathcal{P}}^{2,+}$ is defined in an analogous way to $\overline{J}^{2,+}$:

$$\begin{aligned}
&\overline{\mathcal{P}}_{\mathcal{O}}^{2,+} u(\hat{z}) = \{(\tau, p, X) \in \mathbf{R} \times \mathbf{R}^N \times \mathbf{S}^N;\ \text{there is a} \\
&\text{sequence } (z_j, \tau_j, p_j, X_j) \in Q \times \mathbf{R} \times \mathbf{R}^N \times \mathbf{S}^N \quad (j = 1, 2, \ldots) \\
&\text{such that } (\tau_j, p_j, X_j) \in \mathcal{P}_{\mathcal{O}}^{2,+} u(z_j) \quad \text{and} \\
&(z_j, u(z_j), \tau_j, p_j, X_j) \to (\hat{z}, u(\hat{z}), \tau, p, X) \\
&\text{as } j \to \infty\}. \\
&\overline{\mathcal{P}}_{\mathcal{O}}^{2,-} u(\hat{z}) = -\overline{\mathcal{P}}_{\mathcal{O}}^{2,+}(-u)(\hat{z}).
\end{aligned}$$

4.2.2 *Solutions on Semiclosed Time Intervals*

When we study an evolution equation (4.1.1), it is sometimes convenient to consider solutions in $Q_* = \Omega \times (0, T]$ instead of $Q = \Omega \times (0, T)$, where Ω is an open set in $\mathbf{R}^N$.

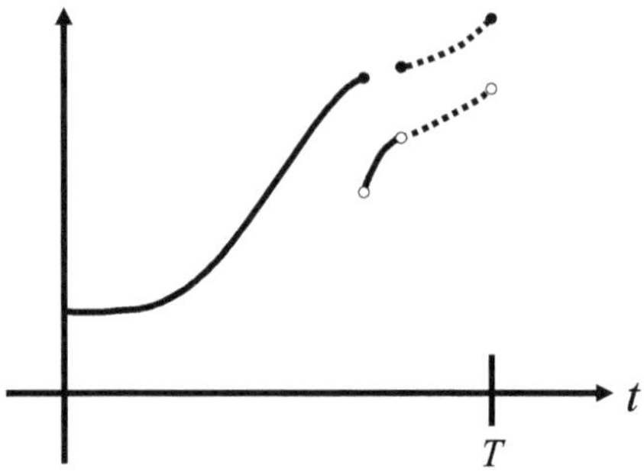

Fig. 4.1 Left accessible function at T

Theorem 4.2.10 (Extension)

(i) *Assume that F is lower semicontinuous in $W = \overline{\Omega} \times [0, T] \times \mathbf{R} \times \mathbf{R}^N \times \mathbf{S}^N$ with values in $\mathbf{R} \cup \{-\infty\}$. Let u be a subsolution of* (4.1.1) *in Q. Then its upper semicontinuous envelope u^* defined in Q_* is a subsolution of* (4.1.1) *in Q_* provided that $u^*(x, T) < \infty$ for all $x \in \Omega$.*

(ii) *Assume that F fulfills* (F1) *and* (F3′) *and that $a\mathcal{F}_\Omega \subset \mathcal{F}_\Omega$ for all $a > 0$. Let u be an $\mathcal{F}_\Omega$-subsolution of* (4.1.1) *in Q. Then u^* is an $\mathcal{F}_\Omega$-subsolution of* (4.1.1) *in Q_* provided that $u^*(x, T) < \infty$ for all $x \in \Omega$.*

Definition 4.2.11 (Left Accessibility) Let (y_0, t_0) be a point in $\mathbf{R}^m \times \mathbf{R}$. A function w defined in $B_r(y_0) \times (t_0 - \delta, t_0]$ is called *left accessible* at (y_0, t_0) if there are sequences $y_\ell \to y_0$, $t_\ell \to t_0 (\ell \to \infty)$ such that $t_\ell < t_0$ and $\lim_{\ell \to \infty} w(y_\ell, t_\ell) = w(y_0, t_0)$.

Remark 4.2.12 We note that by definition our extended function u^* in Q_* is always left accessible at any point (x, T), $x \in \Omega$ (Fig. 4.1). As we state later it turns out that u^* is left accessible at any (x_0, t_0), $t_0 < T$ provided that u is a subsolution (under extra assumptions on F). However, it is clear that u^* for a general function u in Q may not be left accessible at a point (x_0, t_0) for $t_0 < T$.

As usual, the symmetric statement corresponding to Theorem 4.2.10 holds for supersolutions. (The same remark applies following lemmas and its corollary in this section.)

Proof of Theorem 4.2.10

(i) For $\varphi \in C^2(Q_*)$ let $(\hat{x}, \hat{t})$ be maximizer of $u^* - \varphi$ over Q_*. We may assume that $\hat{t} = T$ and $u^* - \varphi$ attains its strict maximum at $(\hat{x}, T)$. Since u^* is left accessible at any point (x, T), $x \in \Omega$, it is easy to see that

$$\limsup_{\alpha \to \infty}{}^* (u^* - \varphi_\alpha) = u^* - \varphi \quad \text{on } Q_*$$

for $\varphi_\alpha(x, t) = \varphi(x, t) + 1/(\alpha(T - t))$, $\alpha > 0$. Let (x_α, t_α) be a maximizer of $u^* - \varphi_\alpha$ over Q_*. Since $\varphi_\alpha = +\infty$ at $t = T$, we see $t_\alpha < T$. By the convergence of maximum points (Sect. 3.2.2) $(x_\alpha, t_\alpha) \to (\hat{x}, T)$ and

$u^*(x_\alpha, t_\alpha) - \alpha/(T - t_\alpha) \to u^*(\hat{x}, T)$ as $\alpha \to \infty$ (without taking subsequences since the convergence of $u^* - \varphi_\alpha$ is monotone). Since $\overline{\lim}_{\alpha\to\infty} u^*(x_\alpha, t_\alpha) \le u^*(\hat{x}, T)$, we now have $u^*(x_\alpha, t_\alpha) \to u(\hat{x}, T)$ as $\alpha \to \infty$, where $\overline{\lim}$ denotes lim sup as usual. Since u is a subsolution in Q,

$$\frac{\partial \varphi_\alpha}{\partial t}(x_\alpha, t_\alpha) + F(x_\alpha, t_\alpha, u^*(x_\alpha, t_\alpha), \nabla\varphi(x_\alpha, t_\alpha), \nabla^2\varphi(x_\alpha, t_\alpha)) \le 0. \tag{4.2.5}$$

Since $\partial\varphi_\alpha/\partial t > \partial\varphi/\partial t$, sending $\alpha \to \infty$ yields

$$\frac{\partial \varphi}{\partial t}(\hat{x}, T) + F(\hat{x}, T, u^*(\hat{x}, T), \nabla\varphi(\hat{x}, T), \nabla^2\varphi(\hat{x}, T)) \le 0 \tag{4.2.6}$$

by lower semicontinuity of F.

(ii) The basic idea of the proof is the same as for (i). We may assume that Ω is bounded and $\mathcal{F}_\Omega \ne \emptyset$ and $a\mathcal{F}_\Omega \subset \mathcal{F}_\Omega$ for $a > 0$. We take $\varphi \in C^2_F(Q_*)$ and argue in the same way to get (4.2.5) if $\nabla\varphi(x_\alpha, t_\alpha) \ne 0$ and

$$\frac{\partial \varphi_\alpha}{\partial t}(x_\alpha, t_\alpha) < 0 \quad \text{if } \nabla\varphi(x_\alpha, t_\alpha) = 0 \tag{4.2.7}$$

since $\varphi_\alpha \in C^2_F(Q_*)$. If $\nabla\varphi(\hat{x}, T) \ne 0$, we get (4.2.6) from (4.2.5) so we may assume that $\nabla\varphi(\hat{x}, T) = 0$. Our goal is now to prove

$$\varphi_t(\hat{x}, T) \le 0. \tag{4.2.8}$$

For φ we take ψ as in (3.2.7) so that $\psi(\cdot, t) \in \mathcal{F}_\Omega$ is of the form

$$\psi(x, t) = \varphi(\hat{x}, T) + \varphi_t(\hat{x}, T)(t - T) + 2f(|x - \hat{x}|) + \theta_1(|t - T|)$$

for some $f \in \mathcal{F}_\Omega$ and some $\theta \in C^2[0, \infty)$, $\theta(0) = \theta'(0) = 0$ and $\theta(r) > 0$ for all $r > 0$ so that

$$\begin{aligned} &\varphi(x, t) < \psi(\hat{x}, T) \quad \text{for} \quad x \ne \hat{x} \quad \text{or} \quad t < T \text{ and} \\ &\varphi(\hat{x}, T) = \psi(\hat{x}, T) \end{aligned}$$

for all (x, t) in a neighborhood $\mathcal{O}' = \operatorname{int} B_\rho(\hat{x}) \times (T - \delta, T]$ of $(\hat{x}, T)$ with some $\rho > 0$ and $\delta > 0$. As argued for φ since $(\hat{x}, T)$ is a strict maximizer of $u^* - \psi$ over $\mathcal{O}'$, a maximizer $(y_\alpha, s_\alpha) \in \mathcal{O}'$ of $u^* - \psi_\alpha$ over $\mathcal{O}'$ converges to $(\hat{x}, T)$ as $\alpha \to \infty$, where $\psi_\alpha = \psi + \alpha/(T - t)$. Thus, instead of (4.2.5) and (4.2.7) we have

$$\frac{\partial \psi_\alpha}{\partial t}(y_\alpha, s_\alpha) + F(y_\alpha, s_\alpha, u^*(y_\alpha, s_\alpha), \nabla\psi(y_\alpha, s_\alpha), \nabla^2\psi(y_\alpha, s_\alpha)) \le 0 \tag{4.2.9}$$

if $\nabla\psi(y_\alpha, s_\alpha) \neq 0$, and

$$\psi_t(y_\alpha, s_\alpha) \leq \frac{\partial \psi_\alpha}{\partial t}(y_\alpha, s_\alpha) < 0 \quad \text{if } \nabla\psi(y_\alpha, s_\alpha) = 0. \tag{4.2.10}$$

We may assume that $\nabla\psi(y_\alpha, s_\alpha) \neq 0$ for sufficiently large α since otherwise (4.2.10) together with $\psi_t(\hat{x}, T) = \varphi_t(\hat{x}, T)$ implies (4.2.8). By definition of ψ the inequality (4.2.9) is of the form

$$\varphi_t(\hat{x}, T) + \theta_1'(T - t_\alpha) + F(y_\alpha, s_\alpha,\ u^*(y_\alpha, s_\alpha), \nabla h(y_\alpha), \nabla^2 h(y_\alpha)) \leq 0, \tag{4.2.11}$$

where $h(y) = 2f(|y - \hat{x}|)$. Since $y_\alpha \to \hat{x}$ and $2f \in \mathcal{F}_\Omega$,

$$\lim_{\alpha\to\infty} F(y_\alpha, s_\alpha, u^*(y_\alpha, s_\alpha),\ \nabla h(y_\alpha),\ \nabla^2 h(y_\alpha)) = 0.$$

Thus letting $\alpha \to \infty$ in (4.2.11) yields (4.2.8).

□

Lemma 4.2.13 (Localization)

(i) *Assume the same hypothesis of Theorem 4.2.10* (i) *concerning F. If u is a subsolution of* (4.1.1) *in $Q = \Omega \times (0, T)$ (resp. $Q_* = \Omega \times (0, T]$), then for any $T' < T$ (resp. $T' \leq T$), u is a subsolution of* (4.1.1) *in $Q' = \Omega \times (0, T']$ (resp. $\Omega \times (0, T')$).*

(ii) *Assume the same hypotheses of Theorem 4.2.10* (ii) *concerning F. If u is an $\mathcal{F}_\Omega$-subsolution of* (4.1.1) *in Q (resp. Q_*), then for any $T' < T$, u is an $\mathcal{F}_\Omega$-subsolution of* (4.1.1) *in Q' (resp. $\Omega \times (0, T')$).*

Proof Suppose that u is a subsolution of (4.1.1) in Q. We may assume that Ω is bounded. Assume that $u^* - \varphi$ attains its strict maximum at (x_0, t_0) over Q' for $\varphi \in C^2(Q')$. Extend φ to a C^2 function on Q (or Q_*) (still denoted φ) and set $\varphi_\delta = \varphi + g(t)/\delta$ for $\delta > 0$, where $g = 0$ for $t < t_0$ and $g = (t - t_0)^3$ for $t \geq t_0$ so that $g \in C^2(\mathbf{R})$. Let (x_δ, t_δ) be a maximizer of $u^* - \varphi_\delta$. Then $t_\delta \leq t_0$ since $g(t) = 0$ for $t \leq t_0$ and $g(t) > 0$ for $t > t_0$. Since

$$\limsup_{\delta\to 0}{}^*(u^* - \varphi_\delta) = \begin{cases} u^* - \varphi & t \leq t_0, \\ -\infty & t > t_0, \end{cases}$$

the convergence of maximum points (Sect. 3.2.2) implies that $(x_\delta, t_\delta) \to (x_0, t_0)$ and $u^*(x_\delta, t_\delta) - g(t_\delta)/\delta \to u^*(x_0, t_0)$ as $\delta \to 0$. Since u^* is upper semicontinuous and $g \geq 0$, this convergence yields $g(t_\delta)/\delta \to 0$, since $g \geq 0$, so that $u^*(x_\delta, t_\delta) \to u^*(x_0, t_0)$. Since u is a subsolution in Q (or Q_*), we get (4.2.5) with α replaced by δ. Since $\partial\varphi/\partial t \leq \partial\varphi_\delta/\partial t$, sending $\delta \to 0$ yields

$$\varphi_t(x_0, t_0) + F(x_0, t_0, u^*(x_0, t_0), \nabla\varphi(x_0, t_0), \nabla^2\varphi(x_0, t_0)) \leq 0.$$

Thus u is a subsolution of (4.1.1) in Q'. The proof for the statement for localization by an open set $\Omega \times (0, T')$ is easier so is omitted.
(ii) The idea of the proof is a combination of that of (i) and Theorem 4.2.10 (ii). It is safely left to the reader as an exercise. □

At this moment, for a subsolution u we wonder whether or not u^* at $t = T' < T$ agrees with the upper semicontinuous envelope of the restriction of u^* on Q'. In other words we wonder whether u^* is left accessible at (x, T') for all $x \in \Omega$, $T' < T$. As already seen in the counterexample in Sect. 3.1.2, there may be a subsolution which is not left accessible for a singular degenerate parabolic equation. We have to restrict F or modify the notion of solutions as in Sect. 3.1.3 to conclude the left accessibility of solutions. The next lemma is essentially found in a paper by Y.-G. Chen et al. [107]. We do not give the proof here since we won't use this result. (The statement for $\mathcal{F}$-solutions are not included in the above article but the proof is easily extended to this case by replacing $|z_i - y_{0i}|^4$ and $|z_i - z_{i0}|^4$ by $f(|z_i - y_{0i}|)$ and $f(|z_i - z_{0i}|)$ for $f \in \mathcal{F}$, respectively without assuming the second inequality of (4.2.12) of course.)

Lemma 4.2.14 (Accessibility) *Let k be a positive integer. Let $T > 0$ and $y_{0i} \in \mathbf{R}^{Ni}$ and let Ω_i be an open set in $\mathbf{R}^{Ni}$ such that $y_{0i} \in \Omega_i$ for $1 \le i \le k$.*

(i) *Assume that $F = F_i : W_i \to \mathbf{R} \cup \{-\infty\}$ is lower semicontinuous and satisfies*

$$\begin{aligned} &F(x, t, r, p, X) > -\infty \ \ \textit{for} \ \ p \not\equiv 0, r \in \mathbf{R}, X \in \mathbf{S}^N, \\ &F(x, t, r, 0, O) > -\infty \ \ \textit{for} \ \ r \in \mathbf{R} \end{aligned} \tag{4.2.12}$$

with $N = N_i$ and $t = T$ for all x near y_{0i} $(1 \le i \le k)$, where $W_i = \overline{\Omega}_i \times [0, T] \times \mathbf{R} \times \mathbf{R}^{Ni} \times \mathbf{S}^{Ni}$. Let u_i be a subsolution of (4.1.1) *with $F = F_i$ in $Q_{i*} = \Omega_i \times (0, T]$. Then the function*

$$w(z, t) = \sum_{i=1}^{k} u_i^*(z_i, t)$$

is left accessible at (y_0, T), where $z = (z_1, \ldots, z_k)$, $z_i \in \Omega_i$ and $y_0 = (y_{01}, \ldots, y_{0k})$.

(ii) *Assume the same hypothesis of Theorem 4.2.10* (ii) *concerning $F = F_i$ with $\Omega = \Omega_i$, $N = N_i$. Then w is left accessible at (y_0, T) provided that u_i is an $\mathcal{F}_{\Omega_i}$-subsolution of* (4.1.1) *with $F = F_i$.*

Corollary 4.2.15 *Let Ω be an open set in $\mathbf{R}^N$.*

(i) *Assume that F satisfies* (4.2.12) *for all $(x, t) \in Q = \Omega \times (0, T)$. If u is a subsolution of* (4.1.1) *in Q, then u^* is left accessible at each $(x, t) \in Q$.*
(ii) *Assume the same hypothesis of Theorem 4.2.10* (ii) *concerning F. If u is an $\mathcal{F}_\Omega$-subsolution of* (4.1.1) *in Q, then u^* is left accessible at each $(x, t) \in Q$.*

This follows from localization and accessibility lemmas (Lemmas 4.2.13 and 4.2.14. In Lemma 4.2.14 the conclusion for general k cannot be reduced to the case of a single function since the sum $u(x,t) + v(y,t)$ of two left accessible (upper semicontinuous) functions u and v may not be left accessible in general. Lemma 4.2.14 implies not only accessibility of subsolution u itself but also accessibility of the sum $u_1(x,t) + u_2(y,t)$ of two subsolutions or difference $u(x,t) - v(y,t)$ where v is a supersolution (with assumptions of F symmetric to (4.2.12)).

4.3 General Idea for the Proof of Comparison Principles

We consider a simple equation of non-evolution type to motivate the idea to establish comparison principles for viscosity sub- and supersolutions.

4.3.1 A Typical Problem

Let $\mathcal{O}$ be a bounded domain in $\mathbf{R}^d$. We consider

$$u + F(Du, D^2u) = 0 \quad \text{in } \mathcal{O}. \tag{4.3.1}$$

This equation has a comparison principle for degenerate elliptic F for example of the following form.

Theorem 4.3.1 *Assume that $F : \mathbf{R}^d \times \mathbf{S}^d \to \mathbf{R}$ is degenerate elliptic and continuous. Let u and v, respectively, be sub- and supersolutions of* (4.3.1). *If $-\infty < u^* \le v_* < \infty$ on $\partial\mathcal{O}$, then $u^* \le v_*$ in $\mathcal{O}$.*

Classical Idea If both u and v are C^2 in $\mathcal{O}$ and continuous in $\overline{\mathcal{O}}$, the proof is simple. We consider the difference $g(z) = u(z) - v(z)$. Assume that g is positive somewhere in $\mathcal{O}$. Since g is continuous and $\overline{\mathcal{O}}$ is bounded, g attains its positive maximum over $\overline{\mathcal{O}}$ at some point $\hat{z} \in \overline{\mathcal{O}}$. Since on the boundary $\partial\mathcal{O}$ we have $g \le 0$, so $\hat{z}$ must be an interior point of $\mathcal{O}$. A classical maximum principle for a C^2 function implies

$$Dg(\hat{z}) = 0,\ D^2g(\hat{z}) \le O,$$

so that

$$Du(\hat{z}) = Dv(\hat{z}),\ D^2u(\hat{z}) \le D^2v(\hat{z}). \tag{4.3.2}$$

Since u and v are classical sub- and supersolutions of (4.3.1), respectively (cf. Proposition 3.1.2), we have

$$u(\hat{z}) + F(Du(\hat{z}), D^2u(\hat{z})) \le 0 \le v(\hat{z}) + F(Dv(\hat{z}), D^2v(\hat{z})).$$

By (4.3.2) and the degenerate ellipticity of F we conclude $g(\hat{z}) = u(\hat{z}) - v(\hat{z}) \le 0$ which contradicts the positivity of the maximum of g.

This argument can be easily generalized if one of u^* and v_* belongs to $C^2(\mathcal{O})$. For example, assume $v_* \in C^2(\mathcal{O})$. Since $g = u^* - v_*$ is upper semicontinuous and $\overline{\mathcal{O}}$ is compact as before, g attains its positive maximum at an interior point $\hat{z} \in \mathcal{O}$. By definition of $J^{2,+}$ instead of (4.3.2) we have

$$(Dv_*(\hat{z}), D^2v_*(\hat{z})) \in J^{2,+}_{\mathcal{O}} u^*(\hat{z}).$$

Since u is a subsolution and v is a classical supersolution,

$$u^*(\hat{z}) + F(Dv(\hat{z}), D^2v(\hat{z})) \le 0 \le v_*(\hat{z}) + F(Dv(\hat{z}), D^2v(\hat{z})),$$

which again yields a contradiction $g(\hat{z}) \le 0$.

Doubling Variables If both u and v are not C^2, at least one of $J^{2,+}u^*(\hat{z})$ and $J^{2,-}v_*(\hat{z})$ may be empty so the classical idea does not work. If we are allowed to consider $J^{2,+}u^*$ and $J^{2,-}v_*$ at different points $\hat{z}, \hat{\zeta}$, there are more chances to have semijets. For this purpose we double the variables and consider

$$w(z, \zeta) = u(z) - v(\zeta) \tag{4.3.3}$$

instead of g; here and hereafter we suppress $*$ in both u and v. However, since we are only interested in the behavior of w where z is close to ζ, we need to give a penalty away from the diagonal set

$$\{(z, \zeta) \in \overline{\mathcal{O}} \times \overline{\mathcal{O}};\ z = \zeta\}.$$

Penalized Function We consider

$$\Phi_\alpha(z, \zeta) = w(z, \zeta) - \alpha\psi(z - \zeta) \tag{4.3.4}$$

for $\psi \in C^2(\mathbf{R}^d)$ satisfying at least $\underline{\lim}_{|z|\to\infty}\psi(z) > 0$, $\psi(0) = 0$, and $\psi(z) > 0$ for $z \ne 0$ and α is a positive parameter so that the term $\alpha\psi(z - \zeta)$ tends to infinity as $\alpha \to \infty$ unless $z = \zeta$; here $\underline{\lim}$ denotes $\liminf$. We have are freedom to choose ψ depending on the situation. The function Φ_α is a penalized function of w in the sense that it has less chance to attain its maximum away from the diagonal set for large α (Fig. 4.2). For (4.3.1) one may take

$$\psi(z) = |z|^2/2.$$

We maximize the function Φ_α over $\overline{\mathcal{O}} \times \overline{\mathcal{O}}$. As $\alpha \to \infty$ the maximizer and maximum approximate those of g. In fact, if $(z_\alpha, \zeta_\alpha) \in \overline{\mathcal{O}} \times \overline{\mathcal{O}}$ is a maximizer of Φ_α over $\overline{\mathcal{O}} \times \overline{\mathcal{O}}$, i.e.,

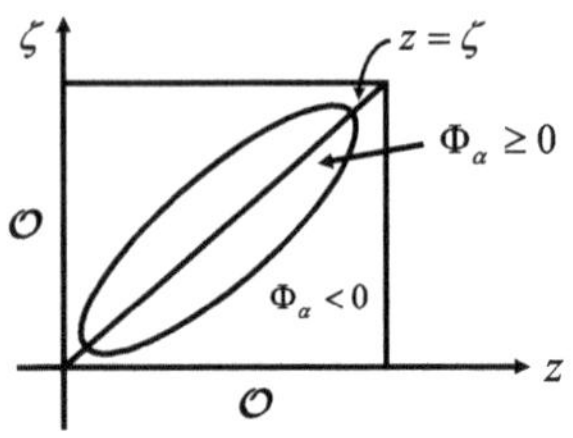

Fig. 4.2 Doubling variables

$$M_\alpha := \max_{\overline{\mathcal{O}} \times \overline{\mathcal{O}}} \Phi_\alpha = \Phi_\alpha(z_\alpha, \zeta_\alpha),$$

then

(i) $\lim_{\alpha\to\infty} \alpha\psi(z_\alpha - \zeta_\alpha) = 0$ and
(ii) $\lim_{\alpha\to\infty} M_\alpha = \max_{\overline{\mathcal{O}}}(u - v) = (u - v)(\hat{z})$ whenever $\hat{z}$ is an accumulation point of $\{z_\alpha\}$ as $\alpha \to \infty$.

(The proof is elementary. Since w is bounded, by definition of maximizers

$$\overline{\lim}_{\alpha\to\infty}\alpha\psi(z_\alpha - \zeta_\alpha) < \infty$$

which in particular implies $z_\alpha - \zeta_\alpha \to 0$ as $\alpha \to \infty$. Let $\hat{z}$ be an accumulation point of $\{z_\alpha\}$ as $\alpha \to \infty$ so that $z_{\alpha_j} \to \hat{z}$, $\zeta_{\alpha_j} \to \hat{z}$ for some subsequence $\{\alpha_j\}$. Since

$$g(z) = g(z) - \alpha\psi(0) \leq w(z_\alpha, \zeta_\alpha) - \alpha\psi(z_\alpha - \zeta_\alpha) \quad \text{for all } z \in \overline{\mathcal{O}}$$

and w is upper semicontinuous, letting $\alpha \to \infty$ yields

$$g(z) \leq g(\hat{z}) - \eta$$

where $\eta = \overline{\lim}_{j\to\infty}\alpha_j\psi(z_{\alpha_j} - \zeta_{\alpha_j})$. We may take $z = \hat{z}$ to conclude $\eta = 0$. This yields $\overline{\lim}_{\alpha\to\infty}\alpha\psi(z_\alpha - \zeta_\alpha) = 0$ and also $\lim_{\alpha\to\infty} M_\alpha = g(\hat{z}) \geq g(z)$ for all $z \in \overline{\mathcal{O}}$.)

Relation of Derivatives Using Φ_α with $\psi(z) = |z|^2/2$ we sketch the proof of Theorem 4.3.1. Assume that g is positive somewhere in $\mathcal{O}$. By approximation of the maximum of g by that of Φ_α, there is $\delta > 0$ such that $M_\alpha > \delta$ for sufficiently large α. Since $g(z) \leq 0$ for $z \in \partial\mathcal{O}$ and the maximizer (z_α, ζ_α) of Φ_α converges to $(\hat{z}, \hat{z})$ (by taking a subsequence), for sufficiently large α, Φ_α does not attain its maximum on $\overline{\mathcal{O}} \times \overline{\mathcal{O}}$ on $\partial(\mathcal{O} \times \mathcal{O})$. We may assume that (z_α, ζ_α) is an (interior) point of $\mathcal{O} \times \mathcal{O}$ and $M_\alpha > 0$ and we fix $\alpha > 0$.

To see relations of derivatives we assume u and v are C^2 around $\hat{z} = z_\alpha$ and $\hat{\zeta} = \zeta_\alpha$ respectively. Since $\Phi = \Phi_\alpha$ attains its maximum at $(\hat{z}, \hat{\zeta})$, a classical maximum principle for functions of $2d$ variables implies

$$Du(\hat{z}) = D_z\varphi(\hat{z}, \hat{\zeta}),\ Dv(\hat{\zeta}) = -D_\zeta\varphi(\hat{z}, \hat{\zeta}), \tag{4.3.5}$$

$$\begin{pmatrix} X & O \\ O & -Y \end{pmatrix} \le (D^2\varphi)(\hat{z}, \hat{\zeta}) \tag{4.3.6}$$

with $\varphi(z,\zeta) = \alpha\psi(z-\zeta)$, where $X = D^2u(\hat{z})$, $Y = D^2v(\hat{\zeta})$. For our choice of $\psi(z) = |z|^2/2$, this yields

$$\begin{pmatrix} X & O \\ O & -Y \end{pmatrix} \le \alpha \begin{pmatrix} I & -I \\ -I & I \end{pmatrix}. \tag{4.3.7}$$

This implies $X \le Y$ since

$$\langle (X-Y)\rho, \rho \rangle = ({}^t\rho, {}^t\rho) \begin{pmatrix} X & O \\ O & -Y \end{pmatrix} \begin{pmatrix} \rho \\ \rho \end{pmatrix} \le \alpha({}^t\rho, {}^t\rho) \begin{pmatrix} I & -I \\ -I & I \end{pmatrix} \begin{pmatrix} \rho \\ \rho \end{pmatrix} = 0$$

for $\rho \in \mathbf{R}^d$. The property $X \le Y$ also follows from (4.3.6) for general φ provided that φ is a function of $z - \zeta$. If φ is a function of $z - \zeta$, (4.3.5) implies $Du(\hat{z}) = Dv(\hat{\zeta})$. Since u and v are classical sub- and supersolutions near $\hat{z}, \hat{\zeta}$ respectively, we see

$$u(\hat{z}) + F(Du(\hat{z}), X) \le 0 \le v(\hat{\zeta}) + F(Dv(\hat{\zeta}), Y).$$

Since $X \le Y$ and $Du(\hat{z}) = Dv(\hat{\zeta})$, one gets $u(\hat{z}) \le v(\hat{\zeta})$ as before by degenerate ellipticity of F which yields a contradiction to $\Phi(\hat{z}, \hat{\zeta}) > 0$.

It is not difficult to extend the property $Du(\hat{z}) = Dv(\hat{\zeta})$ for semijets of functions. In fact, we conclude that there are elements of $J^{2,+}u(\hat{z})$ and $J^{2,-}v(\hat{\zeta})$ whose first derivative part is the same. So this observation yields a rigorous proof of Theorem 4.3.1 for the first order equations. The extension of (4.3.6) for semicontinuous functions is not trivial but surprisingly it is possible with some modification of results.

4.3.2 Maximum Principle for Semicontinuous Functions

We give a maximum principle of type (4.3.5) and (4.3.6), which by now is standard, for semicontinuous functions.

Theorem 4.3.2 *Let $\mathcal{Z}_i$ be a locally compact subset of $\mathbf{R}^{N_i}$ for $i = 1, \dots, k$. Let u_i be an upper semicontinuous functions on $\mathcal{Z}_i$ with values in $\mathbf{R} \cup \{-\infty\}$. Set*

$$w(z) = u_1(z_1) + \cdots + u_k(z_k) \quad \textit{for } z = (z_1, \dots, z_k) \in \mathcal{Z},$$

where

$$\mathcal{Z} = \mathcal{Z}_1 \times \cdots \times \mathcal{Z}_k. \tag{4.3.8}$$

For $\varphi \in C^2(\mathcal{Z})$ suppose that $\hat{z} = (\hat{z}_1, \ldots, \hat{z}_k) \in \mathcal{Z}$ is a point at which a maximum of $w - \varphi$ over $\mathcal{Z}$ is attained, i.e.,

$$\max_{\mathcal{Z}}(w - \varphi) = (w - \varphi)(\hat{z}).$$

Then for each $\lambda > 0$ there is $X_i \in \mathbf{S}^{Ni}$ such that

$$(D_{z_i}\varphi(\hat{z}), X_i) \in \overline{J}^{2,+}_{\mathcal{Z}_i} u_i(\hat{z}_i) \quad \text{for } i = 1, \ldots, k \tag{4.3.9}$$

and the block diagonal matrix with entries X_i satisfies

$$-\left(\frac{1}{\lambda} + |A|\right) I \le \begin{pmatrix} X_1 & & 0 \\ & \ddots & \\ 0 & & X_k \end{pmatrix} \le A + \lambda A^2, \tag{4.3.10}$$

where $A = D^2\varphi(\hat{z}) \in \mathbf{S}^N$, $N = N_1 + \cdots + N_k$, and $|A|$ denotes the operator norm when $\mathbf{R}^N$ is equipped with a standard inner product.

The proof depends on a deep theory of real analysis. There is a nice presentation of the proof in the review paper by M. G. Crandall et al. [128], where u_i is not allowed to take $-\infty$ but this small extension causes no technical problems. The maximum principle has many modified versions so that it applies for example for parabolic problems. The next version, which is an extension of Theorem 4.3.2, is useful for our parabolic problem as pointed out by Dr. M.-H. Giga. We give its proof in Appendix.

Theorem 4.3.3 *Assume the same hypotheses concerning u_i, φ and $\hat{z}$ with orthogonal decomposition*

$$\mathbf{R}^N = \oplus_{i=1}^k V_i, \; V_i = \mathbf{R}^{Ni}, \; \mathcal{Z}_i \subset V_i \quad \text{for } i = 1, \ldots, k,$$

which corresponds to the decomposition of $\mathcal{Z}$ in (4.3.8). *Let $\{P_j\}_{j=1}^{\ell}$ be a family of orthogonal projections on $\mathbf{R}^N$ that satisfies $\sum_{j=1}^{\ell} P_j = I$. Assume that V_i is invariant under the operator of P_j, i.e., $P_j V_i \subset V_i$ for all $i = 1, \ldots, k$, $j = 1, \ldots, \ell$. Assume that $W_j = P_j \mathbf{R}^N$ is invariant under the operation of A, i.e., $AW_j \subset W_j$. Then for each $\lambda_j > 0$ $(1 \le j \le \ell)$ there is $X_i \in \mathbf{S}^{Ni}$ that satisfies* (4.3.9) *and*

$$-\sum_{j=1}^{\ell}\left(\frac{1}{\lambda_j} + |AP_j|\right) P_j \le \begin{pmatrix} X_1 & & 0 \\ & \ddots & \\ 0 & & X_k \end{pmatrix} \le A + \sum_{j=1}^{\ell} \lambda_j A^2 P_j. \tag{4.3.11}$$

The condition (4.3.11) is more useful than (4.3.10) when we would like to handle spatial and time derivatives separately for evolutional problems. In (4.3.11) the W_j-component of the block diagonal matrix of X_i is estimated not only by $|A|$ but also by $|AP_j|$.

Completion of the Proof of Theorem 4.1.1 If the assertion was false, then as observed in Sect. 4.3.1 there is a maximizer $(\hat{z}, \hat{\zeta})$ in $\mathcal{O} \times \mathcal{O}$ of the penalized function $\Phi = \Phi_\alpha$ for fixed α such that the maximum $M = M_\alpha$ of Φ is strictly positive. We apply Theorem 4.3.2 with $k = 2$, $\mathcal{Z}_1 = \mathcal{Z}_2 = \mathcal{O}$, $u_1 = u$, $u_2 = -v$, $\varphi(z, \zeta) = \alpha|z - \zeta|^2/2$. Since

$$D_z\varphi(\hat{z}, \hat{\zeta}) = -D_\zeta\varphi(\hat{z}, \hat{\zeta}) = \alpha(\hat{z} - \hat{\zeta}),\ A = \alpha \begin{pmatrix} I & -I \\ -I & I \end{pmatrix},$$

$$A^2 = 2\alpha A,\ \ |A| = 2\alpha$$

and $\overline{J}_{\mathcal{O}}^{2,-} v = -\overline{J}_{\mathcal{O}}^{2,+}(-v)$, we conclude from (4.3.9) and (4.3.10) that for every $\lambda > 0$ there exist X and $Y \in \mathbf{S}^N$ such that

$$(\alpha(\hat{z} - \hat{\zeta}), X) \in \overline{J}_{\mathcal{O}}^{2,+} u(\hat{z}),\ \ (\alpha(\hat{z} - \hat{\zeta}), Y) \in \overline{J}_{\mathcal{O}}^{2,-} v(\hat{\zeta}) \tag{4.3.12}$$

and

$$-\left(\frac{1}{\lambda} + 2\alpha\right)\begin{pmatrix} I & O \\ O & I \end{pmatrix} \le \begin{pmatrix} X & O \\ O & -Y \end{pmatrix} \le \alpha(1 + 2\lambda\alpha)\begin{pmatrix} I & -I \\ -I & I \end{pmatrix}. \tag{4.3.13}$$

In comparison with (4.3.7) the estimate from above for the matrix containing X and $-Y$ is a little bit different but still our estimate concludes $X \le Y$. Since u and v are sub- and supersolutions of (4.1.1) respectively, thanks to Remark 4.2.9 the relation (4.3.12) yields

$$u(\hat{z}) + F(\alpha(\hat{z} - \hat{\zeta}), X) \le 0 \le v(\hat{\zeta}) + F(\alpha(\hat{z} - \hat{\zeta}), Y).$$

Since $X \le Y$ by (4.3.13), the degenerate ellipticity of F implies that $u(\hat{z}) \le v(\hat{\zeta})$ which contradicts $M = M_\alpha = \Phi_\alpha(\hat{z}, \hat{\zeta}) > 0$. □

We do not use the lower bound in (4.3.13) in the proof. Its presence gives a control for bound for $|X|$ and $|Y|$ which is important in many more complicated problems. We presented Eq. (4.3.1) to motivate the method so Theorem 4.3.1 is by no means optimal for our method. The method presented above already applies to more general equations of form

$$E(z, u, Du, D^2u) = 0$$

as pointed out in the review paper by M. G. Crandall et al. [128]. In the next section we shall prove comparison principles for spatially independent equations of evolution type (4.1.1) but having singularities at zeros of the gradient of solutions. For singular degenerate equations we should be careful in the way of penalizing.

4.4 Proof of Comparison Principles for Parabolic Equations

We now study comparison principles for parabolic equations. We use infinitesimal version (Proposition 4.2.6) definitions of viscosity solutions as in the proof of Theorem 4.3.1. However, since the equation may be singular where the gradient of solutions vanishes, some extra work is necessary.

4.4.1 *Proof for Bounded Domains*

We approximate maximum by penalization. We give its abstract form which also applies to maximizers (z_α, ζ_α) of Φ_α in Sect. 4.3.1.

Lemma 4.4.1 *Let Z_0 be a closed subset of a metric space Z. Let w be an upper semicontinuous function from Z to $\mathbf{R} \cup \{-\infty\}$. Assume that*

$$\overline{\lim}_{\delta \to 0} \sup\{w(z); d(z, Z_0) \le \delta, z \in Z\} \le \sup_{Z_0} w. \tag{4.4.1}$$

Let φ_σ be a nonnegative continuous function on Z parametrized by $\sigma = (\sigma_1, \dots, \sigma_k) \in \mathbf{R}^k, \sigma_i \ge 0$ $(1 \le i \le k)$ such that $\varphi_\sigma = 0$ on Z_0 and that for any $\delta > 0$,

$$\lim_{|\sigma| \to \infty} \inf\{\varphi_\sigma(z);\ d(z, Z_0) \ge \delta\} = \infty. \tag{4.4.2}$$

Let

$$M_\sigma = \sup_Z (w(z) - \varphi_\sigma(z)) \tag{4.4.3}$$

and assume that $M_\sigma < \infty$ for large $|\sigma|$. Let z_σ be such that

$$\lim_{|\sigma| \to \infty} (M_\sigma - (w(z_\sigma) - \varphi_\sigma(z_\sigma))) = 0. \tag{4.4.4}$$

Then

(i) $\lim_{|\sigma| \to \infty} \varphi_\sigma(z_\sigma) = 0$ *so that* $\lim_{|\sigma| \to \infty} d(z_\sigma, Z_0) = 0$ *and*
(ii) $\lim_{|\sigma| \to \infty} M_\sigma = \sup_{Z_0} w = w(\hat{z})$ *and* $\hat{z} \in Z_0$, *whenever* $\hat{z}$ *is an accumulation point of* $\{z_\sigma\}$ *as* $|\sigma| \to \infty$.

Proof Since $M_\sigma < \infty$ for large $|\sigma|$, by definition (4.4.4) of z_σ,

$$\eta = \overline{\lim}_{|\sigma|\to\infty} \varphi_\sigma(z_\sigma) < \infty,$$

which in particular implies $d(z_\sigma, Z_0) \to 0$ (as $|\sigma| \to \infty$) by (4.4.2). By (4.4.1) and (4.4.4) we now conclude that

$$\begin{aligned} w(z) = \lim_{j\to\infty}(w(z) - \varphi_{\sigma_j}(z)) &\le \overline{\lim}_{j\to\infty}(w(z_{\sigma_j}) - \varphi(z_{\sigma_j})) \\ &\le \sup_{Z_0} w - \eta, \ z \in Z_0 \end{aligned}$$

if $\{\sigma_j\}$ is chosen such that $\eta = \lim_{j\to\infty} \varphi_{\sigma_j}(z_{\sigma_j})$. We take sup of the left-hand side over Z_0 which forces $\eta = 0$. This yields (i) and $\overline{\lim}_{\sigma\to\infty} M_\sigma \le \sup_{Z_0} w$. Thus (ii) follows since $M_\sigma \ge \sup_{Z_0} w$ by definition and w is upper semicontinuous at $z = \hat{z}$.
□

Proof of Theorem 4.1.1 By the equivalence of an $\mathcal{F}$-subsolution and usual subsolution (Proposition 3.2.8) under (F1)–(F3) it suffices to prove the case that u and v are, respectively, $\mathcal{F}$-sub- and supersolutions of (4.1.1) in Q under the assumptions (F1), (F2), (F3′) and (F4). We suppress the symbol $\mathcal{F}$ in this proof.

We may assume that u and $-v$ are upper semicontinuous from $\overline{\Omega} \times [0, T)$ to $\mathbf{R} \cup \{-\infty\}$ with $u \le v$ on $\partial_p Q$. Since u and $-v < \infty$ in $\overline{\Omega} \times [0, T)$ the extended function $(u_{T'})^*$ and $(-v_{T'})^* < \infty$ at $t = T'$ and $(u_{T'})^* \le (v_{T'})_*$ on $\partial_p Q$, where $u_{T'}$ and $v_{T'}$ are, respectively, the restriction of u and v on $\overline{\Omega} \times [0, T')$ for $T' < T$. (If we use Corollary 4.2.15, it is easy to see that $(u_{T'})^* = u$, $(-v_{T'})^* = -v$ up to $t = T'$ but we do not need this fact.) By rewriting T' by T we may assume that the extended functions $u^*, -v_*$ to $t = T$ does not take value ∞ on $\overline{\Omega} \times \{t = T\}$ and that $u^* \le v_*$ on $\partial_p \overline{Q}$, where $\partial_p \overline{Q} = \overline{\Omega} \times \{0\} \cup \partial\Omega \times [0, T]$. By Theorem 4.2.10 the extended function u^* and v_* are, respectively, sub- and supersolutions of (4.1.1) in $Q_* := \Omega \times (0, T]$. We shall denote u^* and v_* simply by u and v.

We may also assume that u and v are bounded in $\overline{Q}$. Indeed, since u and $-v$ are upper semicontinuous in $\overline{Q}$, they are bounded from above, i.e., $L = \sup_{\overline{Q}} u < \infty$, $S = \inf_{\overline{Q}} v > -\infty$. Since $-at - 1 + S$ is an $\mathcal{F}$-subsolution of (4.1.1) in Q for large $a > 0$, the function

$$u_S = u \vee (-at - 1 + S)$$

for such an a is a subsolution of (4.1.1) by Lemma 3.4.7. Similarly, $v_L = v \wedge (at + 1 + L)$ is a supersolution for large $a > 0$. We may consider u_S and v_L instead of u and v. So we may assume u and $-v$ are bounded also from below.

In the monotonicity condition (F4) we may assume that $c_0 = 0$ by replacing u by $\tilde{u}e^{c_0 t}$ since the equation for $\tilde{u}$ is

$$\tilde{u}_t + c_0\tilde{u} + e^{-c_0 t} F(x, t, e^{c_0 t}\tilde{u}, e^{c_0 t}\nabla\tilde{u}, e^{c_0 t}\nabla^2\tilde{u}) = 0;$$

the conditions (F1), (F2), (F3′) are invariant under this transform.

As in the proof of Theorem 4.3.1 we double variables for $u - v$ and use a penalizing term with small modifications. We take $f \in \mathcal{F}$ and set

$$\begin{aligned} w_\gamma(x,t,y,s) &= u(x,t) - v(y,s) - \gamma(t+s), \\ \varphi_{\alpha\beta}(x,t,y,s) &= \alpha f(|x-y|) + \beta|t-s|^2 \end{aligned} \tag{4.4.5}$$

for $\alpha, \beta, \gamma > 0$, $(x,t) \in \overline{Q}$, $(y,s) \in \overline{Q}$. Assume that the conclusion was false so that there would exist $(x_0, t_0) \in Q$ such that $u(x_0,t_0) - v(x_0,t_0) > 0$. Then there would exist $\gamma_0 > 0$ and $\delta_0 > 0$ that satisfies

$$\sup_Z w_\gamma \geq \delta_0 \quad \text{for all } 0 < \gamma < \gamma_0, \tag{4.4.6}$$

where $Z = \overline{Q} \times \overline{Q}$. Since $u - v \leq 0$ on $\partial_p \overline{Q}$, we may assume that

$$\sup\{w(x,t,y,s); (x,t,y,s) \in Z_0, (x,t) \in \partial_p \overline{Q}\} \leq \delta_0/2 \tag{4.4.7}$$

for $0 < \gamma < \gamma_0$ by taking γ_0 smaller, where

$$Z_0 = \{(x,t,y,s) \in \overline{Q} \times \overline{Q}; x = y, t = s\}. \tag{4.4.8}$$

We shall fix $\gamma < \gamma_0$ and suppress the subscript γ. Since w is upper semicontinuous and Z is compact, there is a maximizer $z_\sigma \in Z$ of $w - \varphi_\sigma$, where $\sigma = (\alpha, \beta)$, $z_\sigma = (x_\sigma, t_\sigma, y_\sigma, s_\sigma) \in \overline{Q} \times \overline{Q}$. Since Z_0 is compact, the upper semicontinuity of w implies (4.4.1). Since (4.4.2) is trivially fulfilled for φ_σ, we apply Lemma 4.4.1 with (4.4.3) and get $|x_\sigma - y_\sigma| \to 0$, $|t_\sigma - s_\sigma| \to 0$ as $|\sigma| \to \infty$. Since Z is compact, the accumulation point of $\{z_\sigma\}$ always exists so that $x_\sigma \to \hat{x}$, $y_\sigma \to \hat{x}$, $t_\sigma \to \hat{t}$, $s_\sigma \to \hat{t}$ for some $(\hat{x}, \hat{t}) \in \overline{Q}$ by taking subsequences still denoted x_σ, y_σ, s_σ and t_σ. By (ii) of Lemma 4.4.1 and (4.4.6) we see

$$\delta_0 \leq \lim_{|\sigma|\to\infty} M_\sigma = \sup_{Z_0} w = w(\hat{x}, \hat{t}, \hat{x}, \hat{t}).$$

Since (4.4.7) holds, this implies $(\hat{x}, \hat{t}) \notin \partial_p \overline{Q}$. In other words for sufficiently large $\sigma = (\alpha, \beta)$, say $\alpha > \alpha_0$, $\beta > \beta_0$, we observe that $(x_\sigma, t_\sigma, y_\sigma, s_\sigma) \in Q_* \times Q_*$, where $Q_* = \Omega \times (0, T]$.

We shall study behavior of $u(x,t) - v(y,s)$ near $(x_\sigma, t_\sigma, y_\sigma, s_\sigma) \in Q_* \times Q_*$ for $\sigma = (\alpha, \beta)$, $\alpha > \alpha_0$, $\beta > \beta_0$.

Case 1 $x_{\sigma_j} = y_{\sigma_j}$ for some $|\sigma_j| \to \infty$. We fix $\sigma = \sigma_j$. Since

$$(w - \varphi_\sigma)(x, t, y_\sigma, s_\sigma) \leq (w - \varphi_\sigma)(x_\sigma, t_\sigma, y_\sigma, s_\sigma) = M_\sigma$$

for $(x,t) \in Q_*$, we see

$$\max_{Q_*}(u - \varphi^+) = (u - \varphi^+)(x_\sigma, t_\sigma)$$

if

$$\varphi^+(x, t) = \alpha f(|x - y_\sigma|) + \beta(t - s_\sigma)^2 + \gamma t.$$

Similarly, $(w - \varphi_\sigma)(x_\sigma, t_\sigma, y, s) \le M_\sigma$ implies

$$\min_{Q_*}(v - \varphi^-) = (v - \varphi^-)(y_\sigma, s_\sigma),$$

where

$$\varphi^-(y, s) = -\alpha f(|x_\sigma - y|) - \beta(t_\sigma - s)^2 - \gamma s.$$

Our assumption $x_\sigma = y_\sigma$ is equivalent to saying that $\nabla\varphi^+(x_\sigma, t_\sigma) = 0$ and $\nabla\varphi^-(y_\sigma, s_\sigma) = 0$. Since $a\mathcal{F} \subset \mathcal{F}$ for $a > 0$, $\varphi^\pm \in C^2_F(Q_*)$. By the definition of $\mathcal{F}$-solutions in Chap. 3, we observe that

$$\varphi_t^+(x_\sigma, t_\sigma) = 2\beta(t_\sigma - s_\sigma) + \gamma \le 0 \le \varphi_s^-(y_\sigma, s_\sigma) = 2\beta(t_\sigma - s_\sigma) - \gamma.$$

This contradicts $\gamma > 0$ so $x_\sigma = y_\sigma$ does not occur for all $\sigma = (\alpha, \beta)$, $\alpha > \alpha_0$, $\beta > \beta_0$.

Case 2 $x_\sigma \ne y_\sigma$ for sufficiently large σ. We set $\xi = (x, t)$, $\eta = (s, y)$ and $\xi_\sigma = (x_\sigma, t_\sigma)$, $\eta_\sigma = (y_\sigma, s_\sigma)$. Since $w - \varphi_\sigma$ takes its maximum over $Q_* \times Q_*$ at $(\xi_\sigma, \eta_\sigma)$, we see

$$\left(\begin{pmatrix}\hat{\varphi}_\xi \\ \hat{\varphi}_\eta\end{pmatrix}, A\right) \in J^{2,+}w(\xi_\sigma, \eta_\sigma),$$

$$A = \begin{pmatrix}\hat{\varphi}_{\xi\xi} & \hat{\varphi}_{\xi\eta} \\ \hat{\varphi}_{\eta\xi} & \hat{\varphi}_{\eta\eta}\end{pmatrix}$$

where $\hat{\varphi}_\xi = (D_\xi\varphi_\sigma)(\xi_\sigma, \eta_\sigma)$, $\hat{\varphi}_{\xi\xi} = (D_\xi^2\varphi_\sigma)(\xi_\sigma, \eta_\sigma)$ and so on. We apply Theorem 4.3.3 with $k = 2$, $N_1 = N_2 = N + 1$, $\mathcal{Z}_1 = \mathcal{Z}_2 = Q_*$, $\ell = 2$, where projection P_1 and P_2 are defined by

$$P_1 : \mathbf{R}^{N+1} \times \mathbf{R}^{N+1} \to \mathbf{R}^N \times \mathbf{R}^N,\ P_1(x, t, y, s) = (x, y)$$

and $P_2 = I - P_1$. (It is easy to check that A satisfies $AW_j \subset W_j$ with $W_j = P_j\mathbf{R}^{2(N+1)}$.) Then we find that for each $\lambda = (\lambda_1, \lambda_2)$, $\lambda_j > 0$ ($j = 1, 2$) there exists $X_1, -Y_1 \in \mathbf{S}^{N+1}$ such that

$$(\hat{\varphi}_\xi, X_1) \in \overline{J^{2,+}}(u - \gamma t)(x_\sigma, t_\sigma), \tag{4.4.9}$$

$$(-\hat{\varphi}_\eta, Y_1) \in \overline{J^{2,-}}(v + \gamma s)(y_\sigma, s_\sigma),$$

$$-\sum_{j=1}^{2}\left(\frac{1}{\lambda_j} + |AP_j|\right) P_j \le \begin{pmatrix} X_1 & O \\ O & -Y_1 \end{pmatrix} \le A + \sum_{i=1}^{2} \lambda_j A^2 P_j. \quad (4.4.10)$$

If we represent

$$X_1 = \begin{pmatrix} X & \ell_1 \\ {}^t\ell_1 & b_1 \end{pmatrix} \quad Y_1 = \begin{pmatrix} Y & \ell_2 \\ {}^t\ell_2 & b_2 \end{pmatrix}$$

by using $b_i \in \mathbf{R}$, $\ell_i \in \mathbf{R}^N\,(i = 1, 2)$, $X, Y \in \mathbf{S}^N$, (X, Y) can be regarded as a linear operator in $W_1 = P_1\mathbf{R}^{2(N+1)}$. Thus (4.4.10) yields

$$-\left(\frac{1}{\lambda_1} + |A_0|\right) I \le \begin{pmatrix} X & O \\ O & -Y \end{pmatrix} \le A_0 + \lambda_1 A_0^2 \quad (4.4.11)$$

where

$$A_0 = \begin{pmatrix} \hat{\varphi}_{xx} & \hat{\varphi}_{xy} \\ \hat{\varphi}_{yx} & \hat{\varphi}_{yy} \end{pmatrix}.$$

By Remark 4.2.9 the relation (4.4.9) yields

$$(\gamma + \hat{\varphi}_t, \hat{\varphi}_x, X) \in \overline{\mathcal{P}}^{2,+} u(x_\sigma, t_\sigma),$$
$$(-\gamma - \hat{\varphi}_s, -\hat{\varphi}_y, Y) \in \overline{\mathcal{P}}^{2,-} v(y_\sigma, s_\sigma).$$

Since u and v are sub- and supersolutions, respectively and $\hat{\varphi}_x = -\hat{\varphi}_y \neq 0$ by the assumption $x_\sigma \neq y_\sigma$, we have

$$2\beta(t_\sigma - s_\sigma) + \gamma + F(t_\sigma, \hat{u}, \hat{\varphi}_x, X) \le 0 \le 2\beta(t_\sigma - s_\sigma) - \gamma + F(s_\sigma, \hat{v}, -\hat{\varphi}_y, Y), \quad (4.4.12)$$

where $\hat{u} = u(x_\sigma, t_\sigma)$, $\hat{v} = v(y_\sigma, s_\sigma)$. Since $w(x_\sigma, t_\sigma, y_\sigma, s_\sigma) > 0$ by (4.4.6), we see $\hat{u} \ge \hat{v}$. Since $\varphi = \varphi_\sigma$ is a function of $x - y$ and $t - s$, (4.4.11) implies $X \le Y$ as in Sect. 4.3. By monotonicity (F4) with $c_0 = 0$ and (F2) the inequality (4.4.12) yields

$$2\gamma + F(t_\sigma, \hat{u}, \hat{\varphi}_x, X) - F(s_\sigma, \hat{u}, \hat{\varphi}_x, X) \le 0, \quad (4.4.13)$$

since $\hat{\varphi}_x = -\hat{\varphi}_y$. If F is independent of time, we already get a contradiction: $2\gamma \le 0$.

If F depends on t, we send β of $\sigma = (\alpha, \beta)$ to infinity. By a penalty argument as in Lemma 4.4.1, this time we observe that $t_\beta - s_\beta \to 0$ as $\beta \to \infty$. Since we fix α, we write the subscript β instead of σ. We may assume that $x_\beta - y_\beta$ is bounded by the choice of $\alpha > \alpha_0$, $\beta > \beta_0$. There are two cases to discuss.

Case A Assume that $x_\beta - y_\beta$ is bounded away from zero as $\beta \to \infty$. Then by (4.4.11) $X = X_\beta$, $Y = Y_\beta$ is bounded in the space $\mathbf{S}^N$; for an explicit bound see the proof of Theorem 4.1.4, Case 2 in the next subsection Sect. 4.4.2. Since $|x_\beta - y_\beta|$ is bounded from above, $\hat{\varphi}_x$ is bounded as $\beta \to \infty$. Since u is bounded in $\overline{Q}$, $\hat{u}$ is bounded as $\beta \to \infty$. Thus by continuity of $F(t, r, p, X)$ outside $p = 0$, sending $\beta \to \infty$ in (4.4.13) implies $2\gamma \leq 0$ since $t_\beta - s_\beta \to 0$ as $\beta \to \infty$. This contradicts $\gamma > 0$.

Case B Assume that $x_\beta - y_\beta \to 0$ as $\beta \to \infty$. Then we recall $\varphi^\pm$ in Case 1. Since u and v are sub- and supersolutions, respectively, we have

$$\gamma + 2\beta(t_\beta - s_\beta) + F(t_\beta, u, \nabla\varphi^+, \nabla^2\varphi^+) \leq 0 \quad \text{at } (x_\beta, t_\beta),$$
$$0 \leq -\gamma + 2\beta(t_\beta - s_\beta) + F(s_\beta, v, \nabla\varphi^-, \nabla^2\varphi^-) \quad \text{at } (y_\beta, s_\beta).$$

This implies

$$\begin{aligned} &2\gamma + F(t_\beta, u(x_\beta, t_\beta), \nabla\varphi^+(x_\beta), \nabla^2\varphi^+(x_\beta)) \\ &-F(s_\beta, v(y_\beta, s_\beta), \nabla\varphi^-(y_\beta), \nabla^2\varphi^-(y_\beta)) \leq 0; \end{aligned}$$

note that $\nabla\varphi^+, \nabla^2\varphi^+, \nabla\varphi^-, \nabla^2\varphi^-$ is independent of the time variable. By definition of $\mathcal{F}$ sending $\beta \to \infty$ we see the limit involving F equals zero. Again we get a contradiction to positivity of γ.

□

Remark 4.4.2

(i) To get the comparison result stated in Theorem 4.1.1 from the proof we can weaken the definition of class $\mathcal{F}$ by replacing (3.1.11) by

$$\limsup_{p \to 0} \sup_{z \in \overline{\mathcal{O}}} \sup_{|r| \leq M} |F(z, r, \nabla f(|p|), \pm\nabla^2 f(|p|)| = 0$$

for all $M > 0$.

(ii) Instead of setting

$$w(x, t, y, s) = u(x, t) - v(y, s) - \gamma(t + s)$$

we may set

$$w(x, t, y, s) = u(x, t) - v(y, s) - \frac{\gamma}{T - t} - \frac{\gamma}{T - s}.$$

Indeed, M. Ohnuma and K. Sato [368] established results of Theorem 4.1.1 when F is independent of t, r by this choice of w; they also assumed that u^* and $-v^*$ are bounded from above at $t = T$. However, they did not use the fact that u^* and v_* are sub- and supersolutions of Q_* in the proof.

4.4.2 Proof for Unbounded Domains

We shall prove Theorem 4.1.4 by adjusting the proof of Theorem 4.1.1.

Proof of Theorem 4.1.4 As in the proof of Theorem 4.1.1 it suffices to prove the case that u and v are, respectively, $\mathcal{F}$-sub- and supersolutions of (4.1.1) in Q under the assumptions (F1), (F2), (F3′), and (F4) with $c_0 = 0$. We suppress the symbol $\mathcal{F}$ in the proof. We may also assume that u^* and v_* are, respectively, sub- and supersolutions of (4.1.1) on $\Omega \times (0, T]$ and that (4.1.4) holds for $T' = T$ with the property that u^* and v_* are left accessible at $t = T$. We shall denote u^* and v_* simply by u and v.

We set w_γ and $\varphi_{\alpha\beta}$ as in (4.4.5). Assume that the conclusion was false so that

$$M \geq \theta_0 := \lim_{r\downarrow 0} \sup\{u(z) - v(\zeta);\ (z, \zeta) \in Z = \overline{Q} \times \overline{Q},\ |z - \zeta| < r\} > 0,$$

where $M = \sup\{u(z) - v(\zeta);\ (z, \zeta) \in Z\}$; since u and $-v$ are bounded from above, we see that $M < \infty$. Then for sufficiently small $\gamma > 0$ we observe that

$$\mu_0 := \lim_{r\downarrow 0} \sup\{w_\gamma(z, \zeta);\ (z, \zeta) = (x, t, y, s) \in Z,\ |z - \zeta| < r\} > 0.$$

We shall fix γ such that $\mu_0 > 0$ and suppress that subscript γ. We define

$$\Phi_\sigma(x, t, y, s) = w(x, t, y, s) - \varphi_{\alpha\beta}(x, t, y, s) \ \text{ with } \sigma = (\alpha, \beta),\ \theta = \sup_Z \Phi_\sigma,$$

$$\mu_1(r) := \sup\{\Phi_\sigma(x, t, y, s);\ (x, t, y, s) \in Z, |z - \zeta| \leq r\} \leq \theta$$

and observe that $\mu_1(r)$ converges to μ_0 uniformly in σ so that there is $r_1 > 0$ independent of σ such that

$$\mu_1(r) \geq 3\mu_0/4 > 0 \text{ for } r \in (0, r_1).$$

We shall argue as in Lemma 4.4.1. (Unfortunately, (4.4.1) may not hold when Z_0 is unbounded, so Lemma 4.4.1 does not directly apply to our setting.) Since u and $-v$ are bounded from above on Q, $\Phi_\sigma(x, t, y, s) > 0$ implies

$$f(|x - y|) \leq \frac{M}{\alpha} \text{ and } (t - s)^2 \leq \frac{M}{\beta}. \tag{4.4.14}$$

By the hypothesis (4.1.4) on the values of u and v on the boundary there is $r_* > 0$ ($r_* < r_1$) such that

$$\sup\{\Phi_\sigma(z, \zeta);\ (z, \zeta) \in \partial_p\overline{Q} \times \overline{Q} \cup \overline{Q} \times \partial_p\overline{Q},\ |z - \zeta| < r_*\} \leq \frac{\mu_0}{2}.$$

Take α_0, β_0 so large that (4.4.14) with $\alpha > \alpha_0$, $\beta > \beta_0$ always implies $|z - \zeta| < r_*$ with $z = (x, t)$, $\zeta = (y, s)$. By the choice of r_* we see that for all $\alpha > \alpha_0$ and $\beta > \beta_0$ if $z, \zeta \in \overline{Q}$ and $\Phi_\sigma(z, \zeta) > \mu_0/2$, then

$$z, \zeta \in Q_* = \Omega \times (0, T]. \tag{4.4.15}$$

Thus when we approximate the value μ_1, we may assume that z, ζ are away from the parabolic boundary of $\overline{Q}$.

We shall distinguish two cases whether or not the supremum of Φ_σ is approximated at x close to y as $\beta \to \infty$. In the case x close to y the spatial gradient of the test function $\varphi_{\alpha\beta}$ approaches zero so it roughly corresponds to Case 1 in the proof of Theorem 4.1.1; the exactly corresponding classification is slightly different and will be discussed in Remark 4.4.3. We always take σ such that α, β satisfies $\alpha > \alpha_0$, $\beta > \beta_0$. We shall fix α but we shall later send β to infinity. Note that θ and μ_1 depends on β so we sometimes write $\theta(\beta)$, $\mu_1(r, \beta)$ to emphasize its dependence.

Case 1 For each $r \in (0, r_*)$ there is β_r such that $\theta(\beta_r) = \mu_1(r, \beta_r)$ and $\beta_r \to \infty$ as $r \to 0$.

We first fix $r \in (0, r_*)$. By the definition of μ_1 there is a sequence $\{(x_m, t_m, y_m, s_m)\}$ such that

$$\Phi_\sigma(x_m, t_m, y_m, s_m) \geq \theta(\beta) - \frac{1}{m} \quad \text{and } |x_m - y_m| \leq r$$

with $\beta = \beta_r$. By taking a subsequence if necessary we may assume that $(t_m, s_m) \to (\hat{t}, \hat{s})$ and $x_m - y_m \to \hat{\omega}$ for some $\hat{t}, \hat{s} \in [0, T]$ and $\hat{\omega} \in \mathbf{R}^N$ satisfying $|\hat{\omega}| \leq r$. We now consider the function

$$\begin{aligned}
&\psi^+(x, t) = u(x, t) - \varphi^+(x, t),\\
&\varphi^+(x, t) = \alpha f(|x - y_m|) + \beta(t - s_m)^2 + f(|x - y_m - \hat{\omega}|) + (t - \hat{t})^2 + \gamma t.
\end{aligned}$$

Let (ξ_m, τ_m) be a maximum of ψ^+ over $\overline{Q}$. Such a point does exist since $\lim_{r\to\infty} f(r) = \infty$. Since

$$\psi^+(x_m, t_m) \leq \psi^+(\xi_m, \tau_m),$$

subtracting $v(y_m, s_m) + \gamma s_m$ from both sides yields

$$\begin{aligned}
&\Phi_\sigma(x_m, t_m, y_m, s_m) - f(|x_m - y_m - \hat{\omega}|) - (t_m - \hat{t})^2\\
&\leq \Phi_\sigma(\xi_m, \tau_m, y_m, s_m) - f(|\xi_m - y_m - \hat{\omega}|) - (\tau_m - \hat{t})^2.
\end{aligned}$$

Since $\Phi_\sigma \leq \theta$, this implies

$$\begin{aligned}
f(|\xi_m - y_m - \hat{\omega}|) + (\tau_m - \hat{t})^2 \leq{}& \theta(\beta) - \Phi_\sigma(x_m, t_m, y_m, s_m)\\
&+ f(|x_m - y_m - \hat{\omega}|) + (t_m - \hat{t})^2
\end{aligned} \tag{4.4.16}$$

and

$$\Phi_\sigma(x_m, t_m, y_m, s_m) \le \Phi_\sigma(\xi_m, \tau_m, y_m, s_m) + f(|x_m - y_m - \hat{\omega}|) + (t_m - \hat{t})^2. \tag{4.4.17}$$

The inequality (4.4.16) implies that $\tau_m \to \hat{t}$, $\xi_m - y_m \to \hat{\omega}$ since $\Phi_\sigma(x_m, t_m, y_m, s_m) \to \theta(\beta)$, $t_m \to \hat{t}$, $x_m - y_m \to \hat{\omega}$. Since u is bounded, we may assume that $u(\xi_m, \tau_m) \to \hat{u}$ for some $\hat{u} \in \mathbf{R}$. The inequality (4.4.17) implies that

$$\Phi_\sigma(\xi_m, \tau_m, y_m, s_m) > \mu_0/2 \tag{4.4.18}$$

for sufficiently large m; we may assume that (4.4.18) holds for all m.

By (4.4.18) and (4.4.15) we observe that $(\xi_m, \tau_m) \in Q_*$ for all m. Since u is a subsolution in Q_* and since $u - \varphi^+$ is maximized at $(\xi_m, \tau_m) \in Q_*$, we have

$$\varphi_t^+ + F(\tau_m, u, \nabla\varphi^+, \nabla^2\varphi^+) \le 0 \quad \text{at} \quad (\xi_m, \tau_m)$$

if $p_m := \xi_m - y_m \neq 0$ and

$$\varphi_t^+(\xi_m, \tau_m) = 2\beta(\tau_m - s_m) + 2(\tau_m - \hat{t}) + \gamma \le 0$$

if $p_m = 0$. Since $\xi_m - y_m \to \hat{\omega}$, $\tau_m \to \hat{t}$, $s_m \to \hat{s}$, we send m to infinity to get

$$2\beta(\hat{t} - \hat{s}) + \gamma + F\left(\hat{t}, \hat{u}, \alpha\nabla_p f(|p|), \alpha\nabla_p^2 f(|p|)\right) \le 0 \quad \text{at} \quad p = \hat{\omega} \tag{4.4.19}$$

if $\hat{\omega} \neq 0$ and

$$2\beta(\hat{t} - \hat{s}) + \gamma \le 0 \tag{4.4.20}$$

if $\hat{\omega} = 0$ by the definition of $f \in \mathcal{F}$.

In the same way, we consider the function

$$\psi^-(y, s) = -v(y, s) + \varphi^-(y, s),$$
$$\varphi^-(y, s) = -\alpha f(|x_m - y|) - f(|x_m - y - \hat{\omega}|) - \beta(t_m - s)^2 - (s - \hat{s})^2 - \gamma s$$

and study the maximum point of ψ^- over Q_*. By a similar argument as for ψ^+ we use the fact that v is a supersolution to get

$$2\beta(\hat{t} - \hat{s}) - \gamma + F(\hat{s}, \hat{v}, \alpha\nabla_p f(|p|), -\alpha\nabla_p^2 f(|p|)) \ge 0 \quad \text{at} \quad p = \hat{\omega} \tag{4.4.21}$$

if $\hat{\omega} \neq 0$ and

$$2\beta(\hat{t} - \hat{s}) - \gamma \ge 0 \tag{4.4.22}$$

if $\hat{\omega} = 0$. In the latter case (4.4.20) and (4.4.22) yields a contradiction $2\gamma \le 0$ so we may assume that $\hat{\omega} \ne 0$. Subtracting (4.4.21) from (4.4.19) yields

$$2\gamma + F(\hat{t}, \hat{u}, \alpha\nabla_p f(|p|), \alpha\nabla_p^2 f(|p|)) - F(\hat{s}, \hat{v}, \alpha\nabla_p f(|p|), -\alpha\nabla_p^2 f(|p|)) \le 0$$

at $p = \hat{\omega}$. Note that $\hat{\omega}$ depends on r. Since $|\hat{\omega}| \le r$, we now send $r \to 0$ to get $2\gamma \le 0$ by definition of $\mathcal{F}$-solutions, which is a contradiction.

Case 2 There is $r_0 \in (0, r_*)$ such that for sufficiently large β, say $\beta > \beta_1 \ge \beta_0$ for some β_1 the inequality $\theta(\beta) > \mu_1(r_0, \beta)$ holds. We define $\Psi_{\sigma\delta}$ by

$$\Psi_{\sigma\delta}(x, t, y, s) = \Phi_\sigma(x, t, y, s) - \delta|x|^2 - \delta|y|^2$$

for $\delta > 0$. Clearly, $\Psi_{\sigma\delta}$ attains a maximum at some point $(x_{\sigma\delta}, t_{\sigma\delta}, y_{\sigma\delta}, s_{\sigma\delta}) \in Z$ depending on δ and σ. By definition $\sup_Z \Psi_{\sigma\delta} \uparrow \theta$ as $\delta \to 0$. Thus, for sufficiently small δ, say $\delta < \delta_0(\sigma)$, we obtain

$$\sup_Z \Psi_{\sigma\delta} > \mu_2 := \mu_1(r_0, \beta).$$

Since $\Psi_{\sigma\delta}(x_{\sigma\delta}, t_{\sigma\delta}, y_{\sigma\delta}, s_{\sigma\delta}) \ge \sup_Z \Psi_{\sigma\delta}$ and $\mu_2 \ge 3\mu_0/4$, we observe from (4.4.15) that

$$(x_{\sigma\delta}, t_{\sigma\delta}), (y_{\sigma\delta}, s_{\sigma\delta}) \in Q_*.$$

By definition of μ_1 and (4.4.14) we obtain

$$f^{-1}(M/\alpha) \ge |x_{\sigma\delta} - y_{\sigma\delta}| > r_0. \tag{4.4.23}$$

Moreover, since $0 \le \Psi_{\sigma\delta}(x_{\sigma\delta}, t_{\sigma\delta}, y_{\sigma\delta}, s_{\sigma\delta}) \le M - \delta|x_{\sigma\delta}|^2 - \delta|y_{\sigma\delta}|^2$, we have

$$\delta(|x_{\sigma\delta}| + |y_{\sigma\delta}|) \to 0 \text{ as } \delta \to 0. \tag{4.4.24}$$

Since $\Psi_{\sigma\delta} = (w - \delta|x|^2 - \delta|y|^2) - \varphi_\sigma$ attains its maximum at $(\hat{x}, \hat{t}, \hat{y}, \hat{s}) = (x_{\sigma\delta}, t_{\sigma\delta}, y_{\sigma\delta}, s_{\sigma\delta})$ over $Q_* \times Q_*$, we argue as in Case 2 of the proof of Theorem 4.1.1 (Sect. 4.4.1) to get

$$(\gamma + \hat{\varphi}_t, \hat{\varphi}_x + 2\delta x_{\sigma\delta}, X_{\sigma\delta} + 2\delta I) \in \overline{\mathcal{P}}^{2,+} u(x_{\sigma\delta}, t_{\sigma\delta}),$$
$$(-\gamma - \hat{\varphi}_s, -\hat{\varphi}_y - 2\delta y_{\sigma\delta}, Y_{\sigma\delta} - 2\delta I) \in \overline{\mathcal{P}}^{2,-} v(y_{\sigma\delta}, s_{\sigma\delta})$$

for some $X_{\sigma\delta}, Y_{\sigma\delta} \in \mathbf{S}^N$ satisfying (4.4.11) with $X = X_{\sigma\delta}, Y = Y_{\sigma\delta}$. As in the proof of Lemma 4.1.3 a direct calculation yields

$$A_0 = \alpha\frac{f'(\rho)}{\rho}J + \alpha\left(f''(\rho) - \frac{f'(\rho)}{\rho}\right)Q,$$

$$J = \begin{pmatrix} I & -I \\ -I & I \end{pmatrix}, Q = \frac{1}{\rho^2}\begin{pmatrix} p\otimes p & -p\otimes p \\ -p\otimes p & p\otimes p \end{pmatrix}$$

with $p = x_{\sigma\delta} - y_{\sigma\delta}$ and $\rho = |p|$. Since

$$({}^t\xi, {}^t\eta) J \begin{pmatrix} \xi \\ \eta \end{pmatrix} = |\xi - \eta|^2, ({}^t\xi, {}^t\eta) Q \begin{pmatrix} \xi \\ \eta \end{pmatrix} = \frac{1}{\rho^2}|\langle \xi - \eta, p\rangle|^2, \quad \xi, \eta \in \mathbf{R}^N$$

so that $O \le Q \le J \le 2I$, we observe that $A_0 \ge \alpha f''(\rho) Q \ge O$. By $A_0 \ge O$, we see that

$$|A_0| = \sup\{({}^t\xi, {}^t\eta) A_0 \begin{pmatrix} \xi \\ \eta \end{pmatrix}; |\xi|^2 + |\eta|^2 = 1, \quad \xi, \eta \in \mathbf{R}^N\}.$$

Since

$$A_0 \le \alpha \frac{f'(\rho)}{\rho} J + \alpha f''(\rho) Q \le \alpha K_1(\rho) J \le 2\alpha K_1(\rho) I$$

with $K_1(\rho) = f'(\rho)/\rho + f''(\rho)$, we have an estimate $|A_0| \le 2\alpha K_1(\rho)$. Since $J^2 = 2J$, $Q^2 = 2Q$, $JQ = QJ = 2Q$, A_0^2 is of the form

$$A_0^2 = 2\alpha^2 \left\{ \left(\frac{f'(\rho)}{\rho}\right)^2 J + \left(2\frac{f'(\rho)}{\rho} + f''(\rho) - \frac{f'(\rho)}{\rho}\right)\left(f''(\rho) - \frac{f'(\rho)}{\rho}\right) Q \right\}$$

$$\le 2\alpha^2 \left\{ \left(\frac{f'(\rho)}{\rho}\right)^2 J + (f''(\rho))^2 Q \right\} \le 2\alpha^2 K_2(\rho) J$$

with $K_2(\rho) = (f'(\rho)/\rho)^2 + (f''(\rho))^2$. Thus (4.4.11) implies

$$-(\frac{1}{\lambda_1} + 2\alpha K_1(\rho)) I \le \begin{pmatrix} X_{\sigma\delta} & O \\ O & -Y_{\sigma\delta} \end{pmatrix} \le (\alpha K_1(\rho) + 2\alpha^2 \lambda_1 K_2(\rho)) J.$$

This yields $X_{\sigma\delta} \le Y_{\sigma\delta}$ as in Case 2 of the proof of Theorem 4.1.1 (Sect. 4.4.1) and a bound

$$|X_{\sigma\delta}| \le 2\alpha K_1(\rho) + 4\alpha^2 K_2(\rho)\lambda_1 + \frac{1}{\lambda_1}$$

with $\rho = |x_{\sigma\delta} - y_{\sigma\delta}|$. The right-hand side is bounded as $\delta \to 0$ and $\beta \to \infty$ by (4.4.23). We thus have a bound for $X_{\sigma\delta}$ independent of δ and β. By (4.4.23) and (4.4.24) we observe that

$$\hat{\varphi}_x + 2\delta x_{\sigma\delta} \ne 0, \ -\hat{\varphi}_y - 2\delta y_{\sigma\delta} = \hat{\varphi}_x + 2\delta y_{\sigma\delta} \ne 0$$

for sufficiently small $\delta > 0$. Since u and v are sub- and supersolutions, we have

$$2\beta(t_{\sigma\delta} - s_{\sigma\delta}) + \gamma + F(t_{\sigma\delta}, \hat{u}, \hat{\varphi}_x + 2\delta x_{\sigma\delta}, X_{\sigma\delta} + 2\delta I) \leq 0, \quad (4.4.25)$$

$$2\beta(t_{\sigma\delta} - s_{\sigma\delta}) - \gamma + F(s_{\sigma\delta}, \hat{v}, -\hat{\varphi}_y - 2\delta y_{\sigma\delta}, Y_{\sigma\delta} - 2\delta I) \geq 0. \quad (4.4.26)$$

By (F2) and (F4) with $c_0 = 0$, (4.4.26) with $X_{\sigma\delta} \leq Y_{\sigma\delta}$ implies

$$2\beta(t_{\sigma\delta} - s_{\sigma\delta}) - \gamma + F(s_{\sigma\delta}, \hat{u}, -\hat{\varphi}_y - 2\delta y_{\sigma\delta}, X_{\sigma\delta} - 2\delta I) \geq 0. \quad (4.4.27)$$

Sending $\delta \to 0$, we obtain from (4.4.25) and (4.4.27) with a bound for $|X_{\sigma\delta}|$ and (4.4.23) that

$$2\beta(\bar{t}_\sigma - \bar{s}_\sigma) + \gamma + F(\bar{t}_\sigma, \bar{u}_\sigma, \alpha f'(|\bar{p}_\sigma|), \overline{X}_\sigma) \geq 0, \quad (4.4.28)$$

$$2\beta(\bar{t}_\sigma - \bar{s}_\sigma) - \gamma + F(\bar{s}_\sigma, \bar{u}_\sigma, \alpha f'(|\bar{p}_\sigma|), \overline{X}_\sigma) \leq 0 \quad (4.4.29)$$

for some $\bar{t}_\sigma, \bar{s}_\sigma \in [0, T]$, $\bar{u}_\sigma \in \mathbf{R}$, $\overline{X}_\sigma \in \mathbf{S}^N$, $\bar{p}_\sigma \in \mathbf{R}^N$ by taking a subsequence if necessary; here we invoke the assumption that u is bounded so that $\bar{u}_\sigma$ exists. Sending $\beta \to \infty$ so that $\bar{t}_\sigma - \bar{s}_\sigma \to 0$ by (4.4.14), we now obtain from (4.4.28) and (4.4.29) that $2\gamma \leq 0$ since $\overline{X}_\sigma, |\bar{p}_\sigma|, |\bar{p}_\sigma|^{-1}$ are still bounded as $\beta \to \infty$. This contradicts $\gamma > 0$ so we have proved that $\theta_0 \leq 0$. □

Remark 4.4.3 The classification into Case 1 and Case 2 for bounded domains (Theorem 4.1.1) is slightly different from that for unbounded domains (Theorem 4.1.4). If we classify into the following two cases for bounded domains with fixed α, this exactly corresponds to the classification for unbounded domains.

Case 1 For each $r > 0$ there is $\beta_r \to \infty$ (as $r \to 0$) and a maximum $z_{\alpha\beta_r}$ of $w - \varphi_{\alpha\beta_r}$ in $Q_* \times Q_*$ such that $|x_{\alpha\beta_r} - y_{\alpha\beta_r}| \leq r$, where $z_{\alpha\beta} = (x_{\alpha\beta}, t_{\alpha\beta}, y_{\alpha\beta}, s_{\alpha\beta})$.

Case 2 There is $r_0 > 0$ such that for sufficiently large β any maximizer $z_{\alpha\beta}$ of $w - \varphi_{\alpha\beta}$ satisfies $|x_{\alpha\beta} - y_{\alpha\beta}| > r_0$.

4.5 Lipschitz Preserving and Convexity Preserving Properties

We study some general properties of solutions of the Cauchy problem for (4.1.1). As we shall see in Sect. 5.3, for a given $u_0 \in \mathrm{BUC}(\mathbf{R}^N)$ there exists a unique viscosity solution $u \in \mathrm{BUC}(\mathbf{R}^N \times [0, T))$ of (4.1.1) satisfying $u|_{t=0} = u_0$, for example under the assumption of Corollary 4.1.5. Here the space of all bounded, uniformly continuous functions on $D \subset \mathbf{R}^d$ is denoted by $\mathrm{BUC}(D)$. We first observe that global Lipschitz continuity is preserved for spatial homogeneous equations.

Theorem 4.5.1 *Assume that $F(x, t, r, p, X)$ is independent of x and r. Assume that* (F1)–(F2) *and* (F3) *(or* (F3′) *with the assumption that $\mathcal{F}_{\mathbf{R}^N}$ is invariant under* *positive multiplications.) If $u \in \mathrm{BUC}(\mathbf{R}^N \times [0, T))$ is an ($\mathcal{F}$-) solution of* (4.1.1) *in $\mathbf{R}^N \times (0, T)$ with some constant $L > 0$ satisfying*

$$|u(x, 0) - u(y, 0)| \leq L|x - y| \tag{4.5.1}$$

for all $x, y \in \mathbf{R}^n$, then

$$|u(x, t) - u(y, t)| \leq L|x - y| \tag{4.5.2}$$

for all $x, y \in \mathbf{R}^n$, $t \in [0, T)$.

Proof Since F is independent of x and u, we see

$$v(x, t) = u(x + h, t) + L|h|, \quad h \in \mathbf{R}^n$$

is also a viscosity solution in $\mathrm{BUC}(\mathbf{R}^n \times [0, T))$ of (4.1.1) in $\mathbf{R}^n \times (0, T)$. By the assumption of the initial data (4.5.1) and uniform continuity we see that u and v satisfy (4.1.4). By (CP) we have

$$u(x, t) \leq v(x, t)$$

or

$$u(x, t) - u(x + h, t) \leq L|h|$$

for all $x \in \mathbf{R}^N$, $t \in [0, T)$. A symmetric argument comparing $u(x + h, t) - L|h|$ and $u(x, t)$ yields

$$u(x, t) - u(x + h, t) \geq -L|h|.$$

We thus prove (4.5.2). □

We next study whether or not concavity of initial data u_0 is preserved as time develops.

Theorem 4.5.2 *Assume that $F = F(x, t, r, p, X)$ is independent of x and r and satisfies* (F1), (F2). *Assume that F is geometric. Assume that*

$$X \mapsto F(t, p, X) \ \text{ is convex on } \ \mathbf{S}^N \tag{4.5.3}$$

for all $p \in \mathbf{R}^N \setminus \{0\}$, $t \in [0, T]$. Let $u \in C(\mathbf{R}^N \times [0, T))$ be an $\mathcal{F}$-solution of (4.1.1) *in $\mathbf{R}^N \times (0, T)$ satisfying* (4.5.2) *with some $L > 0$. If $u(x, 0)$ is concave, then*

$$u(x, t) + u(y, t) - 2u(z, t) \leq L|x + y - 2z| \tag{4.5.4}$$

holds for all $x, y, z \in \mathbf{R}^N$, $0 \leq t \leq T$. In particular, $x \mapsto u(x, t)$ is concave for all $t \in [0, T)$.

We shall prove Theorem 4.5.2 in several steps. Our first observation is the equivalence of concavity and (4.5.4) type inequality when a function is globally Lipschitz.

Lemma 4.5.3 *Assume that v_0 is globally Lipschitz in $\mathbf{R}^N$ with constant L. Then v_0 is concave if and only if the inequality*

$$v_0(x) + v_0(y) - 2v_0(z) \le L|x + y - 2z| \tag{4.5.5}$$

holds for all $x, y, z \in \mathbf{R}^N$.

Proof If v_0 is concave, then

$$\begin{aligned}
&v_0(x) + v_0(y) - 2v_0(z) \\
&= v_0(x) + v_0(y) - 2v_0\left(\frac{x+y}{2}\right) + 2\left\{v_0\left(\frac{x+y}{2}\right) - v_0(z)\right\} \\
&\le 2\left\{v_0\left(\frac{x+y}{2}\right) - v_0(z)\right\}.
\end{aligned}$$

Since v_0 is globally Lipschitz, the right-hand side is dominated by $2L|\frac{x+y}{2} - z|$. We thus obtain (4.5.5).

The inequality (4.5.5) yields

$$\frac{v_0(x) + v_0(y)}{2} \le v_0\left(\frac{x+y}{2}\right)$$

by taking $z = (x + y)/2$. This says that v_0 is mid-concave. Successive use of this inequality yields

$$\lambda v_0(x) + (1 - \lambda)v_0(y) \le v_0(z), \ z = \lambda x + (1 - \lambda)y$$

for all $\lambda \in (0, 1)$ of the form $\lambda = k2^{-h}$ with positive integers h and k. Since v_0 is continuous, this implies concavity of v_0. □

Proof of Theorem 4.5.2 We suppress the symbol $\mathcal{F}$ in the proof. For $M > 0$ and $R > 0$ let θ^M and θ_R be functions on $\mathbf{R}$ of the form

$$\theta^M(\eta) = \min(M, \eta), \theta_R(\eta) = \max(-R, \eta).$$

Since F is geometric, by the invariance under change of dependent variables proved later in Sect. 5.2.1 we see that

$$\begin{aligned}
u^M(x, t) &= (\theta^M \circ u)(x, t) = \min(M, u(x, t)), \\
u_R(x, t) &= (\theta_R \circ u)(x, t) = \max(-R, u(x, t))
\end{aligned}$$

are solutions of (4.1.1) in $Q = \mathbf{R}^N \times (0, T)$ satisfying (4.5.2). Since u^M is still concave in x, we may assume that u is bounded from *above* in $\mathbf{R}^N \times (0, T)$. As in the proof of Theorem 4.1.1 we may assume that u is a solution of (4.1.1) in $Q_* = \mathbf{R}^N \times (0, T]$ by taking T smaller. By replacing u by u/L we may assume that $L = 1$.

Our goal is to prove

$$u(x,t) + u(y,t) - 2u(z,t) \leq |x + y - 2z| \tag{4.5.6}$$

for all $x, y, z \in \mathbf{R}^N$, $t \in [0, T]$. We take $f \in \mathcal{F}$. For $\kappa > 0$ there is a unique $A(\kappa) > 0$ such that

$$\eta \leq A(\kappa) f(\eta) + \kappa =: N_\kappa(\eta) \quad \text{for all} \quad \eta > 0$$

and the equality holds only at one point $\eta = \eta(\kappa) > 0$. We thus observe that

$$\eta = \inf_{\kappa > 0} N_\kappa(\eta) \tag{4.5.7}$$

for $\eta > 0$. We set

$$w_\gamma^{\kappa R}(x, s_1, y, s_2, z, s_3) = u(x, s_1) + u(y, s_2) - 2u_R(z, s_3) - N_\kappa(|x + y - 2z|) - \gamma s_1$$

for $R > 0$, $\gamma \geq 0$, $\kappa > 0$; we use $w_\gamma^{\kappa\infty}$ when u replaces u_R.

Assume that the conclusion (4.5.6) was false. By (4.5.7) there would exist $\kappa = \kappa_0$ such that

$$\sup\{w_0^{\kappa_0\infty}(x, t, y, t, z, t); x, y, z \in \mathbf{R}^N, t \in [0, T]\} > 0.$$

We may replace $u(z, t)$ by $u_R(z, t)$ for large R and observe that

$$\begin{aligned}\theta_0 := \limsup_{r \downarrow 0} \ &\{w_0^{\kappa_0 R}(x, s_1, y, s_2, z, s_3); x, y, z \in \mathbf{R}^N, s_1, s_2, s_3 \in [0, T],\\ &|s_1 - s_2| \leq r, |s_2 - s_3| < r\} > 0.\end{aligned}$$

Since u is bounded from above on $\overline{Q} = \mathbf{R}^N \times [0, T]$, i.e., $K := \sup_{\overline{Q}} u < \infty$, we see that $\theta_0 \leq 2K + 2R < \infty$. For sufficiently small $\gamma > 0$ we observe that

$$\begin{aligned}\mu_0 := \limsup_{r \downarrow 0} \ &\{w_\gamma^{\kappa_0 R}(x, s_1, y, s_2, z, s_3);\ (x, s_1), (y, s_2), (z, s_3) \in \overline{Q},\\ &|s_1 - s_2| < r, |s_2 - s_3| < r\} > 0.\end{aligned}$$

We shall fix R and γ such that $\mu_0 > 0$ and suppress the subscript γ and the superscript R as well as κ_0. We define

$$\Phi_\beta(x, s_1, y, s_2, z, s_3) := w(x, s_1, y, s_2, z, s_3) - \varphi_\beta(s_1, s_2, s_3),$$

$$\varphi_\beta(s_1, s_2, s_3) := \beta(|s_1 - s_3|^2 + |s_2 - s_3|^2), \quad \theta(\beta) = \sup_W \Phi_\beta,$$

where $W = \overline{Q} \times \overline{Q} \times \overline{Q}$.

We shall prove that for sufficiently large β, say $\beta > \beta_0$, if (x, s_1), (y, s_2), $(z, s_3) \in \overline{Q}$ satisfies

$$\Phi_\beta(x, s_1, y, s_2, z, s_3) > \mu_0/2, \tag{4.5.8}$$

then $s_1 > 0$, $s_2 > 0$ and $s_3 > 0$. By (4.5.5) and (4.5.7) we observe that

$$u_0(x) + u_0(y) - 2u_{R0}(z) - N_\kappa(|x + y - 2z|) \leq 0, \quad x, y, z \in \mathbf{R}^N \tag{4.5.9}$$

for all $\kappa > 0$ where $u_0(x) = u(x, 0)$, $u_{R0}(x) = u_R(x, 0) = \max(-R, u_0(x))$. The left-hand side of (4.5.9) is always negative if either $u_0(x) \leq -R_1$ or $u_0(y) \leq -R_1$ for $R_1 = K + 2R$. Thus (4.5.9) is equivalent to

$$u_{R_10}(x) + u_{R_10}(y) - 2u_{R0}(z) - N_\kappa(|x + y - 2z|) \leq 0, \quad x, y, z \in \mathbf{R}^N \tag{4.5.10}$$

for all $\kappa > 0$. Since u_{R_*0} is in $\mathrm{BUC}(\mathbf{R}^N)$, u_{R_*} is in $\mathrm{BUC}(\mathbf{R}^N \times [0, T))$ for every $R_* > 0$ by the uniqueness of solutions. By this uniform continuity and (4.5.10) there exists a small $r_0 > 0$ such that

$$\begin{aligned}
&\sup\{u_{R_1}(x, s_1) + u_{R_1}(y, s_2) - 2u_R(z, s_3) - N_{\kappa 0}(|x + y - 2z|) \\
&\quad -\gamma s_1 - \varphi_\beta(s_1, s_2, s_3);\ s_1 s_2 s_3 = 0,\ |s_1 - s_3| \leq r_0,\ |s_2 - s_3| \leq r_0, \\
&\quad (x, s_1), (y, s_2), (z, s_3) \in \overline{Q}\} \leq \mu_0/2.
\end{aligned}$$

By the choice of R_1 this implies

$$\begin{aligned}
&\sup\{\Phi_\beta(x, s_1, y, s_2, z, s_3); s_1 s_2 s_3 = 0,\ |s_1 - s_3| \leq r_0,\ |s_2 - s_3| \leq r_0, \\
&\quad (x, s_1), (y, s_2),\ (z, s_3) \in \overline{Q}\} \leq \mu_0/2.
\end{aligned} \tag{4.5.11}$$

Since $\Phi_\beta \leq 2K + 2R$, $\Phi_\beta(x, s_1, y, s_2, z, s_3) > 0$ implies

$$(s_1 - s_3)^2 + (s_2 - s_3)^2 \leq 2(K + R)/\beta. \tag{4.5.12}$$

Take β_0 so large that (4.5.2) with $\beta > \beta_0$ implies that $|s_1 - s_3| \leq r_0$ and $|s_2 - s_3| \leq r_0$. Then from (4.5.11) it follows that (4.5.8) implies $s_1 > 0$, $s_2 > 0$, $s_3 > 0$ if $\beta > \beta_0$.

We shall distinguish two cases whether or not the supremum of Φ_β is approximated at x, y, z with the property that $x + y - 2z$ is close to zero. We shall always assume that $\beta > \beta_0$. We set

$$\mu_1(r,\beta) := \sup\{\Phi_\beta(x,s_1,y,s_2,z,s_3);\ |x+y-2z| \le r,\\ (x,s_1),(y,s_2),(z,s_3) \in \overline{Q}\} \le \theta(\beta). \tag{4.5.13}$$

By the definition of μ_0 there exists $r_* > 0$ (independent of β) such that $\mu_1(r,\beta) \ge 3\mu_0/4$ for $r < r_*$.

Case 1 For each $r \in (0, r_*)$ there is β_r such that $\theta(\beta_r) = \mu_1(r,\beta_r)$ and $\beta_r \to \infty$ as $r \to 0$.

We argue in a similar way to the proof of Case 1 of Theorem 4.1.4 (Sect. 4.4.2). We first fix $r \in (0, r_*)$. By the definition of μ_1 there is a sequence $\{a_m\}$ with $a_m = (x_m, s_m^1, y_m, s_m^2, z_m, s_m^3)$ such that

$$\Phi_\beta(a_m) \ge \theta(\beta) - 1/m \quad \text{and} \quad |x_m + y_m - 2z_m| \le r$$

with $\beta = \beta_r$. We may assume that $(s_m^1, s_m^2, s_m^3) \to (\hat{s_1}, \hat{s_2}, \hat{s_3})$, $x_m + y_m - 2z_m \to \hat{\omega}$ for some $\hat{s_1}, \hat{s_2}, \hat{s_3} \in [0, T]$ and $\hat{\omega} \in \mathbf{R}^N$ with $|\hat{\omega}| \le r$ by taking a subsequence. We now consider the function

$$\begin{aligned}
\psi^+(x,t) &= u(x,t) - \varphi^+(x,t),\\
\varphi^+(x,t) &= Af(|x + y_m - 2z_m|) + \beta(t - s_m^3)^2\\
&\qquad + f(|x + y_m - 2z_m - \hat{\omega}|) + (t - \hat{s_1})^2 + \gamma t,\ A = A(\kappa_0).
\end{aligned}$$

Let (ξ_m, τ_m) be a maximum of ψ^+ over $\overline{Q}$. Such a point exists since $\lim_{r\to\infty} f(r) = \infty$. By definition we observe that

$$\psi^+(x_m, s_m^1) \le \psi^+(\xi_m, \tau_m).$$

Adding $u(y_m, s_m^2) - 2u(z_m, s_m^3) - \beta(s_m^2 - s_m^3)^2$ to both sides yields

$$\begin{aligned}
&\Phi_\beta(a_m) - f(|x_m + y_m - 2z_m - \hat{\omega}|) - (s_m^1 - \hat{s_1})^2\\
&\quad \le \Phi_\beta(b_m) - f(|\xi_m + y_m - 2z_m - \hat{\omega}|) - (\tau_m - \hat{s_1})^2
\end{aligned}$$

with $b_m = (\xi_m, \tau_m, y_m, s_m^2, z_m, s_m^3)$. Since $\Phi_\beta \le \theta$, this implies

$$\begin{aligned}
&f(|\xi_m + y_m - 2z_m - \hat{\omega}|) + (\tau_m - \hat{s_1})^2\\
&\le \theta(\beta) - \Phi_\beta(a_m) + f(|x_m + y_m - 2z_m - \hat{\omega}|) + (s_m^1 - \hat{s_1})^2
\end{aligned} \tag{4.5.14}$$

and

$$\Phi_\beta(a_m) \le \Phi_\beta(b_m) + f(|x_m + y_m - 2z_m - \hat{\omega}|) + (s_m^1 - \hat{s_1})^2. \tag{4.5.15}$$

The inequality (4.5.14) implies that $\xi_m + y_m - 2z_m \to \hat{\omega}$, $\tau_m \to \hat{s_1}$ as $m \to \infty$ since $x_m + y_m - 2z_m \to \hat{\omega}$, $s_m^1 \to \hat{s_1}$. The inequality (4.5.15) implies that

$$\Phi_\beta(b_m) > \mu_0/2 \tag{4.5.16}$$

for sufficiently large m since $\Phi_\beta(a_m) \to \theta(\beta)$. We may assume that (4.5.16) holds for all m.

Since (4.5.16) implies (4.5.8), we observe that $\tau_m > 0$ for all m so that $(\xi_m, \tau_m) \in Q_*$. Since u is a subsolution in Q_* and since $u - \varphi^+$ is maximized at $(\xi_m, \tau_m) \in Q_*$, we have as in the proof of Theorem 4.1.4,

$$\varphi_t^+ + F(\tau_m, \nabla\varphi^+, \nabla^2\varphi^+) \le 0 \quad \text{at} \quad (\xi_m, \tau_m)$$

if $p_m := \xi_m + y_m - 2z_m \neq 0$ and

$$\varphi_t^+(\xi_m, \tau_m) = 2\beta(\tau_m - s_m^3) + 2(\tau_m - \hat{s_1}) + \gamma \le 0$$

if $p_m = 0$. Since $\xi_m + y_m - 2z_m \to \hat{\omega}$, $\tau_m \to \hat{s_1}$, $s_m^i \to \hat{s_i}$, we send m to infinity to get

$$2\beta(\hat{s_1} - \hat{s_3}) + \gamma + F(\hat{s_1}, A\nabla_p f(|p|), A\nabla_p^2 f(|p|)) \le 0 \quad \text{at} \quad p = \hat{\omega} \tag{4.5.17}$$

if $\hat{\omega} \neq 0$ and

$$2\beta(\hat{s_1} - \hat{s_3}) + \gamma \le 0$$

if $\hat{\omega} = 0$ by the definition of $f \in \mathcal{F}$.

In the same way we study the maximum of

$$\begin{aligned} \psi_+(y, s) &= u(y, s) - \varphi_+(y, s), \\ \varphi_+(y, s) &= Af(|x_m + y - 2z_m|) + \beta(s - s_m^3)^2 \\ &\quad + f(|x_m + y - 2z_m - \hat{\omega}|) + (s - \hat{s_2})^2 \end{aligned}$$

and obtain

$$2\beta(\hat{s_2} - \hat{s_3}) + F(\hat{s_2}, A\nabla_p f(|p|), A\nabla_p^2 f(|p|)) \le 0 \quad \text{at} \quad p = \hat{\omega} \tag{4.5.18}$$

if $\hat{\omega} \neq 0$ and $2\beta(\hat{s_2} - \hat{s_3}) \le 0$ if $\hat{\omega} = 0$.

In the same way we study the maximum of

$$\begin{aligned} \psi^-(z, s_3) &= -u(z, s_3) + \varphi^-(z, s_3), \\ 2\varphi^-(z, s_3) &= -Af(|x_m + y_m - 2z|) - \beta(s_m^1 - s_3)^2 - \beta(s_m^2 - s_3)^2 \\ &\quad - f(|x_m + y_m - z + \hat{\omega}|) - (s_3 - \hat{s_3})^2 \end{aligned}$$

and obtain

$$\beta(\hat{s_1} - \hat{s_3}) + \beta(\hat{s_2} - \hat{s_3}) + F(\hat{s_3}, \frac{A}{2}\nabla_p f(|p|), -\frac{A}{2}\nabla_p^2 f(|p|)) \geq 0 \quad \text{at} \quad p = \hat{\omega} \tag{4.5.19}$$

if $\hat{\omega} \neq 0$ and $\beta(\hat{s_1} - \hat{s_3}) + \beta(\hat{s_2} - \hat{s_3}) \leq 0$ if $\hat{\omega} = 0$. If $\hat{\omega} = 0$, we easily get a contradiction: $\gamma \leq 0$ so we may assume that $\hat{\omega} \neq 0$. We consider (4.5.17) + (4.5.18) $-\, 2\times$ (4.5.19) and send $r \to 0$. Since $|\hat{\omega}| \leq r$, this yields a contradiction: $\gamma \leq 0$.

Case 2 There is $r_0 \in (0, r_*)$ such that for sufficiently large β, say $\beta > \beta_1$ for some β_1, the inequality $\theta(\beta) > \mu_1(r_0, \beta)$ holds.

We set $\mu_2 = \mu(r_0, \beta)$ for $\beta > \beta_1$. For $\delta > 0$ we introduce

$$\Psi_{\beta r}(a) = \Phi_\beta(a) - \delta(|x|^2 + |y|^2 + |z|^2)$$

with $a = (x, s_1, y, s_2, z, s_3)$. By definition

$$\sup_W \Psi_{\beta\delta} \uparrow \theta \quad \text{as} \quad \delta \downarrow 0.$$

Thus there is $\delta_0 = \delta_0(\beta) > 0$ that satisfies

$$\sup_W \Psi_{\beta\delta} > \max(\mu_2, \mu_0/2) \quad \text{for} \quad 0 < \delta < \delta_0$$

since $\mu_0 \leq \theta$. Clearly, $\Psi_{\beta\delta}$ attains a maximum at some points $\hat{\xi} = (\hat{x}, \hat{s_1}, \hat{y}, \hat{s_2}, \hat{z}, \hat{s_3})$
$\in W = \overline{Q} \times \overline{Q} \times \overline{Q}$ depending on β and δ. We suppress the dependence with respect to β and δ for simplicity of notation but we shall later send δ to zero first and β to infinity. Since (4.5.8) is fulfilled at $\hat{\xi}$, we see that $\hat{s_i}$ is positive for $i = 1, 2, 3$. By the definition of μ_2 we observe that

$$f^{-1}(2(K + R)/A(\kappa_0)) \geq |\hat{x} + \hat{y} - 2\hat{z}| > r_0; \tag{4.5.20}$$

the left inequality follows from the fact that $\Psi_{\beta\delta}$ is bounded from above by $2(K+R)$ and $\Psi_{\beta\delta}(\hat{\xi}) > 0$. This fact also yields the bound of $\delta(|\hat{x}|^2 + |\hat{y}|^2 + |\hat{z}|^2)$ so in particular we have

$$\delta(|\hat{x}| + |\hat{y}| + |\hat{z}|) \to 0 \quad \text{as} \quad \delta \to 0. \tag{4.5.21}$$

Since $s_i > 0$ for $\delta < \delta_0(\beta)$, $i = 1, 2, 3$, the function

$$\Psi_{\beta\delta} = u(x, s_1) + u(y, s_2) - 2u(z, s_3) - \delta(|x|^2 + |y|^2 + |z|^2) - \psi,$$
$$\psi = N_{\kappa_0}(|x + y - 2z|) + \varphi_\beta(s_1, s_2, s_3)$$

attains its maximum at $\hat{\xi} = (\hat{x}, \hat{s_1}, \hat{y}, \hat{s_2}, \hat{z}, \hat{s_3})$ over $Q_* \times Q_* \times Q_*$ for $\delta < \delta_0(\beta)$. We now apply Theorem 4.3.3 with $k = 3$, $N_1 = N_2 = N_3 = N + 1$, $\mathcal{Z}_1 = \mathcal{Z}_2 = \mathcal{Z}_3 = Q_*$; $\ell = 2$ with projection P_1 and P_2 defined by $P_1(x, s_1, y, s_2, z, s_3) = (x, y, z)$ and $P_2 = I - P_1$. As in Case 2 of the proof of Theorem 4.1.1, for each $\lambda_1 > 0$ there exists $X, Y, Z \in \mathbf{S}^N$ such that

$$-\left(\frac{1}{\lambda_1} + |B_0|\right) \le I \le \begin{pmatrix} X & O & O \\ O & Y & O \\ O & O & -Z \end{pmatrix} \le B_0 + \lambda_2 B_0^2, \tag{4.5.22}$$

$$(\gamma + \hat{\psi}_{s_1}, \hat{\psi}_x + 2\delta\hat{x}, X + 2\delta I) \in \overline{\mathcal{P}}^{2,+} u(\hat{x}, \hat{s_1}), \tag{4.5.23}$$

$$(\hat{\psi}_{s_2}, \hat{\psi}_y + 2\delta\hat{y}, Y + 2\delta I) \in \overline{\mathcal{P}}^{2,+} u(\hat{y}, \hat{s_2}), \tag{4.5.24}$$

$$\left(-\frac{\hat{\psi}_{s_3}}{2}, -\frac{1}{2}(\hat{\psi}_z + 2\delta\hat{z}), \frac{1}{2}(Z - 2\delta I)\right) \in \overline{\mathcal{P}}^{2,-} u(\hat{z}, \hat{s_3}), \tag{4.5.25}$$

with

$$B_0 = \begin{pmatrix} \hat{\psi}_{xx} & \hat{\psi}_{xy} & \hat{\psi}_{xz} \\ \hat{\psi}_{yx} & \hat{\psi}_{yy} & \hat{\psi}_{yz} \\ \hat{\psi}_{zx} & \hat{\psi}_{zy} & \hat{\psi}_{zz} \end{pmatrix}, \tag{4.5.26}$$

where $\hat{\psi}_x$ denotes $\nabla_x \psi$ evaluated at $\hat{\xi}$ and so on. We shall fix λ_1. Note that X, Y, Z may depend on δ and β.

We shall derive a bound for X, Y, Z. As in Case 2 of the proof of Theorem 4.1.4 a direct calculation of (4.5.26) yields

$$B_0 = A\left\{\frac{f'(\rho)}{\rho} S + \left(f''(\rho) - \frac{f'(\rho)}{\rho}\right) R\right\}, \quad A = A(\kappa_0),$$

$$S = \begin{pmatrix} I & I & -2I \\ I & I & -2I \\ -2I & -2I & 4I \end{pmatrix}, \quad R = \frac{1}{\rho^2}\begin{pmatrix} p \otimes p & p \otimes p & -2p \otimes p \\ p \otimes p & p \otimes p & -2p \otimes p \\ -2p \otimes p & -2p \otimes p & 4p \otimes p \end{pmatrix}$$

with $p = \hat{x} + \hat{y} - 2z$, $\rho = |p|$. Since

$$({}^t\xi, {}^t\eta, {}^t\zeta) S \begin{pmatrix} \xi \\ \eta \\ \zeta \end{pmatrix} = |\xi + \eta - 2\zeta|^2,$$

$$({}^t\xi, {}^t\eta, {}^t\zeta)R\begin{pmatrix}\xi\\ \eta\\ \zeta\end{pmatrix} = |\langle \xi + \eta - 2\zeta, p\rangle|^2, \quad \xi, \eta, \zeta \in \mathbf{R}^N$$

so that $O \le R \le S(\le 6I)$, we observe that

$$AK_1(\rho)S \ge B_0 \ge Af''(\rho)R \ge O \tag{4.5.27}$$

with $K_1(\rho) = f'(\rho)/\rho + f''(\rho)$. We thus have an estimate

$$|B_0| \le AK_1(\rho)|S| = 6A(\kappa_0)K_1(\rho) \tag{4.5.28}$$

since $|S| = 6$. Since $S^2 = 6S$, $R^2 = 6R$, $RS = SR = 6R$, as in Case 2 of the proof of Theorem 4.1.4

$$\begin{aligned} B_0^2 &= 6A^2\left\{\left(\frac{f'(\rho)}{\rho}\right)^2 S + \left((f''(\rho))^2 - \left(\frac{f'(\rho)}{\rho}\right)^2\right) R\right\} \\ &\le 6A^2K_2(\rho)S \end{aligned} \tag{4.5.29}$$

with $K_2(\rho) = (f'(\rho)/\rho)^2 + (f''(\rho))^2$. Applying (4.5.27)–(4.5.29) to (4.5.22) we obtain

$$-\left(\frac{1}{\lambda_1} + 6AK_1(\rho)\right) I \le \begin{pmatrix} X & O & O\\ O & Y & O\\ O & O & -Z\end{pmatrix} \le A(K_1(\rho) + 6\lambda_1 AK_2(\rho))S.$$

This inequality implies that $X + Y \le Z$ since

$$({}^t\xi, {}^t\xi, {}^t\xi)S\begin{pmatrix}\xi\\ \xi\\ \xi\end{pmatrix} = |\xi + \xi - 2\xi|^2 = 0.$$

Moreover, it provides a bound for X, Y and Z,

$$|X|, |Y|, |Z| \le 6AK_1(\rho) + 12A^2K_2(\rho)\lambda_1 + \frac{1}{\lambda_1} \tag{4.5.30}$$

with $\rho = |\hat{x} + \hat{y} - 2\hat{z}|$. By (4.5.20) the right-hand side of (4.5.30) is bounded as $\delta \to 0$ and $\beta \to \infty$.

By (4.5.20) and (4.5.21) we observe that

$$\hat{\psi}_x + 2\delta\hat{x} \ne 0, \quad \hat{\psi}_y + 2\delta\hat{y} \ne 0, \quad -\hat{\psi}_z - 2\delta\hat{z} \ne 0$$

for sufficiently small $\delta > 0$. Since u is a solution, the relation (4.5.23) and (4.5.24) yields

$$2\beta(\hat{s_1} - \hat{s_3}) + \gamma + F(\hat{s_1}, \hat{\psi}_x + 2\delta\hat{x}, X + 2\delta I) \leq 0, \tag{4.5.31}$$

$$2\beta(\hat{s_2} - \hat{s_3}) \quad + F(\hat{s_2}, \hat{\psi}_y + 2\delta\hat{y}, Y + 2\delta I) \leq 0. \tag{4.5.32}$$

Since $X + Y \leq Z$, the relation (4.5.25) with degenerate ellipticity yields

$$\beta(\hat{s_1} - \hat{s_3}) + \beta(\hat{s_2} - \hat{s_3}) + F(\hat{s_3}, -\frac{1}{2}(\hat{\psi}_z + 2\delta\hat{z}), \frac{1}{2}(X + Y - 2\delta I)) \geq 0. \tag{4.5.33}$$

Since $\hat{\psi}_x = \hat{\psi}_y = -\hat{\psi}_z/2 = Af'(\rho)p/\rho$, sending $\delta \to 0$ in the inequality (4.5.31) + (4.5.32) $-\, 2\times$ (4.5.33) yields

$$\gamma + F(\overline{s}_1, \mu, \overline{X}) + F(\overline{s}_2, \mu, \overline{Y}) - 2F(\overline{s}_3, \mu, (\overline{X} + \overline{Y})/2)) \leq 0$$

with $\mu = Af'(|\overline{p}|)\overline{p}/|\overline{p}|$ with some $\overline{X}, \overline{Y} \in \mathbf{S}^N$, $\overline{p} \in \mathbf{R}^N$, $\overline{s}_1, \overline{s}_2, \overline{s}_3 \in [0, T]$. Since $\overline{X}, \overline{Y}$ are bounded as $\beta \to \infty$ by (4.5.30) and $|\overline{p}|^{-1}$, $|\overline{p}|$ is bounded as $\sigma \to \infty$ by (4.5.20), sending $\beta \to \infty$ implies

$$\gamma + F(\tilde{s}, \tilde{\mu}, \tilde{X}) + F(\tilde{s}, \tilde{\mu}, \tilde{Y}) - 2F(\tilde{s}, \tilde{\mu}, (\tilde{X} + \tilde{Y})/2) \leq 0$$

by (4.5.12) for some $\tilde{s} \in [0, T]$, $\tilde{X}, \tilde{Y} \in \mathbf{S}^N$, $\tilde{\mu} \in \mathbf{R}^N$. Since $X \mapsto F(t, r, X)$ is convex in X, this yields $\gamma \leq 0$ which is a contradiction. We have thus proved $\theta_0 \leq 0$ to get (4.5.6). □

Corollary 4.5.4 *Assume that $F = F(x, t, r, p, X)$ is independent of x and r and satisfies* (F1), (F2). *Assume that F is geometric. Assume that*

$$X \mapsto F(t, p, X) \quad \textit{is concave on } \mathbf{S}^N$$

for all $p \in \mathbf{R}^N \setminus \{0\}$, $t \in [0, T]$. Let $u \in C(\mathbf{R}^N \times [0, T])$ be an $\mathcal{F}$-solution of (4.1.1) *in $\mathbf{R}^N \times (0, T)$ satisfying* (4.5.2) *with some $L > 0$. If $u(x, 0)$ is convex, then $x \mapsto u(x, t)$ is convex for all $t \in [0, T]$.*

This immediately follows from Theorem 4.5.2 by replacing u by $-u$ since the concavity of $X \mapsto F(t, p, X)$ is equivalent to that of $X \mapsto F(t, p, -X)$.

Remark 4.5.5 (Applicability) The concavity (resp. convexity) of $X \mapsto F_f(p, X)$ is fulfilled if and only if f in (2.6.1) is convex (resp. concave) in $\nabla\mathbf{n}$, since F_f is defined by (2.6.4). Our Theorem 4.5.2 and Corollary 4.5.4 apply to the level set equations of (2.5.2), (2.5.4), and (2.5.6) provided that conditions in Sect. 4.1.3 are fulfilled.

For the affine curvature flow equation (2.5.14) or its generalization $V = (\kappa_+)^\alpha$, $\alpha > 0$ one should take the orientation $\mathbf{n}$ inward for convex curves so that the equation is not a trivial one $V \equiv 0$. For the convex initial curves Γ_0 let u_0 be a function representing Γ_0, i.e., $\Gamma_0 = \partial\{u_0 > 0\} = \partial\{u_0 < 0\}$ with $\mathbf{n} = -\nabla u/|\nabla u|$. One can take u_0 so that it is convex; however it is impossible to take concave u_0. The convexity of the solution u of the level set equation of $V = (\kappa_+)^\alpha$ is preserved if $\alpha \geq 1$, by Corollary 4.5.4.

Since

$$\left[\det\left(\frac{X+Y}{2}\right)\right]^{1/m} \geq \frac{1}{2}((\det X)^{1/m} + (\det Y)^{1/m})$$

for $X \geq O$, $Y \geq O$, $X, Y \in \mathbf{S}^m$ (see e.g. D. S. Mitrinović [354]), for the equation $V = (\kappa_1^+ \cdots \kappa_{N-1}^+)^{1/(N-1)}$ in $\mathbf{R}^N$ the right-hand side is concave in $\nabla\mathbf{n}$ so that $F_f(p, X)$ is convex in X. Unfortunately, we again have to consider convex initial data u_0 for convex initial hypersurface Γ_0 so neither Theorem 4.5.2 nor Corollary 4.5.4 apply to conclude that solution u is convex in x. So it is unlikely that Theorem 4.5.2 and Corollary 4.5.4 apply for surface evolution equations by principal curvatures like (2.6.22), (2.6.23).

4.6 Spatially Inhomogeneous Equations

We present a few versions of comparison principles for (4.1.1) when F also depends on the spatial variable x. We restrict ourselves to the case when Ω is bounded, since the results for unbounded Ω are very complicated to state. We first study the case when inhomogeneity appears only in the first order term.

4.6.1 Inhomogeneity in First Order Perturbation

The next result applies when F is of the form

$$F(x, t, r, p, X) = F_0(t, r, p, X) + F_1(x, t, p) \tag{4.6.1}$$

with F_0 satisfying the assumptions on F of Theorem 4.1.1 (including (F3)) and $F_1 \in C(\overline{\Omega} \times [0, T] \times \mathbf{R}^N)$ that satisfies

$$|F_1(x, t, p) - F_1(y, t, q)| \leq C(1 + |p|)|x - y| \tag{4.6.2}$$

with some constant $C > 0$ independent of $x, y \in \overline{\Omega}$, $p \in \mathbf{R}^N$. In particular, the level set equation of (2.5.2) applies even when the constancy property of c in Sect. 4.1.3 is relaxed as a uniform Lipschitz continuity in x:

$$|c(x, t) - c(y, t)| \leq C|x - y|.$$

Theorem 4.6.1 *Assume that* Ω *is bounded. Assume that* (F1)–(F3) *and* (F4) *hold. Assume that for* $R > 0$ *there is a modulus* ω_R *such that*

$$|F(x, t, r, p, X) - F(y, t, r, p, X)| \le \omega_R(|x - y|(|p| + 1)) \tag{4.6.3}$$

holds for $x, y \in \overline{\Omega}$, $t \in [0, T]$, $|r| \le R$, $p \in \mathbf{R}^N \setminus \{0\}$ *and* $X \in \mathbf{S}^N$. *Then the comparison principle* (BCP) *is valid. If we assume* (F3′) *instead of* (F3), *then* (BCP) *is still valid (by replacing solutions by* $\mathcal{F}_\Omega$*-solutions) provided that*

(i) $\mathcal{F}_\Omega$ *is invariant under positive multiplication;*
(ii) *there exists* $f \in \mathcal{F}_\Omega$ *such that* $f'(\rho)\rho/f(\rho)$ *is bounded on* $[0, 1]$.

Corollary 4.6.2 *Assume that* Ω *is bounded and that* F *satisfies* (F1), (F2) *and* (4.6.3). *Assume that* F *satisfies*

$$|F(x, t, p, \pm I)| \le c_0|p|^{-\eta} \quad \text{on} \quad \overline{\Omega} \times [0, T] \times (B_{r_0}(0) \setminus \{0\}) \tag{4.6.4}$$

for some $r_0 > 0$, $c_0 > 0$, $\eta > 0$. *Then* (BCP) *holds if* F *is geometric.*

From the proof of Lemma 4.1.3 it is clear that the assumption on the existence of c_0 guarantees the existence of $f \in \mathcal{F}_\Omega$ satisfying (ii) of Theorem 4.6.1. The assumption (i) of Theorem 4.6.1 is automatically fulfilled since F is geometric as proved in Lemma 4.1.3. Thus, Corollary 4.6.2 follows from Theorem 4.6.1.

Proof of Theorem 4.6.1 We argue in the same way as in the proof of Theorem 4.1.1 right before formula (4.4.12), where x-dependence should be taken into account. Instead of (4.4.13) we obtain

$$2\gamma + F(x_\sigma, t_\sigma, \hat{u}, \hat{\varphi}_x, X) - F(y_\sigma, s_\sigma, \hat{u}, \hat{\varphi}_x, X) \le 0. \tag{4.6.5}$$

We send $\beta \to \infty$ to get $t_{\alpha\beta} - s_{\alpha\beta} \to 0$. If $x_{\alpha\beta} - y_{\alpha\beta} \to 0$ as $\beta \to \infty$, we get a contradiction as in the Case B of the proof of Theorem 4.1.1. Thus we may assume that $x_{\alpha\beta} \to x_\alpha$, $y_{\alpha\beta} \to y_\alpha$ with $x_\alpha \ne y_\alpha$. We may also assume that

$$\begin{aligned}
&t_{\alpha\beta} \to \hat{t},\ s_{\alpha\beta} \to \hat{t},\ \hat{u} \to u_\alpha \in \mathbf{R},\\
&\hat{\varphi}_x \to q_\alpha := \alpha f'(|p_\alpha|)p_\alpha/|p_\alpha| \quad \text{with} \quad p_\alpha = x_\alpha - y_\alpha;\\
&X \to X_\alpha \in \mathbf{S}^N \quad \text{(by (4.4.11))}
\end{aligned}$$

as $\beta \to \infty$ by taking a subsequence. Thus

$$2\gamma + F(x_\alpha, \hat{t}, u_\alpha, q_\alpha, X_\alpha) - F(y_\alpha, \hat{t}, u_\alpha, q_\alpha, X_\alpha) \le 0. \tag{4.6.6}$$

By (4.6.3) with $R = \sup_{\overline{Q}} |u| + 1$ this inequality yields

$$2\gamma - \omega_R(|x_\alpha - y_\alpha|(|q_\alpha| + 1)) \le 0. \tag{4.6.7}$$

Since $M_\sigma \to \sup_{Z_0} w$ (as $\sigma \to \infty$ by Lemma 4.4.1), we observe that $\alpha f(|x_\alpha - y_\alpha|) \to 0$ as $\alpha \to \infty$. By the assumption (ii),

$$|x_\alpha - y_\alpha||q_\alpha| = \alpha f'(|p_\alpha|)|p_\alpha| \le C\alpha\, f(|p_\alpha|)$$

with some C independent of α. Thus $|x_\alpha - y_\alpha||q_\alpha| \to 0$ as $\alpha \to \infty$. Since $|p_\alpha| \to 0$ as $\alpha \to \infty$, (4.6.7) now yields $2\gamma \le 0$ which is a contradiction. □

Remark 4.6.3

(i) The assumption (4.6.4) is fulfilled for $F = F_f$ of (2.6.19) provided that g is $\hat{e}_m$ or $\hat{e}_m/e_\ell$ ($m > \ell$). If F_0 in (4.6.1) is equal to such an F_f, F of (4.6.1) satisfies the assumption of Corollary 4.6.2 provided that F_1 is geometric and satisfies (4.6.2). For example, Corollary 4.6.2 guarantees that (BCP) holds for the level set equation of

$$V = K + c(x, t)$$

if c satisfies the uniform (in t) Lipschitz continuity in x provided that the Gaussian curvature K is interpreted as its modified form $\hat{e}_{N-1}(\kappa_1, \ldots, \kappa_{N-1})$.

(ii) It is of course possible to prove (CP) when Ω is unbounded. The proof is similar to that of Theorem 4.1.4. Since we shall send $\alpha \to \infty$, we modify the classification Case 1, Case 2 as follows. The details are left to the interested readers.
Case 1 There is a sequence $\alpha_j \to \infty$ such that for each $r \in (0, r_*)$ there is $\beta_r \to \infty$ such that $\theta(\beta_r) = \mu_1(r, \beta_r)$ as $r \to 0$.
Case 2 For sufficiently large α, say $\alpha > \alpha_0$, there is $r_\alpha \in (0, r_*)$ such that $\theta(\beta) > \mu_1(r_\alpha, \beta)$ for all $\beta > \beta_\alpha$ with some β_α.

4.6.2 *Inhomogeneity in Higher Order Terms*

Unfortunately, results in Sect. 4.6.1 do not apply to (4.1.1) when a spatial inhomogeneity appears in higher order terms. For example, it does not apply to $F(x, t, p) = -a(x)$ trace X unless $a \ge 0$ is a constant. The next result applies to such an F.

Theorem 4.6.4 *Assume that Ω is bounded. Assume that* (F1)–(F3) *and* (F4) *hold. Assume that for $R > 0$ there is a modulus ω_R such that*

$$F(x, t, r, p, X) - F(y, t, r, p, Y) \ge -\omega_R(|x - y|(|p| + 1) + \mu|x - y|^2) \tag{4.6.8}$$

for $x, y \in \overline{\Omega}$, $t \in [0, T]$, $p \in \mathbf{R}^N \setminus \{0\}$, $X, Y \in \mathbf{S}^N$, $|r| \le R$, $\mu > 0$ whenever

$$-3\mu I \le \begin{pmatrix} X & O \\ O & -Y \end{pmatrix} \le 3\mu J, \ J = \begin{pmatrix} I & -I \\ -I & I \end{pmatrix}. \tag{4.6.9}$$

Then (BCP) *holds.*

Remark 4.6.5 The assumption (F2) follows from (4.6.8) and (4.6.9) so it is redundant. Indeed,

$$\begin{aligned} {}^t\xi X\xi - {}^t\eta Y\eta &= {}^t(\xi-\eta)X(\xi-\eta) + 2{}^t\eta X(\xi-\eta) + {}^t\eta(X-Y)\eta \\ &\le |X||\xi-\eta|^2 + 2|X||\xi-\eta||\eta| + {}^t\eta(X-Y)\eta \\ &\le |X|(1+\varepsilon^{-1})|\xi-\eta|^2 + \varepsilon|X||\eta|^2 + {}^t\eta(X-Y)\eta \end{aligned}$$

for $\xi, \eta \in \mathbf{R}^n$, $\varepsilon > 0$. If $X \le Y$, this implies

$$\begin{pmatrix} X & O \\ O & -Y \end{pmatrix} \le (1+\varepsilon^{-1})|X|J + \varepsilon|X|I. \tag{4.6.10}$$

(By the way estimate (4.6.10) where $\varepsilon|X|$ replaced by $|X-Y| + \varepsilon|X|$ in front of I holds for arbitrary $X, Y \in \mathbf{S}^N$.) This evidently implies

$$\begin{pmatrix} X_\varepsilon & O \\ O & -Y_\varepsilon \end{pmatrix} \le (1+\varepsilon^{-1})|X|J$$

with $X_\varepsilon = X - \varepsilon|X|I$, $Y_\varepsilon = Y + \varepsilon|X|I$. Thus X_ε, Y_ε satisfies (4.6.9) by a suitable choice of $\mu = \mu_\varepsilon$. By (4.6.8) we have

$$F(x,t,r,p,X_\varepsilon) - F(x,t,r,p,Y_\varepsilon) \ge 0.$$

Sending ε to zero yields (F2).

We give a typical example of F to which Theorem 4.6.4 applies.

Corollary 4.6.6 *Assume that Ω is bounded. Assume that F is of the form* (4.6.1) *with $F_1 \in C(\overline{\Omega} \times [0,T] \times \mathbf{R}^N)$ satisfying* (4.6.2). *Assume that F_0 is of the form*

$$F_0(x,t,r,p,X) = -\mathrm{trace}({}^tA(x,t,p)XA(x,t,p)) \tag{4.6.11}$$

and $A(x,t,p)$ are $\mathbf{S}^N$-valued bounded continuous functions in $\overline{\Omega} \times [0,T] \times (\mathbf{R}^N \setminus \{0\})$. Assume that $A \ge O$ and satisfies

$$|A(x,t,p) - A(y,t,p)| \le C|x-y| \tag{4.6.12}$$

with $C > 0$ independent of $p \in \mathbf{R}^N \setminus \{0\}$, $x, y \in \overline{\Omega}$, $t \in [0,T]$. Then (BCP) *holds for* (4.1.1).

Proof Since (F3) and (F4) are clearly fulfilled, it suffices to prove that (4.6.9) implies (4.6.8). Such a property is closed under addition of F. Thus, we may assume that $F = F_0$, since the assertion is clear for F_1.

From (4.6.9) we see $X \leq Y$, more precisely, ${}^t\xi X \xi - {}^t\eta Y \eta \leq 3\mu|\xi - \eta|^2$. This implies

$$F_0(x, t, r, p, X) - F_0(y, t, r, p, Y) \geq -3\mu||A(x, t, p) - A(y, t, p)||_2^2$$

where $|| \cdot ||_2$ denotes the Hilbert–Schmidt norm. By (4.6.12) the estimate (4.6.8) is fulfilled by choosing $\omega_R(s) = 3NC^2 s$. □

Proof of Theorem 4.6.4 We take $f(\eta) = \eta^4 \in \mathcal{F}$ and argue as in the proof of Theorem 4.6.1 right before the formula (4.6.5); we invoke (F3) to handle Case 1 and Case A. Instead of (4.6.5) we obtain

$$2\gamma + F(x_\sigma, t_\sigma, \hat{u}, \hat{\varphi}_x, X) - F(y_\sigma, s_\sigma, \hat{u}, \hat{\varphi}_x, Y) \leq 0$$

without using $X \leq Y$. We send $\beta \to \infty$ and obtain

$$2\gamma + F(x_\alpha, \hat{t}, u_\alpha, q_\alpha, X_\alpha) - F(y_\alpha, \hat{t}, u_\alpha, q_\alpha, Y_\alpha) \leq 0 \tag{4.6.13}$$

instead of (4.6.6), where $q_\alpha = f'(|p_\alpha|)p_\alpha/|p_\alpha|(= 4|p_\alpha|^2 p_\alpha)$, $p_\alpha = x_\alpha - y_\alpha$.

In the estimate (4.4.11) we may take λ_1, $\lambda_1^{-1} = \alpha K_1(\rho)$, $K_1(\rho) - f'(\rho)/\rho + f''(\rho)$ with $p = x_\sigma - y_\sigma$ to get a bound

$$-3\alpha K_1(\rho_\alpha) I \leq \begin{pmatrix} X_\alpha & O \\ O & -Y_\alpha \end{pmatrix} \leq 3\alpha K_1(\rho_\alpha) J$$

with $\rho_\alpha = |x_\alpha - y_\alpha|$; see the estimate of

$$\begin{pmatrix} X_{\sigma\delta} & O \\ -O & -Y_{\sigma\delta} \end{pmatrix}$$

in the proof of Theorem 4.1.4, Case 2. Thus X_α, $Y_\alpha \in \mathbf{S}^N$ fulfills (4.6.9) with $\mu = \alpha K_1(\rho_\alpha)$. Applying (4.6.8) to (4.6.13) we obtain

$$2\gamma - \omega_R(|x_\alpha - y_\alpha|)(|q_\alpha| + 1) + \alpha K_1(\rho_\alpha)|x_\alpha - y_\alpha|^2) \leq 0 \tag{4.6.14}$$

by taking $R = \sup_{\overline{Q}} |u| + 1$. We have observed that $\alpha f(|x_\alpha - y_\alpha|) = \alpha|\rho_\alpha|^4 \to 0$ as $\alpha \to \infty$ in the proof of Theorem 4.6.1. Thus, the quantity in ω_R tends to zero as $\alpha \to \infty$. Sending $\alpha \to \infty$ in (4.6.14) now yields a contradiction $2\gamma \leq 0$. Thus (BCP) has been proved. □

Example By Corollary 4.6.2 we observe that (BCP) holds for the level set equation of (2.5.2) provided that $\beta > 0$ is continuous and γ in (2.3.7) is convex and C^2 except the origin and that $a \geq 0$ and $c \in C(\overline{\Omega} \times [0, T])$ satisfies the Lipschitz continuity in x uniformly in t, i.e.,

$$|c(x,t) - c(y,t)| \le C|x-y|.$$

We seek a sufficient condition for (BCP) when β and γ also depend on x. We first recall the level set equation of

$$\beta(x,\mathbf{n})V = -a\ \mathrm{div}_{\Gamma_t}\xi(x,\mathbf{n}) + c(x,t,\mathbf{n}). \tag{4.6.15}$$

where $\xi = \nabla_p\gamma$, $p \mapsto \gamma(x,p)$ is convex, positively homogeneous of degree 1. If we write its level set equation in the form of (4.1.1),

$$\begin{aligned} &F(x,t,p,X)\\ &= -\frac{a}{\beta(x,-\frac{p}{|p|})}\{\mathrm{trace}\nabla_p^2\gamma(x,-p)X + \mathrm{trace}\nabla_x\nabla_p\gamma(x,-p)\}|p| - c|p|, \end{aligned} \tag{4.6.16}$$

$$F_0(x,p,X) = -\frac{a}{\beta(x,-\frac{p}{|p|})}(\mathrm{trace}\nabla_p^2\gamma(x,-p)X)|p|, \tag{4.6.17}$$

$$F_1(x,t,p,X) = -\frac{a}{\beta(x,-\frac{p}{|p|})}(\mathrm{trace}\nabla_x\nabla_p\gamma(x,-p))|p| - c|p|. \tag{4.6.18}$$

The condition (4.6.2) for F_1 in (4.6.18) is fulfilled if (and only if)

$$M(x,t,p) = -a\frac{\mathrm{trace}\nabla_x\nabla_p\gamma(x,-p)}{\beta(x,-p/|p|)} - c\,(x,t,-p/|p|) \tag{4.6.19}$$

is Lipschitz continuous in x uniformly in t, p, i.e.,

$$|M(x,t,p) - M(y,t,p)| \le C|x-y| \tag{4.6.20}$$

with C independent of $x, y \in \overline{\Omega}$, $t \in [0,T]$, $p \in \mathbf{R}^N \setminus \{0\}$. To apply Corollary 4.6.6 we shall find a condition that

$$A(x,p) = \left(\left(\nabla_p^2\gamma(x,-p)\right)|p|\right)^{1/2}\frac{a^{1/2}}{\beta^{1/2}} \tag{4.6.21}$$

satisfies (4.6.12) so that F_0 in (4.6.17) is written in the form (4.6.11) with (4.6.12). Since A is positively homogeneous of degree zero in p, it suffices to check the condition that (4.6.12) for $p \in S^{N-1}$. The next result gives a sufficient condition for (4.6.12).

Lemma 4.6.7 *Let Ω be an open set in $\mathbf{R}^N$. Assume that $\gamma = \gamma(x,p)$, $x \in \overline{\Omega}$, $p \in \mathbf{R}^N$ is convex and positively homogeneous of degree 1 on $\mathbf{R}^N$ as a function of p. Assume that $\gamma \in C(\overline{\Omega} \times \mathbf{R}^N)$ is C^2 as a function of p except at the origin and $\Gamma = \nabla_p^2\gamma \in C(\overline{\Omega} \times (\mathbf{R}^N \setminus \{0\})$. Then the square root $\Gamma^{1/2}$ of Γ is Lipschitz* *continuous on* $\overline{\Omega} \times S^{N-1}$ *(respectively, Lipschitz continuous in* $x \in \Omega$ *uniformly in* $p \in S^{N-1}$*) if one of following conditions holds.*

(i) Γ *is* C^1 *and its derivative* $\nabla_p \Gamma$, $\nabla_x \Gamma$ *(resp.* $\nabla_x \Gamma$*) is Lipschitz continuous on* $\overline{\Omega} \times S^{N-1}$ *(resp. Lipschitz continuous in* $x \in \overline{\Omega}$ *uniformly in* $p \in S^{N-1}$*).*
(ii) *All eigenvalues of* $\Gamma(x, p) + p \otimes p$ *are uniformly away from* 0 *in* $\overline{\Omega} \times S^{N-1}$ *and* Γ *has the corresponding Lipschitz regularity.*

Proof If (i) holds, Γ has the Lipschitz property in $\overline{\Omega} \times U$ where U is some tubular neighborhood of S^{N-1}. Since $\Gamma \geq 0$, this implies that $\Gamma^{1/2}$ has the desired Lipschitz continuity; see M. G. Crandall et al. [128, Example 3.6].

By homogeneity the matrix $\Gamma(x, p)$ has a zero eigenvalue with eigenvector $p \in S^{N-1}$. If (ii) holds, $(\Gamma + p \otimes p)^{1/2}$ is positive so it has the desired Lipschitz property. Since

$$(\Gamma(x, p))^{1/2} = Q_p((\Gamma(x, p) + p \otimes p)^{1/2}), \quad p \in S^{N-1}$$

with $Q_p(X) = R_p X R_p$, $R_p = I - p \otimes p/|p|^2$, this yields the Lipschitz property of $(\Gamma(x, p))^{1/2}$. □

Summarizing these results, we conclude that F in (4.6.16) satisfies the assumptions of Corollary 4.6.6 if (i) or (ii) of Lemma 4.6.7 is fulfilled for Γ and β, $\nabla_x \nabla_p \gamma(x, -p)$ and c are continuous in $\overline{\Omega} \times [0, T] \times S^{N-1}$ and Lipschitz continuous in x uniformly in $p \in S^{N-1}$ and $t \in [0, T]$ provided that $\beta \geq \beta_0 > 0$ on $\overline{\Omega} \times S^{N-1}$ with some constant β_0.

Remark 4.6.8 It seems to be nontrivial to allow spatially inhomogeneity in the second order term for (BCP) if the singularity at $\nabla u = 0$ in (4.1.1) is very strong so that (F3) is violated. We consider the level set equation of $V = ((a(x))^2 H)^3$, where $a \in C^\infty(\overline{\Omega})$ such that $\inf_{\overline{\Omega}} a > 0$. It is of the form

$$u_t + F(x, \nabla u, \nabla^2 u) = 0$$

with

$$F(x, p, X) = -(G(x, p, X))^3 / |p|^2, \; G(x, p, X) = \mathrm{trace}((aI) R_p X (aI)).$$

We argue as in the proof of Theorem 4.6.4 for $f(\eta) = \eta^6 \in \mathcal{F}$ to obtain (4.6.13) with the estimate

$$-3\mu I \leq \begin{pmatrix} X_\alpha & O \\ O & -Y_\alpha \end{pmatrix} \leq 3\mu J, \; \mu = \alpha K_1(\rho_\alpha).$$

This relation yields

$$G(y_\alpha, q_\alpha, Y_\alpha) - G(x_\alpha, q_\alpha, X_\alpha) \geq -N\mu |a(x_\alpha) - a(y_\alpha)|^2.$$

It also yields

$$|G(x_\alpha, q_\alpha, X_\alpha)|, |G(y_\alpha, q_\alpha, Y_\alpha)| \le C_1 K_1(\rho_\alpha).$$

Since

$$\begin{aligned} &F(x, p, X) - F(y, p, Y) \\ &= \frac{1}{|p|^2}\{(G(x, p, X))^2 + G(x, p, X)G(y, p, Y) + (G(y, p, Y))^2\} \\ &\quad \times\{G(y, p, Y) - G(x, p, X)\}, \end{aligned}$$

plugging in $x_\alpha, y_\alpha, X_\alpha, Y_\alpha, q_\alpha$ yields

$$F(x_\alpha, q_\alpha, X_\alpha) - F(y_\alpha, q_\alpha, Y_\alpha) \ge -\alpha C_2 K_1(\rho_\alpha)\frac{K_1(\rho_\alpha)^2}{(f'(\rho_\alpha))^2}\rho_\alpha^2 \tag{4.6.22}$$

if a is Lipschitz continuous. Here C_1 and C_2 are constants independent of α. By the form of f we see $f''(\eta)\eta^2/30 = f'(\eta)\eta/6 = f(\eta)$ so $\alpha K_1(\rho_\alpha)\rho_\alpha^2 \to 0$ as $\alpha \to \infty$ since $\alpha f(\rho_\alpha) \to 0$. However, there is no control on $K_1(\rho_\alpha)/q_\alpha$ nor $1/\rho_\alpha$ as $\alpha \to \infty$ so it is not clear that the right-hand side of (4.6.22) tends to zero as $\alpha \to \infty$.

4.7 Boundary Value Problems

We consider the boundary value problem

$$u_t + F(x, t, u, \nabla u, \nabla^2 u) = 0 \quad \text{in} \quad Q = \Omega \times (0, T), \tag{4.7.1}$$

$$B(x, \nabla u) = 0 \quad \text{on} \quad \partial\Omega \times (0, T). \tag{4.7.2}$$

When Ω is convex with C^2 boundary $\partial\Omega$ and (4.7.2) is the homogeneous Neumann boundary condition, i.e., $B(x, p) = \langle \nu(x), p \rangle$, where ν is the outward unit normal of $\partial\Omega$, it is easy to state a condition to guarantee the comparison principle:

(BCPB) Let u and v be sub- and supersolutions of (4.7.1) and (4.7.2) in $\overline{\Omega} \times (0, T)$, where Ω is bounded. If $-\infty < u^* \le v_* < \infty$ at $t = 0$, then $u^* \le v_*$ in Q.

Theorem 4.7.1 *Assume that Ω is a convex domain with C^2 boundary in $\mathbf{R}^N$. Assume that F is independent of x. Assume that* (F1)–(F3) *and* (F4) *holds. Then the comparison principle* (BCPB) *holds. If we assume* (F3′) *instead of* (F3)*, then* (BCPB) *is still valid (by replacing solutions by $\mathcal{F}_\Omega$-solutions) provided that $\mathcal{F}_\Omega$ is invariant under positive multiplication.*

By Lemma 4.1.3 this result yields:

Corollary 4.7.2 *Assume that Ω is a convex domain with C^2 boundary in $\mathbf{R}^N$. Assume that F is independent of x and r. Assume that $F : [0, T]\times(\mathbf{R}^N\setminus\{0\})\times\mathbf{S}^N \to \mathbf{R}$ is continuous and degenerate elliptic. Then* (BCPB) *holds if F is geometric.*

Proof of Theorem 4.7.1 To simplify the presentation we give a proof when F is also independent of r.

We first construct a 'barrier' of the boundary. Let $d(x, \partial\Omega)$ be the distance of x and the boundary $\partial\Omega$. As well known, $d(x, \partial\Omega)$ is C^2 near $\partial\Omega$ for $x \in \overline{\Omega}$; see e.g. D. Gilbarg and N. S. Trudinger [235, Lemma 14.16]. We set $b(x) = -d(x, \partial\Omega)$ near $\partial\Omega$ and extend to Ω in a suitable way so that $b \in C^2(\overline{\Omega})$ and $b < 0$ in Ω. Clearly, b fulfills $\nu(x) = \nabla b(x)$. We set $g(x) = b(x) - \inf_{\overline{\Omega}} b$ so that $g \geq 0$ in $\overline{\Omega}$.

We argue in the same way as in the proof of Theorem 4.1.1. Let $z_\sigma = (x_\sigma, t_\sigma, y_\sigma, s_\sigma) \in \overline{Q} \times \overline{Q}$ be a maximizer of $w - \varphi_\sigma$. We invoke the assumption that $u \leq v$ at $t = 0$ and observe that $t_\sigma > 0$, $s_\sigma > 0$ for large σ, say $\alpha > \alpha_0$, $\beta > \beta_0$. The main difference from the proof of Theorem 4.1.1 is that x_σ and y_σ may be on $\partial\Omega$.

We divide the situation into two cases following Remark 4.4.3. We fix $\alpha > \alpha_0$ and suppress the subscript α.

Case 1 For each $r > 0$ there is $\beta_r \to \infty$ (as $r \to 0$) and a maximizer z_{β_r} of $w - \varphi_{\beta_r}$ in $Q_{**} \times Q_{**}$ such that $|x_{\beta_r} - y_{\beta_r}| \leq r$, where $z_\beta = (x_\beta, t_\beta, y_\beta, s_\beta)$ and $Q_{**} = \overline{\Omega} \times (0, T]$.

We first fix r and denote β_r simply by β.

$$(w - \varphi_\beta)(x, t, y_\beta, s_\beta) \leq (w - \varphi_\beta)(x_\beta, t_\beta, y_\beta, s_\beta)$$

for $(x, t) \in Q_{**}$, we see

$$\max_{Q_{**}}(u - \varphi^+) = (u - \varphi^+)(x_\beta, t_\beta)$$

if

$$\varphi^+(x, t) = \alpha f(|x - y_\beta|) + \beta(t - s_\beta)^2 + \gamma t.$$

Our goal is to prove

$$\varphi_t^+(x_\beta, t_\beta) + F(t_\beta, \alpha\nabla_p f(|p|), \alpha\nabla_p^2 f(|p|)) \leq 0 \quad \text{at} \quad p = p_\beta \tag{4.7.3}$$

if $p_\beta := x_\beta - y_\beta \neq 0$ and

$$\varphi_t^+(x_\beta, t_\beta) = 2\beta(t_\beta - s_\beta) + \gamma \leq 0 \tag{4.7.4}$$

if $p_\beta = 0$. Since u is a subsolution in Q, this is clear if $x_\beta \in \Omega$. So we may assume that $x_\beta \in \partial\Omega$. We modify φ^+ to

$$\tilde{\varphi}^+ = \varphi^+ + f(|x - x_\beta|) + |t - t_\beta|^2$$

so that $u - \tilde{\varphi}^+$ takes a strict maximum at $x = x_\beta$. We set $\varphi_\delta^+ = \tilde{\varphi}^+ + \delta b$. By the convergence of maximum points (Lemma 3.2.5), there is a maximizer (x^δ, t^δ) of $u - \varphi_\delta^+$ in $\overline{Q}$ converging to (x_β, t_β) as $\delta \downarrow 0$.

If $y_\beta \neq x_\beta$, then $\nabla\varphi_\delta^+(x^\delta, t^\delta) \neq 0$ for small $\delta > 0$. If $x^\delta \in \partial\Omega$, then

$$\langle \nu(x^\delta), \nabla\tilde{\varphi}^+(x^\delta)\rangle \geq 0$$

as Ω is convex; we suppress t^δ as $\nabla\tilde{\varphi}^+$ does not depend on t. Since $\langle \nu(x^\delta), \nabla b(x^\delta)\rangle = 1$, this implies

$$\langle \nu(x^\delta), \nabla\varphi_\delta^+(x^\delta)\rangle \geq \delta > 0. \tag{4.7.5}$$

We recall Definition 3.3.7 with $\varphi = \varphi_\delta^+$ near (x_β, t_β). The estimate (4.7.5) guarantees that

$$\varphi_{\delta t}^+ + F(t^\delta, \nabla\varphi_\delta^+, \nabla^2\varphi_\delta^+) \leq 0 \quad \text{at} \quad (x^\delta, t^\delta) \tag{4.7.6}$$

regardless of whether $x^\delta \in \partial\Omega$ or $x^\delta \in \Omega$. Sending $\delta \to 0$ we obtain (4.7.3).

We discuss the case $y_\beta = x_\beta \in \partial\Omega$. Even in this case, if there is a subsequence $\delta_j \to 0$ such that $\nabla\varphi_{\delta_j}^+(x^{\delta_j}, t^{\delta_j}) \neq 0$, then we obtain (4.7.3) either $x_{\delta_j} \in \partial\Omega$ for all j or $x_{\delta_{j_k}} \in \Omega$ for some subsequence $\{j_k\}$ as in the previous paragraph. The remaining case is that $\nabla\varphi_\delta^+(x^\delta, t^\delta) = 0$, i.e.,

$$(\alpha + 1)\nabla_x f\left(|x - y_\beta|\right) + \delta\nabla b(x) = 0 \quad \text{at} \quad x = x^\delta. \tag{**}$$

Since $b(x) = -d(x, \partial\Omega)$ near $\partial\Omega$ for $x \in \bar{\Omega}$ and the identity $f'(0) = 0$, this identity (**) implies that $x^\delta \neq y_\beta = x_\beta$. Moreover, if $x^\delta \in \partial\Omega$, (**) implies that $\nu(x^\delta)$ and $x^\delta - x_\beta$ are parallel; see Fig. 4.3. This is impossible for small $\delta > 0$ since $x^\delta \to x_\beta$ so that $\nu(x^\delta) \to \nu(x_\beta)$ and $x^\delta - x_\beta/(x^\delta - x_\beta)$ with $x^\delta \in \partial\Omega$ converges to a tangent vector at x_β by taking a subsequence. These two vectors are not parallel, so we get a contradiction. We thus conclude that $x^\delta \in \Omega$ for sufficiently small $\delta > 0$. Since $\nabla\varphi_\delta^+(x^\delta, t^\delta) = 0$, by definition of a subsolution, we conclude that

$$\varphi_{\delta t}^+(x^\delta, t^\delta) \leq 0.$$

Sending $\delta \to 0$ yields (4.7.4). We thus conclude that either (4.7.3) or (4.7.4) holds even if $y_\beta = x_\beta$.

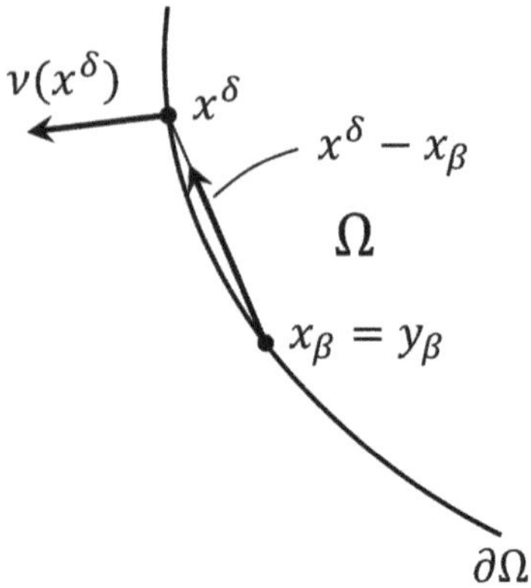

Fig. 4.3 Location of x^δ and x_β

We observe that

$$\min_{Q_{**}}(v - \varphi^-) = (v - \varphi^-)(y_\beta, s_\beta)$$

with $\varphi^-(y, s) = -\alpha f(|x_\beta - y|) - \beta(t_\beta - s)^2 - \gamma s$. We argue in a similar way to obtain

$$\varphi_t^-(y_\beta, s_\beta) + F(s_\beta, \alpha\nabla_p f(|p|), -\alpha\nabla_p^2 f(|p|)) \geq 0 \quad \text{at} \quad p = p_\beta \tag{4.7.7}$$

if $p_\beta \neq 0$ and

$$\varphi_t^-(y_\beta, s_\beta) = 2\beta(t_\beta - s_\beta) - \gamma \geq 0 \tag{4.7.8}$$

if $p_\beta = 0$. If $p_\beta = 0$, (4.7.4) and (4.7.8) yields a contradiction $2\gamma \leq 0$ so we may assume that $p_\beta \neq 0$. Subtracting (4.7.7) from (4.7.3) and sending $r \to 0$ so that $p_{\beta_r} \to 0$ (since $|p_{\beta_r}| < r$), we obtain a contradiction $2\gamma \leq 0$.

Case 2 There is $r_0 > 0$ such that for sufficiently large β any maximizer z_β of $w - \varphi_\beta$ in $Q_{**} \times Q_{**}$ satisfies $|x_{\beta_r} - y_{\beta_r}| > r_0$.

We define $\Psi_{\beta\delta}$ by

$$\Psi_{\beta\delta}(x, t, y, s) = \Phi_\beta(x, t, y, s) - \delta(g(x) + g(y))$$

for $\delta > 0$. Let $\hat{z} = (\hat{x}, \hat{t}, \hat{y}, \hat{s})$ be a maximizer of $\Psi_{\beta\delta}$ on $\overline{Q} \times \overline{Q}$. Clearly, $\sup_Z \Psi_{\beta\delta} \uparrow \sup_Z \Phi_\beta$ as $\delta \downarrow 0$, so as in Sect. 4.4 for sufficiently small δ, say $\delta < \delta_0(\beta)$, we see that $\hat{z} \in Q_{**} \times Q_{**}$ and $|\hat{x} - \hat{y}| > r_0$. (Note that $\hat{z}$ depends on β and δ.)

Since $\Psi_{\beta\delta} = w - \delta(g(x) + g(y)) - \varphi_\beta$ attains its maximum at $\hat{z}$ over $Q_{**} \times Q_{**}$, we apply Theorem 4.3.3 with $k = 2$, $N_1 = N_2 = N + 1$, $\mathcal{Z}_1 = \mathcal{Z}_2 = Q_{**}$, $\ell = 2$ to $w - \varphi = \Psi_{\beta\delta}$. Then we conclude as in the proofs of Theorems 4.1.1 and 4.1.4 (Case 2), that for each $\lambda_1 > 0$ there is $X, Y \in \mathbf{S}^N$ such that (4.4.11) holds and

$$(\gamma + \hat{\varphi}_t, \hat{\varphi}_x + \delta \nabla g(\hat{x}), X + \delta \nabla^2 g(\hat{x})) \in \overline{\mathcal{P}}^{2,+}_{Q_{**}} u(\hat{x}, \hat{t}), \tag{4.7.9}$$

$$(-\gamma + \hat{\varphi}_t, \hat{\varphi}_y - \delta \nabla g(\hat{y}), Y - \delta \nabla^2 g(\hat{y})) \in \overline{\mathcal{P}}^{2,-}_{Q_{**}} v(\hat{y}, \hat{s}). \tag{4.7.10}$$

Since $|\hat{x} - \hat{y}| > r_0$ for small δ, $\hat{\varphi}_x + \delta \nabla g(\hat{x}) \neq 0, \ -\hat{\varphi}_y - \delta \nabla g(\hat{y}) \neq 0$.

If $\hat{x}$ is on the boundary, because of the presence of g and the convexity of the domain

$$\langle \hat{\varphi}_x + \delta \nabla g(\hat{x}), \nu(x) \rangle \geq \delta > 0.$$

Thus the alternate requirement $B \leq 0$ in Definition 3.3.7 is not fulfilled. So (4.7.9) implies

$$\gamma + \hat{\varphi}_t + F(\hat{t}, \hat{\varphi}_x + \delta \nabla g(\hat{x}), X + \delta \nabla^2 g(\hat{x})) \leq 0 \tag{4.7.11}$$

either for $\hat{x} \in \Omega$ or $\hat{x} \in \partial\Omega$ since u is a subsolution in Q_{**}. Similarly for $\hat{y} \in \partial\Omega$,

$$\langle -\hat{\varphi}_y - \delta \nabla g(\hat{y}), \nu(y) \rangle \leq -\delta < 0$$

so (4.7.10) yields

$$-\gamma + \hat{\varphi}_t + F(\hat{s}, -\hat{\varphi}_y - \nabla g(\hat{y}), Y - \delta \nabla^2 g(\hat{y})) \geq 0 \tag{4.7.12}$$

since v is a supersolution. Subtracting (4.7.12) from (4.7.11), we send $\delta \to 0$ and then send $\beta \to \infty$ to get a contradiction $2\gamma \leq 0$ as in the proof of Theorem 4.1.4 Case 2. We omit the detail. □

Remark 4.7.3 There are several directions of extension of Theorem 4.7.1 by removing the convexity or handling more general boundary conditions. However, all such extensions so far exclude very singular equations and forced to assume (F3). We give a typical result.

Theorem 4.7.4 *Assume that Ω is a bounded domain in $\mathbf{R}^N$ with smooth boundary. We consider the level set mean curvature flow equation with the prescribed contact angle boundary condition (in Sect. 2.6.1), i.e.,*

$$\begin{cases} u_t - \Delta u + \sum \frac{u_{x_j} u_{x_j}}{|\nabla u|^2} u_{x_i x_j} = 0 \ \text{in} \ \ \Omega \times (0, T), \\ \frac{\partial u}{\partial \nu} + z|\nabla u| = 0 \qquad \text{on} \ \ \partial\Omega \times (0, T) \end{cases}$$

with a smooth function $z = z(x)$ satisfying $\sup_{\partial\Omega} |z| < 1$. Then (BCPB) *holds.*

The level set mean curvature flow equation can be extended to (2.6.10) provided that $\beta \in C^1(S^{N-1})$ and $\gamma \in C^4(S^{N-1})$ so that the assumption (i) of Lemma 4.6.7 is fulfilled. We do not give the proof of Theorem 4.7.4 because it is very technical. The reader is referred to papers of H. Ishii and M.-H. Sato [283] and G. Barles [47] for more general results as well as the proof of Theorem 4.7.4.

4.8 Notes and Comments

A version of comparison principles for viscosity solutions was first proved by M. G. Crandall and P.-L. Lions [130] and then by M. G. Crandall et al. [126] for first-order equations. Some comparison principles were proved for a special second order equation called the Hamilton–Jacobi–Bellman equation by P.-L. Lions [321–323] by an ad hoc stochastic control method. However, a general theory for the second order equations (even without singularity) remained open for quite a while until R. Jensen [294] developed several key ideas to overcome difficulties. Then H. Ishii [271] extended the theory to include more examples by introducing matrix inequalities of the general form (4.3.7); see also H. Ishii and P.-L. Lions [277]. A maximum principle (Theorem 4.3.2) for semicontinuous functions is due to M. G. Crandall and H. Ishii [127]. This work also introduced the notion of semijets $J^{2,\pm}$ and $\overline{J}^{2,\pm}$. For more detailed information on the development of the theory for comparison principles the reader is referred to the User's Guide by M. G. Crandall et al. [128] and a book by S. Koike [315]. See also B. Kawohl and N. Kutev [300] for further examination of structures of equations.

There are other types of comparison (or maximum) principles based on Alexandrov-Bakelman-Pucci type inequalities. For elliptic problems $F(x, u, Du, D^2u) = 0$ this type of theory often requires uniform ellipticity while the theory developed in the present chapter requires only degenerate ellipticity but strict monotonicity for $r \mapsto E(x, r, p, X)$. The reader is referred to a book of L. A. Caffarelli and X. Cabré [72] for a basic Alexandrov-Bakelman-Pucci inequality as well as a book of S. Koike [315]. See also I. Capuzzo-Dolcetta et al. [86] for further references and extension to unbounded domains. We shall not touch this type of results in this book.

A level set equation has a singularity at $\nabla u = 0$ if it is second order. For such an equation a comparison principle was first proved by Y.-G. Chen et al. [106] and independently by L. C. Evans and J. Spruck [166] (for the level set mean curvature flow equation). Theorem 4.1.1 under (F3) is due to Y.-G. Chen et al. [106]. The results were extended to unbounded domains and spatially inhomogeneous problems by Y. Giga et al. [210] which adjusts the elliptic version by M.-H. Sato [388] to a parabolic one. This work includes Theorem 4.1.4 under (F3). The extension to very singular equations (without (F3)) was established by S. Goto [239] and by H. Ishii and P. Souganidis [284] for geometric equations. The methods are different and the former work seems to be limited to bounded domains while the latter work applies to general domains. Also the latter work introduced the notion of $\mathcal{F}$-solution and it is not limited to geometric equations. In fact M. Ohnuma and K. Sato [368] extended their theory to non-geometric equations. When F is independent of t, Theorem 4.1.4 without (F3) is due to M. Ohnuma and K. Sato [368] and Corollary 4.1.5 without (F3) is due to H. Ishii and P. Souganidis [284].

Contents of Sect. 4.2.1 is essentially taken from User's Guide with special attention to parabolic semijets. Extension of a solution defined in $\Omega \times (0, T)$ to $\Omega \times (0, T]$ is often important as discussed in Sect. 4.2.2. It is stated in several papers including Y.-G. Chen et al. [107] and H. Ishii and P. Souganidis [284] (for $\mathcal{F}$-solutions). Most results in Sect. 4.2.2 are taken from the work of Y.-G. Chen et al. [107].

The contents of Sect. 4.3.1 is essentially taken from User's Guide except Theorem 4.3.3 that is very useful to handle parabolic problems. The proof of Theorem 4.1.1 is more transparent than the original proofs under (F3) of Y.-G. Chen et al. [106] and L. C. Evans and J. Spruck [166] since during their researches Theorem 4.3.2 was not available. The proof of Theorem 4.1.4 without (F3) is slightly different from H. Ishii and P. Souganidis [284] or M. Ohnuma and K. Sato [368] since F depends on t. Even under (F3) it is different from that of Y. Giga et al. [210] since we rather use Theorem 4.3.3 instead of the usual parabolic version of maximum principle (due to M. G. Crandall and H. Ishii [127]).

The Lipschitz preserving property is clear if F is spatially homogeneous, i.e., F is independent of x. The convexity preserving property is more difficult to obtain. When F satisfies (F3) and is independent of t, this property was first proved by Y. Giga et al. [210] by adjusting idea of H. Ishii and P.-L. Lions [277] for singular equations. A statement similar to Theorem 4.5.2 is stated by H. Ishii and P. Souganidis [284] without proof when F is independent of time. We here give a complete proof.

When F depends explicitly on the spatial variable x it is hard to state the results in a simple way. When x-dependence appears in first order terms, it is relatively easy to state (Theorem 4.6.1). Although this is not explicitly stated in the literature without assuming (F3), the proof seems to be standard. When x-dependence appears in the top order term, Theorem 4.6.4 is considered as a variant of comparison results of Y. Giga et al. [210] and of G. Barles et al. [55]. Even for nonsingular equations the power 2 of $\mu|x-y|^2$ in (4.6.8) is optimal in the sense that we cannot replace $\mu|x-y|^2$ by $\mu|x-y|^k$ with $k < 2$ $(k > 0)$. This is already pointed out by H. Ishii [271, Theorem 3.3]. As mentioned in Remark 4.6.8 it is nontrivial to extend spatial inhomogeneity in the second order term when (F3) is violated.

Boundary Value Problems The Neumann type boundary problem, i.e., $B(x,p) = \langle \nu(x), p \rangle$ was proposed for viscosity solutions first by P.-L. Lions [320] and established a comparison principle for first order equations. It was extended to second order equations with more complicated boundary condition by G. Barles [45] and by H. Ishii [273] including oblique type boundary conditions, when the equation has no singularities at $\nabla u = 0$. The first work for singular equation was done by M.-H. Sato [389], where he proved Theorem 4.7.1 under (F3). Extension to $\mathcal{F}$-solutions seems to be nontrivial from his proof so we provide a detailed proof. Under (F3) the convexity assumption on the domain was successfully removed by Y. Giga and M.-H. Sato [230] at the expense of restricting a class of F. Later, this result was generalized to an oblique type problem where $B(x,p) = \langle \nu(x), p \rangle + z|p|$ with a constant $z \in (-1, 1)$ by M.-H. Sato [390] when the domain is a half space. For a general domain and a general boundary condition $B(x, \nabla u) = 0$ the comparison principle has been proved by G. Barles [47] and H. Ishii and M.-H. Sato [283] by a different method and for a different generality. Such a comparison principle is useful not only to the level set method itself but also to stability analysis of stationary solution as presented in S.-I. Ei et al. [153].

Level Set Anisotropic Mean Curvature Flow Equation When we consider the level set equation (2.6.10) of a general evolution equation of an isothermal interface (2.5.2), our theory in this chapter requires at least C^2 regularity of the interfacial energy γ on S^{N-1} as mentioned in Sect. 4.1.3. In applied problems it is sometimes too restrictive. There are several extensions to relax regularity assumptions on γ. For example, M. E. Gurtin et al. [247] and M. Ohnuma and M.-H. Sato [369] independently relax the assumption in the way that $\nabla\gamma$ is Lipschitz but may not be C^1 in a finitely many points on S^1 when $N = 2$. See also Y. Giga [199], for a comparison principle of graph-like solutions. There is also a higher dimensional extension by H. Ishii [276].

When γ is not C^1, the equation has nonlocal nature. A typical example is the case when a Frank diagram of γ is a convex polyhedra. When $N = 2$, such energy is called crystalline and in this case the evolution law is not expected to be local. Nevertheless, when $N = 2$, a comparison principle has been established by introducing an appropriate notion of viscosity solutions by M.-H. Giga and Y. Giga [191]; for announcement see M.-H. Giga and Y. Giga [188]. The theory developed there establishes the level set method as well as several convergence results. In particular it provides convergence of the crystalline algorithm and asserts that crystalline flow is obtained as a limit of motion by smooth approximated interfacial energy as discussed in M.-H. Giga and Y. Giga [190]. See also K. Ishii and H. M. Soner [288] for convergence of the crystalline algorithm for curve shortening equation. Such convergence results were first proved by P. M. Girão and R. V. Kohn [237] (with convergence rate) and T. Fukui and Y. Giga [180] for graph-like functions. For a curve shortening equation P. M. Girão [236] proved the convergence for convex curves; see also a review of P. M. Girão and R. V. Kohn [238]. The crystalline motion was first proposed by S. Angenent and M. E. Gurtin [27] and independently by J. E. Taylor [411].

The theory of M.-H. Giga and Y. Giga [191] is based on the corresponding theory for graph-like functions developed by M.-H. Giga and Y. Giga [187, 189]. An idea of converting the problem to that of graph-like functions may be useful to analyze geometric equations. For the background of these materials the reader is referred to a review article of Y. Giga [202] and references cited there as well as recent development of the theory.

Further Comments for the Second Edition

Progress Related to Crystalline Energy and Nonlocal Flow. In the case γ is not C^1, the theory for higher dimensional cases ($N \geq 3$) has been well developed. As pointed out in the last paragraph the problem has nonlocal nature and definition of a solution itself is not a priori clear even if the evolving hypersurface is given as the graph of a function.

Let us concentrate the level set equation of

$$V = M(\mathbf{n})\left(-\operatorname{div}_{\Gamma_t} \xi(\mathbf{n})\right), \quad \xi = \nabla_p \gamma \tag{4.8.1}$$

with continuous $M(\mathbf{n}) \geq 0$, which is a special case of (2.5.2) when γ is not C^1. If γ is piecewise linear and convex with $\gamma > 0$ outside the origin, we say that γ is *(purely) crystalline*. For such γ, the theory of viscosity solutions has been established by Y. Giga and N. Požár [226–228]; see also a review paper by Y. Giga and N. Požár [229] where several approaches to this problem and related problems are compared. The main idea is extending the notion of viscosity solutions for nonlocal setting.

It is instructive to consider a total variation flow equation

$$u_t - \operatorname{div}(\nabla u/|\nabla u|) = 0 \tag{4.8.2}$$

which is more singular at $|\nabla u| = 0$ than that of the level set mean curvature flow equation

$$u_t - |\nabla u| \operatorname{div}(\nabla u/|\nabla u|) = 0.$$

To simplify the setting we consider (4.8.2) in $\mathbf{R}^n$ to avoid boundary condition. Since (4.8.2) is an L^2-gradient flow of the total variation flow of the total variation energy $\int_{\mathbf{R}^n} |\nabla u|$, it turns out that a solution can be interpreted as a function u satisfying

$$u_t - \operatorname{div} z = 0$$

with some $z = z(x,t)$ satisfying $|z(x,t)| \leq 1$ almost everywhere and $\operatorname{div} z$ is square-integrable in space which satisfies $-\int u \operatorname{div} z \, dx = \int |\nabla u|$ for almost all $t > 0$. Roughly speaking, integrating by parts of the last identity yields $z = \nabla u/|\nabla u|$ provided that $|\nabla u| \neq 0$ by the Schwarz inequality; see a nice book by F. Andreu-Vaillo et al. [16] for analysis of total variation flow equations.

The basic idea of approaches by Y. Giga and N. Požár is to introduce a class of test functions so that $\operatorname{div} z$ is well defined and assign its value as $\operatorname{div} z$. This idea has been introduced by M.-H. Giga and Y. Giga [187], but its analysis was limited for $n = 1$. M.-H. Giga et al. [196, 197] extended this method to higher dimensions to establish a unique solvability of equation with total variation flow type including

$$u_t - \sqrt{1 + |\nabla u|^2} \operatorname{div}(\nabla u/|\nabla u|) = 0 \tag{4.8.3}$$

under periodic boundary conditions with suitable definitions of viscosity solution. The comparison principle has been proved by careful analysis of facet, i.e., the flat part where $|\nabla u| = 0$. The existence is proved by approximation since Perron's method does not work. In fact, the proof for showing that a maximum subsolution should be a solution (Lemma 2.4.2) does not work in this type of nonlocal setting. Y. Giga and N. Požár [226–228] adjusted these ideas to the level set equation of

(4.8.1). Because of several anisotropies, approximation of facets are more involved than that for (4.8.3) to establish the comparison principle.

There is a complete different approach proposed by A. Chambolle et al. [96, 98]. Their approach is based on distance function of an evolving set. As already observed by H. M. Soner [399], if $E \subset \mathbf{R}^N \times (0, T)$ is a closed evolution of the mean curvature flow, the distance function d from $E(t)$ satisfies

$$d_t - \Delta d \geq 0 \quad \text{outside} \quad E.$$

For the wording "closed evolution" see the next chapter. Based on this observation, they introduced a notion of superflow. It is easier to write the case $\gamma = M$ and $\gamma(p) = \gamma(-p)$ for $p \in S^{N-1}$ in (4.8.1). For this case a main assumption of a superflow is

$$d_t^\gamma - \operatorname{div} z \geq 0 \quad \text{outside} \quad E$$
$$z \in \partial\gamma(\nabla d^\gamma)$$

for almost all (x, t), where

$$d^\gamma(x, t) = \inf\left\{\gamma^0(x - y) \mid y \in E\right\},$$

and $\gamma^0(x) = \inf\left\{\lambda \mid x \in \lambda W_\gamma\right\}$. The quantity $\partial\gamma$ denote the subdifferential of γ in $\mathbf{R}^N$, i.e.,

$$\partial\gamma(p) = \left\{v \mid \gamma(p+q) - \gamma(p) \geq v \cdot p \text{ for all } p \in \mathbf{R}^N\right\}.$$

In their approach, the comparison principle is easy to prove while construction of a solution is more involved. They constructed their solution by using what is now called Chambolle's scheme discussed in the added paragraphs of Sect. 6.6. For comparison of applicability of both theories are discussed in the review paper by Y. Giga and N. Požár [229].

We also note that the level set method is extended to nonlocal flow involving fractional mean curvature. Although there are several papers, we just mention a work by A. Chambolle et al. [97] where they introduce a suitable notion of a solution to the level set equation of quite general nonlocal curvature flow equations.

There are also several attempts to extend a level set approach to problems having order preserving structures like Hele-Shaw problem initiated by I. C. Kim [302]. We also note it is sometimes extended to problems which violates order-preserving structure like volume-preserving mean curvature flow developed by I. Kim and D. Kwon [304] and I. Kim et al. [305] by restricting profiles of evolutions like starshaped profiles.

The higher order crystalline flow like crystalline surface diffusion flow is still difficult to analyze and only special case is analyzed by M.-H. Giga and Y. Giga [194].

Boundary Value Problems. In this book, we mainly discuss the Neumann type boundary conditions. However, we note that the boundary condition involving time derivative is also interesting. A typical example is

$$u_t + \langle \nu(x), \nabla u \rangle = 0 \tag{4.8.4}$$

on the boundary $\partial\Omega$ of a domain Ω. If u is a solution of a level set mean curvature flow equation, this boundary condition formally says that

$$v = \langle \nu(x), \mathbf{n}(x) \rangle$$

where v is the normal velocity in $\partial\Omega$ on the contact set to $\partial\Omega$ of each level set of u. This type of problem was proposed by S. Angenent and M. E. Gurtin [28]. The boundary condition involving time derivatives is often called the *dynamic boundary conditions* and there is a large literature. We just mention results related to the theory of viscosity solutions. The theory of viscosity solutions including the comparison principle are established for a general class of fully nonlinear parabolic equations with nonlinear dynamic boundary condition by G. Barles [45, 47] but it excludes level set equations (involving time derivatives). For level set equations, less in known. So far comparison principle is only established by Y. Giga and N. Hamamuki [212], when Ω is a half space and the boundary condition is linear like (4.8.4). A game-theoretic approach to such dynamic boundary problems for level set curvature flow equations and general fully nonlinear parabolic equations is studied by N. Hamamuki and Q. Liu [250, 251]. Further details about the game-theoretic method can be found in Chaps. 7 and 8 of this book. We note that such a class of dynamic boundary problems for the Hamilton-Jacobi equations naturally arises in superconductivity and interface evolution is studied by C. M. Elliott et al. [156], while a degenerate dynamic boundary condition is considered by N. Hamamuki [249] to establish unique existence of viscosity solutions. Several asymptotic limits of time-dependent eikonal equations with dynamic boundary condition were studied by E. S. Al-Aidarous et al. [4].

In addition to recent developments on dynamic boundary problems, there has also been progress on geometric evolution equations with singular boundary conditions. An example, in the level set formulation, is

$$B(x, p) = \langle \nu(x), p \rangle - |p|, \quad x \in \partial\Omega, \ p \in \mathbf{R}^N,$$

which corresponds to a zero contact angle between the evolving surface and the boundary. It generalizes the classical oblique boundary condition, where the contact angle is typically assumed to be nonzero. Such a boundary condition may arise in power-type curvature flows $V = \kappa^\alpha$ with $\alpha > 1$, when the parabolic operator exhibits strong degeneracy on the boundary. In the case of bounded, graph-like motion by power curvature in one spatial dimension, the existence, uniqueness, and large-time behavior of viscosity solutions satisfying such boundary conditions have been studied by T. Kagaya and Q. Liu [296]. See T. Kagaya [295] for further well-posedness results in the case of unbounded solutions with a different regime of α. The large time behavior for the motion by power curvature of strictly convex smooth curves is studied by B. Choi et al. [112]. Similar problems have also been studied extensively in the context of power Gauss curvature flow equations. See for example the works by J. Urbas [421], K. Choi et al. [114] and B. Choi et al. [113].

Extensions to Non-Euclidean Settings. Recent advances in the analysis of curvature flow equations extend beyond the Euclidean framework. A level set method for surface evolution was extended to Riemannian manifolds by T. Ilmanen [267] for the mean curvature flow and later by D. Azagra et al. [39] for more general flows.

Sub-Riemannian differential geometry has also grown into an important area, especially in connection with P. Pansu's celebrated 1982 conjecture regarding the sub-Riemannian isoperimetric profile. We refer to the monograph of L. Capogna et al. [85] for an excellent introduction to the basics of the subject in the Heisenberg group. Motivated by mathematical models of perceptual completion and image processing, L. Capogna and G. Citti [83] initiated the study of the horizontal mean curvature flow equation on sub-Riemannian manifolds, establishing existence and uniqueness results for cylindrical viscosity solutions. While the definition of viscosity solutions is analogous to the Euclidean case, the singularity caused by vanishing horizontal gradients is stronger in the sub-Riemannian context, making the comparison principle and uniqueness significantly more difficult to prove. We refer to F. Ferrari et al. [172] for a comparison theorem for rotationally symmetric solutions in the Heisenberg group. By strengthening the notion of solutions via vanishing viscosity and Riemannian techniques, E. Baspinar and G. Citti [58] established a uniqueness result. The general comparison principle remains open in the sub-Riemannian setting. Related studies regarding geometric evolution in sub-Riemannian manifolds include the works of N. Dirr et al. [145], L. Capogna et al. [84], and A. Kijowski et al. [301] among many others.

Chapter 5
Classical Level Set Method

We introduce a notion of generalized solutions for surface evolution equations which tracks the evolution after singularities develop. For this purpose we solve the level set equations (Chap. 2) globally in time in the sense of viscosity solutions developed in Chaps. 3 and 4. We also prove that each level set of solution is determined by the corresponding level set of initial data and is independent of the choice of initial data for solutions of the level set equations. This uniqueness of level set is fundamental to define a notion of generalized solution by a level set of solutions of the level set equations. We also study various general properties of solutions. In particular we explain what is called *fattening phenomena*; a level set of solutions of the level set equations may have interior points even if the corresponding level set of initial data has no interior points.

5.1 Brief Sketch of a Level Set Method

We consider a surface evolution equation of the form

$$V = f(z, \mathbf{n}, \nabla \mathbf{n}) \quad \text{on} \quad \Gamma_t, \tag{5.1.1}$$

where $f : \mathbf{R}^N \times [0, T] \times \mathbf{E} \to \mathbf{R}$ is a given function as in Chap. 2. Here $\mathbf{E}$ is a bundle defined by

$$\mathbf{E} = \{(p, Q_p(X));\ p \in S^{N-1}, X \in \mathbf{S}^N\},$$

where $Q_p(X) = (I - p \otimes p)X(I - p \otimes p)$ for unit vector p. Examples of surface evolution equations including the Hamilton–Jacobi equations (2.5.6), the mean curvature flow equation (2.5.4) and its anisotropic version, truncated Gaussian

Y. Giga, Q. Liu, *Surface Evolution Equations*, Monographs in Mathematics 99,
https://doi.org/10.1007/978-3-032-20303-8_5

and symmetric curvature flow equation (2.6.22), and the affine curvature flow equations (2.5.14) enjoy at least the following two properties:

(f1) (Continuity) $f : \mathbf{R}^N \times [0, T] \times \mathbf{E} \to \mathbf{R}$ is continuous.
(f2) (Degenerate ellipticity)

$$f(z, p, Q_p(X)) \le f(z, p, Q_p(Y)) \quad \text{whenever } Q_p(X) \ge Q_p(Y).$$

In this book we consider (5.1.1) satisfying (f1) and (f2), i.e., continuous and degenerate parabolic surface evolution equations. We associate the level set equation of (5.1.1):

$$u_t + F(z, \nabla u, \nabla^2 u) = 0, \tag{5.1.2}$$

where F is given by

$$F(z, p, X) = -|p| f(z, -\hat{p}, -Q_{\hat{p}}(X)/|p|), \ \hat{p} = p/|p|. \tag{5.1.3}$$

We would like to solve (5.1.1) with given initial hypersurface globally in time with aid of the level set equation. Our level set method is summarized as follows. To simplify the argument we assume that Γ_t is a compact hypersurface (without boundary) so that $\Gamma_0 = \partial D_0$ for some bounded open set in $\mathbf{R}^N$.

1. **Initial auxiliary function**. We take an auxiliary function u_0 which is at least continuous in $\mathbf{R}^N$ such that

$$\Gamma_0 = \{x \in \mathbf{R}^N; u_0(x) = 0\}, \ D_0 = \{x \in \mathbf{R}^N; u_0(x) > 0\}. \tag{5.1.4}$$

The assumption that u_0 is positive in D_0 gives the orientation of Γ_0. Formally, outward unit normal from D_0 equals $\mathbf{n} = -\nabla u_0/|\nabla u_0|$ by this choice of u_0. For convenience we often arrange that u_0 equals a negative constant $-\alpha$ outside some big ball (Fig. 5.1).

2. **Global unique solvability of level set equations.** We solve (5.1.2) with initial data $u(x, 0) = u_0(x)$ globally in time in the sense of viscosity solutions.

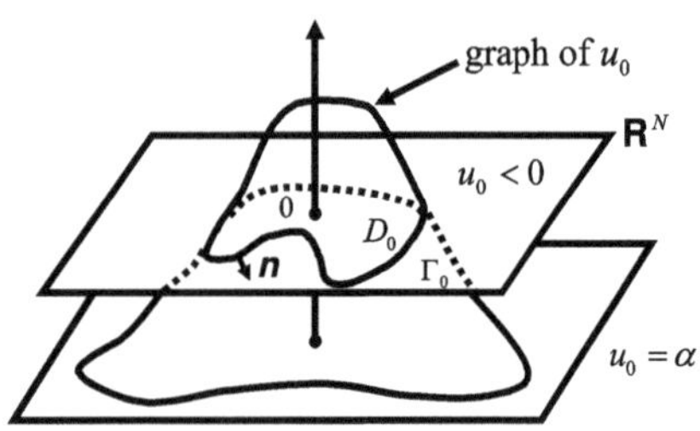

Fig. 5.1 An auxiliary function

3. **Generalized evolution**. For the global solution u of (5.1.2) with initial data u_0 we set

$$\Gamma = \{(x, t) \in \mathbf{R}^N \times [0, T);\ u(x, t) = 0\},$$
$$D = \{(x, t) \in \mathbf{R}^N \times [0, T);\ u(x, t) > 0\}$$

and expect that the cross-section

$$\Gamma(t) = \{x \in \mathbf{R}^N;\ (x, t) \in \Gamma\},$$
$$D(t) = \{x \in \mathbf{R}^N;\ (x, t) \in D\}$$

is a kind of generalized solution. (The orientation $\mathbf{n}$ of $\Gamma(t)$ is formally taken so that it is outward from $D(t)$ and $\mathbf{n} = -\nabla u/|\nabla u|$.) The sets Γ and D are a kind of weak or generalized solutions with initial data Γ_0 and D_0. Since (5.1.2) is invariant in addition of a constant, the value zero in the definition of D and Γ may be replaced by another value c, of course.

Step 1 is easy to be implemented by taking

$$u_0(x) = \max(\mathrm{sd}(x, \partial D_0), -1),$$

where sd denotes the signed distance of ∂D_0 defined by

$$\mathrm{sd}(x, \partial D_0) = \begin{cases} \mathrm{dist}(x, \partial D_0), & x \in D_0, \\ -\mathrm{dist}(x, \partial D_0), & x \notin D_0, \end{cases}$$

where $\mathrm{dist}(x, \partial D_0)$ denotes the distance from a point x to a set ∂D_0. The global solvability of Step 2 is one of main topics of this chapter. Actually, Step 2 is implemented with the aid of the theory of viscosity solutions. Since our definition of Γ and D in Step 3 is extrinsic, to see that our definition is well defined, we should check that Γ and D are determined by Γ_0 and D_0 respectively and independent of the choice of u_0 satisfying (5.1.4). We call this part of the problem "uniqueness of generalized evolutions" by naming Γ and D generalized evolutions. In the next section we discuss the problem of uniqueness of generalized evolutions.

We conclude this section by giving rigorous definitions of generalized evolutions. There are at least two ways to define them. The first one only applies evolution of bounded sets (or its complement) so it is restrictive. However, the proof for the uniqueness depends on comparison principle (BCP) in a bounded domain other than the invariance (Theorem 5.2.1) and it is instructive. The second one is very general but we need the comparison principle in $\mathbf{R}^N$ and the global solvability for (5.1.2) to prove the uniqueness of the generalized evolutions. Moreover the proof is not intuitive. We only give the proof for the first one in Sect. 5.2 and postpone the proof for the second one to Sect. 5.3.

Definition 5.1.1 Let D_0 be a bounded open set in $\mathbf{R}^N$. An open set D in $Z = \mathbf{R}^N \times [0, T)$ is called a *(generalized) open evolution* of (5.1.1) with initial data D_0 if there exist an ($\mathcal{F}_{\mathbf{R}^N}$-)solution $u \in K_\alpha(Z)$ of (5.1.2) in Z that satisfies

$$D = \{(x, t) \in Z; u(x, t) > 0\}, \; D_0 = \{x \in \mathbf{R}^N; u(x, 0) > 0\},$$

for some $\alpha \le 0$, where $K_\alpha(Z)$ denotes the space of all real-valued continuous functions u on Z such that $u \equiv \alpha$ outside $B_R(0) \times [0, T)$ for some $R > 0$.

Let E_0 be a bounded closed set in $\mathbf{R}^N$. A closed set E in Z is called a *(generalized) closed evolution* of (5.1.1) with initial data E_0 if there exists an ($\mathcal{F}_{\mathbf{R}^N}$-)solution $u \in K_\alpha(Z)$ of (5.1.2) in Z that satisfies

$$E = \{(x, t) \in Z; u(x, t) \ge 0\}, \; E_0 = \{x \in \mathbf{R}^N, u(x, 0) \ge 0\}$$

for some $\alpha < 0$. If $E_0 = \overline{D_0}$, the set $E \backslash D = \Gamma$ is called a *(generalized) interface evolution* of (5.1.1) with initial data $\Gamma_0 = E_0 \backslash D_0$.

Definition 5.1.2 We replace $K_\alpha(Z)$ by BUC(Z) to define generalized evolutions for arbitrary open and closed sets in Z. By definition, if D is an open evolution of (5.1.2) with initial data D_0 in Definition 5.1.1, it should be so in the sense of this definition since $K_\alpha(Z) \subset$ BUC(Z).

Remark 5.1.3 We are tempted to use the term 'level set solution' to describe 'generalized evolution'. The reason we did not use this word is that we use 'level set solution' in Chap. 6 in a different sense although it turns out both notions are equivalent (Proposition 6.2.8 and Remark 6.2.9). We often suppress the word 'generalized'. The word 'evolution' to describe D and E was used by S. Altschuler et al. [10].

5.2 Uniqueness of Bounded Evolutions

We shall study whether level set solutions D and E are determined by D_0 and E_0 respectively and are independent of the choice of an auxiliary function u.

5.2.1 Invariance Under Change of Dependent Variables

We first study special invariance of geometric equations. As observed in Sect. 2.6.4 if u solves a geometric equation so does $\theta(u)$ for any θ with $\theta' \ge 0$ at least formally. We state this property in a rigorous way.

Theorem 5.2.1 (Invariance) *Assume that* $F : W_0 = \overline{\Omega} \times [0, T] \times (\mathbf{R}^N \backslash \{0\}) \times \mathbf{S}^N \to \mathbf{R}$ *is continuous and geometric, where* Ω *is an open set in* $\mathbf{R}^N$. *If* u *is an* $\mathcal{F}_\Omega$*-subsolution (resp.* $\mathcal{F}_\Omega$*-supersolution) of*

$$u_t + F(x, t, \nabla u, \nabla^2 u) = 0 \quad in \quad Q = \Omega \times (0, T), \tag{5.2.1}$$

then the composite function $\theta \circ u = \theta(u)$ *is also an* $\mathcal{F}_\Omega$*-subsolution (resp.* $\mathcal{F}_\Omega$*-supersolution) of* (5.2.1) *provided that* $\theta : \mathbf{R} \to \mathbf{R}$ *is continuous and nondecreasing. One may weaken the assumption on continuity of* θ *by upper-semicontinuity (resp. lower-semicontinuity) with values* $\mathbf{R} \cup \{-\infty\}$ *(resp.* $\mathbf{R} \cup \{+\infty\}$*).*

To prove this result we need invariance of class C_F^2 under change of dependent variables.

Lemma 5.2.2 *Assume the same hypothesis of Theorem 5.2.1 concerning* F *and* Ω.

(i) *If* $g \in \mathcal{F}_\Omega$ *and* $\theta : \mathbf{R} \to \mathbf{R}$ *is* C^2 *with* $\theta' > 0$, $\theta'' \geq 0$ *everywhere, then* $\theta \circ g \in \mathcal{F}_\Omega$.
(ii) *If* $\varphi \in C_F^2(Q)$*, then* $\theta \circ \varphi \in C_F^2(Q)$ *for any* $\theta \in C^2(\mathbf{R})$ *with* $\theta' > 0$.

Proof

(i) Since F is geometric, for $f = \theta \circ g$ and g we see by (4.1.2) (in the proof of Lemma 4.1.3)

$$F(z, \nabla_p f(\rho), \pm\nabla_p^2 f(\rho)) = \frac{f'(\rho)}{\rho} F(z, p, \pm I),$$

$$F(z, \nabla_p g(\rho), \pm\nabla_p^2 g(\rho)) = \frac{g'(\rho)}{\rho} F(z, p, \pm I)$$

for $\rho = |p| > 0$. Since $f'(\rho) = \theta'(g(\rho))g'(\rho)$, the preceding two formulas imply that

$$\lim_{p \to 0} \sup_{z \in Q} |F(z, \nabla_p f(\rho), \pm\nabla_p^2 f(\rho))| = 0$$

if the same formula holds for g instead of f. Other conditions $f(0) = f'(0) = f''(0) = 0$, $f''(r) > 0$ for $r > 0$ follow from $\theta' > 0, \theta'' \geq 0$ and the corresponding properties for g.

(ii) This can be proved independent of (i). We set $\psi = \theta \circ \varphi$. We note that $\nabla\varphi(z) = 0$ is equivalent to $\nabla\psi(z) = 0$ since $\theta' > 0$. Thus it suffices to check the behavior of ψ near point $\hat{z}$ where $\nabla\varphi(\hat{z}) = 0$. We may assume that $\theta(0) = 0$, $\varphi(\hat{z}) = 0$, $\varphi_t(\hat{z}) = 0$. Our goal is to prove

$$|\psi(z)| \leq f(|x - \hat{x}|) + \omega_1(|t - \hat{t}|)$$

as $z = (x, t) \to \hat{z} = (\hat{x}, \hat{t})$ in Q for some $f \in \mathcal{F}_\Omega$ and ω_1 with $\omega_1(\sigma)/\sigma \to 0$ as $\sigma \to 0$, $\omega_1(0) = 0$. Since $\varphi \in C_F^2(Q)$, there is $g \in \mathcal{F}_\Omega$ and $\overline{\omega}_1$ with $\overline{\omega}_1(\sigma)/\sigma \to 0$ as $\sigma \to 0$ with $\overline{\omega}_1(0) = 0$ such that

$$|\varphi(z)| \le g(|x-\hat{x}|) + \overline{\omega}_1(|t-\hat{t}|)$$

as $(x,t) \to (\hat{x},\hat{t})$ in Q. This implies

$$|\psi(z)| = |\theta \circ \varphi(z)| \le \theta(g(|x-\hat{x}|) + \overline{\omega}_1(|t-\hat{t}|)).$$

Since $\theta(0) = 0$ and $\theta'(0) = \gamma > 0$, the right-hand side is dominated by

$$2\gamma g(|x-\hat{x}|) + 2\gamma\overline{\omega}_1(|t-\hat{t}|)$$

for (x,t) close to $(\hat{x},\hat{t})$. Since $ag \in \mathcal{F}$, this yields the desired estimate for $|\psi(z)|$ with $f = 2\gamma g$, $\omega_1 = 2\gamma\overline{\omega}_1$.

□

Proof of Theorem 5.2.1 We may assume that $\mathcal{F}_\Omega$ is not empty. The property $(\theta \circ u)^*(z) < \infty$ for $z \in Q$ is trivial since $u^*(z) < \infty$ for $z \in Q$.

1. We first discuss the case when $\theta \in C^2(\mathbf{R})$ and $\theta' > 0$ everywhere. Assume that $\varphi \in C^2_F(Q)$ and $z = (x,t) \in Q$ satisfies

$$\max_Q(\theta \circ u - \varphi) = \theta(u(z)) - \varphi(z) = 0.$$

Since $\theta \in C^2$ and $\theta' > 0$ the inverse function $h = \theta^{-1}$ is also C^2 and $h' > 0$. By definition

$$\max_Q(u - \psi) = u(z) - \psi(z) = 0, \tag{5.2.2}$$

where $\psi = h \circ \varphi$. By Lemma 5.2.2 (ii) $\psi = h \circ \varphi \in C^2_F(Q)$; we do not use Lemma 5.2.2 (i). If $\nabla\varphi(z) = p \neq 0$, geometricity of F implies that

$$F(z, \nabla\psi(z), \nabla^2\psi(z)) = F(z, \mu p, \mu X + \sigma p \otimes p) = \mu F(z, p, X),$$

where $X = \nabla^2\varphi(z)$, $\mu = h'(\varphi(z))$, $\sigma = h''(\varphi(z))$. Since u is a subsolution of (5.2.1), it follows that

$$\begin{aligned} 0 &\ge \frac{\partial\psi}{\partial t}(z) + F(z, \nabla\psi(z), \nabla^2\psi(z)) \\ &= h'(\varphi(z))\left(\frac{\partial\varphi}{\partial t}(z) + F(z, \nabla\varphi(z), \nabla^2\varphi(z))\right), \end{aligned}$$

which implies the condition for subsolution when $\nabla\varphi(z) = p \neq 0$. If $\nabla\varphi(z) = p = 0$ then $\nabla(h \circ \varphi)(z) = 0$. Since u is an $\mathcal{F}$-subsolution, we see

$$0 \geq \frac{\partial \psi}{\partial t}(z) = h'(\varphi(z))\frac{\partial \varphi}{\partial t}(z)$$

so $\theta \circ u$ is now an $\mathcal{F}$-subsolution.

2. For general θ we take an approximate sequence $\theta_m \in C^2(\mathbf{R})$ with $\theta_m' > 0$ such that $\theta = \limsup^* \theta_m$; the existence of such a sequence is guaranteed in Lemma 5.2.3 below. Since $u_m = \theta_m \circ u$ is an $\mathcal{F}$-subsolution and $(\theta \circ u)^* = \limsup^* u_m$, the stability result (Theorem 3.2.1) in Chap. 3 implies that u is an $\mathcal{F}$-subsolution.

The proof of a supersolution is symmetric so is omitted. □

Lemma 5.2.3 *Let θ be an upper semicontinuous nondecreasing function defined on $\mathbf{R}$ with values in $\mathbf{R} \cup \{-\infty\}$. Then there is a smooth θ_m with $\theta_m' > 0$ such that $\theta = \limsup^* \theta_m$. If $\theta : \mathbf{R} \to \mathbf{R}$ is moreover continuous, the sequence θ_m can be arranged so that θ_m converges to θ uniformly in $\mathbf{R}$.*

Proof

1. We approximate θ by nondecreasing piecewise linear functions. We may assume that $\theta \not\equiv -\infty$ since otherwise $\theta_m(t) = -m + t/m$ gives a desired sequence. Let I be the infinite interval such that

$$I = \{t \in \mathbf{R},\ \theta(t) > -\infty\}.$$

For an integer j and positive integer m we set

$$a_j^{(m)} = \inf\{t \in I;\ \theta(t) \geq j/m\}$$

with convention that $a_j^{(m)} = +\infty$ if there is no element $t \in I$ that satisfies $\theta(t) \geq j/m$. Since θ is nondecreasing, $a_j^{(m)} \leq a_{j+1}^{(m)}$ for any integer j and positive integer m and $\{a_j^{(m)}\}_{j=-\infty}^{\infty}$ has no accumulation points in the interior of I. We divide the situation into two cases depending on whether or not $\theta(t) \to -\infty$ as $t \downarrow \gamma = \inf I$.

(i) (the case $\lim_{t\downarrow\gamma} \theta(t) = -\infty$)
We take

$$\vartheta_m(t) = (j_+ + 1)/m \quad \text{at} \quad t = a_j^{(m)} \in I$$

where j_+ is a number such that

$$j_+ = \max\{\ell;\ a_\ell^{(m)} = a_j^{(m)}\}.$$

Fig. 5.2 Construction of ϑ_m

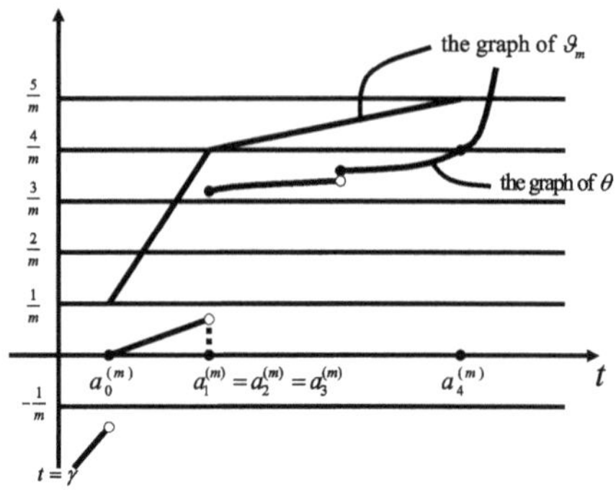

We assign the value of ϑ_m for $t \neq a_j^{(m)}$ such that ϑ_m is continuous, piecewise linear (linear outside $\{a_j^{(m)}\}_{j=-\infty}^{\infty}$) in $I = (\gamma, \infty)$ (Fig. 5.2). To extend ϑ_m outside I we set

$$\theta_m(t) = \begin{cases} \min(-m, \vartheta_m(t)) & t \in I, \\ -m & t \notin I. \end{cases}$$

Since $\lim_{t \downarrow \gamma} \theta(t) = -\infty$, $\theta_m(t)$ is continuous, piecewise linear and nondecreasing. By definition $\vartheta_m \geq \theta$, and since θ is upper semicontinuous it is not difficult to see that

$$\theta = \limsup_{m \to \infty}{}^* \theta_m.$$

(ii) (the case $\lim_{t \downarrow \gamma} \theta(t) = \beta > -\infty$)
We set ϑ_m as in (i) so that ϑ_m is well defined on $I = [\gamma, \infty)$. We set

$$\theta_m(t) = \begin{cases} \vartheta_m(t) & \text{for} \quad t \geq \gamma, \\ m(t - \gamma) + (j_1 + 1)/m & \text{for} \quad t \leq \gamma, \end{cases}$$

where $j_1 = \max\{\ell; a_\ell^{(m)} = \gamma\}$. The function $\theta_m \geq \theta$ is continuous, piecewise linear and nondecreasing. The convergence $\theta = \limsup^* \theta_m$ is not difficult to prove. In both cases (i) and (ii) we remark that the converge $\theta_m \to \theta$ is uniform if $\theta : \mathbf{R} \to \mathbf{R}$ is continuous by this construction.

2. We approximate the nondecreasing piecewise linear function θ_m by a nondecreasing C^2 function from above by mollifying θ_m near nondifferentiable points. We still denote the C^2 approximation of θ_m by θ_m.
3. We approximate the nondecreasing C^2 function θ_m by a C^2 function whose derivative is always positive. We take a positive bounded C^2 function r on $\mathbf{R}$ whose derivative is positive everywhere; a typical example of such an r is $r(t) = \operatorname{Arctan} t + \pi/2$, so we write

$$\tilde{\theta}_m(t) = \theta_m(t) + r(t)/m$$

to get $\theta = \limsup^* \tilde{\theta}_m$ and $\tilde{\theta}_m \geq \theta$. This $\tilde{\theta}_m \in C^2(\mathbf{R})$ now satisfies $\tilde{\theta}'_m(t) > 0$ for all $t \in \mathbf{R}$. We thus obtained the desired sequence. Again if $\theta : \mathbf{R} \to \mathbf{R}$ is continuous, the convergence $\tilde{\theta}_m \to \theta$ is uniform which is inherited from Step 1. □

Remark 5.2.4 The invariance property stated in Theorem 5.2.1 is easily extended to more general geometric equations including boundary value problems. We shall state these results without proof since the proofs are essentially the same as that of Theorem 5.2.1.

Theorem 5.2.5 (Invariance for General Equations) *Let $E(z, \cdot, \cdot)$ be a real-valued function defined on a dense set of W of $\mathbf{R}^d \times \mathbf{S}^d$ for $z \in \mathcal{O}$, where $\mathcal{O}$ is a local compact subset $\mathbf{R}^d$. Assume that the equation*

$$E(z, Du, D^2u) = 0 \quad in \quad \mathcal{O} \tag{5.2.3}$$

is geometric in the sense of Definition 2.6.13. If u is a subsolution (resp. super-solution) of (5.2.3) in $\mathcal{O}$, then $\theta \circ u$ is also a subsolution (resp. supersolution) if θ is nondecreasing, upper semicontinuous (resp. lower semicontinuous) on $\mathbf{R}$ with values in $\mathbf{R} \cup \{-\infty\}$ (resp. $\mathbf{R} \cup \{+\infty\}$).

This results applies to level set equations for surface equations with boundary conditions. For boundary value problems of very singular equations we also get invariance by interpreting their solutions by $\mathcal{F}$-solutions in Definition 3.3.7.

5.2.2 Orientation-Free Surface Evolution Equations

As we already observed in Chap. 2, there are a class of surface evolution equations invariant under the change of orientations of hypersurfaces. Such a surface evolution equation is called orientation-free which is rigorously defined in Definition 5.2.6 below. Examples include the mean curvature flow equation. For such a class of equations we shall show that $\theta \circ u$ is a solution if u is a solution, without assuming that θ is nondecreasing in Theorem 5.2.1, (where θ is assumed to be continuous).

Definition 5.2.6 If f in (5.1.1) fulfills

$$f(z, -p, -Q_p(X)) = -f(z, p, Q_p(X)) \tag{5.2.4}$$

for all $z \in \mathbf{R}^N \times [0, T]$, $(p, Q_p(X)) \in \mathbf{E}$, Eq. (5.1.1) is called *orientation-free*. It is clear that this condition is equivalent to the property

$$F(z, -p, -X) = -F(z, p, X) \tag{5.2.5}$$

for all $z \in \mathbf{R}^N \times [0, T]$, $p \in \mathbf{R}^N \backslash \{0\}$, $X \in \mathbf{S}^N$ where F is in (5.1.3). So if (5.2.5) holds for geometric F, then Eq. (5.2.1) is also called *orientation-free*.

Theorem 5.2.7 *Assume the same hypothesis of Theorem 5.2.1 and* Ω. *Assume that* (5.2.1) *is orientation-free. If* u *is an* $\mathcal{F}_\Omega$*-solution of* (5.2.1), *then so is* $\theta \circ u$ *provided that* $\theta : \mathbf{R} \to \mathbf{R}$ *is continuous.*

Proof

1. We suppress the symbol $\mathcal{F}_\Omega$ in the proof. We first note that $(\theta \circ u)^* < \infty$ and $(\theta \circ u)_* > -\infty$ since $-\infty < u_* \leq u^* < +\infty$. By (5.2.5) and definition of sub- and supersolutions we see that $-u$ is a subsolution (resp. supersolution) if u is a supersolution (resp. subsolution).
2. By Theorem 5.2.1 and Step 1, $\theta \circ u$ is a solution if $\theta : \mathbf{R} \to \mathbf{R}$ is continuous and either nondecreasing or nonincreasing and u is a solution.
3. For general continuous θ we may assume that u is bounded on $Q = \Omega \times (0, T)$. Indeed, for a given $M > 0$, by Theorem 5.2.1, $u_M = \sigma_M \circ u$ for $\sigma_M(t) = (t \wedge M) \vee (-M)$ is a solution if u is a solution. (Here we again use a convention $a \vee b = \max(a, b)$ and $a \wedge b = \min(a, b)$.) If $\theta(u_M)$ is a solution, so is $\theta(u)$ by stability principle (Sect. 3.2.1) since $\theta(u)$ is given as a limit of $\theta(u_M)$ as $M \to \infty$.
4. We approximate θ by a polynomial θ_m uniformly on $[-M - 1, M + 1]$ by Weierstrass' approximation theorem. Thus again by the stability principle we may assume that θ is a polynomial on $[-M - 1, M + 1]$.
5. Since θ is polynomial, there are only a finite number of local maximizers and minimizers of θ in $(-M - 1, M + 1)$. We may assume that θ has either a maximizer or a minimizer since otherwise θ is either nondecreasing or nonincreasing. By the symmetry of the argument we may assume that θ has a maximizer in $(-M - 1, M + 1)$. For given m we truncate θ near a maximizer r; we set
$$\theta_m(t) = \min(\theta(t), \theta(r) - 1/m)$$
for all $t \in (s_-, s_+)$, where (s_-, s_+) is a maximal interval where θ takes no local maximum or minimum in (s_-, s_+) other than at $t = r$. Similarly, we also truncate θ near a minimizer r by the value $\theta(r) + 1/m$ and define new θ_m. Our θ_m has the property that θ_m is constant on some closed interval where θ_m takes either a local maximum or minimum at least for sufficiently large m. Again by stability principle we may assume that θ is constant on some closed interval where θ takes either a local maximum or minimum and the number of such intervals is finite in $[-M - 1, M + 1]$.
6. Assume that $|u| \leq M$ and θ is the truncated function obtained at the end of Step 5. Assume that
$$\max(\theta \circ u - \varphi) = (\theta \circ u - \varphi)(\hat{x}, \hat{t})$$

for some $(\hat{x}, \hat{t}) \in Q$ and $\varphi \in C_F^2(Q)$. By the choice of θ the function θ is either nondecreasing or nonincreasing in a neighborhood $[\hat{u} - 2\delta, \hat{u} + 2\delta]$ for some small $\delta > 0$, where $\hat{u} = u^*(\hat{x}, \hat{t})$. By invariance (Theorem 5.2.1)

$$u_\delta = (u \wedge (\hat{u} + \delta)) \vee (\hat{u} - \delta)$$

is also a solution. We modify θ outside $[\hat{u} - \delta, \hat{u} + \delta]$ to get a nonincreasing continuous function $\tilde{\theta}$ that satisfies $\theta = \tilde{\theta}$ on $[\hat{u} - \delta, \hat{u} + \delta]$, since $\theta \circ u_\delta = \tilde{\theta} \circ u_\delta$.

Step 2 implies that $\theta \circ u_\delta$ is a solution. Thus in particular we have the desired inequality for φ at $(\hat{x}, \hat{t})$ for subsolution since

$$\max(\theta \circ u_\delta - \varphi) = (\theta \circ u_\delta - \varphi)(\hat{x}, \hat{t}).$$

We thus conclude that $\theta \circ u$ is a subsolution. The proof for a supersolution is symmetric so is not repeated. □

5.2.3 Uniqueness

In this subsection we prove the uniqueness of generalized evolutions defined in Definition 5.1.1. As we will see in Sect. 5.3, it is possible to prove the uniqueness of generalized evolutions defined in Definition 5.1.2 and practically and logically speaking this is enough since Definition 5.1.2 is more general than Definition 5.1.1. However, the uniqueness proof for generalized evolutions in the sense of Definition 5.1.2 needs several properties of equations including global solvability of (5.1.2) other than the comparison principle and the invariance property. The proof for evolutions in Definition 5.1.2 is intuitive and instructive and it only depends on the comparison principle in a bounded domain and the invariance property (Theorem 5.2.1). So we present the proof here.

Theorem 5.2.8 *Assume that the level set Eq.* (5.1.2) *of surface evolution equation* (5.1.1) *(satisfying* (f1)*) has the comparison principle* (BCP) *in every ball. (This implicitly assumes* (f2).*) Then there is at most one open (resp. closed) evolution of Definition 5.1.1 for a given initial bounded open (resp. closed) set in* $\mathbf{R}^N$.

For the proof we introduce an elementary fact for level sets.

Lemma 5.2.9 *Let Y be a closed set in $\mathbf{R}^N$. For $u_0, v_0 \in C(Y)$ assume that the set $\{u_0 > 0\}(= \{x \in Y; u_0(x) > 0\})$ is included in $\{v_0 > 0\}$. Assume that the boundary of $\{u_0 > 0\}$ is compact. (If $\{v_0 > 0\}$ includes a neighborhood of infinity, we further assume that $\liminf_{|x|\to\infty} v_0(x) > 0$ and that u_0 is bounded from above.) Then there exists a nondecreasing function $\theta \in C(\mathbf{R})$ such that $\theta(s) = 0$ (for $s \leq 0$) and $\theta(s) > 0$ (for $s > 0$) and*

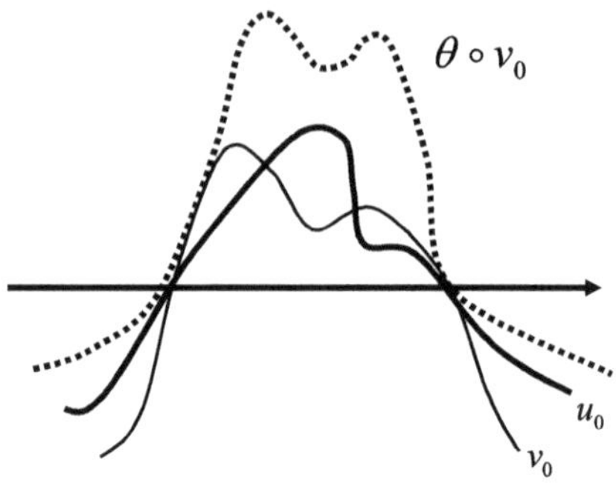

Fig. 5.3 Graphs of u_0, v_0 and $\theta \circ v_0$

$$u_0 \leq \theta \circ v_0 \quad in \quad Y.$$

Proof We define $\overline{\theta} : [0, \infty) \to [0, \infty)$ by

$$\overline{\theta}(\sigma) = \sup\{u_0(y) \vee 0;\ y \in V(\sigma)\},\ V(\sigma) = \{y \in Y;\ 0 \leq v_0(y) \leq \sigma\}.$$

We extend $\overline{\theta}$ outside $[0, \infty)$ by zero and the extended $\overline{\theta}$ is still denoted $\overline{\theta}$. By definition $\overline{\theta}$ is nondecreasing and $u_0 \leq \overline{\theta} \circ v_0$ in Y. Our assumption on v_0 implies that $V(\sigma)$ is compact for small $\sigma \geq 0$. Thus $\overline{\theta} \to 0$ as $\sigma \to 0$ since otherwise it would contradict the assumption that $u_0(y) > 0$ implies $v_0(y) > 0$ for all $y \in Y$. By Lemma 3.1.9 there is a nondecreasing function $\theta \in C(\mathbf{R})$ that satisfies $\theta \geq \overline{\theta}$ in $\mathbf{R}$ and $\theta(\sigma) = \overline{\theta}(\sigma)$ for $\sigma \leq 0$. This θ fulfills all desired properties since $\theta \geq \overline{\theta}$ implies that $u_0 \leq \overline{\theta} \circ v_0 \leq \theta \circ v_0$ in Y (Fig. 5.3). □

Proof of Theorem 5.2.8 Let $u \in K_\alpha(Z)$ and $v \in K_\beta(Z)$ be solutions of (5.1.2), where $\alpha, \beta \leq 0$. Assume that

$$D_0 = \{u_0 > 0\} = \{v_0 > 0\}$$

where $u_0(x) = u(x, 0)$, $v_0(x) = v(x, 0)$ and that D_0 is bounded. By Lemma 5.2.9 there exists a nondecreasing continuous function $\theta : \mathbf{R} \to \mathbf{R}$ such that $\theta(s) = 0$ for $s \leq 0$, $\theta(s) > 0$ for $s > 0$ and that $u_0 \leq \theta \circ v_0$ in $\mathbf{R}^N$. By the invariance (Theorem 5.2.1) the function $w := \theta \circ v \in K_0(Z)$ is a solution of (5.1.2). By definition there is a ball $B_R(0)$ such that $w \equiv 0$ and $u \equiv \alpha$ outside $B_R(0) \times [0, T)$. Since $u_0 \leq \theta \circ v_0$ implies that $u \leq w$ initially, the comparison principle (BCP) for $\Omega = \mathrm{int} B_R(0)$ implies that $u \leq w$ on $B_R(0) \times [0, T)$. It now follows that $\{u > 0\}(= \{(x, t) \in Z; u(x, t) > 0\})$ is included in $\{w > 0\}$. Since the two sets $\{w > 0\}$ and $\{v > 0\}$ agree with each other, $\{u > 0\}$ is included in $\{v > 0\}$. If we exchange the role of u_0 and v_0, the opposite inclusion holds. Thus we see $D = \{u > 0\}$ is determined by D_0 and is independent of the choice of u_0.

The proof for closed evolutions is symmetric if we consider the complement set $\{-u > 0\}$ of $\{u \geq 0\}$. □

Remark 5.2.10 (Orientation-Free Equations) If Eq. (5.1.1) is orientation-free, then $|u|$ solves (5.1.2) if u solves (5.1.2) by Theorem 5.2.7. Thus we may assume that $u \geq 0$ in Definitions 5.1.1 and 5.1.2 to define a generalized interface evolution Γ or zero level set of u. If Γ is bounded, then as in the same way we proved Theorem 5.2.8 we see that the set $\{u > 0\}$ is determined by $\{u_0 > 0\}$ and is independent of the choice of u_0. In other words Γ is uniquely determined by $\Gamma_0 = \Gamma(0)$. Note that we do not need to assume that Γ_0 is contained in a boundary of some bounded set in this argument. (Even if Γ is not bounded, Γ is determined by Γ_0 once uniqueness for arbitrary open evolutions is established.)

5.2.4 Unbounded Evolutions

We now prove the uniqueness of evolutions in the sense of Definition 5.1.2 admitting the global solvability of (5.1.2).

Theorem 5.2.11 *Assume that the level set Eq.* (5.1.2) *of surface evolution equation* (5.1.1) *satisfying* (f1) *has the comparison principle* (CP) *in* $\mathbf{R}^N$. *Assume that for given data* $g \in \mathrm{BUC}(\mathbf{R}^N)$ *there is a solution* $w \in \mathrm{BUC}(Z)$ *of* (5.1.2) *with* $w(x, 0) = g(x)$. *Then there is at most one open (resp. closed) evolution of Definition 5.1.2 for a given initial open (resp. closed) set in* $\mathbf{R}^N$.

To show this statement we need the monotone convergence result stated below. By $a_m \uparrow a$ (as $m \to \infty$) we mean the convergence is monotone, i.e., $a_m \leq a_{m+1}$ and $\lim_{m\to\infty} a_m = a$, where $a_m, a \in \mathbf{R}$.

Lemma 5.2.12 (Monotone Convergence) *Assume the same hypotheses of Theorem 5.2.11 concerning* (5.1.2). *Assume that* $u_{0m} \uparrow u_0$ *where* $u_{0m}, u_0 \in \mathrm{BUC}(\mathbf{R}^N)$. *Let* u_m *and* u *be the* $\mathcal{F}_{\mathbf{R}^N}$*-solutions of* (5.1.2) *with initial data* u_{0m} *and* u_0 *respectively. Then* $u_m \uparrow u$.

We postpone the proof of Lemma 5.2.12 to Sect. 5.6.

Proof of Theorem 5.2.11 Let $u \in \mathrm{BUC}(Z)$ and $v \in \mathrm{BUC}(Z)$ be solutions of (5.1.2). Assume that $\{u_0 > 0\} = \{v_0 > 0\}$, where $u_0(x) = u(x, 0)$, $v_0(x) = v(x, 0)$. Our goal is to prove $\{u > 0\} = \{v > 0\}$. By the invariance (Theorem 5.2.1) $u_+ = \theta \circ u$, $v_+ = \theta \circ v$ where $\theta(\sigma) = \max(\sigma, 0)$ are solutions of (5.1.2) so we may assume that $u \geq 0$ and $v \geq 0$ in Z.

For $m = 1, 2, \ldots$ we set

$$g_m = u_0 \wedge m v_0$$

and let $w_m \in \mathrm{BUC}(Z)$ be a solution with initial data g_m; here the assumption of the global solvability is invoked. Clearly, $g_m(x) \uparrow u_0(x)$ for all $x \in \mathbf{R}^N$. By the monotone convergence lemma (Lemma 5.2.12) we conclude that $w_m(z) \uparrow u(z)$ for all $z \in Z$ as $m \to \infty$. Since $g_m \leq m v_0$ in $\mathbf{R}^N$ and mv is the solution of (5.1.2) (by the invariance) with initial data mv_0, the comparison principle (CP) yields that $w_m \le mv$ in Z.

Since $w_m(z) \uparrow u(z)$ for $z \in Z$, for a given point $z \in \{u > 0\}$ we see $w_m(z) > 0$ for some m. Since $w_m \le mv$, this implies $v(z) > 0$. Thus we have proved that $\{u > 0\}$ is included in $\{v > 0\}$. If we exchange the role of u and v, the opposite inclusion holds. Thus the set $\{u > 0\}$ is determined by $D_0 = \{u_0 > 0\}$ and is independent of the choice of u_0.

The proof of the closed evolutions is symmetric as in Sect. 5.2.3. □

Remark 5.2.13 For an orientation-free equation, thanks to Theorem 5.2.7 it is possible to prove that each ℓ-level set of solution of (5.1.2) is determined by its initial shape independent of $\{u > \ell\}$ and $\{u < \ell\}$.

5.3 Existence by Perron's Method

We shall prove the existence of a global-in-time solution for the Cauchy problem of (5.1.2) or (5.2.1) by Perron's method developed in Sect. 3.4. For $\alpha \in \mathbf{R}$, let $K_\alpha(\mathbf{R}^N)$ be the space of all real-valued continuous functions that equal α outside some ball $B_R(0)$.

Theorem 5.3.1 *Assume that* $F : \mathbf{R}^N \times [0, T] \times (\mathbf{R}^N \setminus \{0\}) \times \mathbf{S}^N \to \mathbf{R}$ $(0 < T < \infty)$ *is continuous and geometric. Assume that* (5.2.1) *satisfies* (BCP) *for arbitrary open ball* Ω. *Assume that there is* $c \in C(0, r_0]$ *with some constant* $r_0 > 0$ *that satisfies*

$$|F(x, t, p, \pm I)| \le c(|p|) \quad \textit{on} \quad \mathbf{R}^N \times [0, T] \times (B_{r_0}(0) \setminus \{0\}). \tag{5.3.1}$$

Then for $\alpha \in \mathbf{R}$ *and each* $u_0 \in K_\alpha(\mathbf{R}^N)$ *there exists a unique* $\mathcal{F}_{\mathbf{R}^N}$*-solution* $u \in K_\alpha(Z)$ *of* (5.2.1) *in* $\mathbf{R}^N \times (0, T)$ *with* $u(x, 0) = u_0(x)$ *for all* $x \in \mathbf{R}^N$, *where* $Z = \mathbf{R}^N \times [0, T)$.

If F is degenerate elliptic and independent of the spatial variables x, then (BCP) holds (Corollary 4.1.2). Moreover, it is easy to check the existence of c in (5.3.1); see the proof of Lemma 4.1.3. Since the existence of initial (Lipschitz continuous) auxiliary function u_0 is clear (Sect. 5.1), we see that level set solutions exist globally for (5.1.1) satisfying (f1), (f2) provided that f is independent of x.

Corollary 5.3.2 *Assume that* $f : [0, T] \times \mathbf{E} \to \mathbf{R}$ *satisfies* (f1) *and* (f2) *with* $T < \infty$. *Then for each bounded open (resp. closed) set* D_0 *(resp.* E_0*) there exists a unique bounded open (closed) evolution* D *(resp.* E*) of* (5.1.1) *with initial data* D_0 *(resp.* E_0*). (The sets* D *and* E *are considered as subsets of* $Z = \mathbf{R}^N \times [0, T)$*.)*

We shall prove Theorem 5.3.1 by constructing suitable sub- and supersolutions and applying Theorem 3.4.9. For this purpose we prepare special, radial sub- and supersolutions.

Lemma 5.3.3 (Radial Solutions) *Assume the same hypotheses of Theorem 5.3.1 concerning F (except* (BCP)*) but with degenerate ellipticity. Then $\mathcal{F}_{\mathbf{R}^N} \neq \emptyset$. Moreover, for $h \in \mathcal{F}_{\mathbf{R}^N}$ with $\sup_{\mathbf{R}} h' < \infty$ and each $A > 0$ there is a constant $M = M(A, T, F, h)$ such that $V^+(x - \xi, t)$ and $v^+(x - \xi, t)$ (resp. $V^-(x - \xi, t)$ and $v^-(x - \xi, t)$) are $\mathcal{F}_{\mathbf{R}^N}$-supersolutions (subsolutions) of* (5.2.1) *in $\mathbf{R}^N \times (0, T)$ if $V^\pm$, $v^\pm$ is of form*

$$V^\pm(x, t) = \pm(Mt + Ah(|x|)), \ v^\pm(x, t) = \pm(Mt - Ah(|x|)). \tag{5.3.2}$$

Proof By Lemma 4.1.3 the set $\mathcal{F}_{\mathbf{R}^N}$ is nonempty since F is geometric and (5.3.1) holds. (By definition of $\mathcal{F}_{\mathbf{R}^N}$ there is $h \in \mathcal{F}_{\mathbf{R}^N}$ such that $\sup_{\mathbf{R}} h' < \infty$ if $\mathcal{F}_{\mathbf{R}^N} \neq \emptyset$.)

We shall prove that c in (5.3.1) can be extended to a continuous function on $(0, \infty)$ so that (5.3.1) holds on $\mathbf{R}^N \times [0, T] \times (\mathbf{R}^N \setminus \{0\})$ and that $c(\rho)/\rho$ is constant on $[r_0, \infty)$. We may assume $r_0 = 1$ by replacing p by p/r_0. Since F is degenerate elliptic and geometric, we see

$$\begin{aligned} F(x, t, p, I) &= |p| F(x, t, p/|p|, I/|p|) \\ &\geq |p| F(x, t, p/|p|, I) \geq -|p| c(1) \quad \text{for} \quad |p| \geq 1. \end{aligned}$$

Similarly, we have

$$\begin{aligned} F(x, t, p, I) &\leq F(x, t, p, O) \leq |p| F(x, t, p/|p|, O) \\ &\leq |p| F(x, t, -(-p/|p|), -I) \leq |p| c(1), \quad p \in \mathbf{R}^N. \end{aligned}$$

These two inequalities yield

$$|F(x, t, p, I)| \leq |p| c(1), \quad |p| \geq 1.$$

A symmetric argument yields $|F(x, t, p, -I)| \leq |p| c(1)$, $|p| \geq 1$. Thus the value of $c(\rho)$ for $\rho \geq 1$ can be defined by $\rho c(1)$ so that (5.3.1) holds on $\mathbf{R}^N \times [0, T] \times (\mathbf{R}^N \setminus \{0\})$.

As in the proof of Lemma 4.1.3 by (4.1.3) we see

$$|F(x, t, \nabla_p(h(\rho)), \pm\nabla_p^2(h(\rho)))| \leq \frac{h'(\rho)}{\rho} c(\rho) = h'(\rho) c(1)$$

for $\rho = |p| \geq 1$. If $h \in \mathcal{F}_{\mathbf{R}^N}$ and $\sup h' < \infty$, the definition of $\mathcal{F}_{\mathbf{R}^N}$ and the above inequality yields

$$B := \sup_{0 \leq t \leq T} \sup_{x, p \in \mathbf{R}^N} |F(x, t, \nabla_p(h(\rho)), \pm\nabla_p^2(h(\rho)))| < \infty.$$

We take $M \geq AB$ to observe that

$$V_t^+(x-\xi,t)+F(x,t,\nabla_x(V^+(x-\xi,t)),\nabla_x^2(V^+(x-\xi,t)))$$
$$\geq M-AB\geq 0.$$

Since $V_t^+ = M > 0$ and V^+ is C^2 in $\mathbf{R}^N \times (0,T)$, we see that V^+ is an $\mathcal{F}_{\mathbf{R}^N}$-supersolution of (5.2.1) in $\mathbf{R}^N \times (0,T)$ by Remark 3.1.10. The proof for v^+, V^-, v^- is similar so is omitted. □

Lemma 5.3.4 *Assume the same hypotheses of Lemma 5.3.3 concerning F. Let u_0 be a uniformly continuous function in $\mathbf{R}^N$. There is an $\mathcal{F}_{\mathbf{R}^N}$-sub- and supersolution u^- and u^+ of (5.2.1) in $\mathbf{R}^N \times (0,T)$ with initial data u_0 which satisfies $u^+(x,t) \geq u_0(x) \geq u^-(x,t)$ for all $x \in \mathbf{R}^N$, $t \in [0,T)$,*

$$\lim_{\substack{t\to 0\\ \delta\to 0}} \sup_{|x-y|\leq\delta} |u^\pm(y,t)-u_0(x)| = 0, \tag{5.3.3}$$

and $u^\pm$ is locally bounded in $\mathbf{R}^N \times [0,T)$. Moreover, $u^\pm$ is bounded in $\mathbf{R}^N \times [0,T)$ if u_0 is bounded.

Proof Since u_0 is uniformly continuous, there is a modulus ω such that

$$u_0(x)-u_0(\xi)\leq\omega(|x-\xi|),\ \ x,\xi\in\mathbf{R}^N.$$

Since $\mathcal{F}_{\mathbf{R}^N} \neq \emptyset$, there is $h \in \mathcal{F}_{\mathbf{R}^N}$ with $\sup_{\mathbf{R}} h' < \infty$. For each $\delta > 0$ there is $A_\delta > 0$ that satisfies $\omega(s) \leq \delta + A_\delta h(s)$ for $s \geq 0$. Thus,

$$u_0(x)\leq u_0(\xi)+\delta+A_\delta h(|x-\xi|).$$

By Lemma 5.3.3 for some M depending on A_δ the function

$$\begin{aligned} w_\delta^\xi(x,t) &= u_0(\xi)+\delta+A_\delta h(|x-\xi|)+Mt \\ &= u_0(\xi)+\delta+V^+(x-\xi,t),\ \ \xi\in\mathbf{R}^N \end{aligned}$$

is an $\mathcal{F}_{\mathbf{R}^N}$-supersolution of (5.2.1) in $\mathbf{R}^N \times (0,T)$. Now we set

$$u^+(x,t)=\inf\{w_\delta^\xi(x,t);\,\delta\in(0,1),\ \xi\in\mathbf{R}^N\} \tag{5.3.4}$$

and observe that u^+ is an $\mathcal{F}_{\mathbf{R}^N}$-supersolution of (5.2.1) in $\mathbf{R}^N \times (0,T)$ by closedness under infimum (Lemmas 3.4.5 and 3.4.7). Since $w_\delta^\xi(x,t) \geq u_0(x)$, $x \in \mathbf{R}^N$, $t > 0$, we have $u^+(x,t) \geq u_0(x)$ for $x \in \mathbf{R}^N$, $t > 0$. In particular, u^+ is locally bounded from below.

We shall prove (5.3.3) and local boundedness of u^+ from above. Since $u^+(x,t) \leq w_\delta^x(x,t)$ for $\delta \in (0,1)$,

$$u^+(x,t)-u_0(x)\leq w_\delta^x(x,t)-u_0(x)\leq\delta+Mt.$$

(This in particular implies that u^+ is locally bounded from above.) Here M depends only on δ; it is independent of x. Thus

$$\limsup_{t\to 0} \sup_{x\in\mathbf{R}^N} (u^+(x,t) - u_0(x)) = 0.$$

Since u_0 is uniformly continuous and $u^+(x,t) \geq u_0(x)$, this yields (5.3.3) for u^+. If u_0 is bounded, it is easy to see that u^+ in (5.3.4) is bounded. Construction of u^- is symmetric. We use V^- in (5.3.2) instead of V^+. The proof of the property for u^- is the same as for u^+. □

Proof of Theorem 5.3.1 For $u_0 \in K_\alpha(\mathbf{R}^N)$ and $h \in \mathcal{F}_{\mathbf{R}^N}$ with $\sup_{\mathbf{R}} h' < \infty$ we take σ large such that

$$u_0(x) < \sigma - h(|x|) \text{ for } x \text{ satisfying } u_0(x) \neq \alpha.$$

Let v^+ be as in (5.3.2) with $A = 1$. By Lemma 5.3.3, v^+ is an $\mathcal{F}_{\mathbf{R}^N}$-supersolution of (5.2.1) so that $v^+ + \sigma$ is also an $\mathcal{F}_{\mathbf{R}^N}$-supersolution. By the invariance (Theorem 5.2.1)

$$w^+(x,t) = \max\{v^+(x,t) + \sigma, \alpha\}$$

is an $\mathcal{F}_{\mathbf{R}^N}$-supersolution of (5.2.1) in $\mathbf{R}^N \times (0,T)$. Evidently, w^+ is continuous in $\mathbf{R}^N \times [0,T)$. By the choice of σ we see

$$u_0(x) \leq w^+(x,0) \quad \text{for all } x \in \mathbf{R}^N.$$

Since $h(s) \to \infty$ as $s \to \infty$,

$$\operatorname{spt}(w^+ - \alpha) \subset \Omega' \times [0,T) \quad \text{with } \Omega' = \operatorname{int} B_R$$

for sufficiently large ball $B_R = B_R(0)$, where spt denotes the support of a function. Similarly, one can construct an $\mathcal{F}_{\mathbf{R}^N}$-supersolution w^- of (5.2.1) in $\mathbf{R}^N \times (0,T)$ that satisfies

$$u_0(x) \geq w^-(x,0) \quad \text{for all } x \in \mathbf{R}^N$$

and spt $(w^- - \alpha) \subset \Omega' \times [0,T)$ by taking R larger if necessary.

Let $u^\pm$ be functions in Lemma 5.3.4 for uniformly continuous function $u_0 \in K_\alpha(\mathbf{R}^N)$. We set

$$U^+(x,t) = \min(u^+(x,t),\ w^+(x,t)),$$
$$U^-(x,t) = \max(u^-(x,t),\ w^-(x,t)).$$

Since u^+, w^+ are $\mathcal{F}_{\mathbf{R}^N}$-supersolutions of (5.2.1), so is U^+ by closedness of infimum for supersolutions (Lemmas 3.4.5 and 3.4.7). A symmetric argument yields that U^- is an $\mathcal{F}_{\mathbf{R}^N}$- supersolution (5.2.1) in $\mathbf{R}^N \times (0, T)$. By the choice of $w^\pm$,

$$U^\pm(x, t) = \alpha \quad \text{on } (B_{2R} \setminus B_R) \times [0, T) \quad \text{and } U^- \leq U^+ \quad \text{on } B_{2R} \times (0, T).$$

Since $U^\pm$ is locally bounded on Z, we apply Perron's method (Theorem 3.4.9) in $\mathcal{O} = \Omega \times (0, T)$ with $\Omega = \mathrm{int} B_{2R}$ to get a solution u of (5.2.1) in $\mathcal{O}$ that satisfies $U^- \leq u \leq U^+$ in $\mathcal{O}$. Since $u = \alpha$ on $(B_{2R} \setminus B_R) \times [0, T)$, we extend u in $(\mathbf{R}^N \setminus B_R) \times [0, T)$ by α and conclude that the extended function (still denoted u) is a solution of (5.2.1) in $\mathbf{R}^N \times (0, T)$.

It remains to prove that $u|_{t=0} = u_0$ and $u \in K_\alpha(Z)$. We shall use (BCP). Since $U^\pm$ is continuous at $t = 0$ by Lemma 5.3.4, and $U^\pm|_{t=0} = u_0$, we see that u is continuous at $t = 0$. In other words $u^* \leq u_*$ on B_{2R} at $t = 0$. Since $u = \alpha$ in $(B_{2R} \setminus B_R) \times [0, T)$, the relation $u^* \leq u_*$ holds on $\partial B_{2R} \times [0, T)$. We now apply (BCP) to get $u^* \leq u_*$ in $B_{2R} \times [0, T)$. Thus $u^* = u_*$ which implies that u is continuous in $B_{2R} \times [0, T)$. Since $u = \alpha$ in $(\mathbf{R}^N \setminus B_R) \times [0, T)$, we conclude $u \in K_\alpha(Z)$. Since $U^- \leq u \leq U^+$ and $U^+|_{t=0} = u_0$, the initial value of u must be u_0. (The uniqueness of solutions follows from (BCP).) The proof is now complete. □

There is another version of existence results with initial data not necessarily in $K_\alpha(\mathbf{R}^N)$ but in a larger space $\mathrm{UC}_*(\mathbf{R}^N) = \{u_0 \in C(\mathbf{R}^N);\ (u_0)_M$ is uniformly continuous in $\mathbf{R}^N$, i.e., $(u_0)_M \in \mathrm{UC}(\mathbf{R}^N)$ for every $M > 0\}$, where $(u_0)_M = (u_0 \wedge M) \vee (-M)$. Such a function u_0 may not be a constant at space infinity. The space $\mathrm{UC}_*(Z)$ is defined by replacing $\mathbf{R}^N$ by Z.

Theorem 5.3.5 *Assume that* $F : \mathbf{R}^N \times [0, T] \times (\mathbf{R}^N \setminus \{0\}) \times \mathbf{S}^N \to \mathbf{R}$ $(0 < T < \infty)$ *is continuous and geometric. Assume that* (5.2.1) *satisfies* (CP) *with* $\Omega = \mathbf{R}^N$. *Assume that there is* $c \in C(0, r_0)$ *with some* $r_0 > 0$ *satisfying* (5.3.1). *Then for* $u_0 \in \mathrm{UC}_*(\mathbf{R}^N)$ *there exists a unique* $\mathcal{F}_{\mathbf{R}^N}$*-solution* $u \in \mathrm{UC}_*(Z)$ *of* (5.2.1) *in* $\mathbf{R}^N \times (0, T)$ *with* $u|_{t=0} = u_0$. *If* $u_0 \in \mathrm{BUC}(\mathbf{R}^N)$, *then* $u \in \mathrm{BUC}(Z)$.

Proof For $u_0 \in \mathrm{UC}_*(\mathbf{R}^N)$ we take $u^\pm$ as in Lemma 5.3.4. Since u_0 is assumed to be uniformly continuous in Lemma 5.3.4, we have to modify construction of u^+ for $u_0 \in UC_*(\mathbf{R}^N)$. We set

$$\mathcal{U}_k = \left\{ x \in \mathbf{R}^N \mid k - 1 < u_0(x) < k + 1 \right\}, \quad k \in \mathbf{Z}.$$

Let ω_k be a modulus of u_0 on $\mathcal{U}_k$. We construct w_δ^ε as in Lemma 5.3.4 and set

$$u_k^+(x, t) = \inf \left\{ w_\delta^\varepsilon(\xi, t),\ \delta \in (0, 1),\ \xi \in \mathcal{U}_k \right\}$$

for $u \in \mathcal{U}_k$, $t \in (0, T)$ which is bounded and uniformly continuous. It is an $\mathcal{F}_{\mathbf{R}^N}$-supersolution in $\mathcal{U}_k \times (0, T)$ and it is uniformly continuous in $\mathcal{U}_k \times [0, T)$. Moreover, u_k^+ is bounded on $\mathcal{U}_k \times (0, T)$ and $u_k^+(x, t) \geq u_0(x)$ on $\mathcal{U}_k \times (0, T)$ with $u_k^+(x, 0) =$

$u_0(x)$ on $\mathcal{U}_k$. We set for each $x \in \mathbf{R}^N$,

$$u^+(x,t) = \inf\left\{u_k^+(x,t) \;\middle|\; k \text{ satisfies } x \in \mathcal{U}_k\right\}.$$

By Lemmas 3.4.5 and 3.4.7, this u^+ is an $\mathcal{F}_{\mathbf{R}^N}$-subsolution in $\mathbf{R}^N \times (0,T)$. By definition, $u^+(x,0) = u_0(x)$ and $u^+(x,t) \geq u_0(x)$, and u^+ is uniformly continuous on $\mathcal{U}_k$. We can similarly construct u^-.

Thus, instead of (5.3.3), we conclude that for any $k \in \mathbf{Z}$

$$\begin{cases} \lim\limits_{\substack{t\to 0\\ \delta\to 0}} \sup\limits_{|x-y|\leq\delta} \left|u^+(y,t) - u_0(x)\right| = 0 \quad \text{for } x \text{ such that } |u_0(x)-k| \leq 1/2 \\ \lim\limits_{\substack{t\to 0\\ \delta\to 0}} \sup\limits_{|x-y|\leq\delta} \left|u^-(y,t) - u_0(x)\right| = 0 \quad \text{for } x \text{ such that } |u_0(x)-k| \leq 1/2. \end{cases} \tag{4.3.3'}$$

We apply Theorem 3.4.9 for (5.2.1) with $\mathcal{O} = \mathbf{R}^N \times (0,T)$ to get an $\mathcal{F}_{\mathbf{R}^N}$-solution u of (5.2.1) in $\mathcal{O}$ satisfying $u^- \leq u \leq u^+$ in $\mathcal{O}$. Since $u^\pm$ satisfies (4.3.3'), we see $u|_{t=0} = u_0$. If u_0 is bounded, then $u^\pm$ is bounded in Z by Lemma 5.3.4 to get u is bounded in Z.

It remains to prove that $u \in \mathrm{UC}_*(Z)$. Here we use the comparison principle (CP). By invariance (Theorem 5.2.1) u_M is an $\mathcal{F}_{\mathbf{R}^N}$-solution of (5.2.1) so we may assume that u is bounded. Since $u^\pm$ satisfies (4.3.3') and $\bar{u} \leq u^+$, we see

$$\lim_{\substack{t\to 0\\ s\to 0\\ \delta\to 0}} \sup_{|x-y|\leq\delta} (u^*(x,t) - u_*(y,s)) = 0.$$

We now apply (CP) to get

$$\lim_{\delta\to 0} \sup\{u^*(x,t) - u_*(y,s);\ |x-y| \leq \delta,\ |t-s| \leq \delta,\ 0 \leq t,\ s < T\} \leq 0.$$

This yields $u \in \mathrm{BUC}(Z)$. The uniqueness of solutions follows from (CP). □

From Theorem 5.3.5 with Lemma 4.1.3 there easily follows an existence result.

Corollary 5.3.6 *Assume that* $f : [0,T] \times \mathbf{E} \to \mathbf{R}$ *satisfies* (f1) *and* (f2) *with* $T < \infty$. *Then for each open (resp. closed) set* D_0 *(resp.* E_0*) there exists a unique open (closed) evolution* D *(resp.* E*) of* (5.1.1) *with initial data* D_0 *(resp.* E_0*).*

Remark 5.3.7 (Condition (5.3.1)**)** If we examine the proof, we realize that the condition (5.3.1) is actually unnecessary when initial data is in K_α. Indeed, by Lemma 4.1.3 there always exists $c_R \in C(0, r_0)$ with some $r_0 > 0$ satisfying

$$|F(x,t,p,\pm I)| \leq c_R(|p|) \quad \text{for} \quad x \in B_R, t \in [0,T], p \in B_{r_0} \setminus \{0\} \tag{5.3.5}$$

for every $R > 0$. This is enough to carry out the proof of Theorem 5.3.1 , although Lemmas 5.3.3 and 5.3.4 should be modified. So Corollary 5.3.2 can be extended to a spatially inhomogeneous equation like (4.6.15) provided it satisfies (BCP). There is no need to assume (5.3.1).

By the way the condition (5.3.1) is interpreted as control of speed of the growth of a unit ball moved by (5.1.1). For the proof of Lemma 4.1.3 it is clear that (5.3.1) is fulfilled if $f : \mathbf{R}^N \times [0, T] \times K$ is bounded for every compact set K of $\mathbf{E}$.

Remark 5.3.8 (Exact Solutions) Exact solutions in Sect. 2.7 are actually an interface evolution. For example, in Sect. 2.7.1 (level set approach) we pointed out that

$$u(x,t) = -(t + |x|^2/2(N-1))(+\text{const.})$$

is a solution of the level set mean curvature flow equation (2.6.5). As anticipated, it is easy to check that this function u solves (2.6.5) in the viscosity sense. So we conclude that the shrinking sphere defined by (2.7.2) and (2.7.3) is actually an interface evolution. In Chap. 6 we shall show that smooth solution $\{\Gamma_t\}$ is an interface evolution so the notion of our generalized solution is consistent with classical solutions.

5.4 Existence by Approximation

Perhaps it is more standard than Perron's method to construct solutions as a limit of solutions of approximate equations. For example, when we are asked to solve first order equation

$$u_t + F(\nabla u) = 0, \quad u|_{t=0} = u_0 \tag{5.4.1}$$

globally in time, we often consider a regularized problem

$$u_t^\varepsilon + F(\nabla u^\varepsilon) = \varepsilon \Delta u^\varepsilon, \quad u^\varepsilon|_{t=0} = u_0 \tag{5.4.2}$$

for $\varepsilon > 0$ and construct the solution u of the original problem (5.4.1) as a limit of solution u^ε of (5.4.2) as $\varepsilon \to 0$. This method is called a vanishing viscosity method since the parameter ε resembles the viscosity in equations of fluid dynamics. The viscosity solution is obtained as the limit of such a problem. The name of 'viscosity solution' stems from this type of a vanishing viscosity method. For more background the reader is referred to a book of P.-L. Lions [320].

For spatially homogeneous level set equations as pointed out by Y.-G. Chen et al. [105], we approximate the equation by a strictly (or uniformly) parabolic equation and observe that a solution of a level set equation is obtained as the local uniform limit of solutions of the approximate problem. In the above paper by Y.-G. Chen et al. [105], the way of approximation was not mentioned. For the level set mean curvature flow equation L. C. Evans and J. Spruck [166] solved a strictly parabolic equation of the form

$$u_t - \sqrt{|\nabla u|^2 + \varepsilon^2}\,\mathrm{div}\left(\frac{\nabla u}{\sqrt{\varepsilon^2 + |\nabla u|^2}}\right) = 0 \tag{5.4.3}$$

or equivalently

$$u_t - \Delta u + \sum_{1 \le i,j \le n} \frac{u_{x_i} u_{x_j}}{\varepsilon^2 + |\nabla u|^2} u_{x_i x_j} = 0 \tag{5.4.4}$$

with suitable initial data. Then they obtained a solution of the level set Eq. (2.6.7) with initial data $u_0 \in K_\alpha(\mathbf{R}^N)$ as a local uniform limit of the solution of (5.4.4).

This method consists of two parts:

(i) **Solvability of approximate problem.** One has to solve (5.4.3).
(ii) **Limiting procedure.** One has to prove that the desired solution of (2.6.7) is obtained as a limit of the approximate solution u^ε of (5.4.4).

For the first part (i) we need the theory of parabolic equations; see e.g. the book of O. A. Ladyženskaja et al. [317] or A. Lunardi [341]. So we do not touch this problem here. We give a precise statement for part (ii).

Theorem 5.4.1 *Assume that* $u^\varepsilon \in C(\mathbf{R}^N \times (0,\infty))$ *is a unique smooth solution of* (5.4.3) *with initial data* $u_0 \in \mathrm{BUC}(\mathbf{R}^N)$. *Then there exists* $u \in C(\mathbf{R}^N \times [0,\infty))$ *such that* $u \in \mathrm{BUC}(\mathbf{R}^N \times [0,T))$ *for every* $T > 0$ *and that* u *is obtained as a local uniform limit of* u^ε *in* $\mathbf{R}^N \times [0,\infty)$. *Moreover,* u *is a viscosity solution of* (2.6.7).

The statement is actually a special version of our convergence result (Theorem 5.6.3). It is easy to check assumptions of Theorem 5.6.3 are fulfilled.

However, there is a more classical way to prove such a statement when u_0 is more regular. Indeed, if $u_0 \in C^2(\mathbf{R}^N) \cap K_\alpha(\mathbf{R}^N)$ by the maximum principle, one gets a uniform bound for $|\nabla u^\varepsilon|$, $|u^\varepsilon|$, $|u^\varepsilon_t|$ in $\mathbf{R}^N \times (0,\infty)$. By Ascoli–Arzèla's compactness theorem, $u^{\varepsilon_j} \to u$ locally uniformly in $\mathbf{R}^N \times [0,\infty)$ with some function u by taking a subsequence $\varepsilon_j \to 0$. By a stability principle with local uniform convergence (Sect. 3.1.2) we conclude that u is a viscosity solution of (2.6.7) with initial data u_0. By the uniqueness of the initial value problem for (2.6.7), the limit u is independent of the choice of subsequence so we obtain a full convergence. The reader is referred to the work of L. C. Evans and J. Spruck [166] for details of this type of argument.

5.5 Various Properties of Evolutions

We study various general properties of open and closed evolutions.

Assumptions on Well-Posedness (W) Assume that $f : \mathbf{R}^N \times [0,T] \times \mathbf{E} \to \mathbf{R}$ satisfies (f1) and (f2). Assume that (5.1.2) with (5.1.3) satisfies (CP) with $\Omega =$

$\mathbf{R}^N$ and (5.3.1) with some c. (The last assumption is fulfilled if for example, f is independent of the space variables.)

By Theorems 5.2.11 and 5.3.1 if we assume (W), then for each open set D_0 in $\mathbf{R}^N$ there is a unique open evolution $D(\subset \mathbf{R}^N \times [0, T))$ if (5.1.1) with $D(0) = D_0$, where $D(t)$ denotes the cross-section of D at time t, i.e.,

$$D(t) = \{x \in \mathbf{R}^N; (x, t) \in D\}.$$

By translation in time under (W) there is a unique open evolution $\tilde{D} \subset \mathbf{R}^N \times [s, T)$ of (5.1.1) with $\tilde{D}(s) = D_0$ where s is a given positive number. Let $U(t, s)$ denote the mapping: $D_0 \mapsto \tilde{D}(t)$. Similarly, let $M(t, s)$ denote the mapping which maps a closed set E_0 to $\tilde{E}(t)$ where $\tilde{E}$ is a closed level set solution with $\tilde{E}(s) = E_0$. By unique existence of evolution we have a semigroup property.

Theorem 5.5.1 (Semigroup Property) *Assume (W). Then*

$$U(t, \tau) \circ U(\tau, s) = U(t, s), \ \ M(t, \tau) \circ M(\tau, s) = M(t, s)$$

for all s, τ, t satisfying $0 \le s \le \tau \le t < T$.

The operators M and U have order preserving properties which follow from (CP). Let $\mathrm{dist}(x, A)$ denote the *distance* between $x \in \Omega$ and a set $A \subset \Omega$, i.e.,

$$\mathrm{dist}(x, A) = \inf\{|x - y|; y \in A\}.$$

We shall often write this quantity simply by $d(x, A)$. For an open set D_0 let $\mathrm{sd}(x, \partial D_0)$ denote the *signed distance function* defined by

$$\mathrm{sd}(x, \partial D_0) = \begin{cases} d(x, \partial D_0) & \text{if } x \in D_0, \\ -d(x, \partial D_0) & \text{if } x \notin D_0. \end{cases}$$

Theorem 5.5.2 (Order Preserving Property) *Assume (W). Let D_0 and D_0' be two open sets in $\mathbf{R}^N$ and let E_0, E_0' be two closed sets in $\mathbf{R}^N$.*

(i) *$D_0 \subset D_0'$ implies $U(t, s)D_0 \subset U(t, s)D_0'$;*
(ii) *$E_0 \subset E_0'$ implies $M(t, s)E_0 \subset M(t, s)E_0'$;*
(iii) *$D_0 \subset E_0$ implies $U(t, s)D_0 \subset M(t, s)E_0$;*
(iv) *if $E_0 \subset D_0$ and $\mathrm{dist}(E_0, \partial D_0) > 0$, then $M(t, s)E_0 \subset U(t, s)D_0$ for all $t, s \in [0, T]$ satisfying $t \ge s$.*

Proof

(i) We take

$$u_0(x) = \max(\mathrm{sd}(x, \partial D_0), -1),$$
$$v_0(x) = \max(\mathrm{sd}(x, \partial D_0'), -1)$$

as initial auxiliary function for D_0 and D'_0 so that $u_0 \le v_0$. Since the solutions u and v starting from u_0 and v_0 at $t = s$ are bounded, uniformly continuous on $\mathbf{R}^N \times [s, T')$ for any $T' < T$, u and v satisfies the assumptions of (CP). Thus $u \le v$ on $\mathbf{R}^N \times [s, T)$ which implies $U(t,s)D_0 \subset U(t,s)D'_0$.
(ii), (iii) The proof is similar to (i).
(iv) We may assume $s = 0$. We take

$$u_0(x) = \begin{cases} \operatorname{dist}(x, \partial D_0)/\{\operatorname{dist}(x, \partial D_0) + \operatorname{dist}(x, E_0)\}, & x \in D_0, \\ -\min\{\operatorname{dist}(x, \partial D_0), 1\} & x \notin D_0, \end{cases}$$

so that $D_0 = \{u_0 > 0\}$ and $E_0 = \{u_0 \ge 1\}$. By the assumption $\operatorname{dist}(E_0, \partial D_0) > 0$, u_0 is a Lipschitz continuous function. Let u be the solution of (5.1.2) with initial data u_0. Since $u - 1$ also solves (5.1.2), $M(t,0)E_0 = \{x \in \mathbf{R}^N; u(x,t) \ge 1\}$. Since $U(t,0)D_0 = \{u(\cdot, t) \ge 0\}$, it is clear that $M(t,0)E_0 \subset U(t,0)D_0$. □

We shall study convergence properties of level set solutions. Below we use following notation. Let $\{A_j\}_{j\ge1}$ be a sequence of sets and B be a set. By $A_j \uparrow B$ we mean that $A_j \subset A_{j+1}$ and $\cup_{j\ge1} A_j = B$. Similarly, by $A_j \downarrow B$ we mean that $A_j \supset A_{j+1}$ and $\cap_{j\ge1} A_j = B$.

Lemma 5.5.3 (Approximation) *Let D be an open evolution. There exist two sequences of open evolution $\{D'_k\}_{k\ge1}$ and closed evolution $\{E'_k\}_{k\ge1}$ such that*

$$D'_k \uparrow D \quad \text{and } D'_k \subset E'_k \subset D'_{k+1}. \tag{5.5.1}$$

Proof Let u be a solution of (5.1.2) such that $\{u > 0\} = D$. Then we define

$$D'_k = \left\{(x,t); u(x,t) > \frac{1}{2^k}\right\}, \ E'_k = \left\{(x,t); u(x,t) \ge \frac{1}{2^k}\right\}.$$

These sets clearly fulfill (5.5.1). Since $u - 2^{-k}$ is a solution of (5.1.2), D'_k and E'_k are open and closed evolutions. □

Theorem 5.5.4 (Monotone Convergence)

(i) *Let D and $\{D_j\}_{j\ge1}$ be open evolutions with initial data D_0 and $\{D_{j0}\}$ respectively. If $D_{j0} \uparrow D_0$, then $D_j \uparrow D$.*
(ii) *Let E and $\{E_j\}_{j\ge1}$ be closed evolution with initial data E_0 and $\{E_{j0}\}$ respectively. If $E_{j0} \downarrow E_0$, then $E_j \downarrow E$.*

Proof

(i) The proof is easy if D_0 is bounded. Let D'_k and E'_k be the approximating open and closed evolutions for D which were constructed in Lemma 5.5.3 . If D_0 is bounded, then $E'_k(0)$ is compact, so there is a $j_k \ge 1$ such that $E'_k(0) \subset D_{j_k0}$. By comparison (Theorem 5.5.2 (iv)), we have $E'_k \subset D_{j_k}$. The sequence E'_k was constructed so that $D'_k \subset E'_k \uparrow D$. Thus $D_{j_k} \uparrow D$, which proves (i) when D_0 is bounded. In general, we use Lemma 5.2.12 with

$$u_{0j}(x) = (\mathrm{sd}(x, D_{j0}) \wedge 1) \vee (-1),$$
$$u_0(x) = (\mathrm{sd}(x, D_0) \wedge 1) \vee (-1)$$

and observe that the solution u_j with initial data u_{0j} satisfies $u_j \uparrow u$, where u is the solution with initial data u_0. Since $D_j = \{u_j > 0\}$ and $D = \{u > 0\}$, $u_j \uparrow u$ implies $D_j \uparrow D$.

(ii) The proof is similar so is omitted.

□

We shall study continuity of $D(t)$ and $E(t)$ for an open and closed evolution as a function of time t. To formulate continuity of open-set valued functions we define the *ε-core* $\mathcal{C}_\varepsilon(W)$ of an open set $W \subset \mathbf{R}^N$ to be

$$\mathcal{C}_\varepsilon(W) = \{x \in \mathbf{R}^N;\ \mathrm{dist}(x, W^c) \geq \varepsilon\},$$

where $W^c = \mathbf{R}^N \setminus W$. This concept is dual to that of an *ε-neighborhood* $\mathcal{N}_\varepsilon(Y)$ of a closed set Y in $\mathbf{R}^N$ in the sense that $\mathcal{C}_\varepsilon(W) = (\mathcal{N}_\varepsilon(W^c))^c$.

Theorem 5.5.5 (Continuity in Time) *Let D and E be open and closed evolutions, respectively.*

(ia) *$D(t)$ is a lower semicontinuous function of $t \in [0, T)$, in the sense that for any $t_0 \geq 0$, and sequence $x_n \in (D(t_n))^c$ with $x_n \to x_0$, $t_n \to t_0$, the limit $x_0 \in (D(t_0))^c$. If $D(0)$ is bounded so that $\mathcal{C}_\varepsilon(D(t_0))$ is compact, this implies that for any $t_0 \geq 0$, $\varepsilon > 0$ there is a $\delta > 0$ such that $|t - t_0| < \delta$ implies $D(t) \supset \mathcal{C}_\varepsilon(D(t_0))$.*

(ib) *$E(t)$ is an upper semicontinuous function of $t \in [0, T)$ in the sense that for any $t_0 \geq 0$ and sequence $x_n \in E(t_n)$ with $x_n \to x_0$, $t_n \to t_0$ the limit $x_0 \in E(t_0)$. If $E(0)$ is bounded so that $\mathcal{N}_\varepsilon(E(t_0))$ is compact, this implies that for any $t_0 \geq 0$, $\varepsilon > 0$, there is a $\delta > 0$ such that $|t - t_0| < \delta$ implies $E(t) \subset \mathcal{N}_\varepsilon(E(t_0))$.*

(iia) *$D(t)$ is left upper semicontinuous in t in the sense that for any $t_0 \in (0, T)$, $x_0 \in (D(t_0))^c$ there is a sequence $x_n \to x_0$ and $t_n \uparrow t_0$ with $x_n \in (D(t_n))^c$. Moreover, for any $t_0 \in (0, T)$, $\varepsilon > 0$ there exists a $\delta > 0$ such that $t_0 - \delta < t < t_0$ implies $\mathcal{C}_\varepsilon(D(t)) \subset D(t_0)$.*

(iib) *$E(t)$ is left lower semicontinuous in t in the sense that for any $t_0 \in (0, T)$, $x_0 \in E(t_0)$ there is a sequence $x_n \to x_0$ and $t_n \uparrow t_0$ with $x_n \in E(t_0)$. Moreover, for any $t_0 \in (0, T)$, $\varepsilon > 0$ there exists a $\delta > 0$ such that $t_0 - \delta < t < t_0$ implies $\mathcal{N}_\varepsilon(E(t)) \supset E(t_0)$ (Fig. 5.4).*

For any two closed sets $C_1, C_2 \subset \mathbf{R}^N$ the *Hausdorff distance* between them is defined by

$$d_H(C_1, C_2) = \inf\{\varepsilon > 0;\ C_1 \subset \mathcal{N}_\varepsilon(C_2) \quad \text{and} \quad C_2 \subset \mathcal{N}_\varepsilon(C_1)\}.$$

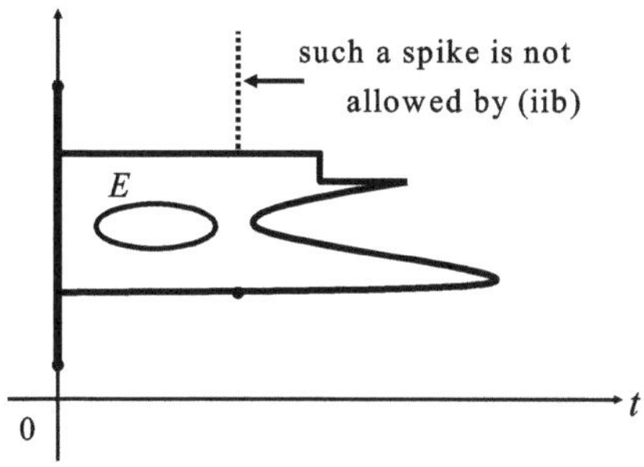

Fig. 5.4 Closed evolution

For open subsets $U_1, U_2 \subset \mathbf{R}^N$ we define

$$d_H^*(U_1, U_2) = \inf\{\varepsilon > 0;\ \mathcal{C}_\varepsilon(U_1) \subset U_2 \quad \text{and} \quad \mathcal{C}_\varepsilon(U_2) \subset U_1\}.$$

By definition $d_H^*(U_1, U_2) = d_H(U_1^c, U_2^c)$ so d_H^* is a metric of open sets in $\mathbf{R}^N$. Theorem 5.5.5 implies that $D(t)$ and $E(t)$ are left continuous functions with respect to the Hausdorff metric d_H^* and d_H, provided that D_0 and E_0 are bounded.

Proof The Properties (ia), (ib) follow from the fact that D, E are open and closed sets in $\mathbf{R}^N \times [0, T)$. Since (iia) and (iib) are dual to each other, we only prove (iia). Let V^- be a radial subsolution defined in (5.3.2) with some M and A. For $\varepsilon > 0$ and $t_0 > 0$ we define $\delta = Ah(\varepsilon)/M$. For given $x_0 \in \mathcal{C}_\varepsilon(D(t))$ we set

$$W_\tau = \{(y, s);\ V^-(y - x_0, s - \tau) > 0\}.$$

For $t \in (t_0 - \delta, t_0)$ we take $\tau > t_0$ such that $\tau - t < \delta$. By the choice of δ we see $W_\tau(t) \subset D(t)$. Since V^- is a subsolution of (5.1.2), by Remark 5.5.14 (ii) on Theorem 5.5.2 (see at the end of Sect. 5.5), we have $W_\tau(t_0) \subset D(t_0)$ so that $x_0 \in D(t_0)$. □

Fattening and Regularity If initial closed set $E_0 = \overline{D_0}$ and D_0 is open, then one would expect that $E = \overline{D}$, where E (resp. D) is a closed (resp. open) evolution with initial data E_0 (resp. D_0). However, unfortunately this is not true in general even if the Eq. (5.1.1) is the curve shortening equation $V = k$ and the initial data is compact. Thus we say an open evolution D is *regular* if $E = \overline{D}$. We say that an interface evolution $\Gamma = E \setminus D$ fattens if $\Gamma(t)$ has an interior point for some $t > 0$ although $\Gamma(0) = E_0 \setminus D_0$ has no interior. If $\Gamma(t_0)$ has an interior at some t_0, then Γ has an interior in $\mathbf{R}^N \times [0, T)$. Indeed, for $x_0 \in \operatorname{int}(\Gamma(t_0))$ we take τ slightly larger than t_0 so that $W_\tau(t_0) \subset \Gamma(t_0)$ and $D \cap W_\tau \cap \{t \geq t_0\} = \emptyset$. Then $W_\tau(t) \subset \Gamma(t)$ for $t (> t_0)$ close to t_0. Thus Γ has an interior in $\mathbf{R}^N \times [0, T)$. If D is regular, then the interface evolution Γ with initial data $\Gamma(0) = \partial D_0$ does not fatten. The converse is not clear by our observations given so far.

Applying the monotone convergence theorem to D_0 with $\Gamma_0 = \partial D_0$ and $E_0 = \overline{D_0}$, we see that evolutions D and E are obtained as a limit of evolutions approximating D_0 from the interior and E_0 from the exterior, respectively. If $\Gamma(t)$

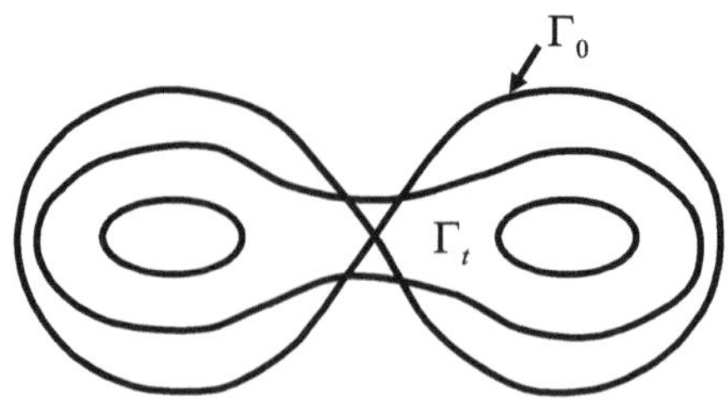

Fig. 5.5 Fattening of Figure 8

has an interior point, these two "solutions" do not agree. In particular, "continuity of solutions with respect to initial data", which usually is expected for differential equations generally, is not valid in this case. The situation $\overline{D} \neq E$ can be also interpreted as a loss of uniqueness (Sect. 6.2.2).

For the curve-shortening equation $V = k$, if the initial data $\Gamma_0 = \partial D_0$ has the shape of figure "8" (embedded in $\mathbf{R}^2$), then Γ fattens instantaneously (Fig. 5.5). This was first observed by L. C. Evans and J. Spruck [166]. It is intuitively clear that a solution approximating from the interior does not agree with one approximated from the exterior. In Sect. 6.2.2 we give a rigorous proof of fattening when Γ_0 consists of two lines crossing at one point with right angle. See also Sect. 8.1.1 for an alternative game-theoretic proof of this fattening phenomenon.

We give several criteria for nonfattening or regularity. Note that $\overline{D}(t)$ does not represent the closure of $D(t)$ in $\mathbf{R}^N$. It is a cross-section of the closure of D in $\mathbf{R}^N \times [0, T)$.

Theorem 5.5.6 (Monotone Motion) *Assume f in* (5.1.1) *is independent of t. Assume (W). Assume that D_0 is a bounded open set. If $M(h,0)\overline{D_0} \subset D_0$ for sufficiently small h, then D is regular, where D is an open evolution with initial data D_0.*

Proof Since the equation is autonomous, $M(t,s) = M(t-s,0)$ $t \geq s \geq 0$ so we write $M_h = M(h,0)$ so that $M(t,s) = M_{t-s}$. We use a similar convention for $U(t,s)$.

By order-preserving property (Theorem 5.5.2) we see

$$M_h(\overline{D_0}) \subset D_0 \quad \text{implies} \quad M_t M_h(\overline{D_0}) \subset U_t(D_0).$$

The semigroup property implies $M_t M_h(\overline{D_0}) = E(t+h)$, when E is a closed evolution with initial data $\overline{D_0}$. Thus $E(t+h) \subset D(t)$ for $t > 0$. Iteration of this argument then shows that $E(t) \subset D(s)$ for all $t > s \geq 0$.

We next note that $\overline{D}(t') \downarrow \overline{D}(t)$ when $t' \uparrow t$ since $\overline{D}(t')$ is decreasing in time and $D(t)$ is left continuous as sets by Theorem 5.5.5. Thus $\cap_{0<t'<t} \overline{D}(t') = \overline{D}(t)$.

Since $E(t) \subset D(s)$, we have

$$E(t) \subset \bigcap_{0<t'<t} D(t) \subset \bigcap_{0<t'<t} \overline{D}(t') = \overline{D}(t).$$

Hence $E(t) \subset \overline{D}(t)$ for all $t > 0$ and $E \subset \overline{D}$. Since the converse inclusion $\overline{D} \subset E$ is true by assumption, this completes the proof. □

Theorem 5.5.7 *Assume that f is independent of t. Assume (W). Let D_0 be a smoothly bounded domain such that $f(x, \mathbf{n}, \nabla\mathbf{n}) < 0$ on ∂D_0. Denote the open and closed evolutions with initial data D_0 and $\overline{D_0}$ by D and E, respectively. Then $E(t) \subset D(s)$ for all $t > s \geq 0$ and $\overline{D} = E$.*

Proof By Theorem 5.5.6 it suffices to prove that $M_h\overline{D_0} \subset D_0$ for small $h > 0$. We set $\sigma = \inf_{\partial D_0}(-f(x, \mathbf{n}, \nabla\mathbf{n})) > 0$ and observe that

$$\psi(x, t) = \mathrm{sd}(x, \partial D_0) - \sigma t$$

is a supersolution of (5.1.2). By comparison (CP),

$$M_h\overline{D_0} \subset \{x \in \mathbf{R}^N;\ \psi(x, h) \geq 0\},$$

which yields $M_h\overline{D_0} \subset D_0$ for $h > 0$. □

Corollary 5.5.8 *Assume that f does not depend on t. Assume that f is independent of $\nabla\mathbf{n}$ (so that* (5.1.1) *is of the first order). Assume (W). If f does not change sign, then D is regular for any bounded open initial data D_0.*

In general one gets several criteria based on invariance of equations. It can be formally written as follows. If the equation is invariant under a semigroup of actions $\{S_h\}$. If $S_h\overline{D_0} \subset D_0$ for small $h > 0$, then D is regular. The condition $S_h\overline{D_0} \subset D_0$ is fulfilled if the generator of S_h is negative on ∂D_0.

We give another example of such a situation. For a set G in $\mathbf{R}^N \times [0, \infty)$, let $\mathcal{D}_\lambda^m(G)$ denote

$$\mathcal{D}_\lambda^m(G) = \{(\lambda x, \lambda^{m+1}t);\ (x, t) \in G\},$$

where $m \in \mathbf{R}$, $\lambda > 0$. Similarly, for a set G_0 in $\mathbf{R}^N$ let $\mathcal{D}_{\lambda,0}^m(G_0)$ denote

$$\mathcal{D}_{\lambda,0}(G_0) = \{\lambda x, x \in G_0\}.$$

The condition $\mathcal{D}_{1-h,0}(\overline{G_0}) \subset G_0$ for small $h > 0$ is a kind of starshapedness.

Theorem 5.5.9 (Dilation Invariant Equations) *Assume that f in* (5.1.1) *is independent of x and t. Assume that $f(p, Q_p(X))$ is positively homogeneous of degree $m \in \mathbf{R}$ in $Q_p(X)$, i.e.,*

$$f(p, \mu Q_p(X)) = \mu^m f(p, Q_p(X)), \quad \mu > 0$$

for all $(p, Q_p(X)) \in \mathbf{E}$. Assume (W). Let D_0 be a bounded open set in $\mathbf{R}^N$. If $\mathcal{D}_{1-h,0}(\overline{D_0}) \subset D_0$ for sufficiently small $h > 0$, then D is regular, where D is an open evolution with initial data D_0.

Proof Let F be the operator defined by (5.1.3). By homogeneity of f we have

$$F(\mu p, \mu^2 X) = \mu^{m+1} F(p, X), \ \ \mu > 0.$$

Let E be the closed evolution with initial data $\overline{D_0}$. Then by homogeneity of F the set $\mathcal{D}^m_{1-h}(E)$ is a closed evolution with initial data $\mathcal{D}_{1-h,0}(\overline{D_0})$. Sending h to zero yields $E \subset \overline{D}$. Since $\overline{D} \subset E$ is trivial, D is regular. □

Corollary 5.5.10 *Assume the same hypotheses of Theorem 5.5.9 concerning f. Let D_0 be a smoothly bounded domain such that $\langle x, \mathbf{n}\rangle > 0$ on ∂D_0. Let D be the open evolution with initial data D_0. Then D is regular.*

Proof Since $\langle -x, \mathbf{n}\rangle$ is the generator of group $\{\mathcal{D}_{e^{-q},0}\}$ at $q = 0$, it is clear that $\mathcal{D}_{1-h,0}(\overline{D_0}) \subset D_0$ for small $h > 0$. Thus D is regular by Theorem 5.5.9. □

We give another criterion when f is rotationally symmetric in the sense that

$$f({}^t Rx, t, Rp, {}^t R Q_p(X) R) = f(x, t, p, Q_p(X))$$

for any rotation matrix R. As well known, a one parameter group $\{R(\lambda)\}$ of rotation is generated by a skew symmetric matrix so it is of form $\{e^{\lambda A}\}_{\lambda \in \mathbf{R}}$ with skew-symmetric matrix A. Similarly to Theorem 5.5.9 we see that if f is rotationally symmetric, $e^{\lambda A}(\overline{D_0}) \subset D_0$ for small $\lambda > 0$ implies the regularity of D. The condition $e^{\lambda A}(\overline{D_0}) \subset D_0$ is fulfilled if $\langle Ax, \mathbf{n}\rangle < 0$ on ∂D_0 if D_0 is smooth. Below we give a criterion for regularity when f is independent of t, x, dilation invariant and rotationally symmetric.

Corollary 5.5.11 *Assume the same hypotheses of Theorem 5.5.9 concerning f. Assume moreover, f is rotationally symmetric. Assume that D_0 is a smoothly bounded domain. Assume that there is a nonnegative constant c_1, c_2 and a skew-symmetric matrix A such that*

$$c_1 f(\mathbf{n}, \nabla \mathbf{n}) + \langle Ax, \mathbf{n}\rangle - c_2 \langle x, \mathbf{n}\rangle < 0 \ \ on \ \ \partial D_0.$$

Then an open evolution D with initial data D_0 is regular.

Proof The condition on ∂D_0 guarantees that

$$(M_{c_1 h} \mathcal{D}_{1-c_2 h,0} e^{hA})(\overline{D_0}) \subset D_0$$

for small $h > 0$. By invariance of the equation $E_h = \mathcal{D}^m_{1-c_2 h} e^{hA}(E)$ is a closed evolution if E is the closed evolution with initial data $\overline{D_0}$. By comparison $E_h(t + h) \subset D(t)$, $t \geq 0$ for small $h > 0$. In other words

$$E_h(t) \subset D(t-h) \ \ \text{for} \ \ t - h \geq 0$$

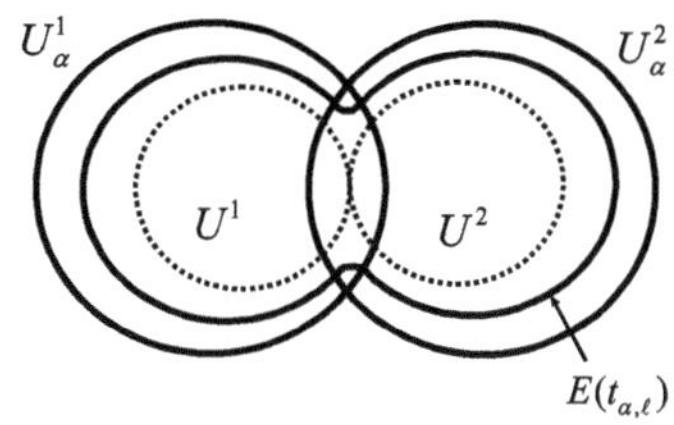

Fig. 5.6 Regularity criterion

for small h. Since $\overline{D}$ is left continuous and $E_h(t) \to E(t)$ as $h \to 0$, we see $E \subset \overline{D}$. Thus D is regular. □

Finally, we give another type of regularity criterion, illustrated in Fig. 5.6, for orientation-free equations when an open evolution is a disjoint sum of regular open evolutions.

Lemma 5.5.12 *Assume (W) and* (5.1.1) *is orientation-free. Let U be a bounded open set in $\mathbf{R}^N$ that may be written as the union of a finite number of disjoint open sets $U^1, \ldots, U^k$. Denote the open evolutions with initial data U and U^i by D and D^i, respectively* ($1 \le i \le k$)*. Let E^i be the closed evolution with initial data $\overline{U}^i$ and assume that $E^i = \overline{D^i}$.*

Assume that there is a sequence of open covers $\{U_\alpha^i, \ldots, U_\alpha^k\}_{\alpha \ge 1}$ of U which satisfies:

(i) $U_\alpha^i \supset U_{\alpha+1}^i$ *and* $\overline{U}^i = \cap_{\alpha \ge 1} \overline{U}_\alpha^i$ *for* $i = 1, \ldots, k$;
(ii) *the sets* $U_\alpha^1, \ldots, U_\alpha^k$ *are pairwise disjoint for* $\alpha \ge 1$.

Finally, let E be a closed evolution for which a double sequence $\{t_{\alpha,\ell}\}_{\alpha,\ell \ge 1}$ exists such that $t_{\alpha,\ell} \downarrow 0$ as $\ell \to \infty$ and

$$E(t_{\alpha,\ell}) \subset U_\alpha^1 \cup \cdots \cup U_\alpha^k.$$

Then $\overline{D} = E$.

Proof Let D_α^i be an open evolution with initial data U_α^i and let D_α be the open evolution with initial data $\cup_{i=1}^k U_\alpha^i$. Since the equation is orientation-free, we apply the separation lemma (stated after this proof as Lemma 5.5.13) to obtain that $D_\alpha = \cup_{i=1}^k D_\alpha^i$.

By order preserving property (Theorem 5.5.2 (iv)) we see

$$E(t) \subset D_\alpha(t - t_{\alpha,\ell}) \subset \cup_{i=1}^k D_\alpha(t - t_{\alpha\ell}) \subset \cup_{i=1}^k \overline{D}_\alpha(t - t_{\alpha,\ell})$$

for $t > t_{\alpha,\ell}$. By continuity, letting $\ell \to \infty$ yields

$$E(t) \subset \cup_{i=1}^k \overline{D}_\alpha^i(t) \ \text{ for } \ t > 0.$$

On the other hand $\overline{D}^i_\alpha \subset E^i_\alpha$, where E^i_α is the closed evolution with initial data $\overline{U}^i_\alpha$. Since $\overline{U}^i_\alpha \downarrow U^i$, by monotone convergence (Theorem 5.5.4) we have $E^i_\alpha \downarrow E^i$, where E^i is the closed evolution with initial data $\overline{U}^i$.

Since $E(t) \subset \cup_{i=1}^k \overline{D}^i_\alpha(t)$, letting $\alpha \to \infty$, we get $E(t) \subset \cup_{i=1}^k E^i(t)$ for all $t > 0$. Since we have assumed $E^i = \overline{D}^i$, we get

$$E(t) \subset \cup_{i=1}^k \overline{D}^i(t) \ \text{for}\ t > 0.$$

By the order preserving property, $D^i \subset D$ for all $i = 1, \dots, k$, so we now obtain $E(t) \subset \overline{D}(t)$ for all $t \geq 0$, i.e., $E \subset \overline{D}$. Thus $\overline{D} = E$ since $D \subset E$ is trivial. □

Lemma 5.5.13 (Separation) *Assume (W) and* (5.1.1) *is orientation-free. Let* D_i *be an open evolution* ($i = 1, 2$).

(i) *If* $D_1(0)$ *and* $D_2(0)$ *are disjoint so are* D_1 *and* D_2.
(ii) *Let* D *be the open evolution with initial data* $D_1(0) \cup D_2(0)$. *If* $D_1(0)$ *and* $D_2(0)$ *are disjoint then* $D = D_1 \cup D_2$.

Proof

(i) Let $d(\geq 0)$ be the distance between $D_1(0)$ and $D_2(0)$ and let $d_j(x)$ denote the distance from $x \in \mathbf{R}^N$ to $\partial(D_j(0))(j = 1, 2)$. Define

$$u_0(x) = \begin{cases} d_1(x) + d/2 & \text{for} \quad x \in D_1(0), \\ -d_2(x) - d/2 & \text{for} \quad x \in D_2(0), \\ (d_2(x) \wedge \frac{d}{2}) - (d_1(x) \wedge \frac{d}{2}) & \text{otherwise.} \end{cases}$$

We claim that u_0 is Lipschitz with constant 1. In the open sets $D_j(0)(j = 1, 2)$ this is clear since the distance to a set is always Lipschitz with constant 1. Outside of the closure of the $D_i(0)$ we always have $d_1(x) + d_2(x) \geq d$, by the triangle inequality. So the open sets

$$V_i = \{x \in \mathbf{R}^N \setminus (\overline{D_1(0)} \cup \overline{D_2(0)});\, d_i(x) < d/2\}$$

are disjoint. On these sets $u_0(x) = \pm(d_i(x) - d/2)$ so that u_0 is also Lipschitz with constant 1 on these sets. Finally $u_0(x)$ vanishes outside $D_1(0) \cup D_2(0) \cup V_1 \cup V_2$. Collecting these separate Lipschitz estimates we find that u_0 is Lipschitz with constant 1 in $\mathbf{R}^N$.

Since u_0 is Lipschitz, there is a unique solution u of (5.1.2) which is uniformly continuous in $\mathbf{R}^N \times [0, T']$ for every $T' < T$. Since $\pm u - c (c \in \mathbf{R})$ is also a solution of (5.1.2) by invariance and the orientation-free property,

$$D_1 = \{(x, t) \in [0, T') \times \mathbf{R}^N;\ u(x, t) > d/2\},$$

$$D_2 = \{(x, \tau) \in [0, T') \times \mathbf{R}^N;\ u(x, t) < -d/2\}$$

are open evolutions with initial data $D_1(0)$ and $D_2(0)$. Clearly D_1 and D_2 are disjoint. Since u is uniformly continuous, the distance of $D_1(t)$, $D_2(t)$ is uniformly positive for all $t \in [0, T']$ if $d > 0$. This property will be used in the proof of (ii).

(ii) There are sequences of open sets $D_{i0}^j \uparrow D_i(0)(i = 1, 2)$ such that the distance from D_{i0}^j to $\partial(D_i(0))$ is positive. Let u_i^j be the solution of (5.1.2) such that $u_i^j|_{t=0} = u_{0i}^j$ is uniformly continuous and that $u_{0i}^j(x) > 0$ for $x \in D_{i0}^j$ and $u_{0i}^j(x) = 0$ for $x \notin D_{i0}^j$. The open evolution D_i^j with initial data D_{i0}^j is given by

$$D_i^j = \{(x,t);\, u_i^j(x,t) > 0\}$$

and $u_i^j \geq 0$ everywhere. Since the distance between D_{10}^j and D_{20}^j is positive, the distance between D_1^j, D_2^j is positive in $\mathbf{R}^N \times [0, T']$. Thus $u_1^j \vee u_2^j$ is also a solution of (5.1.2) so the open evolution D^j with initial data $D_{10}^j \cup D_{20}^j$ equals $D_1^j \cup D_2^j$. By monotone convergence we see $D^j \uparrow D$ and $D_i^j \uparrow D_i$. Thus $D = D_1 \cup D_2$.

□

Remark 5.5.14

(i) If we consider only bounded evolutions D and E in the sense that $D(0)$ and $E(0)$ are bounded, we may replace (CP) by (BCP) in the assumption (W). By Remark 5.3.7 the requirement of (5.3.1) is unnecessary since (5.3.5) is always fulfilled.

(ii) The order preserving property of Theorem 5.5.2 can be strengthened by introducing super level sets of sub- and supersolutions instead of solutions. We just give below a typical result corresponding to Theorem 5.5.2 (i) when D and D' are bounded. Assume that bounded open sets D and D' in $Z = \mathbf{R}^N \times [0, T)$ are super level sets of the form

$$\begin{aligned} D &= \{(x,t) \in Z;\ u(x,t) > 0\}, \\ D' &= \{(x,t) \in Z;\ v(x,t) > 0\} \end{aligned} \tag{5.5.2}$$

for some subsolution u and supersolution v of (5.1.2) and (5.1.3) such that u and v are *continuous* in Z and $u|_{t=0}, v|_{t=0} \in K_\alpha(\mathbf{R}^N)$ for some $\alpha < 0$. If (BCP) holds, then $D \subset D'$ holds provided that $D(0) \subset D'(0)$. This is easy to prove by (BCP) since we may assume that $u \leq v$ at $t = 0$ by the invariance Theorem 5.2.1 and Lemma 5.2.9.

We are tempted to introduce a notion of subsolution for sets D if it is of the form (5.5.2). However, this notion is not the same as the notion of set-theoretic subsolution defined in Definition 6.1.1. It agrees with the level set subsolution in Definition 6.2.2 as proved in Proposition 6.2.3. The difference of these two notions stems from fattening phenomena, as explained in Chap. 6.

5.6 Convergence Properties for Level Set Equations

We shall study whether solutions of approximate equation

$$V = f_\varepsilon(z, \mathbf{n}, \nabla\mathbf{n}) \tag{5.6.1}$$

converges to the solution of

$$V = f(z, \mathbf{n}, \nabla\mathbf{n}) \tag{5.6.2}$$

when f_ε tends to f as $\varepsilon \to 0$ in a certain sense. We first discuss convergence of solutions of level set equations.

Theorem 5.6.1 (Convergence) *Assume that* F_ε, $F : \mathbf{R}^N \times [0, T] \times (\mathbf{R}^N \setminus \{0\}) \times \mathbf{S}^N \to \mathbf{R}$ $(0 < T < \infty, 0 < \varepsilon < 1)$ *are continuous and geometric. Assume that there is* $c \in C(0, r_0)$ *that satisfies* (5.3.1). *Assume that there is a constant* C *such that*

$$\begin{aligned} &|F_\varepsilon(x, t, p, \pm I)| \le C|F(x, t, p, \pm I)| \\ &\text{for } (x, t, p) \in \mathbf{R}^N \times [0, T] \times (B_{r_1}(0) \setminus \{0\}) \end{aligned} \tag{5.6.3}$$

for some $r_1 \in (0, r_0)$. *Assume that* $F_\varepsilon \to F$ *locally uniformly in* $\mathbf{R}^N \times [0, T] \times (\mathbf{R}^N \setminus \{0\}) \times \mathbf{S}^N$. *Assume that* $u_{0\varepsilon} \in \mathrm{BUC}(\mathbf{R}^N)$ *converges to* $u_0 \in \mathrm{BUC}(\mathbf{R}^N)$ *uniformly in* $\mathbf{R}^N$ *as* $\varepsilon \to 0$. *Let* $u_\varepsilon \in \mathrm{BUC}(\mathbf{R}^N \times [0, T))$ *be the* $\mathcal{F}_{\mathbf{R}^N}$*-solution of*

$$u_t + F_\varepsilon(z, \nabla u, \nabla^2 u) = 0 \tag{5.6.4}$$

with initial data $u_{0\varepsilon}$. *Then* $\lim_{\varepsilon\to 0} u_\varepsilon = u \in \mathrm{BUC}(\mathbf{R}^N \times [0, T))$ *exists and* u *is the* $\mathcal{F}_{\mathbf{R}^N}(F)$*-solution of the limit equation*

$$u_t + F(z, \nabla u, \nabla^2 u) = 0 \tag{5.6.5}$$

with initial data u_0 *provided that* (5.6.4), (5.6.5) *satisfies (CP) with* $\Omega = \mathbf{R}^N$. *Moreover, the convergence is locally uniform in* $\mathbf{R}^N \times [0, T)$.

Proof By (5.6.3) and (4.1.2) we see that $\mathcal{F}_{\mathbf{R}^N}(F_\varepsilon) \supset \mathcal{F}_{\mathbf{R}^N}(F)$ and (3.2.4) is fulfilled. (The existence of c guarantees that $\mathcal{F}_{\mathbf{R}^N}(F) \ne \emptyset$ by Lemma 4.1.3.) We now apply the stability results (Theorem 3.2.1) and observe that

$$\overline{u} = \limsup{}^* u_\varepsilon \ \text{ and } \ \underline{u} = \liminf{}_* u_\varepsilon$$

are $\mathcal{F}_{\mathbf{R}^N}(F)$-sub- and supersolutions of (5.2.1) provided that $\overline{u} < \infty$ and $\underline{u} > -\infty$.

We shall prove that

$$\overline{u} < \infty \ \text{ and } \ \overline{\lim_{t\downarrow 0}} \sup_\xi (\overline{u}(\xi, t) - u_0(\xi)) \le 0.$$

Since u_0^ε converges to u_0 uniformly in $\mathbf{R}^N$ and u_0^ε, $u_0 \in \mathrm{BUC}(\mathbf{R}^N)$, there is a modulus of continuity independent of $\varepsilon \in (0, 1)$ such that

$$u_0^\varepsilon(x) - u_0^\varepsilon(\xi) \le \omega(|x - \xi|),\ \varepsilon \in (0, 1),\ x \in \mathbf{R}^N.$$

By the assumptions (5.3.1) and (5.6.3) there is $h \in \mathcal{F}_{\mathbf{R}^N}(F) \subset \mathcal{F}_{\mathbf{R}^N}(F_\varepsilon)$ such that $\sup h' < \infty$. Moreover,

$$B_\varepsilon := \sup_{0 \le t < T} \sup_{x, p \in \mathbf{R}^N} |F_\varepsilon(x, t, \nabla_p(h(\rho)), \pm\nabla_p^2 h(\rho))| < \infty, (\rho = |\rho|) \tag{5.6.6}$$

is bounded for $\varepsilon \in (0, 1)$, i.e.,

$$B = \sup_{0<\varepsilon<1} B_\varepsilon < \infty. \tag{5.6.7}$$

For each $\delta > 0$ there is $A_\delta > 0$ such that $\omega(s) \le \delta + A_\delta s^2$, $s \ge 0$. Thus

$$u_0^\varepsilon(x) \le u_0^\varepsilon(\xi) + \delta + A_\delta h(|x - \xi|).$$

Since B is independent of ε, by Lemma 5.3.3 and its proof the function $V^+(x) = A_\delta h(|x|) + M_\delta t$ is an $\mathcal{F}_{\mathbf{R}^N}(F_\varepsilon)$-supersolution of (5.6.4) (independent of ε). So the function

$$w_\delta^\varepsilon(x, t) = u_0^\varepsilon(\xi) + \delta + V^+(x - \xi, t)$$

is also an $\mathcal{F}_{\mathbf{R}^N}(F_\varepsilon)$-supersolution of (5.6.4). Since u^ε and w_δ^ε is uniformly continuous on $\mathbf{R}^N \times [0, T)$, by (CP) we have $u^\varepsilon \le w_\delta^\varepsilon$. In particular, $\overline{u} < \infty$ on $\mathbf{R}^N \times [0, T)$. Since $u^\varepsilon \le w_\delta^\varepsilon$, we see

$$\overline{u}(\xi, t) \le \lim_{\varepsilon \to 0} u_0^\varepsilon(\xi) + \delta + Mt.$$

Thus

$$\overline{\lim_{t \downarrow 0}} \sup_{\xi \in \mathbf{R}^N} (\overline{u}(\xi, t) - u_0(\xi)) \le \delta.$$

Since $\delta > 0$ is arbitrary, we get

$$\overline{\lim_{t \downarrow 0}} \sup_{\xi \in \mathbf{R}^N} (\overline{u}(\xi, t) - u_0(\xi)) \le 0.$$

A symmetric argument yields $\underline{u} > -\infty$ and

$$\overline{\lim_{t \downarrow 0}} \sup_{\xi \in \mathbf{R}^N} (u_0(\xi) - \underline{u}(\xi, t)) \le 0.$$

In particular $\overline{u}|_{t=0} = \underline{u}|_{t=0} = u_0$. Since u_0 is uniformly continuous, these two inequalities yield

$$\sup_{0\le t,s\le\varepsilon}\ \sup_{|\xi-\eta|\le\varepsilon} (\overline{u}(\xi,t) - \underline{u}(\xi,s)) \to 0 \text{ as } \varepsilon \to 0.$$

We are now in position to apply (CP) to get $\overline{u} \le \underline{u}$. The converse inequality is trivial so $\overline{u} = \underline{u}$. By definition of $\limsup^*$, $\liminf_*$ this implies that u^ε converges to u locally uniformly in $\mathbf{R}^N \times [0,T)$ as $\varepsilon \to 0$. Moreover $u|_{t=0} = u_0$ and u solves (5.6.5) since $u = \underline{u} = \overline{u}$ is both $\mathcal{F}_{\mathbf{R}^N}(F)$-sub- and supersolution. □

Remark 5.6.2

(i) We may replace (5.6.3) by $\mathcal{F}_{\mathbf{R}^N}(F) \subset \mathcal{F}_{\mathbf{R}^N}(F_\varepsilon)$ in Theorem 5.6.1. If f_ε and f are independent of x, t in (5.6.1), (5.6.2) and

$$|f_\varepsilon(p, Q_p(X))| \le A(1 + |Q_p(X)|), \quad \varepsilon \in (0,1)$$

with constant $A > 0$, then $\mathcal{F}(F_{f_\varepsilon}) = \mathcal{F}(F_f)$ $(= \{h \in C^2(\mathbf{R}); h(0) = h'(0) = h''(0) = 0, h''(\sigma) > 0 \text{ for } \sigma > 0\})$. So in this situation we have convergence of solutions of level set equations.

(ii) If we fully use (4.1.5) for $\overline{u}$ and $\underline{u}$, we can also prove that $u^\varepsilon \to u$ uniformly in $\mathbf{R}^N \times [0, T']$ for $T' < T$.

(iii) If $u_0^\varepsilon \in K_{\alpha_\varepsilon}(\mathbf{R}^N)$ and $\operatorname{spt}(u_0^\varepsilon - \alpha_\varepsilon) \subset \operatorname{int} B_R$ for some R independent of $\varepsilon \in (0,1)$, then there is $R' \ge R$ (independent of ε) that satisfies

$$\operatorname{spt}(u^\varepsilon - \alpha_\varepsilon) \subset \operatorname{int} B_{R'} \times [0,T).$$

(In particular, the convergence $u^\varepsilon \to u$ is uniform in $\mathbf{R}^N \times [0, T']$, $T' < T$.) This uniform estimate of the support $u^\varepsilon - \alpha_\varepsilon$ can be proved by constructing a suitable super- and subsolution as in the proof of Theorem 5.3.1. Instead of w^+ we use here

$$w_\varepsilon^+(x,t) = \max\{v^+(x,t) + \sigma, \alpha_\varepsilon\},$$

where σ is chosen such that $u_0^\varepsilon(x) \le \sigma - h(|x|)$ for x satisfying $u_0^\varepsilon(x) \ne \alpha_\varepsilon$. We estimate $w_\varepsilon^- \le u^\varepsilon \le w_\varepsilon^+$ by constructing w_ε^- similarly and observe that $u^\varepsilon - \alpha_\varepsilon = \alpha_\varepsilon$ outside $\Omega' \times [0,T)$, where Ω' is some ball.

(iv) Geometricity of F_ε and F is not necessary if there exists $V^\pm$ (independent of ε) in the proof of Theorem 5.6.1 and if assumptions in Theorem 3.2.1 are fulfilled. We give another version of convergence results whose proof parallels that of Theorem 5.6.1.

Theorem 5.6.3 (Convergence Without Geometricity) *Assume that* $F, F_\varepsilon : \mathbf{R}^N \times [0,T] \times (\mathbf{R}^N \setminus \{0\}) \times \mathbf{S}^N \to \mathbf{R}$ $(0 < T < \infty, 0 < \varepsilon < 1)$ *are continuous. Assume that* $\mathcal{F}_{\mathbf{R}^N}(F) \subset \mathcal{F}_{\mathbf{R}^N}(F_\varepsilon)$ *and fulfills* (3.2.4). *Assume that* B *in* (5.6.6), (5.6.7) *is finite for some* $h \in \mathcal{F}_{\mathbf{R}^N}(F)$. *Assume that* $F_\varepsilon \to F$ *locally uniformly in*

$\mathbf{R}^N \times [0, T] \times (\mathbf{R}^N \setminus \{0\}) \times \mathbf{S}^N$. *Assume that* $u_{0\varepsilon} \in \mathrm{BUC}(\mathbf{R}^N)$ *converges to* $u_0 \in \mathrm{BUC}(\mathbf{R}^N)$ *uniformly in* $\mathbf{R}^N$ *as* $\varepsilon \to 0$. *Let* $u_\varepsilon \in \mathrm{BUC}(Z)$ *be the* $\mathcal{F}_{\mathbf{R}^N}(F_\varepsilon)$-*solution of* (5.6.4) *with initial data* $u_{0\varepsilon}$. *Then* $\lim_{\varepsilon\to 0} u^\varepsilon = u \in \mathrm{BUC}(Z)$ *exists and* u *is the* $\mathcal{F}_{\mathbf{R}^N}(F)$-*solution of* (5.6.5) *with initial data* u_0 *provided that* (5.6.4), (5.6.5) *satisfies (CP) with* $\Omega = \mathbf{R}^N$. *Moreover the convergence is locally uniform in* $Z = \mathbf{R}^N \times [0, T)$.

We take this opportunity to prove Lemma 5.2.12. We suppress the symbol $\mathcal{F}_{\mathbf{R}^N}$ in the proof.

Proof of Lemma 5.2.12 Since $u_{0m} \in \mathrm{BUC}(\mathbf{R}^N)$ and the solution u_m with initial data u_{0m} also belongs to $\mathrm{BUC}(Z)$, we see (CP) is applicable to conclude $u_m \le u_{m+1}(\le u)$ in $Z = \mathbf{R}^N \times [0, T)$. We set

$$v(z) = \sup_{m\ge 1} u_m(z)(= \lim_{m\to\infty} u_m(z)), \quad z \in Z.$$

Since u_m is continuous, v is lower semicontinuous in Z. Thus

$$\begin{aligned} v(z) &= \liminf_{r\downarrow 0}\{v(\zeta);\ \zeta \in Z,\ |\zeta - z| \le r\} \\ &\ge \liminf_{r\downarrow 0}\left\{u_k(\zeta);\ \zeta \in Z,\ |\zeta - z| \le r,\ k \ge \frac{1}{r}\right\} \\ &\ge \liminf\{u_m(\zeta);\ \zeta \in Z, |\zeta - z| \le r\}, \quad m = 1, 2, \ldots. \end{aligned}$$

The last quantity equals $u_m(z)$ so sending $m \to \infty$ yields

$$v(z) \ge \liminf_{r\downarrow 0}\left\{u_k(\zeta);\ \zeta \in Z; |\zeta - z| \le r,\ k \ge \frac{1}{r}\right\} \ge v(z).$$

Thus we conclude

$$v(z) = (\liminf_{m\to\infty}{}_* u_m)(z).$$

By the stability (Theorem 3.2.1) v is a supersolution of (5.2.1).

We shall prove that

$$\limsup_{r\downarrow 0}\{u(x, t) - v(y, s);\ |x - y| \le r,\ t \vee s \le r\} \le 0.$$

Let $h \in \mathcal{F}_{\mathbf{R}^N}$ with $\sup h' < \infty$. As in the proof of Lemma 5.3.4 for each $\delta > 0$ there are positive constants A_δ and $M_\delta = M_{A_\delta}$ such that

$$w_\delta^\xi(x, t) = u_0(\xi) - \delta - A_\delta h(|x - \xi|) - M_\delta t$$

is a subsolution of (5.2.1) in $\mathbf{R}^N \times (0, T)$ and

$$u_0(x) - \delta \geq w_\delta^\xi(x, 0).$$

By Dini's theorem $u_{m0} \to u_0$ locally uniformly in $\mathbf{R}^N$. Since $\{u_{m0}\}$ is bounded from below, there is $\ell = \ell(\delta, \xi)$ such that

$$u_{m0} \geq w_\delta^\xi(\cdot, 0) \text{ in } \mathbf{R}^N \text{ for } m \geq \ell.$$

Since both w_δ^ξ and u_m are uniformly continuous in $\mathbf{R}^N \times [0, T']$, $T' < T$, we see, by (CP), that $u_m \geq w_\delta^\xi$ for all $m \geq \ell$. This yields $v \geq w_\delta^\xi$ in $\mathbf{R}^N \times [0, T)$ for $\delta \in (0, 1)$, $\xi \in \mathbf{R}^N$. In particular, $v(y, s) \geq w_\delta^y(y, s)$. Thus

$$u(x, t) - v(y, s) \leq u(x, t) - w_\delta^y(y, s) \leq u(x, t) - u_0(y) + \delta + M_\delta s.$$

Since u is uniformly continuous in Z, it is clear that

$$\lim_{r \downarrow 0} \sup\{u(x, t) - u_0(y);\ |x - y| \leq r,\ t \leq r\} \leq 0.$$

We thus obtain that

$$\lim_{r \downarrow 0} \sup\{u(x, t) - v(y, s);\ |x - y| \leq r,\ t \vee s \leq r\} \leq \delta.$$

Since $\delta > 0$ is arbitrary, we conclude

$$\lim_{r \downarrow 0} \sup\{u(x, t) - v(y, s);\ |x - y| \leq r,\ t \vee s \leq r\} \leq 0$$

and apply (CP) to get $u \leq v$ in Z. Since $v \leq u$, this implies $u \equiv v$. Thus we have proved that $u_m \uparrow u$. □

The convergence result (Theorem 5.6.1) yields the convergence of evolutions of (5.6.1) to (5.6.2) provided that fattening does not occur.

Theorem 5.6.4 (Convergence of Evolutions) *Assume that* (5.6.1) *and* (5.6.2) *fulfill (W). Assume that f_ε converges to f as $\varepsilon \to 0$ locally uniformly on $\mathbf{R}^N \times [0, T] \times \mathbf{E}$. Assume that F_{f_ε} and F_f satisfy $\mathcal{F}_{\mathbf{R}^N}(F_f) \subset \mathcal{F}_{\mathbf{R}^N}(F_{f_\varepsilon})$ for $\varepsilon \in (0, 1)$. Let E_0^ε and E_0 be compact sets in $\mathbf{R}^N$. Let E^ε and E be a closed evolution of* (5.6.1) *and* (5.6.2) *with initial data E_0^ε and E_0, respectively. Assume that $d_H(E_0^\varepsilon, E_0) \to 0$ as $\varepsilon \to 0$. Assume that* $\operatorname{int} E$ *is regular. Then*

(i) *$d'_H(E^\varepsilon, E) \to 0$ as $\varepsilon \to 0$, where d'_H denotes the Hausdorff distance in $\mathbf{R}^N \times [0, T']$, $T' < T$.*
(ii) *Assume that E is strongly regular in the sense that $E(t) = \overline{D(t)}$ for all $t \in [0, T']$ where D is the open evolution of* (5.6.1) *with initial data* $\operatorname{int} E_0$. *Then $d_H(E^\varepsilon(t), E(t)) \to 0$ as $\varepsilon \to 0$ and the convergence is uniform in $t \in [0, T']$.*

For the proof we need a lemma on level sets.

Lemma 5.6.5 *Let Z and Y be compact metric spaces. Assume that $f_\lambda^\varepsilon \in C(Z)$ converges to $f_\lambda \in C(Z)$ as $\varepsilon \to 0$ uniformly in $\lambda \in Y$, where $\varepsilon \in (0, 1)$. For a given $\ell \in \mathbf{R}$ assume that $\{z \in Z;\ f_\lambda(z) \geq \ell\} := \{f_\lambda \geq \ell\} = \overline{\{f_\lambda > \ell\}}$. Assume that the sets $\{f_\lambda \geq \ell\}$ and $\{f_\lambda^\varepsilon \geq \ell\}$ are compact in Z for all $\lambda \in Y$. Assume that for small $\varepsilon > 0$ the set $\{f_\lambda^\varepsilon \geq \ell\}$ is continuous in $\lambda \in Y$ with respect to the Hausdorff metric d_H in Z. Then $d_H(\{f_\lambda^\varepsilon \geq \ell\}, \{f_\lambda \geq \ell\}) \to 0$ as $\varepsilon \to 0$ uniformly in $\lambda \in Y$.*

Proof We first note that for each $\eta > 0$ there is $\varepsilon_0 \in (0, 1)$ such that $\{f_\lambda^\varepsilon \geq \ell\} \subset \{f_\lambda \geq \ell\}_\eta$ for all $\varepsilon \in (0, \varepsilon_0)$, $\lambda \in Y$, where $A_\eta = \{z \in Z; \operatorname{dist}(z, A) \leq \eta\}$ for a subset A of Z. If not, for some $\eta > 0$ there is a sequence $\varepsilon_j \to 0$, $\{\lambda_j\} \subset Y$, $\{z_j\} \subset Z$ that satisfies dist $(z_j, \{f_{\lambda_j} \geq \ell\}) > \eta$ and $z_j \in \{f_{\lambda_j}^{\varepsilon_j} \geq \ell\}$. We may assume that $\lambda_j \to \lambda_0$, $z_j \to z_0$ as $j \to \infty$ by taking a subsequence. Since f_λ^ε converges to f_λ uniformly in Z and for $\lambda \in Y$, we have $f_{\lambda_j}^{\varepsilon_j}(z_j) \to f_{\lambda_0}(z_0)$. Since $f_{\lambda_j}^{\varepsilon_j}(z_j) \geq \ell$, this implies $f_{\lambda_0}(z_0) \geq \ell$. This is absurd since dist $(z_0, \{f_{\lambda_0} \geq \ell\}) = \lim_{j\to\infty}$ dist $(z_j, \{f_{\lambda_j} \geq \ell\}) \geq \eta$ by continuity of $\{f_\lambda \geq \ell\}$ in $\lambda \in Y$. Thus, $\{f_\lambda^\varepsilon \geq \ell\} \subset \{f_\lambda \geq \ell\}_\eta$ for small $\varepsilon > 0$ uniformly in $\lambda \in Y$.

It remains to prove that $\{f_\lambda \geq \ell\} \subset \{f_\lambda^\varepsilon \geq \ell\}_\eta$ for sufficiently small ε (uniformly in $\lambda \in Y$). If not, there is $r > 0$ and λ_j, z_j with $z_j \in \{f_\lambda \geq \ell\}$ such that $\{f_{\lambda_j}^{\varepsilon_j} \geq \ell\} \cap B_r(z_j) = \emptyset$ for some sequence $\varepsilon_j \to 0$. By continuity of $\{f_\lambda \geq \ell\}$ in λ we may assume that $\lambda_j \to \lambda_0$ and $z_j \to z_0 \in \{f_{\lambda_0} \geq \ell\}$ so that $\{f_{\lambda_j}^{\varepsilon_j} \geq \ell\} \cap B_{r/2}(z_0) = \emptyset$. Since $f_\lambda^\varepsilon \to f_\lambda$ uniformly in Z and for $\lambda \in Y$, this implies that $f_{\lambda_0} \leq \ell$ on $B_{r/2}(z_0)$. This contradicts $\overline{\{f_{\lambda_0} > \ell\}} = \{f_{\lambda_0} \geq \ell\}$. □

Lemma 5.6.6 (Strongly Regular Evolution) *Assume (W). Assume that $E_0 = E(0)$ is compact. If E is strongly regular, then $t \mapsto E(t)$ is continuous as a set-valued function on $[0, T)$.*

Proof Since $t \mapsto E(t)$ is left continuous and upper semicontinuous, it suffices to prove that $E(t)$ is right lower semicontinuous in t.

Assume that $E(t)$ is not right continuous at some point $t_0 \in [0, T)$. Then there is a point $x_0 \in E(t_0)$ and a ball $B_r(x_0)$ such that $E(t_j) \cap B_r(x_0) = \emptyset$ for some $t_j \downarrow t_0$. Since $\overline{D(t_0)} = E(t_0)$ there is a ball $B \subset D(t_0) \cap E(t_0)$. Comparing a special subsolution we conclude that an open evolution starting with $\operatorname{int} B$ at $t = t_0$ contains a center of B for $[t_0, t_0 + \delta)$ for some $\delta > 0$. Thus by comparison we see $D(t_j) \cap B_r(x_0) \neq \emptyset$ which contradicts $E(t_j) \cap B_r(x_0) = \emptyset$. Therefore $E(t)$ is right continuous on $[0, T)$. □

Proof of Theorem 5.6.4 For E_0^ε we set

$$u_0^\varepsilon(x) = (\operatorname{sd}(x, \partial E_0^\varepsilon) \wedge 1) \vee (-1).$$

Since $d_H(E_0^\varepsilon, E_0) \to 0$ as $\varepsilon \to 0$, we see that u_0^ε converges to

$$u_0(x) = (\operatorname{sd}(x, \partial E) \wedge 1) \vee (-1).$$

Let u^ε be the solution of (5.6.4) with $u_0^\varepsilon|_{t=0} = u_0^\varepsilon$ and let u be the solution of (5.6.5) with $u|_{t=0} = u_0$. Our assumption $f_\varepsilon \to f$ guarantees the convergence $F_{f_\varepsilon} \to F_f$ of Theorem 5.6.1. By the convergence result (Theorem 5.6.1 and Remark 5.6.2 (i)) u^ε converges to u uniformly in $\mathbf{R}^n \times [0, T']$ for every $T' < \infty$. Since $E^\varepsilon = \{u^\varepsilon \geq 0\}$, $E = \{u \geq 0\}$, we apply Lemma 5.6.5 with $Z = B_R \times [0, T']$ so that $\{u^\varepsilon \geq 0\} \subset Z$ and $Y = \emptyset$ and $f^\varepsilon = u^\varepsilon$ to get (i). To show (ii) it suffices to take $Z = B_R$, $Y = [0, T']$ and $f_\lambda^\varepsilon = u^\varepsilon(, \lambda)$, since the strong regularity implies the continuity of $E(t)$ by Lemma 5.6.6. □

Remark 5.6.7

(i) As an application of Theorem 5.6.4 (i) we have the convergence of the extinction time T^ε of E^ε is that of E as $\varepsilon = 0$. Here we define

$$T_\varepsilon = \sup\{t;\ E_\varepsilon(t) \neq \emptyset\},\ T_0 = \sup\{t;\ E(t) \neq \emptyset\}.$$

(ii) It is possible to replace E^ε by the interface evolution $\Gamma^\varepsilon = E^\varepsilon \setminus D^\varepsilon$ in Theorem 5.6.4 with trivial modifications.

(iii) In Theorem 5.6.4 (ii) the strong regularity assumption on $[0, T']$ is actually stronger than $E = \overline{D}$. For example if $E(t)$ is a shrinking disk solving the mean curvature flow equation $V = H$ (see Sect. 2.7 and Remark 5.3.8), then at the extinction time T the set $E(T)$ is a singleton but $D(T) = \emptyset$. This $E(T)$ is not right continuous at $t = T$, so continuity in t does not follow from $E = \overline{D}$. (By definition it is clear that strong regularity implies $E = \overline{D}$.)

5.7 Instant Extinction

As we shall see later in Sect. 6.4 a classical solution with smooth closed initial hypersurface agrees with our generalized interface evolution. One advantage of our level set method is that one can take any closed set as an initial data. In this section we would like to give several examples that the interface evolution becomes empty instantaneously for a nonempty initial set.

We begin with an orientation-free problem like the mean curvature flow equation.

Theorem 5.7.1 *Assume* (*W*) *and that* (5.1.1) *is orientation-free. Assume that* (5.1.1) *is strictly parabolic and that* f *is smooth in each variable. Assume that* Γ_0 *is a closed set on a closed smooth (connected) hypersurface* Γ_{10} *in* $\mathbf{R}^N$. *Assume that* $\Gamma_0 \neq \Gamma_{10}$. *Let* Γ *be the interface evolution of* (5.1.1) *with* $\Gamma(0) = \Gamma_0$. *Then* $\Gamma(t) = \varnothing$ *for* $t > 0$.

For example, if the initial data is a hypersurface with a hole, the interface evolution disappears instantaneously (Fig. 5.7).

Proof Let D_{10} be the bounded open set enclosed by Γ_{10}. By assumption of initial data there is another closed hypersurface Γ_{20} ($\neq \Gamma_{10}$) such that $\Gamma_0 \subset \Gamma_{10} \cap \Gamma_{20}$ and

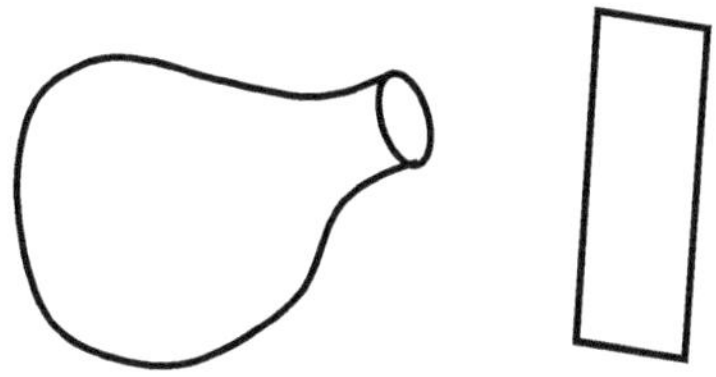

Fig. 5.7 Initial data for instant extinction

the bounded open set D_{20} enclosed by Γ_{20} does not intersect D_{10}. As we mentioned in the last part of Sect. 2.8 there are local smooth solutions Γ_i for (5.1.1) starting from Γ_{i0} $(i = 1, 2)$; see Y. Giga and S. Goto [208] where f is assumed to be independent of space variables but the method actually allows x-dependence. Since the problem is orientation-free, by the strong maximum principle (see e.g. a standard textbook by M. H. Protter and H. F. Weinberger [381]), we see that $\Gamma_1(t) \cap \Gamma_2(t) \neq \varnothing$ for all $t \in (0, T)$, where T is the minimum of life spans of smooth solutions Γ_1 and Γ_2. By consistency (Corollary 6.4.5) it must agree with the interface evolution.

By Remark 5.2.10 for orientation-free equations there are nonnegative auxiliary functions u and u_i solving (5.2.1) in $\mathbf{R}^N \times (0, T)$ such that its zero level set agrees with Γ and Γ_i $(i = 1, 2)$. Then by (BCP) $\Gamma \subset \Gamma_1 \cap \Gamma_2$. Since $\Gamma_1(t) \cap \Gamma_2(t) = \varnothing$ for $t > 0$, this implies that $\Gamma(t) = \varnothing$ for $t \in (0, T)$. Once $\Gamma(t_0) = \varnothing$ for some $t_0 \geq 0$, it is clear that $\Gamma(t) = \Gamma(t_0) = \varnothing$ for $t \geq t_0$. □

If the problem is not orientation-free, such an argument does not work. However, we expect a similar result provided that the problem is strictly parabolic, although a general result is not yet available.

If the problem is not strictly parabolic, typically the first order problem like the Hamilton–Jacobi equation (2.5.6), then the phenomenon is quite different. We give a typical result.

Theorem 5.7.2 *Assume that c is a constant and that β is continuous and positive in* (2.5.6). *Let E_0 be a bounded, closed, convex set with no interior in $\mathbf{R}^N$. Let E be a closed evolution of* (2.5.6) *with initial data E_0. Then*

$$E(t) = E_0 + ctW_{1/\beta} \ (c \geq 0) \quad \text{and} \quad E(t) = \varnothing \ (c < 0)$$

for $t > 0$, where $W_{1/\beta}$ is the Wulff shape defined by

$$W_{1/\beta} = \cap_{|q|=1} \left\{ x \in \mathbf{R}^N; \langle x, q \rangle \leq 1/\beta(q) \right\}.$$

So for $c \geq 0$ even if E_0 is a finite segment or a point, $E(t)$ is not empty for $t > 0$ (Fig. 5.8).

If the curvature term comes in, the situation drastically changes. For example, consider $V = aH + 1$ $(N \geq 3)$ with $a > 0$. Assume that E_0 is a line of the form

$$E_0 = \left\{ (x', x_N); |x_N| \leq L, x' = 0 \right\} \quad (N \geq 3).$$

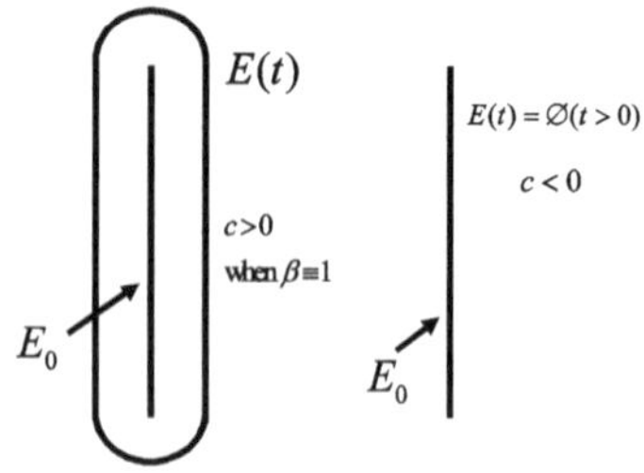

Fig. 5.8 Evolution of a line

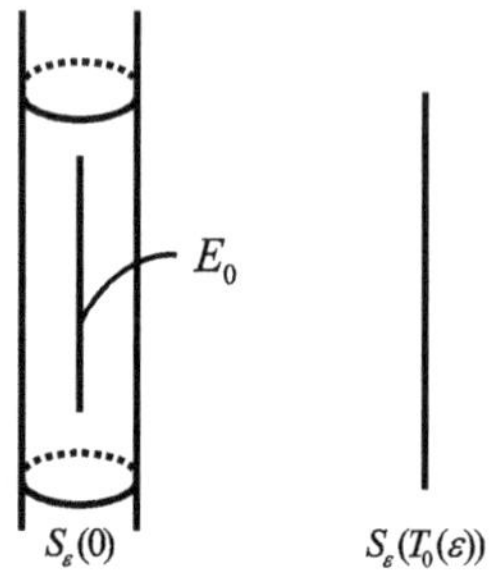

Fig. 5.9 $V = H + 1$

Then there is a cylindrical solution

$$S_\varepsilon(t) = \left\{(x', x_N, t); |x'| \geq R(t)\right\} \quad \text{with}$$
$$dR/dt = -a(N-2)/R + 1,\; R(0) = \varepsilon > 0$$

for $V = aH + 1$. If ε is taken small, then $R(t)$ becomes zero in finite time $T_0(\varepsilon)$ (Fig. 5.9). Moreover, it is easy to see that the time $T_0(\varepsilon)$ when $S_\varepsilon(t)$ shrinks to a line tends to zero as $\varepsilon \to 0$. Since $S_\varepsilon(0) \supset E_0$ for $\varepsilon > 0$, by comparison (CP) we have $E(t) \subset S_\varepsilon(t)$ for $t \in (0, T_0(\varepsilon))$, where E is the closed evolution with initial data E_0. Since $T_0(\varepsilon) \to 0$ as $\varepsilon \to 0$, we observe that $E(t) = \varnothing$ for $t > 0$. In this setting the same instant extinction result holds even if E_0 is an infinite line like the x_N-axis. When $N = 2$ this argument unfortunately does not apply. In fact, if E_0 is the x_2-axis, then the solution agrees with that of $V = 1$. Nevertheless for a finite segment, the instant extinction seems to be valid by taking a suitably shrinking supersolution. The finite interval can be interpreted to have a huge curvature at the end. We do not pursue this problem further in this book.

5.8 Notes and Comments

We take this opportunity to review a scope of equations to which our theory applies.

Unique Existence of Generalized Evolutions Corollaries 5.3.2 and 5.3.6 cover all spatially homogeneous surface evolution equations satisfying (f1) and (f2). This class of equation is the same as in Sect. 4.1.3 so it includes examples mentioned there.

Corollary 5.3.2 can be extended to (5.1.1) when f depends on x provided that (BCP) holds. Several classes of equations satisfying (BCP) have been studied in Sect. 4.6.

For a similar extension of Corollary 5.3.6 we further need the uniform control (5.3.1) as well as (CP) when the equation is spatially inhomogeneous.

Corollary 5.3.2 can be also extended to (5.1.1) in a domain (not necessarily bounded) with prescribed contact angle on the boundary provided that (BCPB) holds. Several examples are provided in Sect. 4.7.

Although we do not mention the one corresponding to (CP) for boundary value problems, Corollary 5.3.6 also can be extended to some boundary value problems. However, it is not explicit what kind of equations admit such a comparison principle for the boundary value problem in the literature.

Since the assumption (W) is essentially fulfilled under the situation mentioned above (see Remark 5.5.14 (i)), results in Sect. 5.5 apply to these problems.

Orientation-Free Equations We list examples of orientation-free equations. Equation (2.5.2) is orientation-free if $\beta(p) = \beta(-p), \gamma(p) = \gamma(-p)$ for $p \in S^{N-1}$ and $c \equiv 0$. In particular the mean curvature flow equation (2.5.4) is orientation-free. Although the Gaussian curvature flow equation (2.5.9) as well as (2.5.12) is orientation-free, our modified Eqs. (2.6.22) and (2.6.23) are not orientation-free. Equation (2.5.13) is orientation-free if $h(\sigma, p) = -h(-\sigma, p), h(\sigma, p) = h(\sigma, -p)$, $\gamma(p) = \gamma(-p)$ for $p \in S^{N-1}$, $\sigma \in \mathbf{R}$. Equation (2.5.14) is not orientation-free although the equation $V = |k|^{\alpha-1}k$ for $\alpha > 0$ is orientation-free.'

Convergence We give a typical example to which Theorem 5.6.4 applies. We consider (2.5.2) parametrized by ε:

$$\beta_\varepsilon(\mathbf{n})V = -a_\varepsilon \mathrm{div}_{\Gamma_t}(\xi_\varepsilon(\mathbf{n})) - c_\varepsilon. \tag{5.8.1}$$

Here c_ε is assumed to be independent of x and t for simplicity. Assume that $c_\varepsilon \to c$ in $\mathbf{R}$ and $a_\varepsilon \to a \in [0, \infty)$ as $\varepsilon \to 0$. If $\beta_\varepsilon \to \beta$, $\partial_p^\alpha \gamma_\varepsilon \to \partial_p^\alpha \gamma$ for $|\alpha| \leq 2$ as $\varepsilon \to 0$ and the convergence is uniform on S^{N-1}, then f_ε converges to f locally uniformly in $\mathbf{E}$ if (5.8.1) is written in the form $V = f_\varepsilon(\mathbf{n}, \nabla\mathbf{n})$ and

$$\beta(\mathbf{n})V = -a \ \mathrm{div}_{\Gamma_t}\xi(\mathbf{n}) - c \tag{5.8.2}$$

is written in the form $V = f(\mathbf{n}, \nabla\mathbf{n})$. In this example as in Remark 5.6.2 it is easy to see that $\mathcal{F}(F_{f_\varepsilon}) = \mathcal{F}(F_f)$. So if both (5.8.1) and (5.8.2) fulfill (W), Theorem 5.6.4 is applicable.

We next point out that there are three review articles on mathematical analysis for the level set method—the article by Y. Giga [200], the article by L. C. Evans published in the lecture note by M. Bardi et al. [41] and the article by L. Ambrosio in the lecture notes by L. Ambrosio and N. Dancer [12]. These articles present main ideas only and do not give details except the one by Ambrosio, where only typical equations are treated. The present chapter gives details and clarifies the class of equations to which the method applies.

The invariance (Theorem 5.2.1) goes back to the work of Y.-G. Chen et al. [106] where they consider conventional viscosity solutions instead of $\mathcal{F}$-solutions. Theorem 5.2.7 for the level set mean curvature flow equation is due to L. C. Evans and J. Spruck [166]. The proofs given here are based on simplification by Y.-G. Chen et al. [108]. Theorem 5.2.8 is essentially due to Y.-G. Chen et al. [106], where they assumed that $f(x, t, \mathbf{n}, \nabla\mathbf{n})$ is independent of x and satisfies (f1), (f2) so that (BCP) holds. For the case of the level set mean curvature flow equation such uniqueness with Remark 5.2.13 is proved by L. C. Evans and J. Spruck [166]. Extension to unbounded evolutions (Sect. 5.2.4) is due to H. Ishii and P. Souganidis [284] as well as $\mathcal{F}$-solutions.

The existence by Perron's method (Sect. 5.3) is essentially due to Y.-G. Chen et al. [106] except extensions to $\mathcal{F}$-solutions and unbounded evolutions. Theorem 5.3.5 is due to H. Ishii and P. Souganidis [284]. The existence by approximation (Sect. 5.4) is due to L. C. Evans and J. Spruck [166].

Semigroup properties were first stated explicitly by L. C. Evans and J. Spruck [166]. The results from Theorem 5.5.2 to Corollary 5.5.8 and in Lemma 5.5.12 are taken from the work of S. Altschuler et al. [10], where they discussed only the mean curvature flow equations; extension to general equations is straightforward. Lemma 5.5.12 is in particular useful to conclude that the fattening does not occur for rotationally symmetric closed hypersurfaces moved by the mean curvature as proved by S. Altschuler et al. [10]; this paper also includes a rigorous proof that the fattening occurs for figure eight; see also a review paper by Y. Giga [200]. For fattening phenomena more references are given in Sect. 6.6. The results from Theorem 5.5.9 to Corollary 5.5.11 are due to G. Barles et al. [55].

The strategy to prove convergence of viscosity solutions only by bound for maximum norm without estimating derivatives goes back to G. Barles and B. Perthame [52, 53] and independently by H. Ishii [272]. A weaker version of convergence is stated in the work of F. Camilli [80], who also discussed the convergence of level sets. His proof is different from ours. It seems that Theorem 5.6.1 was not stated in the literature. Derivation of Theorem 5.6.4 from Theorem 5.6.1 presented here is due to M.-H. Giga and Y. Giga [191]. When $N = 2$, for (5.8.1) and (5.8.2) they proved a stronger result without assuming convergence of the derivative of γ; they only assume the uniform convergence of γ_ε to γ on S^{N-1}.

The phenomenon 'instant extinction' is first observed by L. C. Evans and J. Spruck [166] for the mean curvature flow equation. Theorem 5.7.1 is a general result not stated in the literature. Theorem 5.7.2 is taken from the work of Y. Giga and M.-H. Sato [231]. Such a type of results is later generalized by M. Bardi and Y. Giga [43] when the equation depends on the spatial variables.

We conclude this section by introducing a few recent topics on applications of the level set method not mentioned elsewhere in this book.

A Representation of Solutions as Value Functions The notion of viscosity solutions was introduced to describe value functions of control problems or differential games. See for example, a book of M. Bardi and I. Capuzzo-Dolcetta [40]. A viscosity solution of the level set mean curvature flow equation (2.6.5) can be regarded as a value function of a stochastic differential games as proved by R. Buckdahn et al. [70] and independently by H. M. Soner and N. Touzi [403, 404]. According to the paper by R. V. Kohn and S. Serfaty [312] it can be also regarded as the continuous-in-time limit of value functions of a deterministic two-person game. We will review this game-theoretic approach in Chaps. 7 and 8. There are several ways to derive level set equations as infinitesimal descriptions of various 'filters' in image processing; see for example a book by F. Cao [82]. Such an idea provides an approximation scheme for calculating solutions of level set equations. In fact, the differential game proposed by R. V. Kohn and S. Serfaty [312] has some similarity to the morphological scheme for mean curvature motion developed by F. Catté et al. [93].

Large Time Behavior of a Generalized Solution We just restrict our attention to generalized evolutions because there are too many topics. We consider an anisotropic mean curvature flow equation of the form (4.6.15). We assume that β and c depend only on $\mathbf{n}$ and are independent of x and t as well as assumptions of the example in Sect. 4.6.2. If c is positive and initial compact interface Γ_0 encloses a sufficiently large ball so that the solution grows as time develops, we wonder what is an asymptotic shape as $t \to \infty$. Our level set method is suitable to handle such a problem since a solution may develop singularities during its evolution. Here is an available answer to this equation. Let Γ_t be the interface evolution. Then

$$\Gamma_t/t \to W_{c/\beta} \qquad \text{as} \quad t \to \infty \tag{5.8.3}$$

in the sense of the Hausdorff distance. Here $W_{c/\beta}$ is the Wulff shape defined as in Theorem 5.7.2. This was proved by H. M. Soner [399] for constant $c > 0$ by comparing with self-similar expanding solutions so the result is limited to the case when γ and $1/\beta$ are a constant multiple of each other and W_γ is strictly convex with $a > 0$. If $a = 0$, (5.8.3) was proved by P. Soravia [405] and S. Osher and B. Merriman [375]. Even if γ, β and c are unrelated, the convergence (5.8.3) is still valid as proved in H. Ishii et al. [282]. As pointed out by Y. Giga [202], their proof extends when γ is not C^1 so that W_γ has a flat portion for a planar evolution.

Even if $a = 0$, the case when c or β depends on spatial variables is difficult to treat. If c/β depends only x periodically and independent of $\mathbf{n}$ and t, a large time behavior is an important topic as for example studied by G. Barles and J.-M. Roquejoffre [54].

A large time behavior of the mean curvature flow in a cylinder with homogeneous Neumann condition has been studied by Y. Giga et al. [225] by establishing a strong maximum principle for level sets for level set minimal surface equations. By the way a strong maximum principle for viscosity solutions is not a direct consequence of the classical one, as well discussed in a famous book by M. H. Protter and

H. F. Weinberger [381]. For a strong maximum principle for viscosity solutions and its extensions the reader is referred to a paper by M. Bardi and F. Da Lio [42] and papers by Y. Giga and M. Ohnuma [224] and H. Ishii and Y. Yoshimura [285].

Gaussian Curvature Flow Equations When one considers the Gaussian curvature flow equation (2.5.9) we are forced to modify the equation in the form (2.6.22) to get well-posedness. However, when we consider wearing of a stone, this model is unrealistic, since no nonconvex part of an initial hypersurface moves for a moment, except at its boundary. Even if the initial hypersurface is convex but not strictly convex, the boundary of nonstrictly convex part stays for a moment as estimated by D. Chopp et al. [116]. In a series of works, H. Ishii and T. Mikami [278–281] proposed a nonlocal model which is an extension of the Gaussian curvature flow and established its level set approach as well as its stochastic approximation.

Level Set Approach for Spirals The motion of a spiral-like curve appearing in a surface of crystal during its evolution is modeled by the surface evolution equation typically of the form $V = k + 1$ in the plane. However, the curve may not divide the domain in two parts so the conventional level set method in this chapter does not apply. Fortunately, T. Ohtsuka [371] overcame this difficulty by considering problems in a covering space and established a level set method.

Level Set Method for Calculation of the Graph of a Solution Developing Discontinuities We consider a first order equation

$$u_t + H(u, \nabla u) = 0$$

and interpret this equation as an evolution equation for the hypersurface consisting of the graph of solution. If H is positively homogeneous of degree 1 in ∇u and $r \mapsto H(r, p)$ is nondecreasing, we are able to establish a level set method to track the graph evolution of a possibly discontinuous solution; see a paper by Y. Giga and M.-H. Sato [231]. If one drops the monotonicity assumption in r, then the problem is substantially different since a solution may develop jump discontinuities even if it is initially smooth like a solution of the Burgers equation $u_t + uu_x = 0$. In this situation we have to introduce a singular vertical diffusion to track evolution of the graph of a solution with jump discontinuities in the framework of the level set method. This idea is proposed by Y. Giga [205] and numerically confirmed by a level set method by Y.-H. R. Tsai et al. [419]. Some analytic foundation is given by M.-H. Giga and Y. Giga [192].

Interface in a Thin Plate When an interface is in a thin strip domain with right angle boundary condition, one is tempted to approximate by an evolution of the curve. Such a problem is studied by M. Arisawa and Y. Giga [33].

Further Comments for the Second Edition On the second last topic "level set method for calculation of the graph of a solution developing discontinuities" the reader is referred to a textbook by M.-H. Giga and Y. Giga [195], where several topics of scalar first order evolution equations are discussed.

If one considers an evolution of a hypersurface $\Gamma_t \subset \mathbf{R}^N$ with prescribed geometric boundary Σ even if for orientation-free evolution like the mean curvature flow. The level-set method provided by P. Sternberg and W. P. Ziemer [406] requires that Γ_t is situated in a mean-convex domain Ω with $\Sigma \subset \partial\Omega$. It turns out that this type of Dirichlet problem should be interpreted as an obstacle problem with obstacle Σ. In fact, X. Bian et al. [64] proved that the solution of the obstacle problem is consistent with the solution by P. Sternberg and W. P. Ziemer [406]. Moreover, it is shown that it is consistent with classical solutions. The theory of viscosity solution with obstacles has a long history. We need bilateral obstacles as discussed by N. Yamada [432]. The level set method with bilateral obstacle has been established by G. Mercier [348] and behavior of radial solution was studied by Y. Giga et al. [232]. For a variational approach to the obstacle problem for the mean curvature flow see G. Mercier and M. Novaga [349].

Let us comment on evolution of a spiral. If one considers $V = k+1$, the evolution now depends on orientation. Assume that a curve Γ_t moves in the complement of a finite family non-overlapping closed disks with positive radius in the plane. Assume that the endpoints of Γ_t is situated on the boundary of two of disks. We also impose Γ_t intersect the disks with the right angle. Under these settings S. Goto et al. [240] established a level set method for $V = k + 1$ by showing the uniqueness of their evolution on a covering space. Their modified level set method works well numerically as shown by T. Ohtsuka et al. [372, 373]. However, it is not clear what happens the size of disks becomes zero. It is likely that such problems can be interpreted as an obstacle problem on a covering space but it is not yet carried out at this moment. The topic for a level set method for spirals still attract many researchers. We just mention the work of H. Mitake and H. V. Tran [353] on large time behavior for spiral growth.

Large Time Asymptotics and Homogenization The large-time behavior of viscosity solutions is an important topic that has attracted considerable attention. For convex Hamilton-Jacobi equations with periodic boundary conditions, we refer to A. Fathi [170] and G. Namah and J.-M. Roquejoffre [363]. For developments involving Neumann-type boundary conditions, see G. Barles et al. [51] and the references therein. Related to the level set evolution discussed in this chapter, a notable application arises in the mathematical analysis of the so-called birth-and-spread model for crystal growth. In this context, in addition to the horizontal level set evolution, one also needs to incorporate a source term that accounts for the vertical growth and reflects the effect of nucleation on the crystal surface. See Y. Giga et al. [218, 219] for studies of crystal growth modeled by non-coercive Hamilton-Jacobi equations. A typical second order example is the following level set equation in two space dimensions:

$$u_t - v_\infty |\nabla u| \left(\rho_c \operatorname{div}(\nabla u/|\nabla u|) + 1\right) = c 1_E,$$

where $v_\infty \in \mathbb{R}$ and $\rho_c \geq 0$ denote the step velocity and critical radius respectively, $E \subset \mathbb{R}^2$ represents a set of nucleation centers, and $c > 0$ is the vertical growth rate of the crystal on the set E.

In the case $\rho_c = 0$, a modified notion of viscosity solutions to handle the discontinuous source term was introduced by Y. Giga and N. Hamamuki [211] in general space dimensions, and its large time behavior was later studied by N. Hamamuki [248]. In particular, it was shown by N. Hamamuki [248] that $u(x,t)/t \to c$ as $t \to \infty$ for each x, where $c \in \mathbb{R}$ is a constant called the asymptotic growth rate. The case $\rho_c \neq 0$ is more involved. We refer to Y. Giga et al. [221] for well-posedness results and asymptotic analysis in the case $\rho_c = v_\infty = 1$. It turns out that in general the asymptotic growth rate c depends on the shape and size of the source term. Further discussions on this topic can be found in Y. Giga et al. [222] and Y. Giga et al. [220]. We also refer to a review paper by Y. Giga [206] for a survey of these results. Similar asymptotic behavior has also been studied for the Neumann-type boundary problem. See J. Jang et al. [293] as well as J. Jang [292]. The large time asymptotics for level set equations in the case of spiral growth was discussed in H. Mitake and H. V. Tran [353], as mentioned previously.

A closely related subject is the homogenization of level set curvature flow equations. Beginning with the pioneering works of P.-L. Lions et al. [325] and of L. C. Evans [161], the homogenization of fully nonlinear partial differential equations has become a central theme for viscosity solution theory. For level set geometric evolution equations, homogenization has also been extensively studied within the framework of viscosity solutions. We refer to P.-L. Lions and P. E. Souganidis [326] for homogenization results for quasilinear degenerate elliptic and parabolic equations including the level set mean curvature flow equation. See also P. Cardaliaguet et al. [90], G. Barles et al. [48], L. A. Caffarelli and R. Monneau [74] and S. Armstrong and P. Cardaliaguet [34], among others. Homogenization for certain classes of geometric evolution equations also plays a significant role in turbulent combustion, particularly in understanding the curvature effect on the prediction of burning velocity. For recent developments in this direction, we refer to the works of H. Gao et al. [186] and of H. Mitake et al. [352]. The approach by H. Gao et al. [186] is based on the game theoretic interpretation presented in Chaps. 7 and 8 of this book.

Chapter 6
Set-Theoretic Approach

For surface evolution equations we have introduced generalized solutions as a level set of auxiliary functions. In this chapter we introduce various notions of solutions for surface evolution equations without using auxiliary functions. It turns out that the notion of solutions for evolving sets does not even need level set equations. It only needs surface evolution equations. Since the notion is directly related to evolving sets rather through auxiliary functions, this approach is called intrinsic or set-theoretic. We moreover compare our new notions of solutions with generalized solutions defined in the preceding chapters. The last part of this chapter is devoted to the study of barrier solutions. It turns out that the notion is important to prove the comparison principle via local existence of classical solutions for surface evolution equations without using comparison results in Chap. 4.

6.1 Set-Theoretic Solutions

We consider a surface evolution equation

$$V = f(z, \mathbf{n}, \nabla \mathbf{n}) \quad \text{on} \quad \Gamma_t. \tag{6.1.1}$$

Here $f(z, \cdot, \cdot)$ for $z \in \mathbf{R}^N \times [0, T]$ is a given function defined in

$$\mathbf{E} = \{(p, Q_p(X));\ p \in S^{N-1},\ X \in \mathbf{S}^N\},$$

where $Q_p(X) = (I - p \otimes p)X(I - p \otimes p)$. We recall some of the assumptions on f of the preceding chapter which we still use in this chapter.

(f1) f is continuous in each variable, i.e., $f : \mathbf{R}^N \times [0, T] \times \mathbf{E} \to \mathbf{R}$ is continuous.
(f2) f is degenerate elliptic in the sense that

Y. Giga, Q. Liu, *Surface Evolution Equations*, Monographs in Mathematics 99,
https://doi.org/10.1007/978-3-032-20303-8_6

$$f(z, p, Q_p(X)) \leq f(z, p, Q_p(Y)) \quad \text{whenever} \quad Q_p(X) \geq Q_p(Y).$$

If (f2) is fulfilled, (6.1.1) is called a degenerate parabolic equation. We associate the level set equation of (6.1.1):

$$\begin{aligned} &u_t + F(z, \nabla u, \nabla^2 u) = 0, \\ &F(z, p, X) = -|p| f(z, -\hat{p}, -Q_{\hat{p}}(X)/|p|), \ \hat{p} = p/|p|. \end{aligned} \tag{6.1.2}$$

As already observed, (f1) is equivalent to the condition that F is continuous in $\mathbf{R}^N \times [0, T] \times (\mathbf{R}^N \backslash \{0\}) \times \mathbf{S}^N$; (f2) is equivalent to the degenerate ellipticity of F. Since (6.1.2) is a level set equation, the set $\mathcal{F}_\Omega$ in Chap. 3 is nonempty for each bounded open set Ω of $\mathbf{R}^N$ if (f1) is fulfilled (cf. Lemma 6.6). Moreover $\mathcal{F}_\Omega$ is invariant under positive multiplication. Thus the stability property as well as Perron's method is available for $\mathcal{F}_\Omega$-solutions just as for the usual viscosity solutions.

We often suppress $\mathcal{F}$ of an $\mathcal{F}_\Omega$-solution since we only treat level set equations in this chapter.

6.1.1 *Definition and Its Characterization*

Let χ_D denote the characteristic function of a set D, i.e.,

$$\chi_D(z) = \begin{cases} 1 & z \in D, \\ 0 & \text{otherwise.} \end{cases}$$

If D is a set in a metric space, it is easy to see that

$$(\chi_D)^* = \chi_{\overline{D}}, \ (\chi_D)_* = \chi_{\text{int } D}.$$

Definition 6.1.1 Let G be a set in $\mathbf{R}^N \times J$, where J is an open interval in $(0, T)$. We say that G is a *set-theoretic subsolution* (resp. *supersolution*) of (6.1.1) if χ_G is a subsolution (resp. supersolution) of (6.1.2) in $\mathbf{R}^N \times J$. (If the condition (5.3.1) is not fulfilled, the set $\mathcal{F}_{\mathbf{R}^N}$ may be empty. In such situation by a subsolution we mean an $\mathcal{F}_\Omega$-subsolution of (6.1.2) in $\Omega \times J$ for every open ball Ω in $\mathbf{R}^N$. If G is both a set-theoretic sub- and supersolution of (6.1.1), G is called a *set-theoretic solution* of (6.1.1).

Note that we consider an enclosed set by evolving surface Γ_t rather than by Γ_t itself. This is because we have to fix orientations. In our definition it turns out that the normal $\mathbf{n}$ is taken outward from $G(t)$, where $G(t)$ is the cross-section of G at time t:

$$G(t) = \{x \in \mathbf{R}^N; \ (x, t) \in G\}.$$

To define a set-theoretic subsolution we have used the level set equation (6.1.2). There is a nice characterization of set-theoretic subsolutions without using (6.1.2).

Theorem 6.1.2 *Let J be an open interval in $(0, T)$. Under the continuity assumption (f1) a set G in $\mathcal{O} = \mathbf{R}^N \times J$ is a set-theoretic subsolution of* (6.1.1) *if and only if the following two conditions are fulfilled.*

(i) *Assume that a smoothly evolving hypersurface $\{S_t\}$ around $(x_0, t_0) \in \overline{G}$ has only intersection with $\overline{G}(t)$ at $x_0 \in (\partial G)(t_0)$ around (x_0, t_0). Let $\mathbf{n}_{S_t}$ denote the smooth unit normal vector field of S_t such that $\mathbf{n}_{S_t}(x_0)$ directs outward from $\overline{G}(t)$ at $t = t_0$. Let V_{S_t} denote the normal velocity of S_t in the direction of $\mathbf{n}_{S_t}$. Then*

$$V_{S_t} \le f(x, t, \mathbf{n}_{S_t}, \nabla \mathbf{n}_{S_t}) \quad at \quad x = x_0,\ t = t_0. \tag{6.1.3}$$

(ii) *(Left accessibility) For each $(x_0, t_0) \in \overline{G}$ there is a sequence (x_j, t_j) converging to (x_0, t_0) as $j \to \infty$ with $t_j < t_0$ and $(x_j, t_j) \in \overline{G}$. Here $\overline{G}$ and ∂G denote, respectively, the closure and the boundary of G in $\mathcal{O} = \mathbf{R}^N \times J$.*

Remark 6.1.3 In general $\overline{G}(t)$ does not agree with $\overline{G(t)}$, the closure of $G(t)$ in $\mathbf{R}^N$.

Proof We first prove (i) and (ii) if G is a subsolution. By rotation we may assume that S_t is represented as the graph of a smooth function ψ near (x_0, t_0) of the form

$$x_N = \psi(x', t), \quad x = (x', x_N),$$
$$x_{0N} = \psi(x'_0, t_0), \quad \nabla'\psi(x'_0, t_0) = 0,$$

where ∇' denotes the gradient in x' variables. We may assume that $\mathbf{n}_{S_t}(x_0) = (0, \dots, 0, 1)$ at $t = t_0$. We then set $\varphi(x, t) = \psi(x', t) - x_N$ and observe that $\chi_{\overline{G}} - \varphi$ takes its strict local maximum 1 at (x_0, t_0) since S_t has an intersection with $\overline{G}(t)$ only at x_0 with $t = t_0$. Since $\chi_{\overline{G}}$ is a subsolution and $\nabla\varphi(z_0) \ne 0$, we see

$$\varphi_t(z_0) + F(z_0, \nabla\varphi(z_0), \nabla^2\varphi(z_0)) \le 0 \quad \text{with} \quad z_0 = (x_0, t_0)$$

by Proposition 3.2.2. Since S_t is given by $\varphi = 0$ near (x_0, t_0), the last inequality is a level set representation of (6.1.3). We thus obtain (i). It remains to prove (ii). If (ii) were false, there would exist a point $(x_0, t_0) \in \overline{G}$, $\delta > 0$ and a small ball $B_r(x_0)$ that does not intersect $\overline{G}(t)$ for $t_0 - \delta < t < t_0$ (Fig. 6.1). Thus for any $M > 0$ the function $\chi_{\overline{G}} - \varphi$ with $\varphi(t) = M(t - t_0)$ would take its maximum 1 at (x_0, t_0) over some neighborhood of (x_0, t_0) contained in $B_r(x_0) \times (t_0 - \delta,\ t_0 + \delta)$. We set $\mathcal{O}' = \operatorname{int} B_r(x_0) \times (t_0 - \delta,\ t_0 + \delta)$. Since $\chi_{\overline{G}}$ is an $\mathcal{F}$-subsolution and $\varphi \in C^2_F(\mathcal{O}')$ with $\nabla\varphi \equiv 0$, $\nabla^2\varphi \equiv O$, by definition we have $\varphi_t(x_0, t_0) \le 0$ which contradicts $M > 0$.

We next prove that $\chi_{\overline{G}}$ is an $\mathcal{F}$-subsolution if (i) and (ii) are fulfilled. Suppose that for $(\varphi, z_0) \in C^2_F(\mathcal{O}') \times \mathcal{O}'$

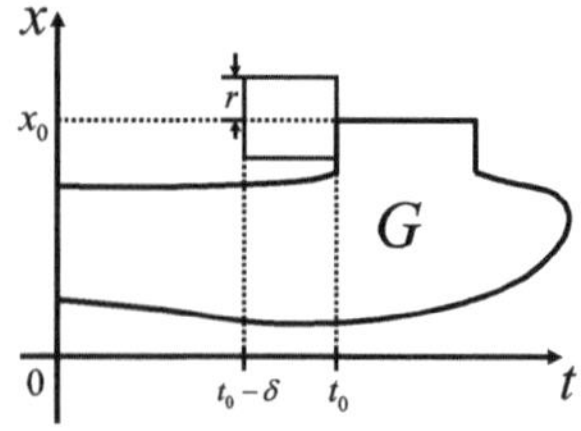

Fig. 6.1 Loss of left accessibility

$$\max_{\mathcal{O}'}(\chi_{\overline{G}} - \varphi) = (\chi_{\overline{G}} - \varphi)(z_0) = 0.$$

We may assume that $z_0 \in \partial G$ since otherwise $\nabla\varphi(z_0) = 0$ and $\varphi_t(z_0) = 0$. Since F is geometric, we may assume that $\varphi \in C^\infty(\mathcal{O}')$ (see Lemma 6.6 and Proposition 3.2.3). We may also assume that $\chi_{\overline{G}} - \varphi$ attains its strict maximum at z_0 by Proposition 3.2.2 since $\mathcal{F}$ is invariant under positive multiplication by geometricity of F. If $\nabla\varphi(z_0) \neq 0$, then by the implicit function theorem

$$S_t = \{x \in \mathbf{R}^N;\ \varphi(x,t) = \varphi(z_0)\}$$

is a smoothly evolving hypersurface around $z_0 = (x_0, t_0)$. Since $z_0 \in \partial G$ is a strict maximum point of $\chi_{\overline{G}} - \varphi$, $\{S_t\}$ has only an intersection with ∂G at z_0 near z_0. Since S_t satisfies (6.1.3) at z_0, as its level set representation we obtain

$$\varphi_t(z_0) + F(z_0,\ \nabla\varphi(z_0),\ \nabla^2\varphi(z_0)) \leq 0;$$

note that $\mathbf{n}_{S_t} = -\nabla\varphi(z_0)/|\nabla\varphi(z_0)|$. It remains to prove that $\varphi_t(z_0) \leq 0$ for $z_0 \in \partial G$ when $\nabla\varphi(z_0) = 0$. Suppose that $\varphi_t(z_0) > 0$. Then $\varphi(x,t) < \varphi(z_0) = 1$ near (x_0, t_0) if $t < t_0$. Since $\chi_{\overline{G}} - \varphi \leq 0$, this would imply

$$\chi_{\overline{G}} = 0 \quad \text{on} \quad B_r(x_0) \times (t_0 - \delta, t_0)$$

for small $r > 0$, $\delta > 0$. This contradicts (ii). □

Remark 6.1.4

(i) Of course there is an equivalent characterization for a set-theoretic supersolution corresponding to Theorem 6.1.2. One should replace (6.1.3) by

$$V_{S_t} \geq f(x,\ t,\ \mathbf{n}_{S_t},\ \nabla\mathbf{n}_{S_t}) \quad \text{at} \quad x = x_0,\ t = t_0,$$

where S_t has only an intersection with $\overline{\mathcal{O}\backslash G}(t)$ at $x_0 \in (\partial(\mathcal{O}\backslash G))(t_0)$ and $\mathbf{n}_{S_t}(x_0)$ directs outward from $\mathrm{int}G(t)$ at $t = t_0$. In (ii) one should replace $\overline{G}$ by $\overline{\mathcal{O}\backslash G}$. The proof for supersolutions parallels that for subsolutions.

(ii) We have defined a notion of set-theoretic solutions only for sets in $\mathbf{R}^N \times J$ but of course one may replace $\mathbf{R}^N$ by an open set Ω in $\mathbf{R}^N$. The statements of Theorem 6.1.2 are still valid if one replaces $\mathbf{R}^N$ by Ω. It is also possible to define a notion of set-theoretic solutions for the boundary value problem. For example if we impose the right angle boundary condition

$$\Gamma_t \perp \partial\Omega$$

to (6.1.1) in Ω, we say that G in $\overline{\Omega} \times J$ is a *set-theoretic subsolution* if $\chi_{\overline{G}}$ is a subsolution of (6.1.2) in $\Omega \times J$ with $\partial u/\partial \nu = 0$. We have a characterization of a set-theoretic subsolution corresponding to Theorem 6.1.2. The statement is almost the same with $\mathbf{R}^N$ replaced by $\overline{\Omega}$ except in condition (ii) at $x_0 \in \partial\Omega \cap \partial G(t_0)$ we only require either (6.1.3) or

$$\langle -\mathbf{n}_{S_t}(x_0),\ \nu \rangle \leq 0 \quad \text{with} \quad t = t_0.$$

(iii) From the proof and Proposition 3.2.3 it is easy to observe that smoothness of S_t in Theorem 6.1.2 may be replaced by the condition that S_t is a $C^{2,1}$ hypersurface as definition of viscosity solutions. We may also replace the assumption that $\{S_t\}$ has only an intersection with $G(t)$ at $x_0 \in (\partial G)(t_0)$ by the condition that $\{S_t\}$ is in $(\mathcal{O} \setminus \operatorname{int} G)(t)$ near $t = t_0$ that satisfies

$$V_{S_t} \leq f(x_0,\ t_0,\ \mathbf{n}_{S_t}(y_0),\ \nabla \mathbf{n}_{S_t}(y_0)) \quad \text{at} \quad t = t_0$$

for $y_0 \in S_{t_0}$ with

$$d(y_0,\ \partial G(t_0)) \leq d(y,\ \partial G(t)) \quad \text{for all } y \in S_t,$$

where t is close to t_0, provided that y_0 is not a geometric boundary point of S_{t_0}.

(iv) As we observed in Chap. 4, u is a subsolution of (6.1.2) in $\mathbf{R}^N \times (t_1, t_2]$ for all $t_1 < t_2$ satisfying $(t_1, t_2] \subset J$. It is easy to observe from the proof of Theorem 6.1.2 that this remark relaxes the assumption on test surface $\{S_t\}$ in Theorem 6.1.2 (i). We require only for $t \leq t_0$ that $\{S_t\}$ has only an intersection with $\overline{G}(t)$ at $x_0 \in \partial G(t_0)$ around (x_0, t_0); $\{S_t\}$ is allowed to intersect $\overline{G}(t)$ for $t > t_0$.

Corollary 6.1.5 *Assume* (f1) *and* (f2). *Let G be a subset of $\mathcal{O} = \mathbf{R}^N \times J$. If ∂G is a smoothly evolving hypersurface in $(0, T)$ and fulfills*

$$V \leq f(x,\ t,\ \mathbf{n},\ \nabla \mathbf{n})$$

on ∂G, where $\mathbf{n}$ *directs outward to G. Then G is a set-theoretic subsolution of* (6.1.1).

This is easy to prove once we admit the characterization (Theorem 6.1.2). We leave the proof to the reader.

6.1.2 Characterization of Solutions of Level Set Equations

The notion of set-theoretic solutions is important to characterize solutions of level set equations. Here we give only its simplest form for a function in $\mathbf{R}^N \times (0, T)$ but as remarked in Remark 6.1.4 its extension to the boundary value problem is straightforward.

Theorem 6.1.6 *Assume the continuity* (f1) *of f in* (6.1.1). *Let J be an open interval in $(0, T)$. Let $u : \mathbf{R}^N \times J \to \mathbf{R} \cup \{-\infty\}$ (resp. $\mathbf{R} \cup \{+\infty\}$) satisfy $u^* < \infty$ (resp. $u_* > -\infty$) on $\mathbf{R}^N \times J$. Then u is a subsolution (resp. supersolution) of* (6.1.2) *in $\mathbf{R}^N \times J$ if and only if each super level set*

$$G_c = \{(x,t) \in \mathbf{R}^N \times J;\ u^*(x,t) \geq c \quad (resp. \quad u_*(x,t) > c)\}$$

is a set-theoretic subsolution (resp. supersolution) of (6.1.1) *in $\mathbf{R}^N \times J$ for all $c \in \mathbf{R}$.*

Proof We have proved several fundamental properties of the level set equation (6.1.2) in Chaps. 3 and 5. We list a number of them for further citation in this section.

(SP) Stability Principle Assume that u_ε is a subsolution (resp. supersolution) of (6.1.2) in $\mathbf{R}^N \times J$ for $\varepsilon > 0$. Then $\overline{u} = \limsup_{\varepsilon \to 0}{}^* u_\varepsilon$ (resp. $\underline{u} = \liminf_{\varepsilon \to 0}{}_* u_\varepsilon$) is a subsolution (resp. supersolution) in $\mathbf{R}^N \times J$ provided that $\overline{u} < \infty$ (resp. $\underline{u} > -\infty$) on $\mathbf{R}^N \times J$.

(CL) Closedness Under Supremum and Infimum Assume that S is a set of subsolutions (resp. supersolutions) of (6.1.2) in $\mathbf{R}^N \times J$. Then

$$\begin{aligned} u(x,t) &= \sup\{v(x,t),\ v \in S\} \\ (\text{resp. } u(x,t) &= \inf\{v(x,t),\ v \in S\}) \end{aligned}$$

is a subsolution (resp. supersolution) of (6.1.2) provided that $u^* < \infty$ (resp. $u_* > -\infty$) in $\mathbf{R}^N \times J$.

(I) Invariance Assume that θ is a continuous and nondecreasing function from $\mathbf{R}$ into $\mathbf{R}$. If u is a subsolution (resp. supersolution) of (6.1.2) in $\mathbf{R}^N \times J$, so does $\theta \circ u = \theta(u)$.

The properties (SP) and (CL) have been proved in Chap. 3; these properties also hold for nongeometric equations. The property (I) has been proved in Chap. 5 and it reflects the property that the level set equation is geometric. Assume that u is a subsolution so that u^* is a subsolution. We approximate the Heaviside function by

$$\theta_\varepsilon(\zeta) = \begin{cases} 1, & \zeta \geq 0, \\ (\zeta + \varepsilon)/\varepsilon, & -\varepsilon \leq \zeta \leq 0, \\ 0, & \zeta \leq -\varepsilon \end{cases} \qquad \text{for} \quad \varepsilon > 0.$$

For $c \in \mathbf{R}$ we set

$$v_\varepsilon(x,t) = \theta_\varepsilon(u^*(x,t) - c)$$

and observe that

$$\limsup_{\varepsilon\to 0}{}^* v_\varepsilon = \chi_{G_c}.$$

By the invariance (I) v_ε is a subsolution of (6.1.2) in $\mathbf{R}^N \times J$. By the stability (SP), χ_{G_c} is now a subsolution of (6.1.2), i.e., G_c is a set-theoretic subsolution.

The converse is easy to prove. We first note that for a closed set G a function

$$I_G(x,t) = \begin{cases} 0 & (x,t) \in G, \\ -\infty & \text{otherwise} \end{cases}$$

is a subsolution of (6.1.2) if χ_G is a subsolution. (Indeed if we take

$$w_\varepsilon(x,t) = \tilde{\theta}_\varepsilon(\chi_G(x,t)) \quad \text{with} \quad \tilde{\theta}_\varepsilon(\zeta) = \begin{cases} 0, & \zeta \ge 1, \\ -(\zeta - 1)/\varepsilon, & \zeta \le 1, \end{cases}$$

then

$$\limsup_{\varepsilon\to 0}{}^* w_\varepsilon = I_G.$$

By (I) and (SP), we see I_G is a subsolution.) Since u^* is upper semicontinuous so that G_c is closed, I_{G_c} is a subsolution of (6.1.2) for each $c \in \mathbf{R}$. Let $u^* - \varphi$ take its maximum at $(\hat{x},\hat{t}) \in \mathbf{R}^N \times J$, where $\varphi \in C^2_{F_f}(\mathbf{R}^N \times J)$. We may assume that $(u^* - \varphi)(\hat{x},\hat{t}) = 0$ and set $c = u^*(\hat{x},\hat{t})$. Since $I_{G_c} - \varphi$ takes its maximum at $(\hat{x},\hat{t}) \in \mathbf{R}^N$, we obtain a desired inequality of φ at $(\hat{x},\hat{t})$, i.e.,

$$\begin{aligned} &\varphi_t + F_f(\hat{x},\hat{t},\nabla\varphi,\nabla^2\varphi) \le 0 \quad \text{at} \quad (\hat{x},\hat{t}) \quad \text{if} \quad \nabla\varphi(\hat{x},\hat{t}) \ne 0, \\ &\varphi_t(\hat{x},\hat{t}) \le 0 \quad \text{if} \quad \nabla\varphi(\hat{x},\hat{t}) = 0. \end{aligned}$$

We give another proof. It is easy to observe that

$$u^*(x,t) = \sup_c (I_{G_c}(x,t) + c)$$

with interpretation that $-\infty + r = -\infty$ for $r \in \mathbf{R}$. Since $I_{G_c} + c$ is a subsolution of (6.1.2) by (I), the closedness (CL) under supremum implies that u^* is a subsolution of (6.1.2) in $\mathbf{R}^N \times J$.

The proof for supersolutions parallels that for subsolutions so is omitted. □

6.1.3 Characterization by Distance Functions

We shall characterize a set-theoretic subsolution (resp. supersolution) by a signed distance function. For a given set A in $\mathbf{R}^N$ we associate the signed distance function

$$\mathrm{sd}(x, A) = \begin{cases} d(x, A^c), & x \in A, \\ -d(x, A), & x \in A^c \end{cases}$$

where A^c denotes the complement of A. We use the convention that $\mathrm{sd}(x, A) \equiv -\infty$ if A is empty and $\mathrm{sd}(x, A) \equiv \infty$ if A^c is empty.

Theorem 6.1.7 *Assume the continuity (f1) of f in* (6.1.1). *Let J be an open interval in $(0, T)$. Let G be a set in $\mathcal{O} = \mathbf{R}^N \times J$. Then G is a set-theoretic subsolution (resp. supersolution) of* (6.1.1) *if and only if $u \equiv \mathrm{sd} \wedge 0$ (resp. $u \equiv \mathrm{sd} \vee 0$) is a subsolution (resp. supersolution) of*

$$u_t + F(x - u\nabla u,\ t,\ \nabla u,\ \nabla^2 u) = 0 \quad in \quad \mathcal{O}, \tag{6.1.4}$$

where $\mathrm{sd(x, t) = sd(x, G(t))}$.

Proof Assume that G is a set-theoretic subsolution. We shall prove that u^* is a subsolution of (6.1.4). We may assume that G is closed so that $u^* = u$ since $u^*(x, t) = \mathrm{sd}(x, \overline{G}(t)) \wedge 0$. Note that $u(x, t) = 0$ is equivalent to $(x, t) \in G$. Suppose that $(\varphi, (\hat{x}, \hat{t})) \in C_F^2(\mathcal{O}') \times \mathcal{O}'$ satisfies

$$\max_{\mathcal{O}'}(u - \varphi) = (u - \varphi)(\hat{x}, \hat{t}) = 0,$$

with $\mathcal{O}' = \Omega \times (0, T)$, where Ω is a bounded open set.

Case 1 If $(\hat{x}, \hat{t}) \in G$, then φ is an upper test function of χ_G at $(\hat{x}, \hat{t})$. Since χ_G is a subsolution of (6.1.2) and $u(\hat{x}, \hat{t}) = 0$, it follows that

$$\varphi_t + F(\hat{x}, \hat{t},\ \nabla\varphi,\ \nabla^2\varphi) \le 0 \quad \text{at} \quad (\hat{x}, \hat{t})$$

or

$$\varphi_t(\hat{x}, \hat{t}) + F(\hat{x} - u^*(\hat{x}, \hat{t})\nabla\varphi(\hat{x}, \hat{t}),\ \hat{t}, \nabla\varphi(\hat{x}, \hat{t}),\ \nabla^2\varphi(\hat{x}, \hat{t})) \le 0 \tag{6.1.5}$$

provided that $\nabla\varphi(\hat{x}, \hat{t}) \neq 0$. If $\nabla\varphi(\hat{x}, \hat{t}) = 0$, then we get $\varphi_t(\hat{x}, \hat{t}) \le 0$ instead of (6.1.5).

Case 2 If $(\hat{x}, \hat{t}) \notin G$, then $u(\hat{x}, \hat{t}) = \mathrm{sd}(\hat{x}, \hat{t})(= -\delta < 0)$. We first observe that $\nabla\varphi(\hat{x}, \hat{t}) \neq 0$. Indeed, by definition of $(\hat{x}, \hat{t})$ we see

$$\mathrm{sd}(x, G(t)) - \varphi(x, t) \le \mathrm{sd}(\hat{x}, G(\hat{t})) - \varphi(\hat{x}, \hat{t}).$$

near $(\hat{x}, \hat{t})$. Setting $t = \hat{t}$ and expanding $\varphi(x, \hat{t})$ near $\hat{x}$ yields

$$d(\hat{x}, G(\hat{t})) - d(x, G(\hat{t})) \leq \langle \nabla\varphi(\hat{x}, \hat{t}),\ x - \hat{x}\rangle + o(|x - \hat{x}|)$$

as $x \to \hat{x}$. Let x_0 be a point in $G(\hat{t})$ that satisfies

$$\delta = d(\hat{x}, G(\hat{t})) = |x_0 - \hat{x}|.$$

We take $x = \hat{x} + \sigma(x_0 - \hat{x})$ for σ close to zero and observe that $d(x, G(\hat{t})) = (1 - \sigma)\delta$. For this choice of x we arrive at

$$\delta - (1 - \sigma)\delta \leq \langle \nabla\varphi(\hat{x}, \hat{t}),\ \sigma(x_0 - \hat{x})\rangle + o(\sigma\delta)$$

as $\sigma \to 0$. This implies that

$$\nabla\varphi(\hat{x}, \hat{t}) = (x_0 - \hat{x})/|x_0 - \hat{x}| \neq 0. \tag{6.1.6}$$

We next observe that $\psi(x, t) < 0$ implies $u(x, t) < 0$ if we set

$$\psi(x, t) = \varphi(x + \hat{x} - x_0,\ t) + \delta.$$

Indeed, the triangle inequality implies

$$-u(x + \hat{x} - x_0,\ t) + u(x, t) \leq |\hat{x} - x_0| = \delta.$$

Since $u - \varphi$ takes its zero maximum at $(\hat{x}, \hat{t})$, $\psi(x, t) < 0$ implies

$$-u(x + \hat{x} - x_0,\ t) \geq -\varphi(x + \hat{x} - x_0,\ t) > \delta.$$

Combining these two inequalities yields $u(x, t) < 0$.

We are now in position to prove (6.1.5) for $(\hat{x}, \hat{t}) \notin G$. Since $\psi(x, t) < 0$ implies $u(x, t) < 0$ and $\psi(x_0, \hat{t}) = 0$, $\chi_G - \psi$ takes its maximum at $(x_0, \hat{t})$. Since χ_G is a subsolution of (6.1.2) and $\nabla\psi(x_0, \hat{t}) = \nabla\varphi(\hat{x}, \hat{t}) \neq 0$, we have

$$\psi_t(x_0, \hat{t}) + F(x_0, \hat{t},\ \nabla\psi(x_0, \hat{t}),\ \nabla^2\psi(x_0, \hat{t})) \leq 0.$$

By (6.1.6) we see

$$x_0 = \hat{x} + \delta\nabla\psi(x_0, \hat{t}) = \hat{x} - u(\hat{x}, \hat{t})\nabla\varphi(\hat{x}, \hat{t}).$$

We thus obtain (6.1.5) for φ.

It remains to prove that G is a set-theoretic subsolution of (6.1.1) if u is a subsolution of (6.1.4). We may again assume that G is closed. The proof is already contained in Theorem 6.1.6 if f is independent of x, since

$$G = \{(x,t) \in \mathbf{R}^N \times (0,T);\ u(x,t) \geq 0\}.$$

If f depends on x, $\theta(u)$ for a nonincreasing function θ may not be a subsolution of (6.1.4). So Theorem 6.1.6 does not apply. However, as in the proof of the invariance lemma in Chap. 5, $w = \theta(u)$ is a subsolution of

$$w_t + F(x - \rho(w)\rho'(w)\nabla w,\ t,\ \nabla w,\ \nabla^2 w) = 0 \quad \text{in} \quad \mathcal{O}$$

for $\rho = \theta^{-1}$ provided that $\theta \in C^1(\mathbf{R})$ with $\theta' > 0$. We take $\tilde{\theta}_\varepsilon$ as in the proof of Theorem 6.1.6 and observe that $w_\varepsilon = \overline{\theta}_\varepsilon(u)$ with $\overline{\theta}_\varepsilon(\sigma) = \tilde{\theta}_\varepsilon(\sigma+1)$ is a subsolution of

$$w_t + F(x - \rho_\varepsilon(w)\rho_\varepsilon'(w)\nabla w,\ t,\ \nabla w,\ \nabla^2 w) = 0 \quad \text{in} \quad \mathcal{O},$$

where $\rho_\varepsilon = \overline{\theta}_\varepsilon^{\,-1}$ implies that $\rho_\varepsilon(\sigma) = 0$ for $\sigma \geq 0$ so that $\rho_\varepsilon\rho_\varepsilon'$ is continuous at $\sigma = 0$. Since $\overline{\theta}_\varepsilon$ is not C^1 at zero, we should approximate $\overline{\theta}_\varepsilon$ by a C^1 function to get this formula by the stability results. By using the stability result we conclude that $I_G = \limsup_{\varepsilon\to 0}{}^* w_\varepsilon$ is a subsolution of (6.1.2) since $\rho_\varepsilon(\sigma) = \varepsilon\sigma \to 0$ as $\varepsilon \to 0$ for $\sigma < 0$ and $\rho_\varepsilon(w) = 0$ for $w > 0$.

The proof for supersolution is similar so is omitted. □

Corollary 6.1.8 *Assume the continuity (f1) of f in* (6.1.1). *The function* $u = \mathrm{sd} \wedge 0$ *is a subsolution of* (6.1.4) *in $\mathcal{O}$ if and only if condition (i) for G of Theorem 6.1.2 is fulfilled and u^* satisfies the left accessibility property : for each $(x_0, t_0) \in \mathbf{R}^N \times (0,T)$ there is a sequence (x_j, t_j) converging to (x_0, t_0) as $j \to \infty$ with $t_j < t_0$ such that $u(x_j, t_j) \to u(x_0, t_0)$. (A similar assertion holds for* $\mathrm{sd} \vee 0$.*)*

Proof Since the left accessibility of $\overline{G}$ is equivalent to the left accessibility of u^*, this follows from Theorems 6.1.2 and 6.1.7. □

6.1.4 Comparison Principle for Sets

We shall review the comparison principle obtained in Chap. 4 from the point of set-theoretic solutions. When F in (6.1.2) is degenerate elliptic, we have proved under reasonable assumptions the comparison principle. We give a slightly different version of (CP) and (BCP) stated in Chap. 4.

(CP) Let u and v be sub- and supersolutions of (6.1.2) in $\mathcal{O} = \mathbf{R}^N \times (0,T)$, respectively.

(i) Assume that u and $-v$ are bounded from above on $\mathcal{O}$. Assume that

$$\lim_{\delta\downarrow 0} \sup\{u^*(x,t) - v_*(y,s);\ (x,t),(y,s) \in \mathbf{R}^N \times [0,T),\ |x-y| \leq \delta,\ t \leq \delta,\ s \leq \delta\} \leq 0, \tag{6.1.7}$$

for each $T' \in (0, T)$ and $u^* > -\infty$, $v_* < \infty$ on $\partial_p \mathcal{O}$. Then

$$\lim_{\delta \downarrow 0} \sup \{u^*(x,t) - v_*(y,s);\ (x,t),(y,s) \in \mathbf{R}^N \times [0, T'],\ |x - y| \le \delta,\ |t - s| \le \delta\} \le 0, \tag{6.1.8}$$

for each $T' < T$.

(ii) If $u^* \le v_*$ at $t = 0$, then $u^* \le v_*$ on $\mathbf{R}^N \times [0, T)$, provided that $u(x,t)$ and $v(x,t)$ are constant outside $B_R(0) \times (0, T)$ for some large $R > 0$.

Of course, the second property follows from the first. The first property is nothing but (CP) in Chap. 4 when $\Omega = \mathbf{R}^N$. The property (CP (i), (ii)) holds for (6.1.2) for example when (6.1.1) is degenerate parabolic with continuous f independent of the space variable x. The second property essentially follows from (BCP). See Chap. 4 for more details.

We give a comparison principle for bounded set-theoretic solutions.

(BCPS) Let E and D be set-theoretic sub- and supersolutions of (6.1.1) in $\mathcal{O}$, respectively. If $E^*(0) \subset D_*(0)$, then $E^* \subset D_*$ provided that D (or $\mathcal{O} \backslash D$) and E (or $\mathcal{O} \backslash E$) are bounded in $\mathcal{O}$. Here E^* denotes the closure of E as a set in $\mathbf{R}^N \times [0, T)$ and D_* denotes the complement of $(\mathcal{O} \backslash D)^*$ in $\mathbf{R}^N \times [0, T)$.

This follows from (CP (ii)) by setting $u = \chi_E$, $v = \chi_D$. It turns out that (BCPS) is equivalent to (CP (ii)).

Lemma 6.1.9 *Assume the continuity* (f1) *of* f *in* (6.1.1). *The property* (BCPS) *for* (6.1.1) *holds if and only if* (CP (ii)) *holds for* (6.1.2), *where* $\mathcal{O} = \mathbf{R}^N \times (0, T)$.

Proof It suffices to prove that (BCPS) implies (CP (ii)). For $c, d \in \mathbf{R}$ we set

$$E_c = \{(x,t) \in \mathbf{R}^N \times [0, T);\ u^*(x,t) \ge c\},$$
$$D_d = \{(x,t) \in \mathbf{R}^N \times [0, T);\ v_*(x,t) > d\}.$$

Since u (resp. v) is constant outside $B_R(0) \times (0, T)$, E_c or $\mathcal{O} \backslash E_c$ (resp. D_d or $\mathcal{O} \backslash D_d$) is bounded in $\mathcal{O}$. By Theorem 6.1.6, E_c and D_d are set-theoretic sub- and supersolutions in $\mathcal{O}$, respectively. By definition of u^* and v_* we see that $E_c = (E_c \cap \mathcal{O})^*$ and $D_d = (D_d \cap \mathcal{O})_*$. Since $u^* \le v_*$ at $t = 0$, $E_c(0) \subset D_{c-\delta}(0)$ for all $c \in \mathbf{R}$, $\delta > 0$. By (BCPS) we see

$$E_c \subset D_{c-\delta}$$

for all $c \in \mathbf{R}$, $\delta > 0$. This implies that $u^* \le v_*$ in $\mathbf{R}^N \times [0, T)$, since otherwise there would exist a point $(x_0, t_0) \in \mathcal{O}$ with $u^*(x_0, t_0) = c > v_*(x_0, t_0) = c - \delta$ for some $c \in \mathbf{R}$, $\delta > 0$ so that $(x_0, t_0) \in E_c$ but $(x_0, t_0) \notin D_{c-\delta}$. □

We now derive a comparison principle for set-theoretic solutions corresponding to (CP (i)) which is not necessarily bounded. It is not difficult to prove that (6.1.7) with $u = \chi_E$ and $v = \chi_D$ is equivalent to

$$\inf\{\mathrm{dist}(E_*(t),\ (\mathcal{O}\backslash D)^*(s));\ 0 \le t \le \varepsilon_0,\ 0 \le s \le \varepsilon_0\} \ge \varepsilon_0 \tag{6.1.9}$$

for some small $\varepsilon_0 > 0$. Similarly, the condition (6.1.8) for $u = \chi_E$ and $v = \chi_D$ is equivalent to saying that for each $0 \le T' < T$,

$$\inf\{\mathrm{dist}(E_*(t),\ (\mathcal{O}\backslash D)^*(s));\ |t - s| \le \varepsilon_1,\ 0 \le t,\ s \le T'\} \ge \varepsilon_1 \tag{6.1.10}$$

for some $\varepsilon_1 > 0$ (which may depend on T'), where

$$\mathrm{dist}(A, B) = \inf\{|x - y|;\ x \in A,\ y \in B\}$$

and the closure is taken in $\mathbf{R}^N \times [0, T)$ not in $\mathcal{O}$.

We now propose a comparison principle for general set-theoretic solutions.

(CPS) Let E and D be set-theoretic sub- and supersolutions of (6.1.1) in $\mathcal{O}$, respectively. If (6.1.9) holds for some $\varepsilon_0 > 0$, then (6.1.10) holds for some ε_1.

Lemma 6.1.10 *Assume the continuity* (f1) *of f in* (6.1.1). *The property* (CPS) *for* (6.1.1) *holds if and only if* (CP (i)) *for* (6.1.2) *holds.*

Proof It suffices to prove that (CPS) implies (CP (i)). We set E_c and D_d as in the proof of Lemma 6.1.9. If (6.1.8) were false, then there would exist a sequence $c_j \in \mathbf{R}$, $(x_j, t_j),\ (y_j, s_j) \in \mathbf{R}^N \times [0, T']$ and a constant $\eta > 0$ that satisfies

$$u^*(x_j, t_j) = c_j > v_*(y_j, s_j) + \eta,$$
$$|x_j - y_j| \to 0,\ |t_j - s_j| \to 0 \quad \text{as} \quad j \to \infty.$$

In particular $x_j \in E_{c_j}(t_j)$ and $y_j \notin D_{c_j - \eta/2}(s_j)$, so that

$$\mathrm{dist}(E_{c_j}(t_j),\ (Z\backslash D_{c_j-\eta/2})(s_j)) \to 0 \tag{6.1.11}$$

as $j \to \infty$, where $Z = \mathbf{R}^N \times [0, T']$. Since (6.1.7) holds, there is $\varepsilon_0 > 0$ independent of c_j that satisfies

$$\inf\{\mathrm{dist}(E_{c_j}(t),\ (Z\backslash D_{c_j-\eta/2})(s)),\ 0 \le t,\ s \le \varepsilon_0\} \ge \varepsilon_0. \tag{6.1.12}$$

By definition of upper and lower semiconvergence there is a closed set E in Z and an open set D in Z that satisfies

$$\chi_E = \limsup_{j\to\infty}{}^* \chi_{E_{c_j}},\ \chi_D = \liminf_{j\to\infty}{}_* \chi_{D_{c_j}-\eta/2}.$$

By (6.1.11) we would obtain

$$\operatorname{dist}(E(\hat{t}),\ (Z\backslash D)(\hat{t})) = 0 \tag{6.1.13}$$

for some $\hat{t} \in [0, T']$ which is an accumulation point of t_j, we also observe that $s_j \to \hat{t}$ since $t_j - s_j \to 0$. By (6.1.12) we see

$$\inf\{\operatorname{dist}(E(t),\ (Z\backslash D)(s)),\ 0 \le t,\ s \le \varepsilon_0\} \ge \varepsilon_0 \tag{6.1.14}$$

which in particular implies $\hat{t} > \varepsilon_0$.

Since the stability principle (SP) holds, E and D are still set-theoretic sub- and supersolutions of (6.1.1), respectively. By (CPS) (6.1.14) implies

$$\inf\{\operatorname{dist}(E(t),\ (Z\backslash D)(s));\ |t - s| \le \varepsilon_1\} \ge \varepsilon_1$$

for some $\varepsilon_1 > 0$ which contradicts (6.1.13). □

Remark 6.1.11

(i) In the comparison principle (CPS) for set-theoretic solutions we have assumed (6.1.9). This assumption needs to include properties of solutions not only at time zero but also near time zero. However, if we assume uniform upper semicontinuity of $E_*(t)$ and $(\mathcal{O}\backslash D)^*(t)$ at $t = 0$, (6.1.9) follows from a simple condition

$$\operatorname{dist}(E^*(0),\ ((\mathcal{O}\backslash D)^*)(0)) > 0. \tag{6.1.15}$$

Here we say that $E^*(t)$ is *uniformly upper semicontinuous* at $t = 0$ if

$$\lim_{t\to 0} \sup\{d(x,\ E^*(0));\ x \in E^*(t)\} = 0.$$

Of course this property always holds if $E^*(0)$ is compact in $\mathbf{R}^N$.

(ii) So far we have taken initial data at $t = 0$ in (CPS) and (BCPS). Sometimes it is necessary to take initial data at $t = t_0 \in (0, T)$. We replace 0 in (CPS) and (BCPS) by t_0 and refer to these conditions as (CPS t_0) and (BCPS t_0), respectively. Of course the statements of Lemmas 6.1.9 and 6.1.10 are still valid for (BCPS t_0) and (CPS t_0) if we replace (CP) in an appropriate way.

6.1.5 Convergence of Sets and Functions

In previous subsections we compared various aspects of set-theoretic solutions and solutions of level set equations. Here we compare the upper relaxed limit of functions and its level set.

Lemma 6.1.12 *Let u^ε $(\varepsilon > 0)$ be an upper semicontinuous function on X with values in $\{-\infty\} \cup \mathbf{R}$. For $c \in \mathbf{R}$ let E_c^ε denote*

$$E_c^\varepsilon = \{x \in X;\ u^\varepsilon(x) \geq c\}.$$

Let $\overline{u}$ and E_c be defined by

$$\overline{u} = \limsup_{\varepsilon \to 0}{}^* u^\varepsilon,\ \ E_c = Limsup_{\varepsilon \to 0, \ell \to c}\, E_\ell^\varepsilon.$$

Then

$$E_c = \{x \in X;\ \overline{u}(z) \geq c\}.$$

Proof The condition $\chi_{\overline{u} \geq c}(z) = 1$ is equivalent to saying that there exists a sequence $\varepsilon_j \to 0$, $z_j \to z$ such that $\liminf_{j \to \infty} u^{\varepsilon_j}(z_j) \geq c$. We observe that the last inequality is equivalent to $\lim_{j \to \infty} \chi_{E_{\ell_j}^{\varepsilon_j}}(z_j) = 1$ with $\ell_j \to c$ by setting $\ell_j = \min(u^{\varepsilon_j}(z_j), c)$. Thus the result follows. □

This lemma together with Theorem 6.1.6 implies that the stability results in Chap. 3 are equivalent to the stability of sets in $X = \mathbf{R}^N \times (0, T)$ or characteristic functions. We leave its explicit form to the reader.

6.2 Level Set Solutions

For a given set G_0 in $\mathbf{R}^N$ we seek a set-theoretic solution $G \subset \mathbf{R}^N \times [0, T)$ of (6.1.1) in $\mathbf{R}^N \times (0, T)$ with initial data G_0. To be precise we say that the initial data of G equals G_0 if $G^*(0) = \overline{G_0}$ and $G_*(0) = \operatorname{int} G_0$. Unfortunately, for a given initial data there may be several set-theoretic solutions, so solutions for the initial value problem of (6.1.1) may not be unique. For this reason we shall introduce the notion of level set sub- and supersolutions.

6.2.1 Nonuniqueness

We first give a simple example of nonuniqueness for the curve shortening equation $V = k$ in $\mathbf{R}^2$.

Example 6.2.1 Let $h = h(z, t)$ be the unique smooth solution of

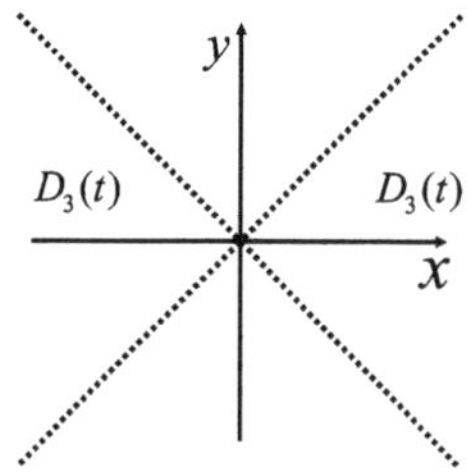

Fig. 6.2 Standing set-theoretic solution

$$\begin{cases} \partial_t h = \dfrac{h_{zz}}{1+h_z^2}, & t > 0,\ z > 0, \\ h_z(0,t) = 0, & t > 0, \\ \lim\limits_{\zeta \to \infty} h_z(\zeta, t) = 1, & t > 0, \\ h(z,0) = z, & z > 0. \end{cases}$$

Such a solution exists globally in time. It can be constructed by an approximation argument; see for example the work of K. Ecker and G. Huisken [152]. Define

$$D_1 = \{(x, y, t) \in \mathbf{R}^2 \times [0, \infty);\ |y| < h(|x|, t)\},$$
$$D_2 = \{(x, y, t) \in \mathbf{R}^2 \times [0, \infty);\ |x| > h(|y|, t)\},$$
$$D_3 = \{(x, y, t) \in \mathbf{R}^2 \times [0, \infty);\ |y| < |x|\}. \text{ (Fig. 6.2)}$$

Since ∂D_j solves $V = k$ in the classical sense, by Corollary 6.1.5, D_1 and D_2 are set-theoretic solutions of $V = k$. By Theorem 6.1.2 and Remark 6.1.4 (i) the set D_3 is also a set-theoretic solution. Indeed, for example to see that $\overline{D_3}$ is a set-theoretic subsolution we observe that all $\{S_t\}$ in Theorem 6.1.2 fulfills (6.1.3) with $V_{S_t} = 0$. Clearly,

$$D_j^*(0) = D_0, \quad D_{j*}(0) = D_0 \quad (j = 1, 2, 3)$$

with $D_0 = \{(x, y) \in \mathbf{R}^2;\ |x| > |y|\}$. Thus all D_j is a set-theoretic solution of $V = k$ with initial data D_0 (Fig. 6.3). See Remark 6.2.12. There is also a similar type example of nonuniqueness for bounded initial data. The uniqueness relates to nonfattening of level sets. In fact, if an open evolution is regular in the sense of Sect. 5.5, then there is at most one set-theoretic solution as stated in Corollary 6.2.10. There are several criteria for uniqueness or nonfattening as already discussed in Sect. 5.5.

6.2.2 Definition of Level Set Solutions

We introduce the notion of a level set subsolution for an open set in $\mathcal{O} = \mathbf{R}^N \times (0, T)$.

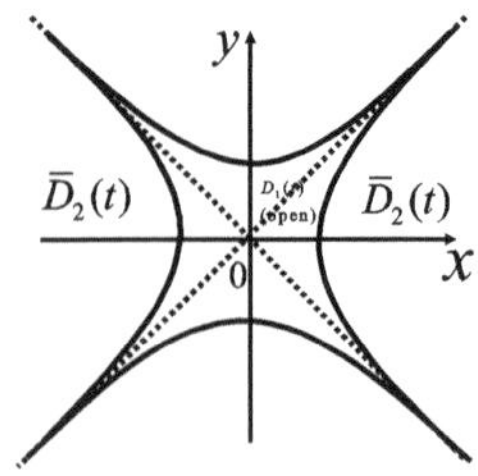

Fig. 6.3 Level set solutions

Definition 6.2.2 Let D be an open set in $Z = \mathbf{R}^N \times [0, T)$. We say that D is a *level set subsolution* of (6.1.1) in $\mathcal{O} = \mathbf{R}^N \times (0, T)$ if there is a sequence $\{E_j\}_{j=1}^{\infty}$ of closed sets in Z that satisfies

(i) $E_j \subset E_{j+1}$ for $j \geq 1$ and $\bigcup\limits_{j=1}^{\infty} E_j = D$,
(ii) E_j is a set-theoretic subsolution of (6.1.1) in $\mathcal{O}$ for $j \geq 1$,
(iii) $\inf\{\mathrm{dist}(E_j(t), (\partial D)(t));\ 0 \leq t < T\} > 0$ for each $j \geq 1$.

For an open set D in $\mathcal{O}$ we say that D is an *(open) level set subsolution* of (6.1.1) in $\mathcal{O}$ if there is an open set D' in Z which is a level set subsolution of (6.1.1) in $\mathcal{O}$ with $D = \mathcal{O} \cap D'$. If D in $\mathcal{O}$ fulfills (i), (ii), (iii) with $\{E_j\}_{j=1}^{\infty} \subset \mathcal{O}$ and $0 \leq t$ in (iii) is replaced by $0 < t$, then such D' always exists by setting $D' = \cup_{j=1}^{\infty} E_j^*$, where E_j^* is the closure of E_j in Z. Note that D' is not uniquely determined by D.
Similarly for a closed set E in Z we say that E is a *(closed) level set supersolution* of (6.1.1) in $\mathcal{O}$ if there is a sequence $\{D_j\}_{j=1}^{\infty}$ of open sets in Z that satisfies

(i) $D_{j+1} \subset D_j$ for $j \geq 1$ and $\bigcap\limits_{j=1}^{\infty} D_j = E$,
(ii) D_j is a set-theoretic supersolution of (6.1.1) for $j \geq 1$,
(iii) $\inf\{\mathrm{dist}((\partial D_j)(t)), E(t));\ 0 \leq t < T\} > 0$ for each $j \geq 1$. We define a closed level set supersolution for a closed set in $\mathcal{O}$ in a similar way as defining a level set subsolution.

Since

$$\chi_{\overline{D}} = \limsup_{j \to \infty}{}^{*} \chi_{E_j},$$

we see that $\overline{D}$ is a subsolution by the stability principle (SP). This justifies our using the name 'subsolution' since D is a set-theoretic subsolution. The same remark applies to level set supersolutions.

It is also possible to define a level set solution in $\mathbf{R}^N \times [t_0, t_1)$ for any $[t_0, t_1) \subset [0, T)$ by replacing $[0, T)$ by $[t_0, t_1)$. We next characterize the solutions as a level set of an auxiliary function.

Proposition 6.2.3 *Assume the continuity* (f1) *of* f *in* (6.1.1). *Let* D *(resp.* E*) be a(n) open (resp. closed) set in* $\mathcal{O} = \mathbf{R}^N \times (0, T)$. *Then* D *(resp.* E*) is a level set subsolution (resp. supersolution) if and only if there is an upper semicontinuous subsolution (resp. a lower semicontinuous supersolution)* u *of* (6.1.2) *in* $\mathcal{O}$ *with properties:*

(i) $D = \{(x,t) \in \mathcal{O}; u(x,t) > 0\}$, $\overline{D} = \{(x,t) \in \mathcal{O}; u(x,t) \geq 0\}$, *(resp.* $E = \{(x,t) \in \mathcal{O}; u(x,t) \geq 0\}$, $\operatorname{int} E = \{(x,t) \in \mathcal{O}; u(x,t) > 0\}$*).*
(ii) $u(x,t)$ *is uniformly continuous in* x *on its zero level set in* $\overline{D}(t)$ *(resp.* $(\mathcal{O} \setminus \operatorname{int} E)(t)$*) uniformly in* $t \in (0, T)$, *i.e., there is a modulus* m *that satisfies*

$$u(x, t_0) \leq m(|x - x_0|)$$

for $x, x_0 \in \overline{D}(t_0)$, $t_0 \in (0, T)$ *with* $u(x_0, t_0) = 0$, $u(x, t_0) \geq 0$ *(resp.* $u(x, t_0) \leq 0$*).*

Proof We only give the proof for D since the proof for E is similar. Assume first that D is a level set subsolution and that an approximate sequence $\{E_j\}_{j=1}^{\infty}$ is taken as in Definition 6.2.2. We set

$$v(x,t) = \sup\left\{\left(\frac{1}{j} + 1\right)\chi_{E_j} - 1;\ j = 1, 2, \ldots\right\}$$

and observe that v is a subsolution of (6.1.2) by the closedness under supremum (CL) since $(j^{-1}+1)\chi_{E_j} - 1$ is a subsolution by the invariance (I). Since $E_j \subset E_{j+1}$ for $j \geq 1$, v is upper semicontinuous in $D = \bigcup_{j=1}^{\infty} E_j$ and moreover

$$v^*(x,t) = \begin{cases} v(x,t), & (x,t) \in D, \\ 0, & (x,t) \in \partial D, \\ -1, & (x,t) \in \mathcal{O} \setminus \overline{D}. \end{cases}$$

If we set $u = v^*$, the property (i) is fulfilled. The uniform continuity (ii) follows from Definition 6.2.2 (iii). Indeed, let $m_0 : [0, \infty) \to [0, \infty)$ be a nondecreasing function of form

$$m_0(\sigma) = \sup\{u(x,t);\ d(x, \partial D(t)) \leq \sigma,\ (x,t) \in \overline{D}\}.$$

By (iii) of Definition 6.2.2 and $\cup_{j=1}^{\infty} E_j = D$ we see $m_0(\sigma) \to 0$ as $\sigma \to 0$. By Lemma 3.1.9 there is a modulus $m \geq m_0$. Thus $u(x, t_0) \leq m(|x - x_0|)$ for $x_0 \in \partial D(t_0)$, $x \in \overline{D}(t_0)$, $t_0 \in (0, T)$.

We now prove the converse. Let u be an upper semicontinuous subsolution u of (6.1.2) satisfying (i) and (ii). We set

$$E_j = \{(x,t) \in \mathcal{O};\ u(x,t) \geq 1/j\}$$

and observe that E_j is a subsolution of (6.1.1) by Theorem 6.1.6. By definition, $E_j \subset E_{j+1}$ and $\cup_{j=1}^{\infty} E_j^* = D_*$. Since

$$1/j \le u(x, t_0) \le m(|x - x_0|)$$

for all $x_0 \in \partial D(t_0)$, $t_0 \in (0, T)$, $x \in E_j(t_0)$, the property (iii) of Definition 6.2.2 now follows. □

Remark 6.2.4 If f in (6.1.1) is independent of x so that F in (6.1.2) is independent of x, then by Theorem 6.1.7 $\mathrm{sd} \wedge 0$ is a subsolution of (6.1.2) for $\mathrm{sd}(x, t) = \mathrm{sd}(x, D(t))$ where D is a level set subsolution. We take

$$w(x, t) = \max(\mathrm{sd} \wedge 0,\ v)$$

with v in the proof of Proposition 6.2.3 and observe from (CL) that w is an upper semicontinuous subsolution and that $w(x, t)$ is uniformly continuous in x on its zero level set (uniformly in $t \in (0, T)$) without restricting x in $\overline{D}(t)$. This is because sd is always Lipschitz continuous.

6.2.3 Uniqueness of Level Set Solutions

We give a comparison principle for level set sub- and supersolutions. For a set D in $\mathbf{R}^N \times [0, T)$ we often denote its intersection with $\mathbf{R}^N \times (0, T)$ still by D.

Theorem 6.2.5 *Assume that* (CPS) *holds. Let G be an open (resp. closed) set in $Z = \mathbf{R}^N \times [0, T)$. Let G be a set-theoretic supersolution (resp. subsolution) of* (6.1.1) *in $\mathcal{O} = \mathbf{R}^N \times (0, T)$. Let D be an open (resp. closed) set in Z. Assume that $(Z \setminus G)(t)$ and $\overline{D}(t)$ are (resp. $G(t)$ and $(Z \setminus \mathrm{int}\, D)(t)$) uniformly upper semicontinuous. Assume that D is a level set subsolution (resp. supersolution) of* (6.1.1) *in $\mathcal{O}$. If $D(0) \subset G(0)$ (resp. $D(0) \supset G(0)$), then $D \subset G$ (resp. $G \subset D$).*

Proof Again we only give the proof when G is a set-theoretic supersolution and D is a level set subsolution since the proof for the other case is similar.

Let $\{E_j\}_{j=1}^{\infty} \subset Z = \mathbf{R}^N \times [0, T)$ be a sequence approximately D in Definition 6.2.2. Since $\overline{D}(t)$ is uniformly upper semicontinuous at $t = 0$, the property (iii) in Definition 6.2.2 yields

$$\inf\{\mathrm{dist}\,(E_j(t),\ (Z\backslash D)(0));\ 0 \le t \le \varepsilon_1\} > 0$$

for some small $\varepsilon_1 > 0$. Since $(Z\backslash G)(t)$ is uniformly upper semicontinuous at $t = 0$, and $D(0) \subset G(0)$, the preceding estimate yields

$$\inf\{\mathrm{dist}\,(E_j(t),\ (Z\backslash G)(s));\ 0 \le t \le \varepsilon_0,\ 0 \le s \le \varepsilon_0\} \ge \varepsilon_0$$

for some small $\varepsilon_0 > 0$. Since E_j is a set-theoretic subsolution and G is a set-theoretic supersolution, the comparison principle (CPS) yields $E_j \subset G$. Since $D = \bigcup_{j=1}^{\infty} E_j$, this implies that $D \subset G$. □

Corollary 6.2.6 *Assume that* (CPS) *holds. Let D_0 (resp. E_0) be a(n) open (resp. closed) set in $\mathbf{R}^N$. There is at most one level set subsolution D (in Z) (resp. supersolution E) which is also a set-theoretic supersolution (resp. subsolution) of* (6.1.1) *in $\mathcal{O} = \mathbf{R}^N \times (0, T)$ with $D(0) = D_0$ (resp. $E(0) = E_0$) provided that $(Z \setminus D)(t)$ and $\overline{D}(t)$ (resp. $E(t)$ and $(Z \setminus \operatorname{int} E)(t)$) are uniformly upper semicontinuous at $t = 0$.*

This follows from Theorem 6.2.5. Indeed, if $\tilde{D}$ and D are both a level set subsolution and a set-theoretic supersolution (with $D(0) = \tilde{D}(0) = D_0$) that satisfies the uniform upper semicontinuity at $t = 0$, then applying Theorem 6.2.5 with $G = \tilde{D}$ yields $\tilde{D} \supset D$. A symmetric argument yields $D \supset \tilde{D}$.

Definition 6.2.7 Let D_0 (resp. E_0) be a(n) open (resp. closed) set in $\mathbf{R}^N$. A(n) open set D (resp. closed set E) in Z is called a *level set solution* of (6.1.1) with initial data D_0 (resp. E_0) if D (resp. E) is simultaneously a level set subsolution (resp. supersolution) and a set-theoretic supersolution (resp. subsolution) with $D(0) = D_0$ (resp. $E(0) = E_0$) and $(Z \setminus D)(t)$ and $\overline{D}(t)$ (resp. $E(t)$ and $(Z \setminus \operatorname{int} E)(t)$) are uniformly upper semicontinuous at $t = 0$.

By Corollary 6.2.6 for D_0 (resp. E_0) a level set solution D (resp. E) with $D(0) = D_0$ (resp. $E(0) = E_0$) is unique. Note that the uniform upper semicontinuity is always fulfilled if one of $Z \setminus D$ or $\overline{D}$ (resp. E and $Z \setminus \operatorname{int} E$) is bounded. We shall compare a generalized evolution of (6.1.1) with a level set solution.

Proposition 6.2.8 *Let D (resp. E) in $Z = \mathbf{R}^N \times [0, T)$ be a generalized open (resp. closed) evolution of* (6.1.1) *with initial data $D(0) = D_0$ (resp. $E(0) = E_0$). Then D (resp. E) is a level set solution (see Definitions 5.1.1 and 5.1.2) of* (6.1.1) *with $D(0) = D_0$ (resp. $E(0) = E_0$).*

Proof Again we only discuss open evolution D since the proof for E is similar. By definition there is a solution $u : Z \to \mathbf{R}$ (of (6.1.2)) that belongs to $\mathrm{BUC}(\mathbf{R}^N \times [0, T'])$ for every $T' < T$ and it satisfies

$$D = \{(x, t) \in Z;\ u(x, t) > 0\}.$$

Since u is a supersolution, D is a set-theoretic supersolution by Theorem 6.1.6. If we take

$$E_j = \{(x, t) \in Z;\ u(x, t) \geq 1/j\},\ \ j = 1, 2, \ldots,$$

then E_j fulfills properties (i), (ii), (iii) of Definition 6.2.2. Indeed, the property (iii) follows from $u \in \mathrm{BUC}(\mathbf{R}^N \times [0, T'])$. The property (ii) follows from Theorem 6.1.6. The property (i) is clear by definition. Thus D is a level set subsolution.

The uniform upper semicontinuity at $t = 0$ follows from the fact that $u \in \mathrm{BUC}(\mathbf{R}^N \times [0, T'])$ so D is a level set solution. □

Remark 6.2.9 By uniqueness (Corollary 6.2.6) there are no level set solutions other than generalized evolution. We may redefine the level set solution to be more restrictive so that Proposition 6.2.8 holds. If we assume that the E_j's in Definition 6.2.2 are bounded, the notion of a level set subsolution comes to be more restrictive than we used above unless D is bounded. However, the statement of comparison and uniqueness (Theorem 6.2.5 and Corollary 6.2.6) still hold for these restrictive level set solutions. As observed in Chap. 5, a generalized open evolution can be approximated by bounded subsolution E_j from inside in many cases, for example the case when (6.1.2) satisfies the comparison principle (CP) as well as the continuity (f1) of f. Thus the statement of Proposition 6.2.8 is still valid for level set solutions in the restrictive sense.

By Proposition 6.2.8 and (CPS) we now observe that a set-theoretic solution of (6.1.1) with a given initial data is unique if (and only if) there is no fattening.

Corollary 6.2.10 (Uniqueness) *Assume that* (CPS) *holds. Let D be an (open) level set solution of* (6.1.1) *with initial data D_0. Assume that $\overline{D}$ is the (closed) level set solution of* (6.1.1) *with initial data $\overline{D_0}$ (i.e., D is regular in the sense of Sect. 5.5). If $G \subset Z = \mathbf{R}^N \times [0, T)$ is a set-theoretic subsolution in $\mathcal{O} = \mathbf{R}^N \times (0, T)$ with $\overline{G}(0) = \overline{D_0}$ and a set-theoretic supersolution in $\mathcal{O}$ with $\operatorname{int} G(0) = D_0$, then $\operatorname{int} G = D$ and $\overline{G} = \overline{D}$ provided that $\overline{G}(t)$ and $(Z \backslash G)(t)$ are uniform upper semicontinuous at time zero. In particular, an open set-theoretic solution with initial data D_0 is unique.*

We characterize a level set solution (open in Z) by the smallest supersolution with the same initial data.

Theorem 6.2.11 *Assume that* (CPS) *holds. For a given open (resp. closed) set D_0 (resp. E_0) in $\mathbf{R}^N$, let S be the set of all open (resp. closed) sets G in $Z = \mathbf{R}^N \times [0, T)$ that satisfy*

(i) *$D_0 \subset G(0)$ (resp. $G(0) \subset E_0$).*
(ii) *G is a set-theoretic supersolution (resp. subsolution) of* (6.1.1) *in $\mathcal{O}$.*
(iii) *$(Z \backslash G)(t)$ (resp. $G(t)$) is uniformly upper semicontinuous at $t = 0$.*

Let D (resp. E) be the level set solution with initial data D_0 (resp. E_0). Then

$$D = \bigcap \{G;\ G \in S\}$$

$$(resp.\quad E = \bigcup \{G;\ G \in S\}).$$

If ∂D_0 (resp. ∂E_0) is compact, the condition (iii) is unnecessary.

Except for the last statement this follows from the comparison principle (Theorem 6.2.5) and $D \in S$ (resp. $E \in S$). If ∂D_0 is bounded, as observed in Chap. 5, there always exists an open set-theoretic supersolution $\hat{G}$ with bounded $\partial\hat{G}$ that satisfies $\hat{G}(0) \supset D_0$. For given open G satisfying (i), (ii), the set $G \cap \hat{G}$ is a set-theoretic supersolution by the closedness under inf operation (CL). Since $\partial(G \cap \hat{G})$ is now bounded, $G \cap \hat{G}$ fulfills (iii) as well as (i), (ii). Thus $\cap\{G; G \in S\}$ is the same as a set even if (iii) in the definition of S is not assumed. The proof for E is symmetric so is omitted.

Remark 6.2.12 In Example 6.2.1 we gave three set-theoretic solutions of the curve shortening equation in $\mathbf{R}^2$ with the same initial data D_0. It turns out that the (open) level set solution with initial data D_0 is D_2 while the (closed) level set solution with initial data $\overline{D_0}$ is $\overline{D_1}$. Indeed, if we set

$$E_j = \{(x, y, t) \in \mathbf{R}^2 \times [0, \infty);\ |x - j^{-1}| > h(|y|,\ t + j^{-1})\},$$

then $\{E_j\}$ satisfies all properties of Definition 6.2.2 with $D = D_2$ since $\{\partial E_j(t)\}$ is a classical solution of the curve shortening equation so that it is a set-theoretic subsolution by Corollary 6.1.5. The uniform upper semicontinuity of the complement of D_2 is easy to check so D_2 is the level set solution with initial data D_0. The proof for $\overline{D_1}$ is similar. Theorem 6.2.11 says that D_2 is the minimal supersolution with $\overline{D_1}(0) \subset \overline{D_0}$ (satisfying the uniform upper semicontinuity at time zero).

6.3 Barrier Solutions

We shall discuss a relation of set-theoretic solutions to evolutions avoiding smooth evolutions.

Definition 6.3.1 Let J be an interval in $[0, T)$. For an interval I let $\mathcal{E}_I$ be a family of sets G in $\mathbf{R}^N \times I$. For $T > 0$ we say that a set B in $\mathcal{O} = \mathbf{R}^N \times J$ is a *barrier supersolution* in $\mathcal{O}$ associated to $\mathcal{E} = \{\mathcal{E}_I;\ I \subset J\}$ if $G \in \mathcal{E}_I$ with $I = [t_0, t_1) \subset J$ and $G(t_0) \subset B(t_0)$ always fulfills $G \subset B \cap (\mathbf{R}^N \times I)$ (Fig. 6.4). The set of all barrier supersolutions associated to $\mathcal{E}$ is denoted by Barr $(\mathcal{E})$.

By definition a restriction $G \cap \mathbf{R}^N \times J'$ for $J' \subset J$ is always a barrier supersolution in $\mathbf{R}^N \times J'$ if $G \in$ Barr $(\mathcal{E})$ in $\mathbf{R}^N \times J$.

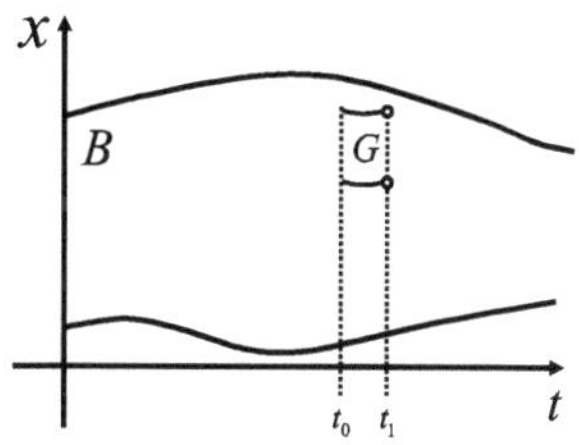

Fig. 6.4 Barrier supersolution

For the surface evolution equation (6.1.1) there are several reasonable choices of $\mathcal{E}_I$. For example

- $\mathcal{E}_I^- = \{G;\ G$ is a bounded closed set $\mathbf{R}^N \times I$ and $\{\partial G(t)\}_{t\in I}$ is a $C^{2,1}$ evolving hypersurface satisfying $V \le f(z, \mathbf{n}, \nabla\mathbf{n})$ on $\partial G(t)$ for $t \in \operatorname{Int} I\}$,
- $\mathcal{E}_I = \{G \in \mathcal{E}_I^-;\ V = f(z, \mathbf{n}, \nabla\mathbf{n})$ on $\partial G(t)$ for $t \in \operatorname{Int} I\}$,
- $\mathcal{E}_I^\infty = \{G \in \mathcal{E}_I;\ \{\partial G(t)\}_{t\in I}$ is a smooth evolving hypersurface$\}$.

By definition

$$\mathrm{Barr}(\mathcal{E}^-) \subset \mathrm{Barr}(\mathcal{E}) \subset \mathrm{Barr}(\mathcal{E}^\infty)$$

with $\mathcal{E}^- = \cup_{I\subset J}\mathcal{E}_I^-$, $\mathcal{E} = \cup_{I\subset J}\mathcal{E}_I$, $\mathcal{E}^\infty = \cup_{I\subset J}\mathcal{E}_I^\infty$. In many cases all these classes become identical although in general the above inclusions may be strict. For a general purpose the set Barr $(\mathcal{E}^-)$ is most useful so we mainly discuss the class Barr $(\mathcal{E}^-)$. We shall compare the notion of a barrier supersolution with a set-theoretic supersolution.

Proposition 6.3.2 *Assume* (f1) *concerning f in* (6.1.1). *Let J be an open interval in $(0, T)$. Let B be an open set in $\mathcal{O} = \mathbf{R}^N \times J$.*

(i) *Assume* (f2) *concerning f in* (6.1.1). *Assume that the comparison principle* (CPS t_0) *holds for* (6.1.1) *with all $t_0 \in [0, T)$. Then B is a barrier supersolution in $\mathcal{O}$ associated to $\mathcal{E}^-$ if B is a set-theoretic supersolution of* (6.1.1) *in $\mathcal{O}$ with the property that $(\mathcal{O}\backslash B)(t)$ is uniformly right upper semicontinuous on J, i.e.,*

$$\lim_{s\downarrow t} \sup\{\mathrm{dist}\,(x,\ (\mathcal{O}\backslash B)(t));\ x \notin B(s)\} = 0 \quad \textit{for } t \in J.$$

(ii) *If B is a barrier supersolution in $\mathcal{O}$ associated to $\mathcal{E}^-$, then B is a set-theoretic supersolution of* (6.1.1) *in $\mathcal{O}$. Assume that f is bounded on $\mathbf{R}^N \times [0, T] \times K$ for every compact set K in $\mathbf{E}$. Then $(\mathcal{O}\backslash B)(t)$ is uniformly right upper semicontinuous at all $t \in J$. If B is a barrier supersolution in $Z = \mathbf{R}^N \times [t_0, t_1]$, $(t_0, t_1) \subset [0, T)$, then $(Z\backslash B)(t)$ is uniformly right upper semicontinuous at all t, $t_0 \le t < t_1$ including $t = t_0$.*

It turns out that the comparison principle is unnecessary to prove the equivalence of barrier supersolutions and set-theoretic supersolutions.

Theorem 6.3.3 *Assume* (f1) *and* (f2) *apply to f in* (6.1.1). *An open set B in $\mathcal{O} = \mathbf{R}^N \times J$ is a barrier supersolution associated to $\mathcal{E}^-$ if and only if B is a set-theoretic supersolution of* (6.1.1) *in $\mathcal{O}$, where J is an open interval in $(0, T)$.*

Corollary 6.3.4 *Assume* (f1) *and* (CPS) *and the uniform continuity of f in Proposition 6.3.2 (ii). Let D_0 be an open set in $\mathbf{R}^N$ and let $D \subset Z = \mathbf{R}^N \times [0, T)$ be an open level set solution of* (6.1.1) *with initial data $D(0) = D_0$. Then*

$$D = \bigcap \{B : B \in \mathit{Barr}(\mathcal{E}^-) \textit{ and } B \textit{ is open in } Z = \mathbf{R}^N \times [0, T) \textit{ with } B(0) \supset D_0\}.$$

The last identity gives a characterization of a level set solution by a minimal barrier containing D_0 at time zero. Since $D = \cap\{G;\ G \in S\}$ by Theorem 6.2.11, Corollary 6.3.4 follows from Proposition 6.3.2 and Theorem 6.3.3.

Remark 6.3.5 The condition of openness of B in Theorem 6.3.3 and Corollary 6.3.4 is unnecessary thanks to the next lemma.

Lemma 6.3.6 *Assume* (f1) *applies to* f *in* (6.1.1). *Let* B *be a set in* $\mathbf{R}^N \times J$, *where* J *is an interval in* $[0, T)$. *Then* $\operatorname{int} B \in \operatorname{Barr}(\mathcal{E}^-)$ *if and only if* $B \in \operatorname{Barr}(\mathcal{E}^-)$.

We postpone the proof of Theorem 6.3.3 and Lemma 6.3.6 to Sect. 6.4.2.

Proof of Proposition 6.3.2

(i) Assume that B is a set-theoretic supersolution of (6.1.1) with uniform right upper semicontinuity. Let G be in $\mathcal{E}_I^-$ with $I = [t_0, t_1) \subset J$. Since G is a classical subsolution, it is a set-theoretic subsolution of (6.1.1) by Corollary 6.1.5; here the degenerate parabolicity of (6.1.1) is invoked. Since G is bounded, $G(t)$ is uniformly right upper semicontinuous. By (CPS t_0) if $G(t_0) \subset B(t_0)$, then $G(t) \subset B(t)$ for all $t \in I$. Thus B is a barrier supersolution associated to $\mathcal{E}^-$.

(ii) Assume now that $B \in \operatorname{Barr}(\mathcal{E}^-)$. We shall prove the right uniform upper semicontinuity of $B(t)$. We may assume $J = [0, T)$. We recall the existence of fundamental subsolutions which is proved essentially in Lemma 5.3.3 as a level set of v^-; note that condition (5.3.1) is fulfilled as pointed out in Remark 5.3.7. Here we use the degenerate ellipticity (f2) of f. Note that the boundedness of f in x is invoked to guarantee that η can be taken independent of (x_0, t_0).

(F) Fundamental Subsolution For sufficiently small $r > 0$, say $r < r_0$, there is $\eta = \eta(r) > 0$ such that for each $(x_0, t_0) \in Z = \mathbf{R}^N \times [0, T)$ there exists $G \in \mathcal{E}_I^-$ for $I = [t_0,\ t_0 + \eta)$ with $G(t_0) = B_r(x_0)$ (the closed ball of radius r centered at x_0) that satisfies $x_0 \in G(t) \subset B_r(x_0)$ for all $t \in I$.

Suppose that $(Z\backslash B)(t)$ were not uniformly right upper semicontinuous at $t_0 \in [0, T)$. Then there would exist $\varepsilon_0 > 0$ and a sequence $\{(x_j,\ t_j)\}_{j=1}^{\infty}$ with $x_j \notin B(t_j)$ $(t_j \downarrow t_0)$ that satisfies $\operatorname{dist}(x_j,\ (Z\backslash B)(t_0)) \geq \varepsilon_0$ for large j, say $j \geq j_0$. We may assume that $\varepsilon_0 < r_0$. By (F) we take $r = \varepsilon_0$ and $G_j \in \mathcal{E}_I^-$ for $I = [t_0,\ t_0 + \eta)$ with $G_j(t_0) = B_r(x_j)$ so that $G_j(t) \ni x_j$ for $t \in I$. Since $G_j(t_0) \subset B(t_0)$ for $j \geq j_0$ and $B \in \operatorname{Barr}(\mathcal{E}^-)$, $G_j(t) \subset B(t)$ for $t \in I$. This contradicts $x_j \notin B(t_j)$. Thus $(Z\backslash B)(t)$ is uniformly right upper semicontinuous at each $t \in [0, T)$.

We shall prove that B is a set-theoretic supersolution in $\mathcal{O} = \mathbf{R}^N \times J$ of (6.1.1), where J is an open interval in $(0, T)$. We shall check the criterion given in Theorem 6.1.2 with Remark 6.1.4 (i). The left accessibility of $\mathcal{O}\backslash B$ follows from (F). Indeed, if not, there would exist a point $(x_0, t_0) \in \mathcal{O}\backslash B$ such that $B_r(x_0) \subset B(t)$ for $t_0 - \delta \leq t \leq t_0$ with some $\delta > 0$, $r > 0$. We may assume that $r < r_0$ and $\delta < \eta = \eta(r)$ for r_0, η given in (F). By (F) there is $G \in \mathcal{E}_I^-$, $I = [t_0 - \delta,\ t_0 - \delta + \eta)$ that satisfies $G(t_0) = B_r(x_0)$, $G(t) \ni x_0$ for all $t \in I$. Since B is a barrier supersolution with $B_r(x_0) \subset B(t_0 - \delta)$, we see $G(t) \subset B(t)$ for $t \in I$. Since

$t_0 - \delta + \eta > t_0$, this contradicts $x_0 \notin B(t_0)$. Note that we only need a weaker version of (F) where η may depend on x_0. Since this version only needs the continuity (f1) and (f2) of f, we do not need the boundedness of f to prove the left accessibility.

It remains to check a version of the definition of supersolution corresponding to Theorem 6.1.2 (i). Let $\{S_t\}$ be a smoothly evolving hypersurface around (x_0, t_0) and $\{S_t\}$ has only an intersection with $(\mathcal{O} \backslash B)(t)$ at $x_0 \in \partial(\mathcal{O} \backslash B)(t_0)$ around (x_0, t_0). By extending $\{S_t\}$ we may assume that there is a bounded closed set G in $B \cap (\mathbf{R}^N \times I) \cup \{(x_0, t_0)\}$ with the property that $\partial G(t) = S_t$ is a smoothly evolving hypersurface for some small interval $I = [t_0 - \delta,\ t_0 + \delta)$, $\delta > 0$. Assume that B does not satisfy the supersolution version of Theorem 6.1.2 (i). Then there would exist $\{S_t\}$ and (x_0, t_0) satisfying the above properties with

$$V_{S_t} < f(x,\ t,\ \mathbf{n}_{S_t},\ \nabla \mathbf{n}_{S_t}) \quad \text{at} \quad x = x_0,\ t = t_0.$$

By continuity of f and $\{S_t\}$ we may assume for some $\varepsilon > 0$,

$$V_{S_t} \le f(x,\ t,\ \mathbf{n}_{S_t},\ \nabla \mathbf{n}_{S_t}) - \varepsilon \quad \text{on} \quad S_t \cap B_r(x_0)$$

for $t \in I = [t_0 - \delta,\ t_0 + \delta)$ by taking δ smaller if necessary. We shall modify S_t so that this type of inequality holds for every point of S_t not necessarily near x_0. As we see in the next lemma, there is a closed set $\hat{G} \in \mathcal{E}_I^-$ such that $\hat{G}(t_0 - \delta) = G(t_0 - \delta)$ and $\hat{G} = G$, a neighborhood of (x_0, t_0) in $\mathcal{O}$ by taking δ smaller. Since $B \in$ Barr $(\mathcal{E}^-)$ and $\hat{G}(t_0 - \delta) \subset B(t_0 - \delta)$, we see $\hat{G}(t_0) \subset B(t_0)$. Since $\hat{G}(t_0) \ni x_0$ and B is open, this contradicts the choice of $x_0 : x_0 \notin B(t_0)$. □

Lemma 6.3.7 (Modification) *Assume the continuity* (f1) *of* f *in* (6.1.1). *Let* G *be a compact set in* $\mathbf{R}^N \times I$ *with* $I = [t_0 - \delta,\ t_0 + \delta] \subset (0, T)$ *for some* $\delta > 0$. *Assume that* $S_t = (\partial G)(t)$ *is a smoothly evolving hypersurface on* I. *If there is* $\eta > 0$ *that satisfies*

$$V_{S_t} \le f(x,\ t,\ \mathbf{n}_{S_t},\ \nabla \mathbf{n}_{S_t}) - \eta \text{ on } S_t \tag{6.3.1}$$

for all $x \in B_\delta(x_0)$, $t \in I$, *then there is* $\delta_1 > 0$ $(\delta_1 < \delta)$ *and a compact set* $\hat{G}$ *in* $\mathbf{R}^N \times J$ *with* $J = [t_0 - \delta_1,\ t_0 + \delta_1]$ *with the property that* $\hat{G}(t_0 - \delta_1) = G(t_0 - \delta_1)$ *and* $\hat{G} = G$ *near* (x_0, t_0) *and that* $\hat{G} \in \mathcal{E}_J^-$.

Proof For G there is a function $u \in C(\mathbf{R}^N \times I)$ such that

$$G = \{(x, t) \in \mathbf{R}^N \times I;\ u(x, t) \ge 0\}$$

and u is smooth near ∂G with $\nabla u \ne 0$. By taking $\delta > 0$ smaller we may assume that $G \subset U \times I$ with a bounded open set U containing $B_\delta(x_0)$ and that $u \in C^\infty(U \times I)$ and $\nabla u \ne 0$ on $U \times I$. We rewrite our assumption (6.3.1) to get

$$u_t + F(z,\ \nabla u,\ \nabla^2 u) \le -\eta |\nabla u| \quad \text{in} \quad (B_\delta(x_0) \times I) \cap \partial G.$$

Since $|\nabla u| > 0$ in $B_\delta(x_0) \times I$ and F is continuous, there is a constant $\lambda > 0$ that satisfies

$$u_t + F(z, \nabla u, \nabla^2 u) \le -\lambda \quad \text{in} \quad B_\delta(x_0) \times I \tag{6.3.2}$$

by taking δ smaller if necessary. We set

$$v(x,t) = u(x,t) - \mu\alpha(x)(t - t_1).$$

Here $\alpha \in C^\infty(\mathbf{R}^N)$ is taken so that $\alpha \equiv 1$ on $\mathbf{R}^N \backslash B_\delta(x_0)$ and $\alpha \equiv 0$ on $B_{\delta/2}(x_0)$ with $\alpha \ge 0$. The constant $t_1 \in (t_0 - \delta,\ t_0)$ and $\mu > 0$ are to be determined later. Since U is bounded we take μ large enough so that

$$u_t + F(z, \nabla u, \nabla^2 u) \le \mu \quad \text{in} \quad U \times I.$$

Since $\alpha \equiv 1$ on $\mathbf{R}^N \backslash B_\delta(x_0)$, this choice of μ yields

$$v_t + F(z, \nabla v, \nabla^2 v) \le 0 \tag{6.3.3}$$

for $x \in \mathbf{R}^N \backslash B_\delta(x_0),\ t \in I$. Since $\alpha \equiv 0$ on $B_{\delta/2}(x_0)$, (6.3.3) is also valid for $x \in B_{\delta/2}(x_0)$ by (6.3.2). Taking $t_1 = t_0 - \delta_1$ close to t_0 we obtain (6.3.3) for $x \in B_\delta(x_0) \backslash B_{\delta/2}(x_0)$ and for $t \in [t_1,\ t_0 + \delta_1)$ since (6.3.2) holds. We may assume that $\nabla v \ne 0$ on $U \times J,\ J = [t_1,\ t_0 + \delta_1]$ and that the zero level set is contained in U by taking δ_1 smaller. We now set

$$\hat{G} = \{(x,t) \in U \times J;\ v(x,t) \ge 0\}$$

and observe that $\hat{G} \in \mathcal{E}_J^-$ by (6.3.3). From definition of v it follows that $\hat{G} = G$ near (x_0, t_0) and $\hat{G}(t_1) = G(t_1)$. □

Remark 6.3.8 (Choice of $\mathcal{E}$) If the comparison principle holds for smooth sub- and supersolutions of (6.1.1) and the local smooth solution exists for every C^∞ initial closed hypersurface (whose existence time interval may depend on smoothness of its initial surface), then Barr $(\mathcal{E}^-)$ = Barr $(\mathcal{E}^\infty)$. Indeed for $B \notin$ Barr $(\mathcal{E}^-)$ with $B \in$ Barr $(\mathcal{E}^\infty)$ there is $G \in \mathcal{E}_I^-$ with some $I = [t_0,\ t_1)$ that satisfies $G(t_0) \subset B(t_0)$ but $G(t) \backslash B(t) \ne \emptyset$ for some t, $t_0 < t < t_1$. We may assume that $\{\partial G(t)\}$ is a smooth evolving hypersurface by approximation. Since $\overline{G}$ is compact, we see

$$t_* = \inf\{t \in I;\ G(t) \backslash B(t) \ne \phi\} > t_0.$$

By the local existence of solutions for each $\delta \ge 0$ there is $\eta > 0$ and $W^\delta \in \mathcal{E}_J^\infty$ with $J = [t_* - \delta,\ t_* - \delta + \eta)(\subset I)$ and $W^\delta(t_* - \delta) = G(t_* - \delta)$. By uniform regularity of $G(t)$ in t we may assume that η is independent of δ and take $\delta > 0$ so that $\eta - \delta > 0$. By comparison of smooth sub- and supersolutions, we see

$$G(t) \subset W^{\delta}(t) \quad \text{for} \quad t \in J.$$

Since $W^{\delta}(t_* - \delta) = G(t_* - \delta) \subset B(t - \delta)$ and $B \in \text{Barr}\ (\mathcal{E}^{\infty})$ it follows that $G(t) \subset W^{\delta}(t) \subset B(t)$ for $t \in J$. This contradicts the definition of t_* since $\eta - \delta > 0$.

Remark 6.3.9 (Subsolutions) There is of course a notion of barrier subsolutions corresponding to subsolutions. Let $\mathcal{E}_I^+$ be the set defined by

$$\begin{aligned}\mathcal{E}_I^+ = \{U \subset \mathbf{R}^N \times I; \quad & U \quad \text{is a bounded open set and} \\ & \{\partial U(t)\}_{t \in I} \quad \text{is a } C^{2,1} \text{ evolving} \\ & \text{hypersurface satisfying} \\ & \qquad V \geq f(z, \mathbf{n}, \nabla \mathbf{n}) \quad \text{on} \quad \partial U(t)\},\end{aligned}$$

where $\mathbf{n}$ is taken outward from $U(t)$. For an interval J in $[0, T)$ a set B in $\mathbf{R}^N \times J$ is called a *barrier subsolution* associated to $\mathcal{E}^+ = \{\mathcal{E}_I^+; I \subset J\}$ if $U \in \mathcal{E}_I^+$ with $I = [t_0, t_1) \subset J$ and $U(t_0) \supset B(t_0)$ always fulfills $U \supset B \cap (\mathbf{R}^N \times I)$. The set of all barrier subsolutions associated to $\mathcal{E}^+$ is denoted by barr $(\mathcal{E}^+)$. It is easy to obtain statements as in Proposition 6.3.2–Lemma 6.3.6 for barrier subsolutions by an argument symmetric to the case of supersolutions. For example, the statement corresponding to Theorem 6.3.3 reads: a closed set B in $\mathcal{O} = \mathbf{R}^N \times J$ (J: open interval) is a barrier subsolution if and only if B is a set-theoretic subsolution of (6.1.1) in $\mathcal{O}$ under the assumptions (f1) and (f2). The statement corresponding to Lemma 6.3.6 reads: $B \in$ barr $(\mathcal{E}^+)$ if and only if $\overline{B} \in$ barr $(\mathcal{E}^+)$ under the assumption of (f1) and (f2). The right uniform upper semicontinuity of $\overline{B}$ follows if $B \in$ barr $(\mathcal{E}^+)$ provided that f satisfies the uniform continuity in Proposition 6.3.2 (ii).

6.4 Consistency

We shall prove that our notion of a level set subsolution is consistent with classical subsolutions. As we observed in Corollary 6.1.5 a classical subsolution is always a set-theoretic subsolution so the question is whether or not there is an approximate family of set-theoretic subsolutions in Definition 6.2.2. We shall also prove Theorem 6.3.3 and Lemma 6.3.6.

6.4.1 *Nested Family of Subsolutions*

Theorem 6.4.1 *Assume the continuity* (f1) *of f in* (6.1.1). *Let F be as in* (6.1.2). *Let G be in $\mathcal{E}_I^-$ with a closed interval $I \subset [0, T)$. Then there is a tubular neighborhood*

W of ∂G in $\mathbf{R}^N \times I$ and $u_1 \in C^{2,1}(W\backslash G) \cap C^1(W\backslash \mathrm{int}\ G)$ (resp. $u_2 \in C^{2,1}$ (int $G \cap W) \cap C^1(G \cap W)$) that satisfies

$$u_t + F(z,\ \nabla u,\ \nabla^2 u) \leq 0 \quad in \quad W\backslash G \quad (resp.\ \mathrm{int}\ G \cap W) \tag{6.4.1}$$

with $u_1 < 0$ in $W\backslash G$, $u_1 = 0$ on ∂G (resp. $u_2 > 0$ in int $G \cap W$, $u_2 = 0$ on ∂G) and that $\nabla u_1 \neq 0$ in $W\backslash$ int G (resp. $\nabla u_2 \neq 0$ in $G \cap W$.) The inequality in (6.4.1) *may be replaced by the strict inequality.*

As level sets of u_1 or u_2 there is a nested family of classical subsolutions $G_s = \{(x, t) \in W;\ u_i(x, t) \geq s\}$ ($i = 1$ for $s < 0$ and $i = 2$ for $s \geq 0$) approximating G from inside or outside of G. As a corollary of Theorem 6.4.1 we have

Theorem 6.4.2 *Assume the same hypothesis of Theorem 6.4.1 concerning f and G. Then there is a nested family $\{G_s\}$ in $\mathcal{E}_I^-$ for $s \in (-\delta,\ \delta)$ with some $\delta > 0$ with the property that:*

(i) $G_0 = G$, $G_s \subset \mathrm{int}\ G_\tau$ *if* $\tau < s$, $\tau, s \in (-\delta,\ \delta)$.
(ii) *For each $\varepsilon > 0$ there is a $\delta_0 > 0$ that satisfies*

$$\sup\{\mathrm{dist}\ (x,\ G(t));\ x \in G_s(t),\ 0 \leq s \leq \delta_0,\ t \in I\} < \varepsilon,$$

$$\sup\{\mathrm{dist}\ (x, (W\backslash\ \mathrm{int}\ G)(t));\ x \in G_s(t),\ -\delta_0 \leq s \leq 0,\ t \in I\} < \varepsilon.$$

(iii) *The strict inequality*

$$V < f(z, \mathbf{n}, \nabla\mathbf{n})$$

is fulfilled on ∂G_s if $s \neq 0$.

Proof of Theorem 6.4.1 We first remark that the construction of u_1 satisfying (6.4.1) is very easy if f is independent of the spatial variable x. In this case we take $u_1 = \mathrm{sd} \wedge 0$ with $\mathrm{sd}(x, t) = \mathrm{sd}(x, G(t))$ and observe by Theorem 6.1.7 that u_1 is a subsolution of (6.1.2) outside G. Since $u_1 \in C^{2,1}(W\backslash G)$ by regularity of G (see e.g. D. Gilbarg and N. S. Trudinger [235]) for some tubular neighborhood W of ∂G, u_1 is a function satisfying (6.4.1) in $W\backslash G$ in a classical sense. Unfortunately, if f depends on x, u_1 is no longer a subsolution of (6.1.2); it is a subsolution of (6.1.4) We need to use another argument to construct u_1 for the general case as well as to construct u_2.

Since $\{\partial G(t)\}$ is a $C^{2,1}$ evolving hypersurface on I, there are a tubular neighborhood W of ∂G and $\varphi \in C^{2,1}(W)$ that satisfies $\nabla\varphi \neq 0$ in $\overline{W}$ with

$$G \cap W = \{(x, t) \in W;\ \varphi(x, t) \geq 0\}.$$

Since $G \in \mathcal{E}_I^-$, φ satisfies

$$\varphi_t + F(z,\ \nabla\varphi,\ \nabla^2\varphi) \leq 0 \quad \text{on} \quad \partial G.$$

Since we may assume that $\overline{W}$ is compact in $\mathbf{R}^N \times I$ and $\nabla\varphi \neq 0$ in $\overline{W}$,

$$\omega_0(\sigma) = \sup\{(\varphi_t + F(z, \nabla\varphi, \nabla^2\varphi))(x,t), (x,t) \in W, |\varphi(x,t)| \leq \sigma\}$$

is nondecreasing and $\omega_0(\sigma) \to 0$ as $\sigma \to 0$. By Lemma 3.1.9 there is a modulus $\omega \in C^\infty(0, \infty) \cap C^0[0, \infty)$ with $\omega_0 \leq \omega$ that satisfies

$$\varphi_t + F(z, \nabla\varphi, \nabla^2\varphi) \leq \omega(|\varphi|) \quad \text{in} \quad W. \tag{6.4.2}$$

We may assume that $\omega(\sigma) \geq \sigma$. If we set

$$\theta(s) = \exp\left(-\int_s^1 \frac{d\sigma}{\omega(\sigma)}\right), \quad s > 0,$$

then $\theta \in C^1[0, \infty) \cap C^\infty(0, \infty)$ with $\theta'(s) > 0$ for $s > 0$. Since F in (6.4.2) is geometric, from (6.4.2) it follows that the function $w = -\theta(-\varphi)(\in C^1(W\backslash \text{int } G) \cap C^{2,1}(W\backslash G))$ solves

$$w_t + F(z, \nabla w, \nabla^2 w) \leq \theta'(-\varphi)\omega(|\varphi|) \quad \text{in} \quad W\backslash G. \tag{6.4.3}$$

By the choice of θ we see

$$\theta'(-\varphi)\omega(|\varphi|) = -w \quad \text{in} \quad W\backslash \text{int } G.$$

From (6.4.3) the function $u_1 = e^{\lambda t} w$ with $\lambda > 1$ solves

$$u_t + F(z, \nabla u, \nabla^2 u) \leq e^{\lambda t}(\lambda - 1)w < 0 \quad \text{in} \quad W\backslash \text{ int } G.$$

Since the construction of u_2 is similar, it is omitted. □

Remark 6.4.3 Of course, statements similar to Theorems 6.4.1 and 6.4.2 hold for classical supersolutions although we do not state them here explicitly.

6.4.2 *Applications*

Theorem 6.4.4 *Assume the continuity* (f1) *and the degenerate ellipticity* (f2) *of* f *in* (6.1.1). *If* $G \in \mathcal{E}_I^-$ *with an interval* $I = [t_0, t_1) \subset [0, T)$, *then* G *is a level set subsolution of* (6.1.1) *in* $\mathbf{R}^N \times [t_0, t_1)$.

This follows from Corollary 6.1.5 and Theorem 6.4.2 since G is compact. Applying Corollary 6.2.10 we have

Corollary 6.4.5 *Assume* (f1) *and* (f2) *apply to* f *in* (6.1.1). *Let* G *be a subset of* $Z = \mathbf{R}^N \times [0, T)$. *Assume that* $G(t)$ *is bounded. If* ∂G *is a smoothly evolving*

hypersurface in $[0, T)$ *and fulfills* (6.1.1) *on* ∂G *for* $t > 0$, *where* $\mathbf{n}$ *directs outward to* G, *then* $\operatorname{int} G$ *and* $\overline{G}$ *are the level set solutions with initial data* $G(0)$ *and* $\overline{G}(0)$, *respectively. In particular, for initial data* $G(0)$ *and* $\overline{G}(0)$ *an open and closed set-theoretic solution is unique, respectively.*

Proof of Lemma 6.3.6 Suppose that $\operatorname{int} B$ does not belong to Barr $(\mathcal{E}^-)$. Then there would exist $I = [t_0, t_1)$ and $G \in \mathcal{E}_I^-$ that satisfies $G(t_0) \subset \operatorname{int} B(t_0)$, $G(t) \subset B(t)$ for $t \in I$ but $G(t_*) \cap \partial B(t_*) \neq \emptyset$ at some $t_* \in (t_0, t_1)$. Since dist $(G(t_0), \mathbf{R}^N \setminus (\operatorname{int} B)(t_0)) = d_0 > 0$, by Theorem 6.4.2 there is $G_s \in \mathcal{E}_I^-$ that satisfies $\operatorname{int} G_s \supset G$ and $G_s(t_0) \subset \operatorname{int} B(t_0)$. Since $\operatorname{int} G_s \supset G$, we see $G_s(t_*)$ contains a point outside $B(t_*)$. This contradicts $B \in$ Barr $(\mathcal{E}^-)$.

Suppose that B does not belong to Barr $(\mathcal{E}^-)$. Then there would exist $I = [t_0, t_1)$ and $G \in \mathcal{E}_I^-$ that satisfies $G(t_0) \subset B(t_0)$ but $G(t_2)\setminus B(t_2) \neq \emptyset$ for some $t_2 \in I$. As before by Theorem 6.4.2 there is $G_s \in \mathcal{E}_I^-$ that satisfies $\operatorname{int} G \supset G_s$ and $G_s(t_0) \subset \operatorname{int} B(t_0)$ and $G_s(t_2) \setminus (\operatorname{int} B)(t_2) \neq \emptyset$ which contradicts $\operatorname{int} B \in$ Barr $(\mathcal{E}^-)$. □

As another application of existence of a nested family of subsolutions we prove that a supersolution is always a barrier supersolution without assuming a comparison principle.

Proof of Theorem 6.3.3 Suppose that B were not a barrier supersolution. There would exist $G \in \mathcal{E}_I^-$ with some $I = [t_0, t_1)$ in $(0, T)$ that satisfies $G(t_0) \subset B(t_0)$ but $G(t)\setminus B(t) \neq \emptyset$ for some $t \in I$. We may assume that $I = [t_0, t_1]$ by taking $t_1 (> t_0)$ smaller. Since G is compact, the time

$$t_* = \inf\{t \in I;\ G(t)\setminus B(t) \neq \emptyset\} \quad (< T)$$

is strictly larger than t_0. By Theorem 6.4.2 there is $\tilde{G} \in \mathcal{E}_I^-$ that satisfies

$$\begin{gathered}\tilde{G}(t_0) \subset B(t_0),\ (\operatorname{int} \tilde{G})(t) \supset G \cap (\mathbf{R}^N \times I),\\ V < f(x,\ \mathbf{n},\ \nabla \mathbf{n}) \quad \text{on} \quad \partial \tilde{G}.\end{gathered} \tag{6.4.4}$$

Since $\tilde{G}$ is compact, we see

$$t_0 < s_* = \inf\{t \in I;\ \tilde{G}(t)\setminus B(t) \neq \emptyset\} < t_*.$$

Since χ_B is a supersolution in $\mathbf{R}^N \times (0, s_*]$ as proved in Chap. 4, there is a set-theoretic characterization corresponding to Theorem 6.1.2 with Remark 6.1.4 (iv). Since $\tilde{G}(t)$ touches $\partial B(t)$ for the first time at s_*, $\partial \tilde{G}(t)$ is regarded as a 'test surface' at $t = s_*$. Since B is a supersolution we must have

$$V \geq f(z,\ \mathbf{n},\ \nabla \mathbf{n}) \quad \text{on } \partial \tilde{G}(s_*) \cap \partial B(s_*)$$

which evidently contradicts (6.4.4). The proof of the converse is included in the proof of Proposition 6.3.2 (ii). □

6.4.3 Relation Among Various Solutions

We have defined several notions of solutions of (6.1.1) for a set in $\mathbf{R}^N \times J$ with an interval J in $[0, T)$. We summarize the relation of these sets. We first assume that J is open in $(0, T)$.

(i) A barrier supersolution $\Rightarrow$ A set-theoretic supersolution. To prove this implication we have used the existence (F) of fundamental subsolutions and the modification (Lemma 6.3.7) as presented in the proof of Proposition 6.3.2 (ii).

(ii) A set-theoretic supersolution $\Rightarrow$ A barrier supersolution. To prove this implication we have used the existence of a nested family of smooth subsolutions (Theorem 6.4.2) as presented in the proof of Theorem 6.3.3 in Sect. 6.4.2. In both (i) and (ii) we first assume the set is open but thanks to Lemma 6.3.6 such an assumption turns to be unnecessary. In the proof of Lemma 6.3.6 in Sect. 6.4.2 we have again used the existence of a nested family of smooth subsolutions. Of course, both (i) and (ii) are still valid if a subsolution replaces a supersolution.

(iii) Both implications (i) and (ii) can be extended to various settings although we only state a few examples without detailed proofs. From the proof one may take J to be a semi-closed interval $[t_0, t_1)$ in both (i) and (ii). Also (i) and (ii) apply to boundary value problems with appropriate modifications.

(iv) Perron's method (cf. Sect. 3.4) is easy to establish for barrier solutions without using results in Sect. 3.4. Here is a version.

Theorem 6.4.6 *Let* Barr $(\mathcal{E}^-)$ *(resp.* barr $(\mathcal{E}^+)$*) be the set of barrier supersolutions (resp. subsolutions) associated to* $\mathcal{E}^-$ *(resp.* $\mathcal{E}^+$*) in* $\mathcal{O} = \mathbf{R}^N \times J$ *with an open interval in* $(0, T)$.

(i) *(Closedness) Let S be a subset in* Barr $(\mathcal{E}^-)$. *Then* $\cap B_{B \in S} \in$ Barr $(\mathcal{E}^-)$.

(ii) *(Minimality) Assume that $U \in \mathcal{E}_I^+$ is always a barrier supersolution associated to $\mathcal{E}^-$ in $\mathbf{R}^N \times$* int I. *Assume that $B_0 \in$* Barr $(\mathcal{E}^-)$ *and $B_1 \in$* barr $(\mathcal{E}^+)$ *fulfills $B_0 \supset B_1$. Then*

$$D = \cap\{B;\ B \in \mathrm{Barr}\,(\mathcal{E}^-),\ B_1 \subset B \subset B_0\}$$

is a barrier sub- and supersolution in $\mathcal{O}$ associated to $\mathcal{E}^+$ and $\mathcal{E}^-$ respectively.

Proof (i) is trivial by definition. To prove (ii) it suffices to prove that $D \in$ barr $(\mathcal{E}^+)$. If not, there is an interval $I = [t_0, t_1) \subset J$ and $U \in \mathcal{E}_I^+$ that satisfies $U(t_0) \supset D(t_0)$ but $D(t) \backslash U(t) \neq \emptyset$ for some $t \in I, t > t_0$. Since $B_1 \in$ barr $(\mathcal{E}^+)$, $U(t) \supset B_1(t)$ for all $t \in I$. As in (i) it is easy to see that

$$\tilde{D} = (D \cap U \cap (\mathbf{R}^N \times I)) \cap (D \cap (\mathbf{R}^N \times (J \backslash I)))$$

belongs to Barr ($\mathcal{E}^-$) since U is a barrier supersolution. Since $U(t) \supset B_1(t)$ for all $t \in I$, $\tilde{D} \supset B_1$. This contradicts the minimality of D. □

The assumption that $U \in \mathcal{E}_I^+$ is always a barrier subsolution is fulfilled if (f2) in (6.1.1) is assumed. So this assumption gives some parabolicity of the problem. Using implication of (i), (ii), our Theorem 6.4.6 establishes Perron's method for set-theoretic solutions under (f1) and (f2), which also can be proved by applying results in Sect. 3.4 for characteristic functions.

(v) As observed in (iv) the notion of barrier solutions simplifies several properties of set-theoretic solutions. The reader might be curious whether there is a notion of barrier solution corresponding to viscosity solution of

$$u_t + F(z, u, \nabla u, \nabla^2 u) = 0 \tag{6.4.5}$$

in $\mathcal{O} = \mathbf{R}^N \times (0, T)$. At least for continuous F we propose to say that u is a barrier supersolution if $v < u$ on $\partial_p Q$ always implies $v < u$ in $\overline{Q}$ whenever $v \in C^{2,1}(\overline{Q})$ is a $C^{2,1}$ subsolution of (6.4.5) in Q and Q is of form $Q = D \times I$ with a bounded domain D in $\mathbf{R}^N$ and an open interval I in $(0, T)$. Our notion is localized but we still have equivalence results as in (i) and (ii). We leave the details to the reader.

We next compare level set solutions with other solutions. Let J be a semi-closed interval $[t_0, t_1)$ in $[0, T)$.

(vi) A level set subsolution $\Rightarrow$ a set-theoretic subsolution in $\mathbf{R}^N \times (t_0, t_1)$ for an open set D in $\mathbf{R}^N \times [t_0, t_1)$. However, the converse may not be true as explained below.

(vii) A level set solution D with (open) initial data D_0 at $t = t_0$ is a minimal open set-theoretic supersolution G whose initial data $G(t_0)$ contains D_0 (at least when ∂D_0 is bounded and for general D_0 the uniform upper semicontinuity of G at $t = t_0$ is assumed) (Theorem 6.2.11). This is based on the comparison principle for set-theoretic sub- and supersolutions. A level set solution with given initial data D_0 is unique while the set-theoretic solution may not be unique as explained in Example 6.2.1 and Remark 6.2.12. In particular the converse of (vi) is not always true.

(viii) A generalized open evolution with initial data $D_0 \Rightarrow$ a level set solution with the same initial data (Proposition 6.2.8). By uniqueness of generalized open evolution (based on the comparison principle for (6.1.2)) the converse is also true.

6.5 Separation and Comparison Principle

We give an alternate way to prove the comparison principle (BCPS) for set-theoretic solutions based on local existence of classical solutions of (6.1.1). The key ingredients are the separation lemma and the equivalence of barrier solutions and set-theoretic solutions.

Lemma 6.5.1 (Separation) *Let E_0 be a compact set contained in an open set D_0 in $\mathbf{R}^N$. Then there is a bounded open set U_0 with C^{1+1} boundary (i.e., the unit normal vector field is Lipschitz) such that*

(i) $E_0 \subset U_0$, $\overline{U}_0 \subset D_0$,
(ii) $\operatorname{dist}(E_0, \partial U_0) + \operatorname{dist}(\overline{U}_0, \partial D_0) = \operatorname{dist}(E_0, \partial D_0)$.

This lemma is due to T. Ilmanen [267]. We do not give the proof here. A simpler proof is given by P. Cardaliaguet [88].
A typical local existence result we need is

(LE) Let U_0 be a bounded open set in $\mathbf{R}^N$ with C^∞ boundary, where the unit normal is taken outward from U_0. Then for a given $t_0 \in [0, T)$ there are $I = [t_0, t_1)$ and $U \in \mathcal{E}_I^- \cap \mathcal{E}_I^+$ with $U(t_0) = U_0$ for some $t_1 (> t_0)$ depending on initial data U_0 only through a bound for the second fundamental form of ∂U_0.

We present a version of the comparison principle when f of (6.1.1) does not depend on the spatial variables.

Theorem 6.5.2 *Assume that f in* (6.1.1) *satisfies* (f1) *and is independent of $x \in \mathbf{R}^N$. Assume that* (LE) *holds. Then* (BCPS) *holds.*

Proof Let E and D be set-theoretic sub- and supersolutions of (6.1.1) in $\mathbf{R}^N \times (0, T)$. We may assume that D and E are bounded since the case that the complements of D and E are bounded can be handled in a similar way. Assume that $E_0 = E^*(0)$ is contained in $D_0 = D_*(0)$. Our goal is to prove $E^* \subset D_*$. We may assume that E is closed and D is open in $\mathbf{R}^N \times [0, T)$.

We shall prove that

$$\operatorname{dist}(E(t),\ D^c(t)) \geq \operatorname{dist}(E_0,\ D_0^c) \quad \text{for} \quad t \in (0, T)$$

where c denotes the complements in $\mathcal{O} = \mathbf{R}^N \times (0, T)$. (This evidently implies that $E \subset D$ but it is actually equivalent to (BCPS) since (6.1.1) is translation invariant in space.) We set

$$t_* = \sup\{t \in [0, T);\ \operatorname{dist}(E(\tau), D^c(\tau)) \geq \operatorname{dist}(E_0, D_0^c) \quad \text{for all} \quad 0 \leq \tau \leq t\}.$$

By left accessibility of E and D^c at t_*,

$$\operatorname{dist}(E(t_*), D^c(t_*)) \geq \operatorname{dist}(E_0, D_0^c).$$

Assume that $t_* < T$. Then by the separation (Lemma 6.5.1) there is an open set U_0 with C^{1+1} boundary such that

(i) $E(t_*) \subset U_0$, $\overline{U}_0 \subset D(t_*)$,
(ii) $\operatorname{dist}(E(t_*), \partial U_0) + \operatorname{dist}(\overline{U}_0, D^c(t_*)) = \operatorname{dist}(E(t_*), D^c(t_*))$.

Unfortunately, U_0 is not quite smooth enough to apply (LE). Let $U_{0\varepsilon}$ be an open set with smooth boundary satisfying (i) with U_0 replaced by $U_{0\varepsilon}$ and

(ii)$'$ $\operatorname{dist}(E(t_*), \partial U_{0\varepsilon}) + \operatorname{dist}(\overline{U}_{0\varepsilon}, D^c(t_*)) \geq \operatorname{dist}(E(t_*), D^c(t_*)) - \varepsilon$;

the principal curvatures of $U_{0\varepsilon}$ are chosen to be bounded uniformly for small $\varepsilon > 0$. By (LE) there is $U_\varepsilon \in \mathcal{E}_I^- \cap \mathcal{E}_I^+$ with $U_\varepsilon(t_*) = U_{0\varepsilon}$ for some $I = [t_*, t_1)$ with $t_1 > t_*$ independent of ε.

Now we recall that $E \in \operatorname{barr}(\mathcal{E}^+)$ and $D \in \operatorname{Barr}(\mathcal{E}^-)$ by Theorem 6.3.3 as subsets of $\mathbf{R}^N \times (0, T)$. Since we consider E^* and D_* as subsets of $\mathbf{R}^N \times [0, T)$, it is easy to see that $E^* \in \operatorname{barr}(\mathcal{E}^+)$, $D_* \in \operatorname{Barr}(\mathcal{E}^-)$ as subsets of $\mathbf{R}^N \times [0, T)$. By this observation one can compare D and E with $\mathcal{E}_I^-$ for I starting at t_*. By definition of barrier solutions and translation invariance we see

$$\operatorname{dist}(E(t), \partial U_\varepsilon(t)) \geq \operatorname{dist}(E(t_*), \partial U_{0\varepsilon}),$$

$$\operatorname{dist}(\overline{U}_\varepsilon(t), D^c(t)) \geq \operatorname{dist}(\overline{U}_{0\varepsilon}, D^c(t_*))$$

for $t \in I$. This together with (ii)$'$ yields

$$\operatorname{dist}(E(t), D^c(t)) \geq \operatorname{dist}(E(t_*), D^c(t_*)) - \varepsilon.$$

Since I is independent of ε, this yields

$$\operatorname{dist}(E(t), D^c(t)) \geq \operatorname{dist}(E(t_*), D^c(t_*))$$

for $t \in [t_*, t_1)$. This contradicts the definition of t_* so $t_* = T$; the proof is now complete. □

Remark 6.5.3

(i) In various situations a local existence of solutions for (6.1.1) has been proved. For the mean curvature flow equation it has been proved by G. Huisken [259]. Existence time t_1 has been estimated by

$$t_1 \geq t_0 + c_N/M^2,$$

where M is a bound of the second fundamental form and c_N depends only on the dimension. Our theorem apparently does not assume parabolicity (f2) but to establish (LE) we need some parabolicity.

(ii) Theorem 6.5.2 together with Lemma 6.1.9 yields (CP (ii)) for the level set equations without using comparison results in Chap. 4. Note that our local existence (LE) does not require uniqueness of smooth solutions. Conceptually speaking, existence result (LE) implies the comparison principle, which yields the existence of a continuous solution by Perron's method.

6.6 Notes and Comments

A set-theoretic interpretation of solutions was introduced by H. M. Soner [399] for the anisotropic curvature flow equation with no explicit dependence on the spatial variables and the time variable. However, his definition is based on (signed) distance functions instead of characteristic functions used in Definition 6.1.1. Our definition seems new although characteristic functions are often used to describe typical properties of generalized evolutions for example in G. Barles et al. [55]. It turns out that our definition agrees with the one by distance functions as stated in Theorem 6.1.7. For the mean curvature flow equation K. Ishii and H. M. Soner [288] called a set satisfying Theorem 6.1.2 (i) a *weak solution*. They compared weak subsolutions with set-theoretic subsolutions based on distance functions. In fact, based on results by H. M. Soner [399, §7, §14] they stated Corollary 6.1.8 for the mean curvature flow equation. Our Theorem 6.1.2 gives a geometric interpretation of our set-theoretic solutions. Although Theorem 6.1.2 follows from Theorem 6.1.7 once Corollary 6.1.8 is proved, it is not stated in the literature as far as the author knows. We give a direct proof without appealing to distance functions.

Consistency It is clear that smooth solution $\Gamma \subset \mathbf{R}^N \times [0, T_0)$ is a set-theoretic solution of (6.1.1) thanks to Corollary 6.1.5. However, it is not very obvious that Γ is also a level set solution or a generalized interface evolution, since one has to prove that there is no fattening. The consistency of a smooth solution (with smooth initial data) with an interface evolution has been proved first by L. C. Evans and J. Spruck [166] for the mean curvature flow equation and later by Y. Giga and S. Goto [208] for general equations (6.1.1) satisfying (f1) and (f2) when F is independent of x and grows linearly in $\nabla \mathbf{n}$. The proof can be extended for all (6.1.1) satisfying (f1) and (f2) without further restrictions. The major task is to construct sub- and supersolutions of the level set equation whose zero level set is the evolving surface by using a distance function. We do not present the detail.

Our characterization of solutions of level set equation (Theorem 6.1.6) is important to bridge the set-theoretic approach to the level set approach. The proof of the 'only if' part is standard and this implication is often used in the literature, for example, in G. Barles et al. [55]. However, the converse seems to be unfamiliar.

Our Theorem 6.1.7 characterizes set-theoretic solutions by distance functions. The 'only if' part has been proved by G. Barles et al. [55] by representing distance functions by sup-convolutions. The proof given here is direct and does not appeal to sup-convolutions. The converse is contained in Theorem 6.1.6 when F in (6.1.4) is independent of the spatial variable. Otherwise we need a little bit of extra work as mentioned in the proof.

It is clear that the comparison principle for level set equations implies the comparison principle for set-theoretic solutions. We here note the converse is also easy to prove. We have given two kinds of such statements (Lemmas 6.1.9 and 6.1.10) for bounded sets and general sets. It seems that there is no literature containing both lemmas.

Several similar problems for evolution equations for h in Example 6.2.1 were studied for example by N. Ishimura [289] (the Gaussian curvature flow equation) and by Y. Kohsaka [314] (the curve-shortening-like equation with free boundaries).

Nonuniqueness of set-theoretic solutions of $V = k$ with a given initial data was first observed by L. C. Evans and J. Spruck [166], where they described this phenomena as 'fattening' of level sets. Example 6.2.1 is taken from H. M. Soner [399]. Several criteria on initial data for uniqueness, described in Sect. 5.5, were given in G. Barles et al. [55], H. M. Soner and P. E. Souganidis [402], S. Altschuler et al. [10]. The latter two works show that axisymmetric evolution for the mean curvature flow is regular. Examples of nonuniqueness were given for various equations by several authors including Y. Giga et al. [209], G. Barles et al. [55], G. Bellettini and M. Paolini [62] and Y. Giga [201] even if initial data has a smooth boundary. For the mean curvature flow equations with $N \geq 3$, examples of nonuniqueness for smooth initial data is also proved by S. B. Angenent et al. [30] for $4 \leq N \leq 7$, $N = 8$ and numerically conjectured by S. Angenent et al. [29]. A survey article by B. White [427] includes an example of nonuniqueness for $N = 3$.

The notion of level set solutions is introduced to describe generalized evolutions constructed in Chap. 5 from the set-theoretic point of view.

The notion of barrier solutions was first introduced by T. Ilmanen [267] and by E. De Giorgi [138], and developed by G. Bellettini and M. Paolini [63] and L. Ambrosio and H. M. Soner [14]. These works include the statement that minimal barrier is equivalent to generalized open evolution (Corollary 6.3.4) for several equations. (For general geometric equations see the work of G. Bellettini and M. Novaga [61]). See also a review article by L. Ambrosio in the lecture note by L. Ambrosio and N. Dancer [12]. These results depend on comparison principles. However, the equivalence of a barrier supersolution and a set-theoretic supersolution seems to be new. We note we did not use the separation lemma (Lemma 6.5.1) nor comparison principles for the proof of this last statement. This characterization provides a method to prove the comparison results for level set equations via local existence of classical solutions of surface evolution equations.

The notion of barrier solutions is also useful to prove convergence of internal layers of the Allen–Cahn equation to the mean curvature flow. In fact, G. Barles and P. E. Souganidis [56] introduced a kind of barrier solutions to solve such problems. In particular, they proved the global-in-time convergence for the homogeneous Neumann problem which was only known for convex domains [298]. This idea is further developed by G. Barles and F. Da Lio [49] to handle oblique type boundary conditions.

Approximation There are several ways to approximate solutions of level set equations or generalized interface evolution itself. We list several related articles (which apply the mean curvature flow equation) without detailed explanation.

(i) **Finite difference approximations.** M. G. Crandall and P.-L. Lions [131] proves the convergence of their schemes (see a paper by K. Deckelnick [135] for further development). The stability of a more intuitive scheme has been proved in Y.-G. Chen et al. [109]. There is also a finite element approximation to calculate solutions of level set equations. The reader is referred to papers by K. Deckelnick and G. Dziuk [136, 137] for convergence of the scheme and A. Handlovičová and K. Mikula [254] for related schemes.

(ii) **Bence–Merriman–Osher scheme.** This is a kind of filter proposed by B. Merriman et al. [350]. The convergence has been proved by G. Barles and C. Georgelin [50] and independently by L. C. Evans [162]. It is further developed by H. Ishii [275], F. Cao [81] and by H. Ishii et al. [282]. Convergence rate is also studied by K. Ishii [287].

(iii) **A variational scheme.** This is proposed by F. Almgren et al. [9] and by I. Fonseca and M. A. Katsoulakis [178]. For further development see a paper by S. Luckhaus and T. Sturzenhecker [338].

(iv) **Approximation by interacting particle systems.** This relates the description of macroscopic model and microscopic models. It has been studied by M. A. Katsoulakis and P. E. Souganidis [299].

The reader is referred to a review by C. M. Elliott [154] and a book by F. Cao [82] for other approximations. Note that the above approximation works for generalized evolution which is not necessarily smooth.

Regularity There are several attempts to prove some regularity of generalized interface evolutions for the mean curvature flow equation. For example, L. C. Evans and J. Spruck [168] proved that the evolution fulfills the clearing-out lemma which is important for the Brakke flow. As an application they estimate the extinction time from above by using the area of initial data. An estimate from below is given by Y. Giga and K. Yama-uchi [233]. They also prove that if a part of interface evolution is given as the graph of a continuous function, the part of evolution is represented as a graph-like solution of the equation. Such a solution has been proved by them to be regular. The proof of this reduction is simplified by M.-H. Giga and Y. Giga [191].

Furthermore, L. C. Evans and J. Spruck [169] prove that almost every level set of solutions of the level set mean curvature flow equation is a Brakke type flow. This is generalized by Y. Tonegawa [414] for anisotropic curvature flow equations. For monotone evolutions the size of the singular sets of generalized interface evolution Γ is well-estimated. In fact, B. White [426] proved that the singular set of $\Gamma \subset \mathbf{R}^N \times [0, \infty)$ has the parabolic Hausdorff dimension at most $N - 2$. This is optimal since a torus shrinking to a ring has $N - 2$ dimensional singularity.

Nonlocal Problems In this book we restrict ourselves to interface equations (2.5.1) whose normal velocity is determined by infinitesimal quantities of an evolving hypersurface, except the notion by crystalline curvature; see a paper by M.-H. Giga and Y. Giga [191]. We list here several other motions whose normal velocity is determined by nonlocal quantities of an evolving hypersurface. If the problem has some order preserving structure, there is a chance to establish a level set method for that problem. The examples include works of D. Adalsteinsson et al. [1], B. Andrews and M. Feldman [25], P. Cardaliaguet and E. Rouy [91], P. Cardaliaguet [87, 88], P. Cardaliaguet and O. Ley [89], F. Da Lio et al. [133], I. C. Kim [302, 303],

H. Ishii and T. Mikami [279]. The scope of problems include etching and deposition problems, the Hele–Shaw problem as well as some shape optimization problems. The barrier approach is somewhat useful, especially when the usual comparison principle is difficult to establish directly. See for example a paper by M.-H. Giga and Y. Giga [188] and a paper by P. Cardaliaguet and O. Ley [89].

Further Comments for the Second Edition We first note that a nice book by G. Bellettini [59] includes a detailed explanation of barrier solutions and their comparisons with level set solutions. This book includes many other interesting topics, for example, approximation by the Allen-Cahn equation.

The variational approximation scheme originally proposed by F. Almgren et al. [9] became very familiar. Let us recall their scheme by recalling Almgren-Taylor-Wang functional. Let $F_0 \subset \mathbf{R}^N$ be a Caccioppoli set. More precisely, the total variation $\int_{\mathbf{R}^N} |\nabla \chi_{F_0}|$ of χ_{F_0} which is called the perimeter of F_0 is finite. For $h > 0$, Almgren-Taylor-Wang functional is defined by

$$\mathcal{A}(F, F_0, h) = \int_{\mathbf{R}^N} |\nabla \chi_F| + \frac{1}{h} \int_{F \Delta F_0} \mathrm{dist}(\cdot, \partial F_0)\, dx$$

for each Caccioppoli set $F \subset \mathbf{R}^N$, where $F \Delta E = (F \backslash E) \cup (E \backslash F)$. For a given Caccioppoli set E_0, find a minimizer $T_h(E)$ which minimizes $E \mapsto \mathcal{A}(E, E_0, h)$. The main idea is to get an approximate flow by repetition. More precisely for a given $h > 0$ we set

$$E^h(t) = T_h^{[t/h]}(E_0) = \overbrace{T_h \circ \cdots \circ T_h}^{[t/h]}(E_0),$$

where $[k]$ denotes the largest integer less than or equal to $k > 0$. F. Almgren et al. [9] derived several properties of this approximate solution and called the limit a *flat Φ curvature flow*. They proved the convergence of $\{E^h(t)\}$ (as $h \downarrow 0$) to a smooth mean curvature flow up to the time when the latter exists. Later, S. Luckhaus and T. Sturzenhecker [338] showed its convergence to a distributional solution (called a BV-solution) χ to the mean curvature flow under *no mass loss condition*:

$$\int_0^T \int_{\mathbf{R}^N} |\nabla \chi_h| \to \int_0^T \int_{\mathbf{R}^N} |\nabla \chi|,$$

where $\chi_h(t) = \chi_{E^h}$. G. De Philippis and T. Laux [141] proved that this no mass loss condition is unnecessary for the convergence whenever initial data E_0 is *outward minimizing*, i.e.,

$$\int_{\mathbf{R}^N} |\nabla \chi_{E_0}| \le \int_{\mathbf{R}^N} |\nabla \chi_F| \quad \text{if} \quad E_0 \subset F.$$

If E_0 is mean convex, this condition is fulfilled.

One problem of this scheme is that the functional $\mathcal{A}$ is not convex, it is not easy to find a minimizer and it may not be unique. To overcome this difficulty, A. Chambolle [94] introduced a new scheme which is convex. To simplify the issue, we concentrate the case when E_0 is bounded. Let Ω be a bounded convex domain containing the closure of E_0 in its interior. For a given closed set E_0 in Ω, the Chambolle functional for $u \in L^2(\Omega)$ is of the form

$$\mathcal{E}_h(u) = \int_\Omega |\nabla u|\, dx + \frac{1}{2h} \int_\Omega (u - d_{E_0})^2\, dx,$$

where d_{E_0} is the signed distance function from E_0. This is a convex minimizing problem and it is easy to find a unique minimizer. Let $w^h_{E_0}$ be its minimizer. As proved by A. Chambolle [94], the set $\left\{w^h_{E_0}(x) \le 0\right\}$ is a minimizer $\mathcal{A}(E, E_0, h)$. Thus, from now on, we set

$$T_h(E_0) := \left\{x \in \Omega \mid w^h_E(x) \le 0\right\}.$$

Then, it turns out that the approximate set $E^h = T_h^{[t/h]}(E_0)$ actually converges to the level-set flow up to fattening as proved by A. Chambolle [94]. It is known that T_h conserved the convexity as proved by V. Caselles and A. Chambolle [92]. A simple proof of the convergence with help of super and subflows which are a kind of set-theoretic solutions in this chapter is given by A. Chambolle and M. Novaga [99]. This scheme is also effective for anisotropic curvature flow as shown in V. Caselles and A. Chambolle [92], T. Eto et al. [160]. We note that the paper by T. Eto et al. [160] includes the case when E_0 is unbounded. This Chambolle type minimizing movement scheme is by now well-studied in many directions. Let us consider a general evolution equation discussed in Sect. 2.5.1. We allow spatial dependence. It is of the form

$$V = M\left(x, \mathbf{n}(x)\right)\left\{-\operatorname{div}_{\Gamma_t}\left(\xi(x, \mathbf{n}(x)\right) + f(x,t)\right\} \quad \text{for} \quad x \in \Gamma_t,$$

where $\xi = \nabla_p \gamma$ and $\gamma(x, \mathbf{n})$ denote a surface energy density. In a wider class of M and γ, A. Chambolle et al. [95] constructed the Chambolle type minimizing movement scheme and prove that the limit must be a *flat flow* and a BV solution provided that no mass loss condition holds. They also prove the convergence of their scheme. However, their results exclude crystalline γ, i.e., the case $\gamma(x, p)$ is not C^1 for p with $|p| = 1$. For spatially homogeneous γ which may not be C^1, A. Chambolle et al. [98] constructed a crystalline flow based on minimizing movement scheme; see also the work by A. Chambolle et al. [96]. We also note that for homogeneous case K. Ishii [286] proved that the minimizing movement scheme converges to the level-set flow for crystalline mean curvature flow provided that the corresponding level-set flow equation is well-posed.

The Chambolle type minimizing movement scheme is also effective to construct a level-set flow solution to the mean curvature flow equations with prescribed contact angle. We consider a Caccioppoli set F_0 in a half space $\mathbf{R}_+^N$ and it touches the boundary $\partial\mathbf{R}_+^N$ G. Bellettini and S. Y. Kholmatov [60] considered a functional

$$\mathcal{A}_\beta(F, F_0, h) = \int_{\mathbf{R}_+^N} |\nabla\chi_F| + \int_{\partial\mathbf{R}_+^N} \beta\gamma(\chi_F)\, d\mathcal{H}^{N-1} + \frac{1}{h}\int_{\mathbf{R}_+^N\cap(F\Delta F_0)} \operatorname{dist}(\cdot, \partial F_0)\, dx$$

for each Caccioppoli set $F \subset \mathbf{R}_+^N$ instead of $\mathcal{A}$. The quantity β defined on $\partial\mathbf{R}_+^N$ is related to constant angle as $\beta = \cos\theta$. The operator γ denotes the trace and $\mathcal{H}^{N-1}$ denotes the dimensional Hausdorff measure. The approximate solution E^h can be constructed as before. G. Bellettini and S. Y. Kholmatov [60] proved that the convergence of their relation to a distributed solution provided that $(N-1)$-dimensional measure of discrete solution to that of the distributional solution which corresponds to no mass loss condition. They also showed regularity of the limit up to the boundary provided that β is Lipschitz continuous. T. Eto and Y. Giga [158, 159] proposed a Chambolle type functional for this problem. Instead of $\mathcal{E}_h$, they introduced

$$\mathcal{E}_h^\beta(u) = \mathcal{E}_h(u) + \int_{\partial\Omega} \beta\gamma u\, d\mathcal{H}^{N-1}.$$

They proved that the approximate solution converges to the level-set mean curvature flow with contact angle θ as $h \downarrow 0$ (up to fattening) provided that Ω is a smooth bounded convex set and $\beta \in C^1(\partial\Omega)$ will $\sup_{\partial\Omega} |\beta(x)| < 1$ and the tangential gradient $\nabla_{\partial\Omega}\beta$ satisfies $|\nabla_{\partial\Omega}\beta(x)| \le k(x)$ for all $x \in \partial\Omega$, where $k(x)$ denotes the minimal nonnegative inward principal curvature.

Chapter 7
Deterministic Game Interpretations

In this chapter, we introduce a more recent approach to surface evolution equations based on deterministic discrete two-person games. It turns out that viscosity solutions of various level set equations can be approximated by the value functions of a family of deterministic games. We first introduce the game for the Cauchy problem of mean curvature flow equation and then discuss several variants on the stationary case and boundary value problems.

7.1 Cauchy Problems

We next give a game interpretation for the curve shortening equation:

$$\text{(MCF)}\begin{cases} u_t - |\nabla u| \operatorname{div}\left(\dfrac{\nabla u}{|\nabla u|}\right) = 0 & \text{in } \mathbf{R}^2 \times (0, \infty) \qquad (7.1.1) \\ u(\cdot, 0) = u_0 & \text{in } \mathbf{R}^2, \qquad (7.1.2) \end{cases}$$

where u_0 is assumed to be Lipschitz in $\mathbf{R}^2$. Note that (7.1.1) is only a special case of (2.6.7) in two space dimensions, which represents one of the most important examples of the nonlinear evolution (2.6.3), with F_f given by (2.6.6).

7.1.1 Finite Horizon Games

Game 1-a We start a two-person game from any given $x \in \mathbf{R}^2$ with any duration $t \in [0, \infty)$. Let the step size be $\varepsilon > 0$. The total of the game steps is $\ell = [t/\varepsilon^2]$. We obtain a sequence $y_k \in \mathbf{R}^2$ $(k = 0, 1, 2, \dots, \ell)$ with $y_0 = x$, following the rules below.

Y. Giga, Q. Liu, *Surface Evolution Equations*, Monographs in Mathematics 99,
https://doi.org/10.1007/978-3-032-20303-8_7

At the k-th step,

- Player I chooses a direction from S^1, i.e., a vector $v_k \in \mathbf{R}^2$ with $|v_k| = 1$;
- Player II then chooses $b_k = \pm 1$;
- The game token moves from y_{k-1} to $y_k = y_{k-1} + \sqrt{2}\varepsilon b_k v_k$.

The goal of Player I is to minimize the value $u_0(y_\ell)$ while Player II attempts to maximize $u_0(y_\ell)$. Then we can define the *value function* of this game by

$$u^\varepsilon(x, t) = \min_{|v_1|=1} \max_{b_1=\pm 1} \ldots \min_{|v_\ell|=1} \max_{b_\ell=\pm 1} u_0(y_\ell). \tag{7.1.3}$$

It is easily seen that u^ε satisfies, for any $(x, t) \in \mathbf{R}^2 \times [\varepsilon^2, \infty)$,

$$u^\varepsilon(x, t) = \min_{|v|=1} \max_{b=\pm 1} u^\varepsilon(x + \sqrt{2}\varepsilon b v, t - \varepsilon^2), \tag{7.1.4}$$

which is referred to as the *dynamic programming principle* (DPP) in the context of optimal control and differential game theory. The main result concerning the game-theoretic representation of solutions to (MCF) is as below.

Theorem 7.1.1 *Assume that u_0 is Lipschitz continuous in $\mathbf{R}^2$. Let u^ε be the value function defined as in* (7.1.3). *Then $u^\varepsilon \to u$ locally uniformly in $\mathbf{R}^2 \times [0, \infty)$ as $\varepsilon \to 0$, where u is the unique viscosity solution of* (MCF).

Remark 7.1.2 This type of deterministic discrete game was introduced by R. V. Kohn and S. Serfaty [312]. However, their game setting is slightly different from ours. In our formulation, the variable t denotes the terminal time of the game. In contrast, in [312] the game starts at time t and ends at a fixed terminal time $T \geq 0$. As a consequence, their approach leads to the following terminal value problem:

$$\begin{cases} u_t + |\nabla u| \operatorname{div}\left(\dfrac{\nabla u}{|\nabla u|}\right) = 0 & \text{in } \mathbf{R}^2 \times (0, \infty), \\ u(\cdot, T) = u_0 & \text{in } \mathbf{R}^2. \end{cases}$$

In what follows, we work with the initial value formulation rather than the terminal value formulation used in [312].

Example 7.1.3 We present a concrete example of the game-theoretic approximation. Let $L > 0$ and

$$u_0(x) = L|x| \tag{7.1.5}$$

for any $x \in \mathbf{R}^2$. In this setting, at any position $y \in \mathbf{R}^2$ Player I may choose $v \in S^1$ satisfying $\langle y, v \rangle = 0$; this choice is actually optimal. A straightforward calculation then shows that, regardless of the choices made by Player II, the outcome of the game is $L(|x|^2 + 2\varepsilon^2\ell)^{1/2}$. This implies

$$u^\varepsilon(x,t) \le L(|x|^2 + 2\varepsilon^2\ell)^{\frac{1}{2}}. \tag{7.1.6}$$

On the other hand, for any unit vector $v \in \mathbf{R}^2$ that Player I may choose at any step, Player II can always select $b = \pm 1$ so that $b\,\langle y, v\rangle \ge 0$. Another direct computation then gives

$$u^\varepsilon(x,t) \ge L(|x|^2 + 2\varepsilon^2\ell)^{\frac{1}{2}}. \tag{7.1.7}$$

We refer to the strategies employed by Player I and Player II above as *strategies of concentric circles*. These strategies will play an important role in later applications.

Combining (7.1.6) and (7.1.7), we have

$$u^\varepsilon(x,t) = L(|x|^2 + 2\varepsilon^2\ell)^{\frac{1}{2}}. \tag{7.1.8}$$

It is then clear that $u^\varepsilon \to u$ locally uniformly in $\mathbf{R}^2 \times (0,\infty)$, where

$$u(x,t) = L(|x|^2 + 2t)^{\frac{1}{2}}$$

for any $(x,t) \in \mathbf{R}^2 \times (0,\infty)$. It is not difficult to verify that u is the unique solution of (MCF) with u_0 given in (7.1.5).

Before rigorously proving Theorem 7.1.1, we first provide a heuristic argument. Assuming that u^ε are smooth and do converge in a strong sense to a certain smooth function u with $|\nabla u| \neq 0$, we show that u solves the problem (MCF). Clearly, $u^\varepsilon(x,0) = u_0(x)$ for any $x \in \mathbf{R}^2$. Applying a Taylor expansion to the right hand side of the dynamic programming equation (7.1.4), we obtain, for $\varepsilon > 0$ small,

$$0 = \min_{|v|=1}\max_{b=\pm 1}\left\{\sqrt{2}\varepsilon b\left\langle \nabla u^\varepsilon(x,t), v\right\rangle + \varepsilon^2\left\langle \nabla^2 u^\varepsilon(x,t)v, v\right\rangle - \varepsilon^2 u_t^\varepsilon(x,t)\right\} + o(\varepsilon^2).$$

It follows that

$$0 = \min_{|v|=1}\left\{\sqrt{2}\varepsilon\left|\left\langle \nabla u^\varepsilon(x,t), v\right\rangle\right| + \varepsilon^2\left\langle \nabla^2 u^\varepsilon(x,t)v, v\right\rangle - \varepsilon^2 u_t^\varepsilon(x,t)\right\} + o(\varepsilon^2). \tag{7.1.9}$$

For $\varepsilon > 0$ sufficiently small, the leading-order term in the minimum forces $\langle \nabla u^\varepsilon(x,t), v\rangle$ to be small. This suggests that any minimizer v is close to either $\nabla^\perp u^\varepsilon(x,t)/|\nabla u^\varepsilon(x,t)|$ or its negative, where $\nabla^\perp u = (-\partial u/\partial x_2, \partial u/\partial x_1)$. Dividing (7.1.9) by ε^2 and sending $\varepsilon \to 0$, we are then led to

$$u_t - \left\langle \nabla^2 u \frac{\nabla^\perp u}{|\nabla u|}, \frac{\nabla^\perp u}{|\nabla u|}\right\rangle = 0,$$

which is equivalent to (7.1.1).

7.1.2 Convergence of Game Values

We next give a rigorous proof of Theorem 7.1.1, following the heuristic idea above. The strategy of the proof is as follows. We first show that u^ε are locally bounded in $\mathbf{R}^2 \times [0, \infty)$ uniformly in ε, which enables us to define the *upper* and *lower relaxed limits*: :

$$\overline{u} = \limsup_{\varepsilon \to 0}{}^* u^\varepsilon, \quad \underline{u} = \liminf_{\varepsilon \to 0}{}_* u^\varepsilon. \tag{7.1.10}$$

It is clear that $\overline{u} \geq \underline{u}$ in $\mathbf{R}^2 \times [0, \infty)$.

We next show that $\overline{u}$ and $\underline{u}$ are respectively a subsolution and a supersolution of (7.1.1). After verifying that $\overline{u}(\cdot, 0) \leq u_0 \leq \underline{u}(\cdot, 0)$, we use the comparison principle to conclude that $\overline{u} \leq \underline{u}$ in $\mathbf{R}^2 \times [0, \infty)$. Since $\underline{u} \leq \overline{u}$ holds by definition, $\overline{u} = \underline{u}$ is the unique solution of (MCF) and the convergence of u^ε follows immediately. Let us first show local boundedness of u^ε.

Lemma 7.1.4 *Assume that u_0 is Lipschitz in $\mathbf{R}^2$, that is, there exists $L > 0$ such that*

$$|u_0(x) - u_0(y)| \leq L|x - y|$$

for any $x, y \in \mathbf{R}^2$. Let u^ε be the value function given in (7.1.3). *Then u^ε satisfies*

$$u_0(y) - L(|x - y|^2 + 2t)^{\frac{1}{2}} \leq u^\varepsilon(x, t) \leq u_0(y) + L(|x - y|^2 + 2t)^{\frac{1}{2}}. \tag{7.1.11}$$

for all $y \in \mathbf{R}^2$ and $t \geq 0$.

Proof Fix $y \in \mathbf{R}^2$ arbitrarily. By the Lipschitz assumption on u_0, we have $u_0(x) \leq w_0(x)$ for any $x \in \mathbf{R}^2$, where

$$w_0(x) = u_0(y) + L|x - y|.$$

We consider the game with objective function w_0 and denote its value function by w^ε. Noticing that w_0 can be reduced to the initial value as in (7.1.5) by translations, we follow the observations in the derivation of (7.1.8) and get

$$w^\varepsilon(x, t) = u_0(y) + L(|x - y|^2 + 2\varepsilon^2\ell)^{\frac{1}{2}}.$$

Since $u_0 \leq w_0$ in $\mathbf{R}^2$, in the game with objective u_0, Player I may use the same strategies as used in the game for w_0 to guarantee

$$u^\varepsilon(x, t) \leq u_0(y) + L(|x - y|^2 + 2\varepsilon^2\ell)^{\frac{1}{2}} \leq u_0(y) + L(|x - y|^2 + 2t)^{\frac{1}{2}}.$$

A symmetric argument yields the other inequality we aimed to show. □

As an immediate consequence of Lemma 7.1.4, we obtain the local boundedness of $\overline{u}$ and $\underline{u}$ in $\mathbf{R}^2 \times [0, \infty)$. Let $F = F_f$ be as in (2.6.6), that is,

$$F(p, X) = -\operatorname{trace}\left(I - \frac{p \otimes p}{|p|^2}\right) X, \quad p \neq 0, \ X \in \mathbf{S}^2.$$

Let F^*, F_* be respectively the upper and lower semicontinuous envelopes of F. Based on the dynamic programming principle, we define a nonlinear operator $H^\varepsilon : C(\mathbf{R}^2) \to C(\mathbf{R}^2)$ by

$$H^\varepsilon[\psi](x) = \min_{|v|=1} \max_{b=\pm 1} \psi\left(x + \sqrt{2}\varepsilon b v\right), \qquad x \in \mathbf{R}^2, \psi \in C(\mathbf{R}^2). \tag{7.1.12}$$

It is clear that H^ε is monotone, that is, for any $\psi_1, \psi_2 \in C(\mathbf{R}^2)$,

$$H^\varepsilon[\psi_1] \leq H^\varepsilon[\psi_2] \quad \text{whenever } \psi_1 \leq \psi_2 \ in \ \mathbf{R}^2.$$

Moreover, it is easily seen that $H^\varepsilon[\psi + c] = H^\varepsilon[\psi] + c$ in $\mathbf{R}^2$ for any $c \in \mathbf{R}$. Below we present another important property of this operator, which provides a rigorous justification of our formal argument for game convergence.

Lemma 7.1.5 (Consistency) *Let $\psi \in C^2(\mathbf{R}^2)$ and H^ε be the operator defined in* (7.1.12). *Then*

$$H^\varepsilon[\psi](x) - \psi(x) \leq -\varepsilon^2 F_*(\nabla\psi(x), \nabla^2\psi(x)) + o(\varepsilon^2), \tag{7.1.13}$$

$$H^\varepsilon[\psi](x) - \psi(x) \geq -\varepsilon^2 F^*(\nabla\psi(x), \nabla^2\psi(x)) + o(\varepsilon^2), \tag{7.1.14}$$

where the constants implicit in the error terms above are locally uniformly in $\mathbf{R}^2$.

Before proving this lemma, we first present an elementary result, which can be found in Lemma 5.2.2 in Y. Giga and Q. Liu [217].

Lemma 7.1.6 *Suppose that $p \in S^1$ and $X \in \mathbf{S}^2$. Then there exists a constant $M > 0$ depending only on the norm of X such that for any $\xi \in S^1$,*

$$|\langle Xp^\perp, p^\perp\rangle - \langle X\xi, \xi\rangle| \leq M|\langle \xi, p\rangle|, \tag{7.1.15}$$

where $p^\perp$ denotes a unit vector orthogonal to p in $\mathbf{R}^2$.

Proof Let $a = \langle \xi, p\rangle$ and $b = \langle \xi, p^\perp\rangle$. It is clear that $\xi = ap + bp^\perp$ and $a^2 + b^2 = 1$. By direct calculations, we get

$$
\begin{aligned}
\left\langle Xp^\perp, p^\perp\right\rangle - \langle X\xi, \xi\rangle &= \left\langle Xp^\perp, p^\perp\right\rangle - \left\langle X(ap+bp^\perp), (ap+bp^\perp)\right\rangle \\
&= (1-b^2)\left\langle Xp^\perp, p^\perp\right\rangle - 2ab\left\langle Xp, p^\perp\right\rangle - a^2\langle Xp, p\rangle \\
&\le a^2\left\langle Xp^\perp, p^\perp\right\rangle - 2ab\left\langle Xp, p^\perp\right\rangle - a^2\langle Xp, p\rangle .
\end{aligned}
$$

It immediately follows that

$$
|\langle Xp^\perp, p^\perp\rangle - \langle X\xi, \xi\rangle| \le M|a|,
$$

where $M > 0$ depends only on $\|X\|$. This is equivalent to (7.1.15). □

Note that here and in the sequel we do not specify the norm of a matrix $X \in \mathbf{S}^2$. One can take $\|X\|$ to be the Hilbert-Schmidt norm $\|X\|_2$, as in the preceding chapters, but our subsequent arguments and applications are unaffected if $\|\cdot\|$ is replaced by any other equivalent norm on $\mathbf{S}^2$.

Remark 7.1.7 By homogeneity, when $p \in \mathbf{R}^2$ is not a unit vector, the relation (7.1.15) still holds for all $\xi \in \mathbf{R}^2$ provided that $p^\perp$ is orthogonal to p and $|p^\perp| = |p| = |\xi|$.

Proof of Lemma 7.1.5 Let us first show (7.1.13). Let us fix $x \in \mathbf{R}^2$. Then by Taylor expansion, we have

$$
H^\varepsilon[\psi](x) = \psi(x) + \min_{|v|=1}\left\{\sqrt{2}\varepsilon|\,\langle\nabla\psi(x), v\rangle\,| + \varepsilon^2\left\langle\nabla^2\psi(x)v, v\right\rangle + o(\varepsilon^2)\right\}, \tag{7.1.16}
$$

where the constant implicit in the error term depends only on the uniform continuity of $\nabla^2\psi$ near x.

If $\nabla\psi(x) \ne 0$, then by taking v perpendicular to $\nabla\psi(x)$, we get

$$
H^\varepsilon[\psi](x) \le \psi(x) + \varepsilon^2\left\langle\nabla^2\psi(x)\frac{\nabla^\perp\psi(x)}{|\nabla\psi(x)|}, \frac{\nabla^\perp\psi(x)}{|\nabla\psi(x)|}\right\rangle + o(\varepsilon^2),
$$

which immediately yields (7.1.13). If $\nabla\psi(x) = 0$, then (7.1.16) reduces to

$$
H^\varepsilon[\psi](x) = \psi(x) + \varepsilon^2\min_{|v|=1}\left\langle\nabla^2\psi(x)v, v\right\rangle + o(\varepsilon^2), \tag{7.1.17}
$$

which also gives (7.1.13).

The proof of (7.1.14) is still based on (7.1.16). Suppose that $\nabla\psi(x) \ne 0$. Applying Lemma 7.1.6 with $\xi = v$, $p = \nabla\psi(x)/|\nabla\psi(x)|$ and $X = \nabla^2\psi(x)$, we have

$$
\left\langle\nabla^2\psi(x)v, v\right\rangle - \left\langle\nabla^2\psi(x)\frac{\nabla^\perp\psi(x)}{|\nabla\psi(x)|}, \frac{\nabla^\perp\psi(x)}{|\nabla\psi(x)|}\right\rangle \ge -M\frac{|\,\langle\nabla\psi(x), v\rangle\,|}{|\nabla\psi(x)|},
$$

where $M > 0$ depends on $\|\nabla^2\psi(x)\|$. Adopting this estimate in (7.1.16), we are led to

$$H^\varepsilon[\psi](x) - \psi(x) \geq \min_{|v|=1}\left(\sqrt{2}\varepsilon - \frac{M\varepsilon^2}{|\nabla\psi(x)|}\right)|\langle\nabla\psi(x), v\rangle| + \varepsilon^2\left\langle\nabla^2\psi(x)\frac{\nabla^\perp\psi(x)}{|\nabla\psi(x)|}, \frac{\nabla^\perp\psi(x)}{|\nabla\psi(x)|}\right\rangle + o(\varepsilon^2).$$

This clearly implies (7.1.14). When $\nabla\psi(x) = 0$, (7.1.14) is again an immediate consequence of (7.1.17). □

We next use Lemma 7.1.5 to verify that $\overline{u}$ and $\underline{u}$ are respectively a subsolution and a supersolution of (7.1.1).

Proposition 7.1.8 (Subsolution Property) *Assume that u_0 is Lipschitz in $\mathbf{R}^2$. Let u^ε be the value function as given in* (7.1.3) *and $\overline{u}$ be defined as in* (7.1.10). *Then $\overline{u}$ is a subsolution of* (7.1.1).

Proof Suppose that there exist $(x_0, t_0) \in \mathbf{R}^2 \times (0, \infty)$ and $\phi \in C^2(\mathbf{R}^2 \times (0, \infty))$ such that $u - \phi$ attains a strict maximum on $\mathbf{R}^2 \times (0, \infty)$ at (x_0, t_0). In particular, there exists $r > 0$ such that

$$\overline{u}(x_0, t_0) - \phi(x_0, t_0) > \overline{u}(x, t) - \phi(x, t)$$

for any $(x, t) \in B_r(x_0, t_0)$ with $(x, t) \neq (x_0, t_0)$. Here, as in the previous sections, $B_r(z)$ denotes the closed ball with radius r and center at z.

Then by definition of $\overline{u}$, there exist a sequence $(\tilde{x}_\varepsilon, \tilde{t}_\varepsilon) \in \mathbf{R}^2 \times (0, \infty)$, still indexed by ε for simplicity, such that $(\tilde{x}_\varepsilon, \tilde{t}_\varepsilon) \to (x_0, t_0)$, $u^\varepsilon(\tilde{x}_\varepsilon, \tilde{t}_\varepsilon) \to \overline{u}(x_0, t_0)$ as $\varepsilon \to 0$. We may take $(x_\varepsilon, t_\varepsilon)$ such that

$$u^\varepsilon(x_\varepsilon, t_\varepsilon) - \phi(x_\varepsilon, t_\varepsilon) \geq \sup_{B_r(x_0,t_0)}(u^\varepsilon - \phi) - \varepsilon^3. \tag{7.1.18}$$

It follows that $(x_\varepsilon, t_\varepsilon) \to (x_0, t_0)$ as $\varepsilon \to 0$ due to the strict maximum of $\overline{u} - \phi$ at (x_0, t_0).

Using (7.1.18) in the dynamic programming principle (7.1.4) at $(x, t) = (x_\varepsilon, t_\varepsilon)$, we have

$$\phi(x_\varepsilon, t_\varepsilon) \leq \min_{|v|=1}\max_{b=\pm 1}\phi(x_\varepsilon + \sqrt{2}\varepsilon bv, t_\varepsilon - \varepsilon^2) + \varepsilon^3.$$

Now applying (7.1.13) in Lemma 7.1.5 with

$$\psi = \phi(\cdot, t_\varepsilon - \varepsilon^2), \quad x = x_\varepsilon, \tag{7.1.19}$$

we have

$$\phi(x_\varepsilon, t_\varepsilon) - \phi(x_\varepsilon, t_\varepsilon - \varepsilon^2) \le -\varepsilon^2 F_* \left(\nabla\phi(x_\varepsilon, t_\varepsilon - \varepsilon^2), \nabla^2\phi(x_\varepsilon, t_\varepsilon - \varepsilon^2)\right) + o(\varepsilon^2).$$

It follows that

$$\varepsilon^2\phi_t(x_\varepsilon, t_\varepsilon - \varepsilon^2) + \varepsilon^2 F_* \left(\nabla\phi(x_\varepsilon, t_\varepsilon - \varepsilon^2), \nabla^2\phi(x_\varepsilon, t_\varepsilon - \varepsilon^2)\right) \le o(\varepsilon^2).$$

Dividing this inequality by ε^2 and then sending $\varepsilon \to 0$, we are led to

$$\phi_t - |\nabla\phi| \operatorname{div}\left(\frac{\nabla\phi}{|\nabla\phi|}\right) \le 0 \quad \text{at } (x_0, t_0), \tag{7.1.20}$$

as desired. □

Proposition 7.1.9 (Supersolution Property) *Assume that u_0 is Lipschitz in $\mathbf{R}^2$. Let u^ε be the value function as given in* (7.1.3) *and $\underline{u}$ be defined as in* (7.1.10). *Then $\underline{u}$ is a supersolution of* (7.1.1).

Proof Assume that there exist $(x_0, t_0) \in \mathbf{R}^2 \times (0, \infty)$ and $\phi \in C^2(\mathbf{R}^2 \times (0, \infty))$ such that $u - \phi$ attains a strict minimum on $\mathbf{R}^2 \times (0, \infty)$ at (x_0, t_0). We may follow the same argument in the proof of Proposition 7.1.8 and obtain a sequence $(x_\varepsilon, t_\varepsilon) \to (x_0, t_0)$ such that

$$\phi(x_\varepsilon, t_\varepsilon) \ge \min_{|v|=1} \max_{b=\pm 1} \phi(x_\varepsilon + \sqrt{2}\varepsilon b v, t_\varepsilon - \varepsilon^2) + \varepsilon^3.$$

We use Lemma 7.1.5 again with (7.1.19), obtaining

$$\varepsilon^2\phi_t(x_\varepsilon, t_\varepsilon - \varepsilon^2) + \varepsilon^2 F^* \left(\nabla\phi(x_\varepsilon, t_\varepsilon - \varepsilon^2), \nabla^2\phi(x_\varepsilon, t_\varepsilon - \varepsilon^2)\right) \ge o(\varepsilon^2).$$

Dividing it by ε^2 and passing to the limit as $\varepsilon \to 0$, we conclude

$$\phi_t - |\nabla\phi| \operatorname{div}\left(\frac{\nabla\phi}{|\nabla\phi|}\right) \ge 0 \quad \text{at } (x_0, t_0).$$

□

We now show that $\overline{u}$ and $\underline{u}$ satisfy the initial condition.

Proposition 7.1.10 (Initial Value) *Assume that u_0 is Lipschitz in $\mathbf{R}^2$ with Lipschitz constant $L > 0$. Let u^ε be the value function as given in* (7.1.3). *Let $\overline{u}$ and $\underline{u}$ be defined as in* (7.1.10). *Then*

$$u_0(x) - L\sqrt{2t} \le \underline{u}(x, t) \le \overline{u}(x, t) \le u_0(x) + L\sqrt{2t} \tag{7.1.21}$$

holds for any $(x, t) \in \mathbf{R}^2 \times [0, \infty)$. In particular, the following results hold:

$$\overline{u}(\cdot, 0) \leq u_0(\cdot) \leq \underline{u}(\cdot, 0) \quad in\ \mathbf{R}^2; \tag{7.1.22}$$

$$\limsup_{\delta \to 0} \left\{ \overline{u}(x, t) - \underline{u}(y, s) : x, y \in \mathbf{R}^2,\ t, s \geq 0\ with\ |x - y| \leq \delta,\ t, s \leq \delta \right\} \leq 0. \tag{7.1.23}$$

Proof The estimate (7.1.21) is a direct consequence of Lemma 7.1.4. Indeed, since (7.1.11) holds uniformly for any $\varepsilon > 0$ and u_0 is continuous, we have

$$\overline{u}(x, t) \leq u_0(x) + L(|x - y|^2 + 2t)^{\frac{1}{2}},$$

for any $x, y \in \mathbf{R}^2$ and $t \geq 0$, which reduces to $\overline{u}(x, t) \leq u_0(x) + L\sqrt{2t}$ if we let $y = x$. The proof of (7.1.21) on the lower bound of $\underline{u}$ is symmetric.

The relation (7.1.22) follows immediately. We can also use (7.1.21), combined with the Lipschitz continuity of u_0, to get

$$\overline{u}(x, t) - \underline{u}(y, s) \leq L|x - y| + L\sqrt{2t} + L\sqrt{2s} \leq L(\delta + 2\sqrt{2\delta})$$

for all $(x, t), (y, s) \in \mathbf{R}^2 \times [0, \infty)$ satisfying $|x - y| \leq \delta$ and $t, s \leq \delta$, which clearly yields (7.1.23). □

We proved (7.1.23) in order to apply the comparison principle (4.1.4) to $\overline{u}$ and $\underline{u}$ in an unbounded domain. We are now in a position to show Theorem 7.1.1.

Proof of Theorem 7.1.1 By Propositions 7.1.8 and 7.1.9, $\overline{u}$ and $\underline{u}$ are respectively a subsolution and a supersolution of (7.1.1). Thanks to Proposition 7.1.10, we can apply comparison principle Theorem 4.1.4 to deduce $\overline{u} \leq \underline{u}$ in $\mathbf{R}^2 \times [0, \infty)$. Since it is clear that $\underline{u} \leq \overline{u}$, we conclude that $\overline{u} = \underline{u}$ is a continuous solution of (7.1.1) with $\overline{u}(\cdot, 0) = \underline{u}(\cdot, 0) = u_0$ in $\mathbf{R}^2$. The locally uniform convergence of u^ε follows immediately. □

Remark 7.1.11 (Inverse of Game 1-a) We conclude this section by emphasizing that there is an *inverse game* with similar rules.
Game 1-a' We can obtain another value function by keeping the rules of games but switching the goals of both players; in other words, we take

$$\tilde{u}^\varepsilon(x, t) = \max_{|v_1|=1} \min_{b_1=\pm 1} \ldots \max_{|v_\ell|=1} \min_{b_\ell=\pm 1} u_0(y_\ell). \tag{7.1.24}$$

In this case, the DPP is

$$\tilde{u}^\varepsilon(x, t) = \max_{|v|=1} \min_{b=\pm 1} \tilde{u}^\varepsilon(x + \sqrt{2}\varepsilon b v, t - \varepsilon^2), \tag{7.1.25}$$

Following the same arguments in this section, one can show that as $\varepsilon \to 0$, $\tilde{u}^\varepsilon$ still converges to the unique solution of (MCF) locally uniformly in $\mathbf{R}^2 \times [0, \infty)$ under the same assumptions in Theorem 7.1.1.

7.1.3 Generalizations

One can consider different variants and generalizations of the game approach to mean curvature flow by changing the rules of Game 1-a.

7.1.3.1 Motion by Positive/Negative Curvature

As a first example, let us introduce the following game for motion by positive or negative curvature, which will be used in our analysis later.

Game 1-b Let the basic game setting be the same as that of Game 1-a. At the k-th step, instead of picking $v_k \in \mathbf{R}^2$ with $|v_k| = 1$, Player I chooses $v_k \in \mathbf{R}^2$ such that $|v_k| \le 1$. This change of game rules leads to a different game value given by

$$u^\varepsilon(x,t) = \max_{|v_1|\le 1} \min_{b_1=\pm 1} \ldots \max_{|v_\ell|\le 1} \min_{b_\ell=\pm 1} u_0(y_\ell). \tag{7.1.26}$$

It clearly satisfies a new DPP:

$$u^\varepsilon(x,t) = \max_{|v|\le 1} \min_{b=\pm 1} u^\varepsilon(x+\sqrt{2}\varepsilon b v, t-\varepsilon^2), \tag{7.1.27}$$

Following the proof of Theorem 7.1.1, we can show the result below.

Theorem 7.1.12 *Assume that u_0 is Lipschitz continuous in $\mathbf{R}^2$. Let u^ε be the value function associated to Game 1-b, as defined in (7.1.26). Then $u^\varepsilon \to u$ locally uniformly in $\mathbf{R}^2 \times [0,\infty)$ as $\varepsilon \to 0$, where u is the unique viscosity solution of*

$$\text{(PCF)} \quad \begin{cases} u_t - |\nabla u| \operatorname{div}\left(\dfrac{\nabla u}{|\nabla u|}\right)_+ = 0 & \text{in } \mathbf{R}^2 \times (0,\infty), \\[2ex] u(\cdot,0) = u_0 & \text{in } \mathbf{R}^2, \end{cases} \tag{7.1.28}$$

where $a_+ = \max\{a, 0\}$ for any $a \in \mathbf{R}$.

By considering the associated inverse game leading to DPP with $\max_{|v|\le 1} \min_{b=\pm 1}$ replaced by $\min_{|v|\le 1} \max_{b=\pm 1}$, one obtains the convergence of the game values to the unique viscosity solution of the Cauchy problem for

$$u_t - |\nabla u| \operatorname{div}\left(\frac{\nabla u}{|\nabla u|}\right)_- = 0 \quad \text{in } \mathbf{R}^2 \times (0,\infty)$$

instead of (7.1.28), where $a_- = \min\{a, 0\}$ for $a \in \mathbf{R}$.

7.1.3.2 Motion by a Convex/Concave Function of Curvature

It is also possible to adapt the game approach to a more general curvature motion whose normal velocity depends on a function of the curvature; namely, we consider

$$u_t - |\nabla u|\, g\left(\operatorname{div}\left(\frac{\nabla u}{|\nabla u|}\right)\right) = 0 \quad \text{in } \mathbf{R}^2 \times (0, \infty) \tag{7.1.29}$$

where $g \in C(\mathbf{R})$ is a given nondecreasing function that is convex or concave. It is not difficult to see that this parabolic equation satisfies the assumptions in Sect. 4.1 and therefore the comparison principle, Theorem 4.1.4, still applies.

Suppose that the game starts at $x \in \mathbf{R}^2$ with duration $t \geq 0$, and $\varepsilon > 0$ and $\ell \geq 0$ are set in the same manner as in Game 1-a. We take a family of nondecreasing functions $f_\varepsilon : (0, \infty) \to \mathbf{R}$, whose relation with g will be clarified later. We again let y_k denote the game position after the k-th move ($k = 1, 2 \ldots, \ell$). At the k-th step,

- Player I chooses two unit vectors $v_k, w_k \in \mathbf{R}^2$ and $s_k \in \mathbf{R}$.
- Player II chooses $b_k = \pm 1$.
- The game token moves from the position y_{k-1} to

$$y_k = \sqrt{2}\varepsilon b_k s_k v_k + \varepsilon^2 f_\varepsilon(s_k^2) w_k.$$

If Player I and Player II still aim to minimize and maximize $u_0(y_\ell)$ respectively, then the value function u^ε is defined by

$$u^\varepsilon(x, t) = \inf_{v_1, s_1} \sup_{b_1} \ldots \inf_{v_\ell, s_\ell} \sup_{b_\ell} u_0(y_\ell).$$

Note that we here used inf sup instead of min max, since the value is now optimized in an open set. The DPP in this case reads

$$u^\varepsilon(x, t) = \inf_{v, s} \sup_{b} u^\varepsilon\left(x + \sqrt{2}\varepsilon b s v + \varepsilon^2 f_\varepsilon(s^2) w, t - \varepsilon^2\right). \tag{7.1.30}$$

Assuming that u^ε is smooth with $\nabla u^\varepsilon(x, t) \neq 0$, we may apply a formal Taylor expansion for (7.1.30) and get

$$0 = \inf_{s > 0} |\nabla u| \left(s^2 \kappa[u](x, t) + f_\varepsilon(s^2)\right) - u_t^\varepsilon(x, t) + o(1)$$

for $\varepsilon > 0$ small, where $\kappa[u]$ denotes the curvature of level sets of u, i.e.,

$$\kappa[u](x, t) = \frac{1}{|\nabla u(x, t)|} \operatorname{trace}\left[\left(I - \frac{\nabla u(x, t) \otimes \nabla u(x, t)}{|\nabla u(x, t)|^2}\right) \nabla^2 u(x, t)\right]$$

if $\nabla u(x, t) \neq 0$.

In order to show, at least formally, that u^ε converges to a solution u of (7.1.29), it suffices to find f_ε that satisfies

$$\inf_{\tau>0} (c\tau + f_\varepsilon(\tau)) \to g(c) \quad \text{as } \varepsilon \to 0 \tag{7.1.31}$$

locally uniformly for $c \in \mathbf{R}$. Such f_ε actually exists provided that $g \in C(\mathbf{R})$ is nondecreasing and concave. Indeed, in this case we may find a family of functions $g_\varepsilon \in C^1(\mathbf{R})$ that are nondecreasing and strictly concave such that $g_\varepsilon \to g$ locally uniformly in $\mathbf{R}$ as $\varepsilon \to 0$. Note that for any $x \in \mathbf{R}$,

$$g_\varepsilon(x) = \inf_{a\in\mathbf{R}} \left\{g_\varepsilon(a) + g_\varepsilon'(a)(x-a)\right\}$$

holds for all $\varepsilon > 0$. Since g_ε' is strictly decreasing due to the strict concavity of g_ε, there exists an inverse function $l_\varepsilon : (0, \infty) \to (0, \infty)$ of g_ε'. Setting

$$f_\varepsilon(\tau) = g_\varepsilon(l_\varepsilon(\tau)) - l_\varepsilon(\tau)\tau, \tag{7.1.32}$$

we obtain (7.1.31), since g_ε defines a one-to-one mapping $(0, \infty) \to (0, \infty)$.

A symmetric argument works if we consider the inverse version of the game with g assumed to be convex instead. The rigorous details are omitted here.

7.2 Stationary Problems

We next consider an elliptic version of the game-theoretic interpretation. The level-set equation we consider in this section is

$$\text{(SDP)} \qquad \begin{cases} -|\nabla U| \operatorname{div}\left(\dfrac{\nabla U}{|\nabla U|}\right) = 1 & \text{in } \Omega \qquad (7.2.1) \\ U = 0 & \text{on } \partial\Omega, \qquad (7.2.2) \end{cases}$$

where $\Omega \subset \mathbf{R}^2$ is assumed to be a bounded domain. For simplicity, we still consider the two dimensional case.

The solution $U(x)$ of (SDP) can be interpreted as the *arrival time* of the front $\partial\Omega$ at the point $x \in \Omega$ under motion by curvature. Heuristically, if there exists a smooth solution U of (SDP), then by defining

$$u(x, t) = U(x) - t \quad \text{for } (x, t) \in \Omega \times [0, \infty),$$

we see, at least formally, that u satisfies (7.1.1) in $\Omega \times (0, \infty)$. We will show the existence and uniqueness of viscosity solutions $u \in C(\overline{\Omega})$ of (SDP), assuming that Ω is smooth and strictly convex. The well-posedness for a nonconvex Ω is substantially more delicate.

7.2.1 An Exit-Time Game

Game 2-a The game rules for (SDP) are essentially the same as those in Sect. 7.1.1. However, this time we do not have a fixed game duration but instead we consider an exit time problem. For any starting game position x, we still denote by y_k the game position after k-th step. Let

$$\ell = \inf\left\{k;\ y_k \in \mathbf{R}^2 \setminus \Omega \text{ under the game rules}\right\}. \tag{7.2.3}$$

Then $\ell\varepsilon^2$ stands for the first time when the game state exits from Ω under fixed strategies of both players. Suppose Player I attempts to minimize $\ell\varepsilon^2$ and Player II is to maximize the time.

The *value function* for this game is expressed as

$$U^\varepsilon(x) = \min_{|v_1|=1} \max_{b_1=\pm 1} \min_{|v_2|=1} \max_{b_2=\pm 1} \cdots \ell\varepsilon^2. \tag{7.2.4}$$

It is clear that $U^\varepsilon(x) = 0$ if $x \in \mathbf{R}^2 \setminus \Omega$. One can easily show that U^ε is bounded uniformly in ε due to the boundedness of Ω. Indeed, since any bounded domain $\Omega \subset \mathbf{R}^2$ is contained in a ball with radius diam $(\Omega)/2$, where diam (Ω) denotes the diameter of Ω, by the strategy of Player I as described in Example 7.1.3, we have

$$U^\varepsilon(x) \le \left(\frac{\operatorname{diam}(\Omega)}{2}\right)^2 \frac{\varepsilon^2}{2\varepsilon^2} \le \frac{1}{8}\operatorname{diam}(\Omega)^2.$$

The *dynamic programming principle* in this case reads

$$U^\varepsilon(x) = \min_{|v|=1} \max_{b=\pm 1} U^\varepsilon(x + \sqrt{2}\varepsilon b v) + \varepsilon^2 \qquad \text{for any } x \in \Omega. \tag{7.2.5}$$

When Ω is assumed to be strictly convex, this time optimal game provides us with an approximation of the unique viscosity solution of (SDP).

Theorem 7.2.1 *Assume that $\Omega \subset \mathbf{R}^2$ is bounded and strictly convex. Let U^ε be defined as in* (7.2.4). *Then $U^\varepsilon \to U$ uniformly in $\overline{\Omega}$ as $\varepsilon \to 0$, where U is the unique viscosity solution of* (SDP).

We again give a formal proof of Theorem 7.2.1 before proceeding to the rigorous argument. Let us assume that U^ε are smooth and converge to U with $\nabla U \neq 0$. Then by applying Taylor expansion to the right hand side of (7.2.5) at any $x \in \Omega$, we get

$$0 = \min_{|v|=1} \max_{b=\pm 1} \left\{ \sqrt{2}\varepsilon b \langle \nabla U^\varepsilon(x), v \rangle + \varepsilon^2 \left\langle \nabla^2 U^\varepsilon(x) v, v \right\rangle \right\} + \varepsilon^2 + o(\varepsilon^2)$$
$$= \min_{|v|=1} \left\{ \sqrt{2}\varepsilon \left| \langle \nabla U^\varepsilon(x), v \rangle \right| + \varepsilon^2 \left\langle \nabla^2 U^\varepsilon(x) v, v \right\rangle \right\} + \varepsilon^2 + o(\varepsilon^2),$$

which suggests that an optimal choice of v is approximately $\nabla^\perp U^\varepsilon(x)/|\nabla U^\varepsilon(x)|$. Dividing by ε^2 and letting $\varepsilon \to 0$, we are led to

$$-\left\langle \nabla^2 U(x) \frac{\nabla^\perp U(x)}{|\nabla U(x)|}, \frac{\nabla^\perp U(x)}{|\nabla U(x)|} \right\rangle = 1,$$

which is precisely (7.2.1).

7.2.1.1 A Viscosity Argument

As in the parabolic case, a rigorous justification of the formal argument can be carried out in the framework of viscosity solutions. Since U^ε are uniformly bounded in $\overline{\Omega}$, we may introduce the upper and lower relaxed limits of U^ε in $\overline{\Omega}$:

$$\overline{U} = \operatorname*{limsup^*}_{\varepsilon \to 0} U^\varepsilon, \quad \underline{U} = \operatorname*{liminf_*}_{\varepsilon \to 0} U^\varepsilon. \tag{7.2.6}$$

It is not difficult to see that $\overline{U}$ and $\underline{U}$ are respectively upper semicontinuous and lower semicontinuous in $\overline{\Omega}$. Also, it is clear that $\overline{U} \geq \underline{U}$ in $\overline{\Omega}$.

Proposition 7.2.2 *Assume that* Ω *is a bounded domain. Let* U^ε *be the value function as defined in* (7.2.4) *and* $\overline{U}, \underline{U}$ *be as in* (7.2.6). *Then* $\overline{U}$ *and* $\underline{U}$ *are respectively a subsolution and a supersolution of* (7.2.1).

Proof We only prove the supersolution part. As in the parabolic case, the verification of subsolution is easier. Assume that there exist $x_0 \in \Omega$ and $\phi \in C^2(\overline{\Omega})$ such that $\underline{U} - \phi$ attains a strict minimum in $\overline{\Omega}$ at x_0. Then by definition of $\underline{U}$, we can get $r > 0$ and a sequence $x_\varepsilon \in \Omega$ such that $x_\varepsilon \to x_0$ as $\varepsilon \to 0$ and

$$(U^\varepsilon - \phi)(x_\varepsilon) \leq \inf_{B_r(x_0)} (U^\varepsilon - \phi) + \varepsilon^3.$$

In view of (7.2.5), we thus have

$$\phi(x_\varepsilon) \geq \min_{|v|=1} \max_{b=\pm 1} \phi(x_\varepsilon + \sqrt{2}\varepsilon b v) + \varepsilon^2 - \varepsilon^3.$$

Using Lemma 7.1.5 with $\psi = \phi$ and $x = x_\varepsilon$, we deduce that

$$0 \geq -\varepsilon^2 F^*(\nabla \phi(x_\varepsilon), \nabla^2 \phi(x_\varepsilon)) + \varepsilon^2 + o(\varepsilon^2),$$

which yields $F^*(\nabla\phi(x_0), \nabla^2\phi(x_0)) \geq 1$. This is equivalent to saying that

$$-|\nabla\phi| \operatorname{div}\left(\frac{\nabla\phi}{|\nabla\phi|}\right) \geq 1 \quad \text{at } x_0$$

when $\nabla\phi(x_0) \neq 0$ and

$$\max_{|v|=1}\left\{-\left\langle\nabla^2\phi(x_0)v, v\right\rangle\right\} \geq 1$$

when $\nabla\phi(x_0) = 0$. □

We next show that $\overline{U}$ and $\underline{U}$ satisfy the Dirichlet boundary condition (7.2.2) under the assumption that Ω is strictly convex. The key is to establish short-time reachability of game trajectories to the boundary when the starting point is sufficiently close.

Proposition 7.2.3 *Assume that $\Omega \subset \mathbf{R}^2$ is a bounded, smooth and strictly convex domain. Let U^ε be the value function as defined in* (7.2.4) *and $\overline{U}$, $\underline{U}$ be as in* (7.2.6). *Then*

$$\overline{U} \leq 0 \leq \underline{U} \quad \text{on } \partial\Omega.$$

Proof The inequality $\underline{U} \geq 0$ is trivial, since $U^\varepsilon \geq 0$ in $\overline{\Omega}$. Below we only show $\overline{U} \leq 0$.

Fix $x_0 \in \partial\Omega$. Without loss of generality, we assume that $x_0 = (0, 0)$ and $\partial\Omega$ near x_0 is the graph of a nonnegative, strictly convex function $x_2 = g(x_1)$ with $g(0) = 0$. By the strict convexity of g, for any $h > 0$ small, there exists $\delta_h > 0$ such that $g(x_1) > h$ for all $|x_1| > \delta_h$, and $\delta_h \to 0$ as $h \to 0$.

Take a closed ball $B_{R_h}(y_h)$ centered at $y_h = (0, R_h + h/2) \in \mathbf{R}^2$ with

$$R_h = \frac{\delta_h^2}{h} + \frac{h}{4}. \tag{7.2.7}$$

We choose such R_h and y_h so that $\partial B_{R_h}(y_h)$ passes through points $(\pm\delta_h, h) \in \mathbf{R}^2 \setminus \Omega$ and $\operatorname{int} B_r(x_0) \cap \Omega \subset \Omega \setminus B_{R_h}(y_h)$ for some $r > 0$ small. Our choice of R_h and y_h thus yields $|x - y_h| \leq h + R_h$ for any $x \in S_h$, where S_h is set of all points $x = (x_1, x_2) \in \Omega \setminus B_{R_h}(y_h)$ satisfying $|x_1| < \delta_h$, $0 < x_2 < h$. Clearly, $S_h \subset \operatorname{int} B_{R_h+h}(y_h) \setminus B_{R_h}(y_h)$.

If we start the exit-time game at any $x \in \operatorname{int} B_r(x_0) \cap \overline{\Omega}$, then the strategy of Player I, as described in Example 7.1.3, directs the game state along the tangent directions of the concentric circles centered at y_h and forces it to leave $\operatorname{int} B_{R_h+h}(y_h)$ after $((R_h + h)^2 - R_h^2)/(2\varepsilon^2)$ steps. Consequently, the minimal exit time from Ω satisfies

$$U^\varepsilon(x) \leq \frac{(R_h + h)^2 - R_h^2}{2} = hR_h + \frac{h^2}{2}$$

for all $x \in \operatorname{int} B_r(x_0) \cap \overline{\Omega}$ and $\varepsilon > 0$. It follows from (7.2.7) that

$$\overline{U}(x_0) \le \delta_h^2 + \frac{3h^2}{4}.$$

We obtain $\overline{U}(x_0) \le 0$, in view of the convergence $\delta_h \to 0$ as $h \to 0$. □

In order to prove Theorem 7.2.1, we need the following comparison principle for (SDP), where, for broader applications, we allow the subsolution and supersolution for comparison to be unbounded.

Theorem 7.2.4 (Comparison Principle) *Assume that $\Omega \subset \mathbf{R}^n$ is a bounded domain. Let U and V be respectively an upper semicontinuous subsolution and a lower semicontinuous supersolution of* (7.2.1). *Suppose that U and V take values in* $[0, \infty]$. *If $U \le V$ on $\partial\Omega$, then $U \le V$ in $\overline{\Omega}$.*

Theorem 7.2.4 is merely a variant of the parabolic result, Theorem 4.1.1. Indeed, we may apply a Kruzhkov-type transform to convert (7.2.1) to a form that is more convenient for the proof of Theorem 7.2.4. Suppose that U and V satisfy the assumptions of Theorem 7.2.4 and let

$$\hat{U}(x) = e^{-U(x)}, \quad \hat{V}(x) = e^{-V(x)}$$

for any $x \in \overline{\Omega}$. Then by direct calculations, we easily see that $\hat{U}$ is a bounded supersolution of

$$U - |\nabla U| \operatorname{div}\left(\frac{\nabla U}{|\nabla U|}\right) = 0 \quad \text{in } \Omega.$$

Similarly, we can show that $\hat{V}$ is a bounded subsolution of the equation above. Moreover, we have $\hat{U} \ge \hat{V}$ on $\partial\Omega$. Following the same argument in the proof of Theorem 4.1.1, we get $\hat{U} \ge \hat{V}$ in $\overline{\Omega}$, which implies that $U \le V$. We omit the detailed proof of Theorem 7.2.4 and prove Theorem 7.2.1 below.

Proof of Theorem 7.2.1 In view of Proposition 7.2.2, Proposition 7.2.3 and Theorem 7.2.4, we have $\overline{U} \le \underline{U}$ in $\overline{\Omega}$. Since $\overline{U} \ge \underline{U}$ clearly holds in $\overline{\Omega}$, we deduce that $\overline{U} = \underline{U}$ in $\overline{\Omega}$, which implies the uniform convergence of U^ε. □

Remark 7.2.5 One can also consider an *inverse game* for the stationary problem similar to the one in the finite horizon case, as introduced in Remark 7.1.11.
Game 2-a' Suppose Player I and Player II respectively attempt to maximize and to minimize $\ell\varepsilon^2$ and, where ℓ is given by (7.2.3). Let

$$\tilde{U}^\varepsilon(x) = \max_{v_1} \min_{b_1} \max_{v_2} \min_{b_2} \cdots \ell\varepsilon^2. \tag{7.2.8}$$

Then $\tilde{U}^\varepsilon$ solves a new DPP, which reads

$$\tilde{U}^{\varepsilon}(x) = \max_{|v|=1} \min_{b=\pm 1} \tilde{U}^{\varepsilon}(x + \sqrt{2}\varepsilon v) + \varepsilon^2 \quad \text{for any } x \in \Omega. \tag{7.2.9}$$

The proof of the convergence $\tilde{U}^{\varepsilon}$ to the unique solution U of (SDP) is similar to that of Theorem 7.2.1. In particular, similar results to Propositions 7.2.2 and 7.2.3 hold for the inverse game.

7.2.1.2 A Verification Argument

When Ω is smooth, bounded and strictly convex, the unique solution $U(x)$ of (SDP) actually belongs to class $C^3(\overline{\Omega})$ and has only one critical point in the domain.

Lemma 7.2.6 *Assume that $\Omega \subset \mathbf{R}^2$ is a smoothly bounded, strictly convex domain. Let U be the unique solution of* (SDP). *Then the following properties of U hold.*

- *U is of class C^3 in $\overline{\Omega}$.*
- *U has only one critical point $x_* \in \Omega$, where U attains a strict maximum in $\overline{\Omega}$.*
- *$\nabla^2 U(x_*) = -I$ and $\nabla^3 U(x_*) = 0$.*

We refer the reader to Lemma 3.1 in R. V. Kohn and S. Serfaty [312] for the proof. One can make use of this result to show the convergence result in Theorem 7.2.1 by the so-called *verification argument*. In fact, we may even obtain the convergence rate.

Theorem 7.2.7 *Assume that $\Omega \subset \mathbf{R}^2$ is a smoothly bounded, strictly convex domain. Let $U \in C^3(\overline{\Omega})$ be the unique solution of (SDP). Let U^{ε} be defined as in* (7.2.4). *Then there exists $C > 0$ depending on the C^3-norm of U in $\overline{\Omega}$ such that*

$$|U^{\varepsilon}(x) - U(x)| \leq C\varepsilon$$

for all $x \in \overline{\Omega}$.

The proof of Theorem 7.2.7 consists of the applications of Proposition 7.2.8 and Proposition 7.2.9 below.

Proposition 7.2.8 *Suppose that the assumptions in Theorem 7.2.7 hold. Then there exists $C > 0$ depending on the C^3-norm of U in $\overline{\Omega}$ such that*

$$U^{\varepsilon}(x) - U(x) \leq C\varepsilon \quad \textit{for all } x \in \overline{\Omega}.$$

Proof We first show that U^{ε} is lower semicontinuous in $\overline{\Omega}$ for every $\varepsilon > 0$. Since U^{ε} takes values in $\{k\varepsilon^2\}$ $(k \in \mathbb{N})$, it is sufficient to prove that

$$\Sigma_k = \{x \in \overline{\Omega};\, U^{\varepsilon}(x) \leq k\varepsilon^2\}$$

is closed for all $k \geq 0$. The case $k = 0$ is obvious.

Suppose by induction that Σ_k is closed for some $k \geq 0$. Assume that there exists a sequence $x_j \in \Sigma_{k+1}$ such that $x_j \to x_0 \in \overline{\Omega}$. We thus have $U^\varepsilon(x_j) \leq (k+1)\varepsilon^2$ for all j. By DPP in (7.2.5) we obtain

$$U^\varepsilon(x_0) \leq \max\{U^\varepsilon(x_0 + \sqrt{2}\varepsilon v), U^\varepsilon(x_0 - \sqrt{2}\varepsilon v)\} + \varepsilon^2 \tag{7.2.10}$$

for all $v \in S^1$, and

$$U^\varepsilon(x_j) = \max\left\{U^\varepsilon(x_j + \sqrt{2}\varepsilon v^j), U^\varepsilon(x_j - \sqrt{2}\varepsilon v^j)\right\} + \varepsilon^2$$

for some $v^j \in S^1$. This implies that $x_j \pm \sqrt{2}\varepsilon v^j \in \Sigma_k$. Letting $j \to 0$ and assuming by compactness that $v^j \to v^0$ for some $v^0 \in S^1$, we have $x_0 \pm \sqrt{2}\varepsilon v^0 \in \Sigma_k$. Applying (7.2.10) with $v = v^0$, we see that $U^\varepsilon(x_0) \leq (k+1)\varepsilon^2$, which completes the proof of the closedness of Σ_{k+1}.

Let us now fix $x \in \Omega$ with $x \neq x_*$, where x_* is given in Lemma 7.2.6. We start Game 2-a from x. Since $\nabla U(x) \neq 0$, the choice $v_1 = \nabla^\perp U(x)/|\nabla U(x)|$ of Player I yields

$$U^\varepsilon(x) \leq \max_{b=\pm 1} U^\varepsilon(x + \sqrt{2}\varepsilon b v_1) + \varepsilon^2.$$

Then Player II will certainly chooses $b_1 = \pm 1$ such that $U^\varepsilon(x + \sqrt{2}\varepsilon b_1 v_1)$ is larger. Setting $y_1 = x + \sqrt{2}\varepsilon b_1 v_1$, we have

$$U^\varepsilon(x) \leq U^\varepsilon(y_1) + \varepsilon^2.$$

Since U is a C^3 solution of (7.2.1), we also have

$$\begin{aligned} U(x + \sqrt{2}\varepsilon b_1 v_1) &\leq U(x) - \varepsilon^2 F(\nabla U(x), \nabla^2 U(x)) + C\varepsilon^3 \\ &= U(x) - \varepsilon^2 + C\varepsilon^3, \end{aligned}$$

where C depends on the C^3-norm of U in $\overline{\Omega}$. It is thus clear that $y_1 \neq x_*$ since $U(y_1) < U(x_*)$ when ε is small. We may repeat the argument above, choosing $v_2 = \nabla^\perp U(y_1)/|\nabla U(y_1)|$ so that

$$U^\varepsilon(y_1) \leq U^\varepsilon(y_2) + \varepsilon^2,$$

$$U(y_2) \leq U(y_1) - \varepsilon^2 + C\varepsilon^3 \leq U(x) - \varepsilon^2 + 2C\varepsilon^3,$$

where $y_2 = y_1 + \sqrt{2}\varepsilon b_2 v_2$ for some $b_2 = \pm 1$. We continue iterating this argument until the ℓ-th step when $y_\ell \notin \Omega$. We then obtain

$$U^\varepsilon(x) \leq U^\varepsilon(y_{\ell-1}) + (\ell - 1)\varepsilon^2,$$

$$U(y_{\ell-1}) \leq U(x) - (\ell - 1)\varepsilon^2 + C(\ell - 1)\varepsilon^3. \tag{7.2.11}$$

By definition, we have $U^\varepsilon(y_{\ell-1}) \leq \varepsilon^2$. Also, it is easily seen that

$$U(y_{\ell-1}) \geq -C\varepsilon \tag{7.2.12}$$

for $\varepsilon > 0$ small due to the Lipschitz continuity of U and the Dirichlet boundary condition (7.2.2). These estimates imply that

$$\begin{aligned} U^\varepsilon(x) \leq U^\varepsilon(y_{\ell-1}) + (\ell - 1)\varepsilon^2 &\leq U(y_{\ell-1}) + C\varepsilon + \ell\varepsilon^2 \\ &\leq U(x) + \varepsilon^2 + C\varepsilon + C(\ell - 1)\varepsilon^3. \end{aligned} \tag{7.2.13}$$

In view of (7.2.11) and (7.2.12) again, we get

$$(\ell - 1)\varepsilon^2(1 - C\varepsilon) \leq U(x) + C\varepsilon,$$

which by the boundedness of U yields the uniform boundedness of $(\ell - 1)\varepsilon^2$ with respect to x. The estimate (7.2.13) therefore implies the existence of some $C > 0$ such that $U^\varepsilon(x) - U(x) \leq C\varepsilon$ for all $x \neq x_*$. By the lower semicontinuity of $U^\varepsilon - U$, the same estimate holds for all $x \in \Omega$. □

Proposition 7.2.9 *Suppose that the assumptions in Theorem 7.2.7 hold. Then there exists $C > 0$ depending on the C^3-norm of U in $\overline{\Omega}$ such that*

$$U^\varepsilon(x) - U(x) \geq -C\varepsilon \quad \textit{for all } x \in \overline{\Omega}.$$

Proof Fix $x \in \Omega$ arbitrarily. By DPP, there exists $v_1 \in \mathbf{R}^2$ with $|v_1| = 1$ such that

$$U^\varepsilon(x) = \max_{b=\pm 1} U^\varepsilon(x + \sqrt{2}\varepsilon b v_1) + \varepsilon^2.$$

We claim that there exist $C > 0$ depending on C^3-norm of U such that

$$\max_{b=\pm 1} U(x + \sqrt{2}\varepsilon b v_1) \geq U(x) - \varepsilon^2 - C\varepsilon^3 \tag{7.2.14}$$

for all $\varepsilon > 0$ small. Indeed, by Taylor expansion, we have

$$U(x + \sqrt{2}\varepsilon b v_1) \geq U(x) + \sqrt{2}b\varepsilon \left\langle \nabla U(x), v_1 \right\rangle + \varepsilon^2 \left\langle \nabla^2 U(x) v_1, v_1 \right\rangle + C\varepsilon^3,$$

which implies that

$$\max_{b=\pm 1} U(x + \sqrt{2}\varepsilon b v_1) \geq U(x) + \sqrt{2}\varepsilon | \left\langle \nabla U(x), v_1 \right\rangle | + \varepsilon^2 \left\langle \nabla^2 U(x) v_1, v_1 \right\rangle - C\varepsilon^3. \tag{7.2.15}$$

Suppose that $x \neq x_*$. This means that $\nabla U(x) \neq 0$. Then we may apply Lemma 7.1.6 with $\xi = v_1$, $p = \nabla U(x)/|\nabla U(x)|$ and $X = \nabla^2 U(x)$ to get

$$\left\langle \nabla^2 U(x) v_1, v_1 \right\rangle - \left\langle \nabla^2 U(x) \frac{\nabla^\perp U(x)}{|\nabla U(x)|}, \frac{\nabla^\perp U(x)}{|\nabla U(x)|} \right\rangle \geq -M \frac{|\langle \nabla U(x), v_1 \rangle|}{|\nabla U(x)|},$$

where $M > 0$ depends on $\|\nabla^2 U(x)\|$. Then by (7.2.15), we obtain

$$\max_{b=\pm 1} U(x + \sqrt{2}\varepsilon b v_1) - U(x) \geq \min_{|v|=1} \left(\sqrt{2}\varepsilon - \frac{M\varepsilon^2}{|\nabla U(x)|} \right) |\langle \nabla U(x), v \rangle|$$
$$+ \varepsilon^2 \left\langle \nabla^2 U(x) \frac{\nabla^\perp U(x)}{|\nabla U(x)|}, \frac{\nabla^\perp U(x)}{|\nabla U(x)|} \right\rangle - C\varepsilon^3.$$

If $|\nabla U(x)| \geq M\varepsilon$, then we easily deduce (7.2.14) for $\varepsilon > 0$ small, since U is the solution of (7.2.1).

It remains to discuss the case when $|\nabla U(x)| \leq M\varepsilon$, which also includes the case $x = x_*$. In this case, x must be close to x_* where U attains the unique maximum in $\overline{\Omega}$. In fact, since U is C^3 and $\nabla^2 U(x_*) = -I$, we have

$$\frac{1}{2}|x - x_*| \leq |\nabla U(x)| \leq M\varepsilon.$$

Hence, by (7.2.15), we get

$$\max_{b=\pm 1} U(x + \sqrt{2}\varepsilon b v_1) \geq U(x) + \varepsilon^2 \left\langle \nabla^2 U(x_*) v_1, v_1 \right\rangle - 2C\varepsilon^3.$$

Using the fact that $\nabla^2 U(x_*) = -I$, we end up with (7.2.14) again.

Our claim above enables us to find $y_1 = x + \sqrt{2}\varepsilon b_1 v_1$, where $b_1 = \pm 1$ is chosen to achieve the maximum in (7.2.14). In particular, we have

$$U^\varepsilon(x) \geq U^\varepsilon(y_1) + \varepsilon^2, \quad U(x) \leq U(y_1) + \varepsilon^2 + C\varepsilon^3.$$

Let us iterate this argument until it stops at the first exit moment $\ell\varepsilon^2$. We thus get

$$U^\varepsilon(x) \geq U^\varepsilon(y_\ell) + \ell\varepsilon^2 \geq \ell\varepsilon^2, \tag{7.2.16}$$
$$U(x) \leq U(y_{\ell-1}) + (\ell - 1)\varepsilon^2 + C(\ell - 1)\varepsilon^3.$$

Noticing that $U(y_{\ell-1}) \leq C\varepsilon$, we have

$$U(x) \leq U^\varepsilon(x) + C\varepsilon + C(\ell - 1)\varepsilon^3$$

for all $x \in \Omega$. We complete the proof by remarking that $\ell\varepsilon^2$ is bounded uniformly in $\varepsilon > 0$ thanks to (7.2.16) and the uniform boundedness of U^ε. □

7.2.2 *Loss of Comparison Principle*

We next study (SDP) when $\Omega \subset \mathbf{R}^2$ is not necessarily convex. The main difficulty in this case is that $U^\varepsilon(x)$ does not tend to 0 when x gets close to a point on the concave portion of $\partial\Omega$.

7.2.2.1 Discontinuity of Game Values

Because of the loss of convexity of Ω, the half relaxed limits of U^ε as $\varepsilon \to 0$ may not satisfy the Dirichlet boundary condition (7.2.2) in the strict sense. Also, the relation $\overline{U} \le \underline{U}$ may not hold in $\overline{\Omega}$, as is observed in the example below.

Example 7.2.10 Suppose that Ω is a bounded domain in $\mathbf{R}^2$ that is locally expressed near $0 \in \partial\Omega$ by

$$x_2 > h(x_1) := \begin{cases} (x_1 - 1)^2 - 1 & \text{if } 0 \le x_1 < 1, \\ (x_1 + 1)^2 - 1 & \text{if } -1 < x_1 < 0, \end{cases}$$

as shown in Fig. 7.1. Let O be the origin $x = 0$.

It is obvious that $\underline{U} = 0$ by the definition that $U^\varepsilon = 0$ on $\partial\Omega$ and $U^\varepsilon \ge 0$ in $\overline{\Omega}$. We also observe that $\overline{U}(0) > 0$. Indeed, there exists $c > 0$ such that $U^\varepsilon(x) > c$ for any $\varepsilon > 0$ and any $x = (0, x_2)$ with $x_2 > 0$ small. On the other hand, since two strictly convex pieces of $\partial\Omega$ meet at O, we have $\overline{U}(x) = 0$ for any $x = (x_1, h(x_1))$ with $|x_1|$ sufficiently small. This follows from the proof of Proposition 7.2.3. We therefore conclude that the origin O is a discontinuous point of $\overline{U}$, and U^ε does not converge uniformly to any continuous function in $\overline{\Omega}$ in this case.

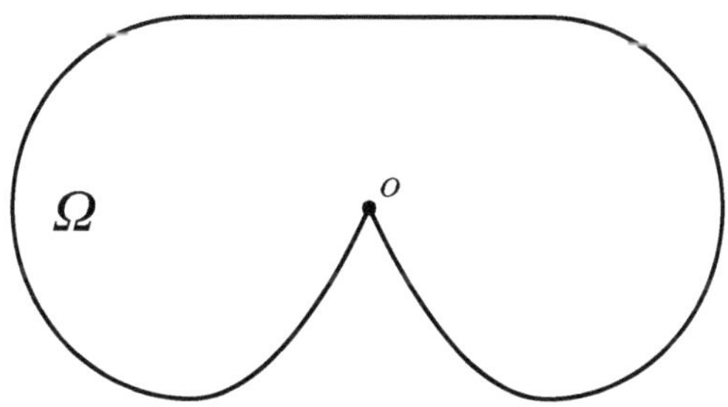

Fig. 7.1 A domain Ω such that $\overline{U}$ and $\underline{U}$ do not agree at the point O

7.2.2.2 Weak Comparison Principle in a Star-Shaped Domain

In order to study the game in a nonconvex domain, we thus need to interpret the Dirichlet condition in a weak sense.

Definition 7.2.11 (Boundary Condition in the Viscosity Sense)

(i) A bounded upper semicontinuous function U in $\overline{\Omega}$ is called a subsolution of (SDP) with the boundary condition (7.2.2) interpreted in the viscosity sense if

(a) U is a subsolution of (7.2.1) and
(b) whenever there exist $x_0 \in \partial\Omega$ and $\phi \in C^2(\mathbf{R}^2)$ such that $U - \phi$ attains a (strict) maximum in $\overline{\Omega}$ at x_0, we have either $U(x_0) \leq 0$ or the following condition holds:

$$-|\nabla\phi| \operatorname{div}\left(\frac{\nabla\phi}{|\nabla\phi|}\right) \leq 1 \quad \text{holds at } x_0$$

provided that $\nabla\phi(x_0) \neq 0$ and

$$\min_{|v|=1}\left\{-\left\langle \nabla^2\phi(x_0)v, v\right\rangle\right\} \leq 1.$$

provided that $\nabla\phi(x_0) = 0$.

(ii) A bounded lower semicontinuous function U in $\overline{\Omega}$ is called a supersolution of (SDP) with (7.2.2) in the viscosity sense if

(a) U is a supersolution of (7.2.1) and
(b) whenever there exist $x_0 \in \partial\Omega$ and $\phi \in C^2(\mathbf{R}^2)$ such that $U - \phi$ attains a (strict) minimum in $\overline{\Omega}$ at x_0, we have either $U(x_0) \geq 0$ or the following condition holds:

$$-|\nabla\phi| \operatorname{div}\left(\frac{\nabla\phi}{|\nabla\phi|}\right) \geq 1 \quad \text{holds at } x_0$$

provided that $\nabla\phi(x_0) \neq 0$ and

$$\max_{|v|=1}\left\{-\left\langle \nabla^2\phi(x_0)v, v\right\rangle\right\} \geq 1.$$

provided that $\nabla\phi(x_0) = 0$.

Proposition 7.2.12 *Assume that* Ω *is a bounded domain. Let* U^ε *be the value function as defined in* (7.2.4) *and* $\overline{U}, \underline{U}$ *be as in* (7.2.6). *Then* $\overline{U}$ *and* $\underline{U}$ *are respectively a subsolution and a supersolution of* (SDP) *with the boundary condition* (7.2.2) *interpreted in the viscosity sense.*

Proof Since we have shown in Proposition 7.2.2 that $\overline{U}$ and $\underline{U}$ are respectively a subsolution and a supersolution of (7.2.1) in Ω, it suffices to verfiy that $\overline{U}$ and $\underline{U}$ satisfy the boundary condition in the viscosity sense. The verification for the part with $\underline{U}$ is actually trivial since $U^\varepsilon \geq 0$ in $\overline{\Omega}$, which implies that $\underline{U} \geq 0$ in $\overline{\Omega}$.

It remains to show that $\overline{U}$ is a subsolution on $\partial\Omega$. Suppose that there exists $x_0 \in \partial\Omega$ and $\phi \in C^2(\mathbf{R}^2)$ such that $\overline{U} - \phi$ attains a strict maximum in $\overline{\Omega}$ at $x_0 \in \partial\Omega$ with $\overline{U}(x_0) > 0$. Then by definition of $\overline{U}$, we may take $r > 0$ and a sequence in $\overline{\Omega}$ still indexed by x_ε such that $x_\varepsilon \to x_0$ as $\varepsilon \to 0$ and

$$(U^\varepsilon - \phi)(x_\varepsilon) \geq \inf_{B_r(x_0)} (U^\varepsilon - \phi) - \varepsilon^3.$$

We thus have $x_\varepsilon \in \Omega$ due to the definition $U^\varepsilon = 0$ on $\partial\Omega$. This observation allows us to obtain (7.2.5) with $x = x_\varepsilon$.

The rest of the proof is omitted, since it resembles the proof of Proposition 7.2.2. □

Remark 7.2.13 It is possible to discuss the inverse game in a nonconvex domain as well. In fact, for the value function $\tilde{U}^\varepsilon$ defined as (7.2.8), one can obtain a result similar to Proposition 7.2.12, that is, the relaxed limits of $\tilde{U}^\varepsilon$ are respectively a subsolution and a supersolution of (SDP).

In general we cannot expect the comparison principle in the usual sense to hold in $\overline{\Omega}$, as shown in Example 7.2.10. We will see in Example 7.2.16 that $\overline{U} \leq \underline{U}$ may fail to hold not only on the boundary of Ω but also in its interior. It is possible to prove a *weak comparison principle* that allows discontinuous solutions. Here follows such a result under a star-shapedness assumption on Ω.

Theorem 7.2.14 (Weak Comparison Theorem) *Assume that $\Omega \subset \mathbf{R}^2$ is a star-shaped domain in the sense that there exist $x_0 \in \Omega$, $0 < \lambda_0 < 1$ and $C > 0$ such that $\overline{\Omega}_\lambda \subset \Omega$ holds for any $\lambda \in (\lambda_0, 1)$, where*

$$\Omega_\lambda := \{\lambda y + (1 - \lambda)x_0;\ y \in \Omega\}. \tag{7.2.17}$$

Let U and V be respectively a subsolution and a supersolution of (SDP) *with the Dirichlet boundary condition* (7.2.2) *interpreted in the viscosity sense. Then $U_* \leq V$ and $U \leq V^*$ in Ω.*

Proof We first claim that $V \geq 0$ in $\overline{\Omega}$. Indeed, since V is lower semicontinuous in $\overline{\Omega}$, it attains a minimum at some $\overline{x} \in \overline{\Omega}$. Taking the test function $\phi \equiv V(\overline{x})$ in Definition 7.2.11(ii), we see that the viscosity inequalities associated to the equation cannot hold at $\overline{x}$. We therefore have $\overline{x} \in \partial\Omega$ and $V(x_0) \geq 0$, which immediately yields $V \geq 0$ in $\overline{\Omega}$. Moreover, since the pointwise maximum of two subsolutions is still a subsolution, we can consider $\max\{U, 0\}$ in place of U and assume that $U \geq 0$ in Ω. By translation, we may also assume that Ω is star-shaped with respect to 0, that is, $\overline{\Omega}_\lambda \subset \Omega$ holds for $\lambda \in (\lambda_0, 1)$ and Ω_λ given by (7.2.17) with $x_0 = 0$.

Take $0 < \delta < 1$ arbitrarily and set $U_\delta = \delta U$ in $\overline{\Omega}$. It is not difficult to show that U_δ is a subsolution of

$$- |\nabla U| \operatorname{div}\left(\frac{\nabla U}{|\nabla U|}\right) = \delta \tag{7.2.18}$$

in Ω with $U_\delta \leq 0$ on $\partial\Omega$ in the viscosity sense. It then suffices to show $U_\delta \leq V$ in $\overline{\Omega}$ for any such δ. For simplicity, we write U instead of U_δ below. For any $\lambda \in (\lambda_0, 1)$, let

$$U^\lambda(x) = \lambda^2 U(x/\lambda) \qquad \text{for } x \in \overline{\Omega}_\lambda.$$

By the star-shapedness assumption on Ω, we have $\overline{\Omega}_\lambda \subset \Omega$. Due to the homogeneity of the operator, U^λ is still a subsolution of (7.2.18) in Ω_λ with $U^\lambda \leq 0$ on $\partial\Omega_\lambda$ in the viscosity sense. Let us extend U^λ to an upper semicontinuous function $\tilde{U}^\lambda$ in $\overline{\Omega}$ by setting

$$\tilde{U}^\lambda(x) = \begin{cases} U^\lambda(x) & x \in \overline{\Omega}_\lambda \\ 0 & x \in \overline{\Omega} \setminus \overline{\Omega}_\lambda. \end{cases}$$

One can show that $\tilde{U}^\lambda$ satisfies the subsolution property of (7.2.18) in Ω. This clearly holds for tests in Ω_λ and $\Omega \setminus \overline{\Omega}_\lambda$. Let us then discuss the case when there exist $\phi \in C^2(\overline{\Omega})$ and $x_0 \in \partial\Omega_\lambda$ such that $\tilde{U}^\lambda - \phi$ attains a maximum in $\overline{\Omega}$ at x_0. If $\tilde{U}^\lambda(x_0) = 0$, then we have $\nabla\phi(x_0) = 0$ and $\nabla^2\phi(x_0) \geq O$ due to the fact that $\tilde{U}^\lambda \geq 0$ in Ω, which immediately yields the desired inequality

$$\min_{|v|=1} \left\{ -\left\langle \nabla^2\phi(x_0)v, v \right\rangle \right\} \leq \delta.$$

If $\tilde{U}^\lambda(x_0) > 0$, then we consider ϕ a test function for $\tilde{U}^\lambda$ in $\overline{\Omega}_\lambda$ at the boundary point $x_0 \in \partial\Omega_\lambda$. Using the condition (b) in Definition (7.2.11)(i), we see that the viscosity inequalities for ϕ still hold at x_0. The proof of the subsolution property of $\tilde{U}^\lambda$ in Ω is complete now.

We next prove that $\tilde{U}^\lambda \leq V$ in $\overline{\Omega}$ for any fixed $\lambda \in (\lambda_0, 1)$. Suppose by contradiction that $\tilde{U}^\lambda - V \geq \mu$ somewhere in $\overline{\Omega}$ for a certain constant $\mu > 0$. Then for any $\varepsilon > 0$,

$$\Phi_\varepsilon(x, y) = \tilde{U}^\lambda(x) - V(y) - \zeta(x, y)$$

attains a positive maximum in $\overline{\Omega} \times \overline{\Omega}$ at some $(x_\varepsilon, y_\varepsilon) \in \overline{\Omega}_\lambda \times \overline{\Omega}$, where we take

$$\zeta(x, y) = \frac{|x - y|^4}{\varepsilon} \quad \text{for } x, y \in \overline{\Omega}.$$

It follows that $\tilde{U}^\lambda(x_\varepsilon) > V(y_\varepsilon) + \mu \geq \mu$ for any $\varepsilon > 0$ and $|x_\varepsilon - y_\varepsilon| \leq O(\varepsilon^{\frac{1}{4}})$ as $\varepsilon \to 0$. By taking a subsequence still indexed by ε, we get $x_\varepsilon, y_\varepsilon \to \hat{x} \in \overline{\Omega}$ as $\varepsilon \to 0$. Since $\tilde{U}^\lambda = 0 \leq V$ on $\overline{\Omega} \setminus \overline{\Omega}_\lambda$, we get $\hat{x} \in \overline{\Omega}_\lambda$, which by the star-shapedness of Ω implies $x_\varepsilon, y_\varepsilon \in \Omega$ for $\varepsilon > 0$ sufficiently small.

By Theorem 4.3.2, for any $\sigma > 0$ there exist $(\xi_\varepsilon, X_\varepsilon) \in \overline{J}^{2,+}\tilde{U}^\lambda(x_\varepsilon)$, $(\eta_\varepsilon, Y_\varepsilon) \in \overline{J}^{2,-}V(y_\varepsilon)$ such that

$$\xi_\varepsilon = \nabla_x \zeta(x_\varepsilon, y_\varepsilon) = -\nabla_y \zeta(x_\varepsilon, y_\varepsilon) = \eta_\varepsilon$$

and

$$\begin{pmatrix} X_\varepsilon & 0 \\ 0 & -Y_\varepsilon \end{pmatrix} \leq A_\varepsilon + \sigma A_\varepsilon^2, \tag{7.2.19}$$

where $A_\varepsilon = \nabla^2 \zeta(x_\varepsilon, y_\varepsilon)$. The latter relation implies that $X_\varepsilon \leq Y_\varepsilon$ by a direct calculation. We next discuss two distinct cases below:

Case 1 Suppose that $x_\varepsilon \neq y_\varepsilon$. Then we have $\xi_\varepsilon = \eta_\varepsilon \neq 0$. Since we have shown that $\tilde{U}^\lambda$ is a subsolution of (7.2.18) in Ω, we obtain

$$-\operatorname{trace}\left[\left(I - \frac{\xi_\varepsilon \otimes \xi_\varepsilon}{|\xi_\varepsilon|^2}\right) X_\varepsilon\right] \leq \delta.$$

On the other hand, applying to V the definition of supersolutions of (7.2.1), we get

$$-\operatorname{trace}\left[\left(I - \frac{\eta_\varepsilon \otimes \eta_\varepsilon}{|\eta_\varepsilon|^2}\right) Y_\varepsilon\right] \geq 1.$$

Taking the difference between these inequalities yields a contradiction.

Case 2 Suppose that $x_\varepsilon = y_\varepsilon$. In this case, $\xi_\varepsilon = \eta_\varepsilon = 0$. Moreover, by (7.2.19), we get $Y_\varepsilon \geq O$. Hence, we can adopt the definition of viscosity supersolutions for V to derive another contradiction.

Hence, we conclude the proof of $\tilde{U}^\lambda \leq V$ in $\overline{\Omega}$ for every $\lambda \in (\lambda_0, 1)$. This implies that $\lambda^2 U(x) \leq V(\lambda x)$ for any $x \in \overline{\Omega}$ and $\lambda \in (\lambda_0, 1)$, and in particular $U(x) \leq V^*(x)$. A symmetric argument yields $U_* \leq V$ in $\overline{\Omega}$. $\square$

As an immediately consequence of Proposition 7.2.12 and Theorem 7.2.14, we conclude that when Ω is star-shaped, the family U^ε admits a unique, possibly discontinuous limit U satisfying $(U^*)_* = U_*$ and $(U_*)^* = U^*$ in $\overline{\Omega}$.

7.2.2.3 Failure of Weak Comparison Principle in General Domains

One may wonder what happens if Ω is not star-shaped. In this case, we cannot expect even the weak comparison principle to hold in general. To see this, we consider an (outer) approximation of Ω given by

$$\Omega^\varepsilon = \{x \in \mathbf{R}^2;\, d(x, \overline{\Omega}) < \varepsilon\}. \tag{7.2.20}$$

We may consider the same exit-time game, or its inverse version, played in Ω^ε. Let us focus on the inverse game in Ω^ε and denote by V^ε (rather than $\tilde{V}^\varepsilon$ for convenience) its value function in Ω^ε. It is easily seen that an analogue of the dynamic programming principle (7.2.5) holds for the modified game, namely,

$$V^\varepsilon(x) = \max_{|v|=1} \min_{b=\pm 1} V^\varepsilon(x + \sqrt{2}\varepsilon v) + \varepsilon^2 \qquad \text{for any } x \in \Omega^\varepsilon. \tag{7.2.21}$$

Define the relaxed limits of V^ε in $\overline{\Omega}$ by

$$\overline{V} = \operatorname*{limsup^*}_{\varepsilon \to 0} V^\varepsilon, \quad \underline{V} = \operatorname*{liminf_*}_{\varepsilon \to 0} V^\varepsilon. \tag{7.2.22}$$

It turns out that, even for a general domain Ω, $\overline{V}$ and $\underline{V}$ are still a subsolution and a supersolution of (SDP) respectively.

Proposition 7.2.15 *Assume that $\Omega \subset \mathbf{R}^2$ is a bounded open set. Let Ω^ε be an open set defined by* (7.2.20). *Let V^ε be the value function of the exit-time game in Ω^ε and $\overline{V}, \underline{V}$ be given by* (7.2.22). *Then $\overline{V}$ and $\underline{V}$ are respectively a subsolution and a supersolution of* (SDP).

Despite the additional approximation of Ω in the game setting, the proof, omitted here, is essentially the same as that of Proposition 7.2.12, taking Remark 7.2.13 into account. The example below then shows that the weak comparison principle fails to hold in general.

Example 7.2.16 Let $\Omega \subset \mathbf{R}^2$ be an open set of a *figure eight* shape, that is,

$$\Omega = \operatorname{int} B_1(P_1) \cup \operatorname{int} B_1(P_2)$$

with $P_1 = (-1, 0)$, $P_2 = (1, 0)$. Note that the exit-time game in Ω is essentially decomposed into two independent games played in the two disks. It follows that the corresponding value function U^ε satisfies $\overline{U}(x) \le \delta$ for any $\delta > 0$ when $|x|$ is sufficiently small.

On the other hand, we may consider the modified game described above in the approximating domain Ω^ε defined by (7.2.20). We first claim that

$$V^\varepsilon(0) \ge \frac{1}{8} \qquad \text{for all } \varepsilon > 0. \tag{7.2.23}$$

Indeed, starting from the origin $x = 0$, Player I may repeatedly choose $v = (1, 0)$ before the distance between the game state and either P_1 or P_2 is less than $\sqrt{2}\varepsilon$. If, for instance, the position moves close to P_1, Player I can then adopt the strategy described in Example 7.1.3, following the tangential directions of the circles centered at P_1. This strategy guarantees that the exit time is close to $1/2$ for all $\varepsilon > 0$ small, independently of the choices made by Player II. This yields (7.2.23).

Moreover, the estimate (7.2.23) can be extended to a cone in $\overline{\Omega}$ near the origin. Indeed, we have

$$V^\varepsilon(x) \geq \frac{1}{8} \quad \text{for any } x \in E \text{ and any } \varepsilon > 0 \text{ small,} \tag{7.2.24}$$

where

$$E = \left\{ x = (x_1, x_2) \in \mathbf{R}^2;\ |x| \leq \frac{1}{2},\ |x_2| \leq |x_1| \right\}.$$

For each $x \in E$ with $x \neq 0$, Player I initially choose directions v parallel to x until the game position crosses the boundary of the square region $\{(x_1, x_2) : |x_1| + |x_2| \leq 1\}$. Once the position leaves this square, Player I can immediately switch to tangential directions along the appropriate concentric circles. Regardless of Player II's responses, the exit time is then bounded below by $1/8$, uniformly for all ε small. The estimate (7.2.24) then implies

$$\overline{V}_*(x) > \underline{U}(x)$$

for all $x \in E$ sufficiently close to the origin. Recall that $\overline{V}$ and $\underline{U}$ are respectively a subsolution and a supersolution of (SDP). This shows that the weak comparison principle fails in such Ω.

In this example, Ω is not a domain in the usual sense, since it is not connected. Without affecting the key estimates above, one can slightly modify the region to make it connected, for instance by adding a thin connecting path between the two disks.

7.2.2.4 A Modified Exit-Time Game

We recall that Game 1-b is a modified version of Game 1-a with $|v| = 1$ replaced by $|v| \leq 1$ in the choices of Player I. Let us make the same change for the exit-time game.

Game 2-b Let the basic game setting be the same as that of Game 2-a. However, at the k-th step, instead of picking $v_k \in R^2$ with $|v_k| = 1$, Player I chooses $v_k \in \mathbf{R}^2$ such that $|v_k| \leq 1$.

This modified game gives rise to an approximation of (SDP) as well. Indeed, the value function in this case is

$$U^{\varepsilon}(x) = \min_{|v_1|\leq 1} \max_{b_1=\pm 1} \min_{|v_2|\leq 1} \max_{b_2=\pm 1} \cdots \ell\varepsilon^2, \tag{7.2.25}$$

which clearly satisfies that

$$U^{\varepsilon}(x) = \min_{|v|\leq 1} \max_{b=\pm 1} U^{\varepsilon}(x + \sqrt{2}\varepsilon b v) + \varepsilon^2. \tag{7.2.26}$$

Based on Taylor expansion of the dynamic programming equation above, we may follow the proofs of Proposition 7.2.2 and Proposition 7.2.12 to show that the relaxed limits $\overline{U}$ and $\underline{U}$ of U^{ε} in (7.2.25) are respectively a subsolution and a supersolution of

$$\begin{cases} -|\nabla U| \operatorname{div}\left(\dfrac{\nabla U}{|\nabla U|}\right)_+ = 1 & \text{in } \Omega \\ U = 0 & \text{on } \partial\Omega, \end{cases} \tag{7.2.27}$$

with the Dirichlet boundary condition interpreted in the viscosity sense. Note that the Dirichlet problem for (7.2.27) above actually coincides with (SDP), since (7.2.27) is clearly equivalent to (7.2.1). We therefore obtain the following result.

Proposition 7.2.17 *Assume that Ω is a bounded domain. Let U^{ε} be the value function defined in* (7.2.25) *and $\overline{U}, \underline{U}$ be its relaxed limits given by* (7.2.6). *Then $\overline{U}$ and $\underline{U}$ are respectively a subsolution and a supersolution of* (SDP) *with the boundary condition* (7.2.2) *interpreted in the viscosity sense.*

Replacing $|v| = 1$ by $|v| \leq 1$ in the original game rules enables us to obtain a subsolution and a supersolution of (SDP). In general, Game 2-a and Game 2-b may yield different value functions. However, when $\Omega \subset \mathbf{R}^2$ is star-shaped, both value functions have the same limit as $\varepsilon \to 0$, thanks to Theorem 7.2.14.

7.2.3 Boundary Regularity and Game Convergence

Example 7.2.16 shows that the comparison principle may fail, even in a weak sense, when Ω is non-convex. However, the domain Ω in that example is not smooth, and it remains unclear whether a weak comparison principle holds in general for smooth non-convex domains. On the other hand, the convergence of game values may be established without relying on such a weak comparison principle. In this section, assuming that $\partial\Omega$ is the graph of a smooth periodic function on $\mathbf{R}$, we show that there exists a continuous solution U of (SDP) that can be approximated by the value function of Game 2-b.

It turns out that Game 2-b is closely connected to the arrival time for motion by positive curvature, which is expressed in the level-set formulation in (PCF). In fact, we have the following observation regarding their relationship.

Lemma 7.2.18 (Connection Between Evolution and Stationary Games) *Let $\Omega \subset \mathbf{R}^2$ be a bounded domain. Let u_0 be a Lipschitz function in $\mathbf{R}^2$ satisfying*

$$\partial\Omega = \{x \in \mathbf{R}^2;\, u_0(x) = 0\} \quad \textit{and} \quad \Omega = \{x \in \mathbf{R}^2;\, u_0(x) < 0\}. \tag{7.2.28}$$

Let u be the unique viscosity solution of (PCF). *Suppose that U^ε is the value function associated to Game 2-b and $\overline{U}, \underline{U}$ are its relaxed limits given by* (7.1.10). *Then for any $x \in \overline{\Omega}$ and $t \geq 0$,*

(i) $u(x,t) > 0$ *implies* $t \geq \overline{U}(x)$;
(ii) $u(x,t) < 0$ *implies* $t \leq \underline{U}(x)$.

Proof Fix $x_0 \in \overline{\Omega}$ and $t \geq 0$ arbitrarily.

(i) Our goal is to show $\overline{U}(x_0) \leq t$ under the assumption that $u(x_0,t) > 0$. By Theorem 7.1.12, we may choose $\delta > 0$ such that $u^\varepsilon(x,t) > \delta$ for all $x \in \operatorname{int} B_\delta(x_0)$ and $\varepsilon > 0$ small. This implies that there exist a strategy of Player I such that for any strategy of Player II, $u_0(y_\ell) > \delta$, and hence $y_\ell \notin \overline{\Omega}$. As a result, by definition, $U^\varepsilon(x) < t$ for all $x \in \operatorname{int} B_\delta(x_0)$ and all $\varepsilon > 0$ small, which yields $\overline{U}(x_0) \leq t$.
(ii) We now prove $\underline{U}(x_0) \geq t$ provided that $u(x_0,t) < 0$. Using (7.1.12) again, we have $\delta > 0$ such that $u^\varepsilon(x,t) < -\delta$ for all $x \in \operatorname{int} B_\delta(x_0)$ and $\varepsilon > 0$. We thus can ensure the existence of a strategy S_{II} of Player II, under which $u_0(y_\ell) < -\delta$ and $y_\ell \in \Omega$ for any strategy of Player I. We claim that $y_\ell \in \Omega$ for all $s \leq t$. Suppose otherwise: there exists $s_0 \leq t$ such that the game position exits Ω. Then Player I may choose the trivial control $v = 0$ for the remainder of the finite-horizon game, which would force $y_\ell \notin \Omega$ at the terminal time. This contradicts the property of S_{II} that $y_\ell \in \Omega$ for any strategy of Player I. Therefore, under the strategy S_{II}, the first exit time satisfies $U^\varepsilon(x_0) > t$ for all $\varepsilon > 0$. Consequently, we conclude that $\underline{U}(x_0) \geq t$.

□

The result above shows that the asymptotic behavior of the exit time $U^\varepsilon(x)$ as $\varepsilon \to 0$ is determined by the first arrival time of the positive curvature flow at x, namely the time t at which $u(x,t) = 0$. For this reason, we now turn to a detailed study of the first arrival time for the parabolic equation (7.1.28).

7.2.3.1 Waiting Time Effect of Positive Curvature Flow

The evolution equation (7.1.28) features a distinctive phenomenon that each level set of the solution u with initial value has the following behavior: its convex parts moves immediately while a concave part stays at the initial position for a while before starting to evolve. The latter phenomenon is called *waiting time* effect. It was conjectured in R. V. Kohn and S. Serfaty [312] that for any nonconvex level curve in the plane,

- there exists a free boundary separating the moving convex part and the stationary concave part;
- the concave part decreases monotonically in size;
- any part of the curve never stops once it starts to move.

It is known that, in general, the answers are negative without regularity assumptions on the initial data beyond Lipschitz continuity; see Example 7.2.10 and another example due to G. Barles and F. Da Lio, presented in Appendix C.3 of R. V. Kohn and S. Serfaty [312].

Let us give a more precise definition of waiting time of the boundary $\partial\Omega$.

Definition 7.2.19 Suppose that u_0 is a Lipschitz function in $\mathbf{R}^2$ satisfying (7.2.28). Let u be the unique viscosity solution of (PCF). We define

$$T_0(x) = \sup\{t \geq 0;\ u(x, t) = 0\} \quad \text{for } x \in \partial\Omega \tag{7.2.29}$$

Since $u(x, T_0(x)) = 0$ due to the continuity of u, by Lemma 7.2.18 we observe that

$$\overline{U}(x) \leq T_0(x) \leq \underline{U}(x)$$

for all $x \in \partial\Omega$, where $\overline{U}$ and $\underline{U}$ are relaxed limits of the value function of Game 2-b. In other words, the limit of game value u^ε essentially satisfies, in a strict sense, a continuous inhomogeneous Dirichlet condition

$$U = T_0 \quad \text{on } \partial\Omega \tag{7.2.30}$$

rather than the homogeneous one (7.2.2) interpreted in the viscosity sense. Hence, in order to show the convergence of U^ε, it suffices to show that the waiting time T_0 is continuous on $\partial\Omega$. Once this continuity is established, the upper and lower relaxed limits of U^ε can be characterized, respectively, as a subsolution and a supersolution of the elliptic equation (7.2.1) with the classical boundary condition (7.2.30), instead of the generalized boundary condition (7.2.2). The convergence of U^ε then follows immediately from the comparison principle, Theorem 7.2.4.

7.2.4 A Special Case with Smooth Graph-Like Boundary

We below show that the continuity of T_0 on $\partial\Omega$ does hold in the following special case. Let us consider the graph-like solution of (7.1.28); in other words, suppose that

$$u(x,t) = w(x_1,t) - x_2, \quad u_0(x,t) = w_0(x_1,t) - x_2$$

for any $x = (x_1, x_2) \in \mathbf{R}^2$. In the rest of this section, we use x instead of x_1 to denote the space variable. Moreover, we impose the following assumptions on w_0.

(A1) w_0 is a periodic Lipschitz function on $\mathbf{R}$ with an open period $I = (a, b)$ satisfying

$$\{x \in [a,b];\, w_0(x) = M\} = \{a, b\}, \tag{7.2.31}$$

where we take $M := \max_{x \in \mathbf{R}} w_0(x)$.

Denote by Ω_0 the epigraph of u_0 in I, i.e.,

$$\Omega_0 := \{(x,y) \in \mathbf{R}^2;\, y > w_0(x),\, x \in I\}. \tag{7.2.32}$$

In this setting, the positive curvature flow is reduced to the following equation:

$$\begin{cases} w_t - \dfrac{(w_{xx})_+}{1 + w_x^2} = 0 & \text{in } \mathbf{R} \times (0, \infty), \qquad (7.2.33) \\ w(x,0) = w_0(x) & \text{in } \mathbf{R}. \qquad (7.2.34) \end{cases}$$

Hereafter, we say that a function w is a subsolution or supersolution of a Cauchy problem such as (7.2.33)–(7.2.34), or of a similar Cauchy–Dirichlet problem, if it satisfies the viscosity inequalities in the interior of the domain as well as the corresponding inequalities on the parabolic boundary. We further impose the following assumption to simplify the exposition.

(A2) Let $I_0 := \{\alpha_0, \beta_0\}$, where $\alpha_0, \beta_0 \in I$ such that

(a) $\alpha_0 < \beta_0$;
(b) w_0 is convex in $I_+ := (\alpha_0, \beta_0)$;
(c) w_0 is of class C^1 in $I_- := I \setminus [\alpha_0, \beta_0]$ with its derivative $(w_0)_x$ strictly decreasing in each interval of I_-;
(d) For $\alpha_0, \beta_0 \in I_0$,

$$k_0^\alpha := \lim_{x \to \alpha_0 -} \frac{w_0(x) - w_0(\alpha_0)}{x - \alpha_0} \le \lim_{x \to \alpha_0 +} \frac{w_0(x) \quad w_0(\alpha_0)}{x - \alpha_0},$$

$$k_0^\beta := \lim_{x \to \beta_0 +} \frac{w_0(x) - w_0(\beta_0)}{x - \beta_0} \ge \lim_{x \to \beta_0 -} \frac{w_0(x) - w_0(\beta_0)}{x - \beta_0};$$

(e) $k_0^\alpha < k_0^\beta$.

Assumptions (a)–(c) yield a partition of the graph according to convexity. We do not explicitly impose any regularity on the convex part I_+ beyond Lipschitz continuity. However, assumption (d) prescribes a compatibility condition between the left and right derivatives at the inflection points, which, together with (a)–(c), implies that u_0 is semiconvex. The assumption (e) further requires that each convex component (α_0^i, β_0^i) contains strictly convex points.

The solution of (7.2.33)–(7.2.34) clearly enjoys monotonicity in time.

Lemma 7.2.20 *Assume* (A1). *Let* w *be the solution of* (7.2.33) *with the initial value* (7.2.34). *Then* $w(x,t) \geq w(x,s)$ *for any* $x \in \mathbf{R}$ *and* $t \geq s \geq 0$.

Since w satisfies $w_t \geq 0$ in $\mathbf{R} \times (0,\infty)$ in the viscosity sense, one can easily show the monotonicity by following Lemma 5.15 in M. Bardi and I. Capuzzo-Dolcetta [40] or Lemma 4.4 in Y. Giga et al. [219]. The large-time behavior for (7.2.33) is also simple for this specific initial value.

Proposition 7.2.21 (Large-Time Behavior) *Assume* (A1). *Let* w *be the solution of* (7.2.33)–(7.2.34). *Then* $w(x,t) \to M = \max_{x\in\mathbf{R}} w_0(x)$ *uniformly for all* $x \in \mathbf{R}$ *as* $t \to \infty$. *Moreover, if* (A2) *is also satisfied, then*

$$w(x,t) < M \quad \text{for all } t \geq 0 \text{ and } x \in (a,b). \tag{7.2.35}$$

Proof One may first take the relaxed limits of u as $t \to \infty$:

$$\overline{w}(x) = \limsup_{t\to\infty}{}^{*} w(x,t), \quad \underline{w}(x) = \liminf_{t\to\infty}{}_{*} w(x,t).$$

Since the constant function M is a supersolution and w is nondecreasing in time, we have $w_0(x) \leq w(x,t) \leq M$, which implies that $w_0 \leq \underline{w} \leq \overline{w} \leq M$ in $\mathbf{R}$. In particular, we have $\overline{w}(a) = \underline{w}(a) = \overline{w}(b) = \underline{w}(b) = M$. By the standard stability theory for viscosity solutions, it follows that $\underline{w}$ in I is a supersolution of

$$\begin{cases} -(w_{xx})_+ = 0 & \text{in } I = (a,b), \\ w(a) = w(b) = M. \end{cases} \tag{7.2.36}$$

We claim that $\underline{w} \equiv M$ in I. Indeed, if there exists $x_0 \in I$ such that $\underline{w}(x_0) < M$, then we can construct a smooth function ϕ in $[a,b]$ satisfying $\phi(a) = \phi(b) = M$, $\underline{w}(x_0) < \phi(x_0) < M$ and $\phi_{xx} > 0$ in I. It is obvious that there exists $\overline{x} \in I$ such that

$$\min_{x\in[a,b]} (w - \phi)(x) = w(\overline{x}) - \phi(\overline{x}) < 0.$$

By the definition of viscosity supersolutions, we obtain $-\phi_{xx}(\overline{x})_+ \geq 0$, which yields $\phi_{xx}(\overline{x}) \leq 0$. This contradicts the fact that $\phi_{xx} > 0$ in I, and hence $\underline{w} \equiv M$ in I. It is now clear that $\overline{w} = \underline{w} = M$, which implies the uniform convergence of $w(\cdot,t)$ to the constant solution of (7.2.36) as $t \to \infty$.

To prove the strict inequality (7.2.35) for all $t \geq 0$ and $x \in \mathbf{R}$ such that $w_0(x) < M$, we construct a supersolution of (7.2.33). For any $\varepsilon > 0$, let

$$W_0^\varepsilon(x) = M + A\sin(\omega x + d)$$

with $A, \omega, d \in \mathbf{R}$ properly chosen such that

$$W_0^\varepsilon(a+\varepsilon) = W_0^\varepsilon(b-\varepsilon) = M$$

and $W_0^\varepsilon(x) > w_0(x)$ for all $x \in [a+\varepsilon, b-\varepsilon]$. Such choices of A, ω, d are possible due to the condition (A2)(c). Let us solve

$$\begin{cases} W_t^\varepsilon - W_{xx}^\varepsilon = 0 & \text{in } (a+\varepsilon, b-\varepsilon) \times (0,\infty), \quad (7.2.37) \\ W^\varepsilon(a+\varepsilon, t) = W^\varepsilon(b-\varepsilon, t) = M & \text{for all } t \in [0,\infty), \quad (7.2.38) \\ W^\varepsilon(\cdot, 0) = W_0^\varepsilon & \text{in } [a+\varepsilon, b-\varepsilon]. \quad (7.2.39) \end{cases}$$

The unique solution is given by

$$W^\varepsilon(x,t) = M + Ae^{-\omega^2 t}\sin(\omega x + d).$$

We thus have $W^\varepsilon(x,t) = W^\varepsilon(x,t) > w_0(x)$ for all $t \geq 0$ and $x \in [a+\varepsilon, t-\varepsilon]$. In addition, $W^\varepsilon(x,t)$ is convex for all $x \in (a+\varepsilon, t-\varepsilon)$ and $t \geq 0$. It follows that in $(a+\varepsilon, b-\varepsilon) \times (0,\infty)$

$$W_t^\varepsilon - \frac{(W_{xx}^\varepsilon)_+}{1+(W_x^\varepsilon)^2} \geq W_t^\varepsilon - W_{xx}^\varepsilon = 0,$$

which, by the comparison principle for the positive curvature flow, yields $W^\varepsilon \geq w$ in $[a+\varepsilon, b-\varepsilon] \times [0,\infty)$. It follows that $w < M$ in $[a+\varepsilon, b-\varepsilon] \times [0,\infty)$. We then obtain (7.2.35) thanks to the arbitrariness of $\varepsilon > 0$. □

As a direct consequence of Lemma 7.2.20 and Proposition 7.2.21 as well as (A1), we obtain $w(a,t) = w(b,t) = M$ for all $t \geq 0$. This amounts to saying that (7.2.33)–(7.2.34) can be considered as a Cauchy-Dirichlet problem in the domain (a,b) with boundary condition $w(a,t) = w(b,t) = M$, $t \geq 0$.

Lemma 7.2.22 (Existence of Waiting Time) *Assume* (A1) *and* (A2). *Let w be the solution of* (7.2.33)–(7.2.34). *For any $x \in I_-$, there exists $\tau_x > 0$ depending on x such that $w(x,t) = w_0(x)$ for all $t \in [0, \tau_x]$.*

Proof Fix an arbitrary $x_0 \in I_-$ and take $W(x) = Bx + C$ with $B, C \in \mathbf{R}$ properly chosen so that there exist $x_1, x_2 \in I_-$ satisfying $x_1 < x_0 < x_2$, $W(x_0) = u_0(x_0)$ and $W(x) > u_0(x)$ for all $x \in [x_1, x_0) \cup (x_0, x_2]$. By continuity of the solution $w(x,t)$, there exists $\tau > 0$ such that $w(x_1, t) < W(x_1)$ and $w(x_2, t) < W(x_2)$ for any $s \in [0, \tau]$. Hence, W is a stationary supersolution of

$$\begin{cases} w_t - \dfrac{(w_{xx})+}{1+w_x^2} = 0 \quad \text{in } (x_1, x_2) \times (0, \tau), \\ w(x_1, t) = w(x_1, t) \text{ and } w(x_2, t) = u(x_2, t) \text{ for all } s \in [0, \tau], \\ w(x, 0) = w_0(x) \text{ for all } x \in [x_1, x_2]. \end{cases} \tag{7.2.40}$$

The comparison theorem yields that $w(x_0, t) \leq W(x_0) \equiv w_0(x_0)$ and therefore $w(x_0, t) = u_0(x_0)$ for any $t \in [0, \tau]$ by the monotonicity of $t \mapsto w(x, t)$. □

The definition of waiting time T_0 in (7.2.29) in this case reduces to

$$T_0(z) = \sup\{t \geq 0 : w(x, t) = w_0(x)\} \quad \text{for any } z = (x, w_0(x)) \in \mathbf{R}^2. \tag{7.2.41}$$

By Proposition 7.2.21, for each $x \in I$, we have $T_0(z) < \infty$ at $z = (x, w_0(x))$.

In what follows, we investigate the behavior of inflection points when t is small. Define

$$\begin{aligned} \alpha_t &:= \sup\{x; w(x, t) = w_0(x),\ a \leq x \leq \alpha_0\}, \\ \beta_t &:= \inf\{x; w(x, t) = w_0(x),\ \beta_0 \leq x \leq b\}. \end{aligned} \tag{7.2.42}$$

Lemma 7.2.22 asserts that the definitions above are valid and $a < \alpha_t \leq \beta_t < b$ when $t > 0$ is sufficiently small.

Let

$$\tau^* := \sup\{t \in [0, \infty); a < \alpha_t \leq \beta_t < b\}. \tag{7.2.43}$$

We shall later prove that $\tau^* = \infty$. Lemma 7.2.20 implies the monotonicity of α_t and β_t; that is, $\alpha_t \leq \alpha_s$ and $\beta_t \geq \beta_s$ for all $0 \leq s \leq t < \tau^*$. We may also define α_{τ^*} and β_{τ^*} by letting $t \to \tau^*$ in (7.2.42). In fact,

$$\alpha_{\tau^*} = \lim_{t \to \tau^*} \alpha_t, \quad \beta_{\tau^*} = \lim_{t \to \tau^*} \beta_t$$

due to the continuity of the solution w.

Remark 7.2.23 Under the assumptions (A1) and (A2), we readily deduce that for any $t < \tau^*$,

$$w(x, s) = w_0(x) \text{ for all } x \in [a, \alpha_t) \cup (\beta_t, b] \text{ and } s \in [0, t). \tag{7.2.44}$$

Indeed, from the definition of α_t and β_t and the continuity of u, we have $w(\alpha_t, t) = w_0(\alpha_t)$ and $w(\beta_t, t) = w_0(\beta_t)$, which by Lemma 7.2.20 implies that w remains stationary at α_t and β_t before time t; in other words, $w(\alpha_t, s) = w_0(\alpha_t)$ and $w(\beta_t, s) = w_0(\beta_t)$ for all $0 \leq s \leq t$. Noting that $w(a, s) = w(b, s) = w_0(a) = w_0(b) = M$ for all $s \geq 0$, we recall the periodic setting and find that $w(x, s)$ is actually a solution of

$$
\begin{cases}
w_s - \dfrac{(w_{xx})_+}{1+w_x^2} = 0 & \text{in } (a, \alpha_t) \cup (\beta_t, b) \times (0, t), \\
w(x, s) = w_0(x) & \text{for all } x \in \{a, b, \alpha_t, \beta_t\} \text{ and } s \in [0, t], \\
w(x, 0) = w_0(x) & \text{for } x \in [a, \alpha_t) \cup (\beta_t, b].
\end{cases}
$$

Since w_0 is clearly the unique solution of this problem, (7.2.44) follows immediately.

Theorem 7.2.24 (Continuity in Time of Inflection Points) *Assume* (A1) *and* (A2). *Let w be the solution of* (7.2.33)–(7.2.34). *Let τ^* be given by* (7.2.43). *For $t \in [0, \tau^*)$, let α_t, β_t be defined as in* (7.2.42). *Then $\alpha_t, \beta_t : [0, \tau^*) \to I$ are both continuous functions.*

Proof Let us only show the continuity for β_t. We fix $t_0 \in [0, \tau^*)$. Our goal is to show $\beta_t \to \beta_{t_0}$ as $t \to t_0$.

We first show the right continuity, i.e., $\lim_{t\to t_0+} \beta_t = \beta_{t_0}$. Indeed, by the monotonicity of β_t in t, we have $\liminf_{t\to t_0+} \beta_t \geq \beta_{t_0}$. On the other hand, by Remark 7.2.23, $w(x, t_0) = w_0(x)$ for all $x \in (\beta_{t_0}, b)$. Set

$$
\tilde{w}(x, t) = w(x, t + t_0)
$$

for $x \in \mathbf{R}$ and $t \in [t_0.t_0 + \varepsilon)$ with $\varepsilon > 0$ small. Then $\tilde{w}$ is the solution of (7.2.33) restricted to $\mathbf{R} \times (t_0, t_0 + \varepsilon)$ with initial data $w(\cdot, t_0)$. Applying Lemma 7.2.22 to $\tilde{u}$ and an arbitrary $x \in (\beta_{t_0}, b)$, we may take $\tau = t - t_0 > 0$ small such that $\tilde{w}(x, s) = w(x, t_0)$ for all $s \in [0, \tau]$, which yields that $w(x, t) = w(x, \tau + t_0) = w_0(x)$. We therefore get $\beta_t < x$ for any $x \in (\beta_{t_0}, b)$ when $t > t_0$ is sufficiently close to t_0. If follows that $\limsup_{t\to t_0+} \beta_t \leq \beta_{t_0}$.

We next consider the left continuity. Thanks to the monotonicity of α_t and β_t in t, we may assume by contradiction that there exists a nondecreasing sequence $t_n \to t_0-$ as $n \to \infty$ such that $\lim_{n\to\infty} \beta_{t_n} < \beta_{t_0} - \delta_0$ for some $\delta_0 > 0$. It follows from Remark 7.2.23 that $w(x, t_n) = w_0(x)$ for all $x \in [\beta_{t_n}, b)$, which, by continuity of u, implies that $w(x, t_n) = w_0(x)$ and therefore $w(x, t_0) = w_0(x)$ for all $x \in [\beta_{t_0} - \delta_0, b)$. This contradicts the definition of β_{t_0}. □

Lemma 7.2.25 (Support Functions at Inflection Points) *Assume* (A1) *and* (A2). *Let w be the solution of* (7.2.33)–(7.2.34). *Let τ^* be given by* (7.2.43). *For $t \in [0, \tau^*)$, let $y = L_t^\alpha(x)$ and $y = L_t^\beta(x)$ be the equations of tangent lines to w_0 respectively at α_t and β_t. Then,*

$$
w(x, t) \geq \max\{L_t^\alpha(x), L_t^\beta(x)\}
$$

for all $x \in [\alpha_t, \beta_t]$ and $t \in [0, \tau^*)$.

Proof The statement clearly holds when $t = 0$ due to (A2). Let us show $w(\cdot, t) \geq L_t^\beta(\cdot)$ in $[\alpha_t, \beta_t]$ for $t \in (0, \tau^*)$. Using a symmetric argument, one can prove $w(\cdot, t) \geq L_t^\alpha(\cdot)$ in $[\alpha_t, \beta_t]$.

Suppose by contradiction that there exists $t > 0$ such that there is $x_1 \in [\alpha_t, \beta_t]$ satisfying $w(x_1, t) < L_t^\beta(x_1)$, which yields that $w(x_1, s) < L_t^\beta(x_1)$ for any $0 \leq s \leq t$. Then we may take $y_1 \in (w(x_1, t), L_t^\beta(x_1))$.

Take another tangent line $y = L^*(x)$ to w_0 passing through (x_1, y_1). We assume that they are tangent at $x = x_0$. Then we must have $x_0 < \beta_t$. We pick x_2 from the interval (β_t, a), which is nonempty due to $t < \tau^*$, and denote $y_2 = L^*(x_2)$. It is obvious that $w(x_2, s) = w_0(x_2) < y_2$ for all $s \in [0, t]$.

It is now clear that $y = L^*(x)$ is a supersolution of the Cauchy-Dirichlet problem (7.2.40) with $\tau = t$. Since w is the solution of the same equation, by comparison theorem for (7.2.40), we have $w(x, s) \leq L^*(x)$ for all $x \in [x_1, x_2]$ and $s \in [0, t]$, and in particular, $w(x_0, t) \leq L^*(x)$. This implies that $w(x_0, t) = w_0(x_0)$. On the other hand, by definition of β_t, we get $w(x_0, t) > w_0(x_0)$, which is a contradiction. □

For any $t \in (0, \tau^*)$, denote

$$\Omega_t := \{(x, y) \in \mathbf{R}^2 : y > w(x, t),\ \alpha_t < x < \beta_t\}. \tag{7.2.45}$$

An immediate consequence of Lemma 7.2.25 is as follows.

Lemma 7.2.26 *Assume* (A1) *and* (A2). *Let τ^* be given by* (7.2.43). *For any $t \in (0, \tau^*)$, let Ω_0 and Ω_t be defined as in* (7.2.32) *and* (7.2.45) *respectively. Then,*

$$\{kz_1 + (1 - k)z_2; k \in [0, 1]\} \subset \overline{\Omega_0}$$

holds for any $z_1, z_2 \in \overline{\Omega_t}$.

Proof Set

$$\mathcal{S} := \{(x, y) \in \mathbf{R}^2; y \geq L_t^\alpha(x) \vee L_t^\beta(x), y \geq u_0(x) \text{ and } \alpha_t \geq x \geq \beta_t\}.$$

Then by definition we have $\mathcal{S} \subset \overline{\Omega_0}$. It is also easily seen that $\mathcal{S}$ is convex. On the other hand, Lemma 7.2.25 and Lemma 7.2.20 implies that $\overline{\Omega_t^i} \subset \mathcal{S}$. Hence the convex hull of $\overline{\Omega_t}$ is contained in $\overline{\Omega_0}$. □

Lemma 7.2.27 (Convexity of Moving Graph Pieces) *Assume* (A1) *and* (A2). *Let w be the solution of* (7.2.33)–(7.2.34). *Let τ^* be given by* (7.2.43). *For any $t \in (0, \tau^*)$, let α_t and β_t be defined as in* (7.2.42). *Then $x \mapsto w(x, t)$ is convex in $[\alpha_t, \beta_t]$ for all $t \in (0, \tau^*)$; namely*

$$w(kx_1 + (1 - k)x_2, t) \leq kw(x_1, t) + (1 - k)w(x_2, t)$$

for any $k \in [0, 1]$ and $x_1, x_2 \in [\alpha_t, \beta_t]$.

Proof We first fix $t \in (0, \tau^*)$. Set for any $x \in [x_1, x_2]$ and $s \in [0, t]$,

$$v(x, s) := w(x_2, s)\frac{x - x_1}{x_2 - x_1} + w(x_1, s)\frac{x_2 - x}{x_2 - x_1}.$$

It follows from Lemma 7.2.26 that $v(x, 0) \geq w_0(x)$. It is also obvious that

$$v(x_1, s) = w(x_1, s), \quad v(x_2, s) = w(x_2, s)$$

for all $s \in [0, t]$. Moreover, due to Lemma 7.2.20, we obtain $v_s(x, s) \geq 0$ in the viscosity sense for any $x \in [x_1, x_2]$ and $s \in (0, t)$. Since v is linear in space variable, the observations above yield that v is a supersolution of (7.2.40) with $\tau = t$. Therefore by comparison principle, we have $v(x, t) \geq w(x, t)$ for all $x \in [x_1, x_2]$. The desired convexity of $x \mapsto w(x, t)$ follows immediately. □

The convexity result yields $\tau^* = \infty$ immediately.

Corollary 7.2.28 (Infinite Convexification Time) *Assume* (A1) *and* (A2). *Let w be the solution of* (7.2.33)–(7.2.34). *Let τ^* be given by* (7.2.43). *Then $\tau^* = \infty$.*

Proof Suppose that $\tau^* < \infty$. Without loss of generality, we assume that $\alpha_{\tau^*} = a$. By Lemma 7.2.27, $w(\cdot, t)$ is convex in $[\alpha_t, \beta_t]$ for any $t \in (0, \tau^*)$. Letting $t \to \tau^*$, we get the convexity of $w(\cdot, \tau^*)$ in $[a, \beta_{\tau^*}]$. Note that $w(\cdot, \tau^*) \geq w_0$ and $(w_0)_x(a) = 0$. We thus have $w(\cdot, \tau^*) \equiv M$ in (a, β_{τ^*}), which is a contradiction to (7.2.35). □

Our next goal is to show the strict expansion of the convex piece as t increases in $[0, \tau^*)$. We first consider a special case.

Lemma 7.2.29 (Immediate Move of a Corner) *Assume* (A1) *and* (A2). *Let w be the solution of* (7.2.33)–(7.2.34) *with initial data w_0. If there is $x_0 \in (a, b)$ such that*

$$-\infty < \lim_{x \to x_0-} \frac{w_0(x) - w_0(x_0)}{x - x_0} < \lim_{x \to x_0+} \frac{w_0(x) - w_0(x_0)}{x - x_0} < \infty, \tag{7.2.46}$$

then $w(x_0, t) > w_0(x_0)$ for any $t > 0$.

Proof It is easy to construct a periodic function $\phi \in C^\infty([x_0 - \delta, x_0 + \delta])$ for some $\delta > 0$ such that $\phi(x_0) = w_0(x_0)$, $\phi(x_0 - \delta) < w_0(x_0 - \delta)$, $\phi(x_0 + \delta) < w_0(x_0 + \delta)$ and $\phi_{xx} \geq c_1$ in $(x_0 - \delta, x_0 + \delta)$ for some $c_1 > 0$. Since $|\phi_x|$ is bounded in $(x_0 - \delta, x_0 + \delta)$, there exists $c > 0$ depending on $c_1 > 0$ such that $\psi(x, t) = \phi(x) + ct$ satisfies

$$\psi_t - \frac{(\psi_{xx})_+}{1 + \psi_x^2} \leq 0 \quad \text{in } (x_0 - \delta, x_0 + \delta) \times (0, t_0)$$

with $t_0 > 0$ small. By taking t_0 further small, we have $\psi \leq w$ on $\{x_0 \pm \delta\} \times [0, t_0)$. By comparison principle, we thus deduce $\psi \leq w$ in $[x_0 - \delta, x_0 + \delta] \times [0, t_0)$. In particular, we have

$$w(x_0, t) \geq w_0(x_0) + ct > w_0(x_0)$$

for all $t \in (0, t_0)$. By Lemma 7.2.20, we conclude that $w(x_0, t) > w_0(x_0)$ for all $t > 0$. □

The argument in the proof above can also be generalized to show a similar situation during the evolution.

Lemma 7.2.30 *Assume* (A1) *and* (A2). *Let w be the solution of* (7.2.33)–(7.2.34) *satisfying*

$$-\infty < \lim_{x \to x_0-} \frac{w(x, t_0) - w(x_0, t_0)}{x - x_0} < \lim_{x \to x_0+} \frac{w(x, t_0) - w(x_0, t_0)}{x - x_0} < \infty$$

for some $x_0 \in \mathbf{R}$ and $t_0 \geq 0$. Then $w(x_0, t) > w(x_0, t_0)$ for all $t > t_0$.

Let $\tilde{w}(x, t) = w(x, t + t_0)$. Then $\tilde{u}$ is the solution of (7.2.33) with initial data $\tilde{w}(x, 0) = w(x, t_0)$. We reach the conclusion by applying Lemma 7.2.29.

Theorem 7.2.31 (Nonstop Moving of Convex Pieces) *Assume* (A1) *and* (A2). *Let w be the solution of* (7.2.33) *and* (7.2.34). *Let α_t, β_t be defined as in* (7.2.42) *for all $t \geq 0$. Then $w(x, t) > w(x, s)$ for all $x \in [\alpha_s, \beta_s]$, $0 \leq s < t$.*

Proof Suppose by contradiction that there are $s \geq 0$, $\delta > 0$ and $x_0 \in [\alpha_s, \beta_s]$ such that $w(x_0, t) = w(x_0, s)$ for all $t \in [s, s + \delta]$. By Lemma 7.2.27, we have

$$-\infty < \lim_{x \to x_0-} \frac{w(x, s) - w(x_0, s)}{x - x_0} \leq \lim_{x \to x_0+} \frac{w(x, s) - w(x_0, s)}{x - x_0} < \infty.$$

We may derive a contradiction immediately by using Lemma 7.2.30 if

$$\lim_{x \to x_0-} \frac{w(x, s) - w(x_0, s)}{x - x_0} < \lim_{x \to x_0+} \frac{w(x, s) - w(x_0, s)}{x - x_0}.$$

We therefore only discuss the case that

$$\lim_{x \to x_0-} \frac{w(x, s) - w(x_0, s)}{x - x_0} = \lim_{x \to x_0+} \frac{w(x, s) - w(x_0, s)}{x - x_0},$$

which means $w(\cdot, s)$ is differentiable at x_0. Let us denote by $\hat{k}$ the quantity above, i.e., the derivative at x_0. By Lemma 7.2.25 and Lemma 7.2.27, we have

$$(w_0)_x(\alpha_s) \leq \hat{k} \leq (w_0)_x(\beta_s).$$

Since the strict concavity of w_0 in (a, α_0) and in (β_0, b) implies that $(w_0)_x(\alpha_s) < (w_0)_x(\beta_s)$, either $\hat{k} > (w_0)_x(\alpha_s)$ or $\hat{k} < (w_0)_x(\beta_s)$ holds. Let us assume the latter without loss of generality. Note that in this case, we may take a smooth and Lipschitz function $\phi : \mathbf{R} \to \mathbf{R}$ which fulfills the following conditions:

(i) Its graph is symmetric about $(x_0, w(x_0, s))$, i.e., $\phi(x) = 2w(x_0, s) - \phi(2x_0 - x)$.
(ii) It serves as a one-sided lower bound in the sense that $\phi(x) < w(x, s)$ for all $x \in (\beta_s, \infty)$ and $\phi(x) = \hat{k}x$ for all $x \in [x_0, x_0 + \delta_1)$ with some $\delta_1 > 0$.
(iii) It is convex in (x_0, β_s) and $\phi_x(x) \equiv \overline{k} > \hat{k}$ for all $x \in (\beta_s, \infty)$.

Set $K = \sup_{\mathbf{R}} |\phi_x|$ and $c = 1/(1 + K^2)$. Let ψ be the unique solution of the heat equation

$$\begin{cases} \psi_t - c\psi_{xx} = 0 & \text{in } \mathbf{R} \times (s, \infty), \\ \psi(x, s) = \phi(x) & \text{in } \mathbf{R}. \end{cases} \tag{7.2.47}$$

We then utilize the symmetry to get $\psi(x_0, t) = w_0(x_0)$ and $\psi_{xx}(x_0, t) = 0$ for all $t \geq s$. We adopt the strong maximum principle for ψ_x to obtain $\psi_x(x_0, t) > \phi_x(x_0)$ for $t \geq s$ and therefore

$$\psi_x(x_0, t) > \hat{k} = \lim_{x \to x_0+} \frac{\psi(x, t) - \psi(x_0, t)}{x - x_0} \quad \text{for all } t > s. \tag{7.2.48}$$

Moreover, the maximum principle for ψ_{xx} implies that $\psi_{xx} \geq 0$ in $(x_0, \infty) \times [s, s+ \delta)$. Besides, there exists $\mu \in (0, \delta)$ small such that $\psi(\beta_s, t) \leq w(\beta_s, t)$ for all $t \in [s, s + \mu]$. It is not difficult to see that ψ also satisfies

$$\begin{cases} \psi_t - \dfrac{(\psi_{xx})_+}{1 + \psi_x^2} \leq 0 & \text{in } (x_0, \beta_s) \times (s, s + \mu), \\ \psi(x, s) \leq w(x, s) & \text{for all } x \in [x_0, \beta_s], \\ \psi(x_0, t) \leq w(x_0, t) \text{ and } \psi(\beta_s, t) \leq w(\beta_s, t) & \text{for all } t \in [s, s + \mu]. \end{cases} \tag{7.2.49}$$

Indeed, since the initial and boundary conditions are satisfied, as proved above, we only need to show that w satisfies the first inequality. Note that the maximum principle for (7.2.47) implies $|\psi_x(x, t)| \leq K$ for all $x \in \mathbf{R}$ and $t \geq 0$. Since ψ stays convex in space in $(x_0, \infty) \times (s, s + \mu)$. Then a direct calculation shows that

$$\psi_t = c\psi_{xx} \leq \frac{(\psi_{xx})_+}{1 + \psi_x^2} \quad \text{in } (x_0, \beta_s) \times (s, s + \mu),$$

which completes the verification of (7.2.49). Hence, $w(x, t) \geq \psi(x, t)$ for all $x \in [x_0, \beta_s]$ and $t \in [s, s + \mu)$, which in turn implies that

$$\lim_{x \to x_0+} \frac{w(x, t) - w(x_0, t)}{x - x_0} \geq \psi_x(x_0, t) > \hat{k} = \lim_{x \to x_0-} \frac{w(x, t) - w(x_0, t)}{x - x_0}$$

for any $t \in (s, s+\mu)$. In particular, the relation above holds for $t = s+\mu$. Since $\mu < \delta$, Lemma 7.2.30 comes into play again for us to obtain $w(x_0, s+\delta) > w(x_0, s)$, which is a contradiction. □

Under the assumptions (A1) and (A2), we have shown that the convex portion in (α_t, β_t) strictly expands for all $t > 0$. The conjecture by R. V. Kohn and S. Serfaty [312] is proved in this particular case.

Corollary 7.2.32 (Behavior of Inflection Points) *Assume* (A1) *and* (A2). *Let* w *be the solution of* (7.2.33) *and* (7.2.34). *Let* α_t, β_t *be defined as in* (7.2.42). *Then,* $\alpha_t < \alpha_s$ *and* $\beta_t > \beta_s$ *hold for any and* $t > s \geq 0$.

7.3 Boundary Value Problems

In this section, we use the game-theoretic method to study various boundary value problems associated to the curvature flow equation.

Let $\Omega \subset \mathbf{R}^2$ be a bounded domain with smooth boundary and $\nu(x)$ be the unit outward normal to $\partial\Omega$ at x. We mainly consider the following Neumann boundary problem

$$
\text{(NP)} \begin{cases} u_t - |\nabla u| \operatorname{div}\left(\dfrac{\nabla u}{|\nabla u|}\right) = 0 & \text{in } \Omega \times (0, \infty), \quad (7.3.1) \\ \langle \nabla u(x,t), \nu(x) \rangle = 0 & \text{for } (x,t) \in \partial\Omega \times (0,\infty), \quad (7.3.2) \\ u(x,0) = u_0(x) & \text{for } x \in \overline{\Omega}. \quad (7.3.3) \end{cases}
$$

Here, the Neumann type boundary condition (7.3.2) is interpreted in the viscosity sense.

7.3.1 Neumann Boundary Condition

Recall that $\mathrm{sd}(\cdot, \partial\Omega)$ denotes the signed distance to $\partial\Omega$; namely,

$$
\mathrm{sd}(x, \partial\Omega) = \begin{cases} \operatorname{dist}(x, \partial\Omega), & x \in \Omega, \\ -\operatorname{dist}(x, \partial\Omega), & x \notin \Omega. \end{cases} \tag{7.3.4}
$$

For every $\sigma \geq 0$, we may take

$$
\Omega_\sigma = \{x \in \Omega;\ \mathrm{sd}(x, \partial\Omega) > \sigma\}. \tag{7.3.5}
$$

The function $\mathrm{sd}(x, \partial\Omega)$ is of class $C^2(\overline{\Omega} \setminus \Omega_\sigma)$ for $\sigma > 0$ small since Ω is of class C^2. We extend the outward unit normal ν by setting

$$\nu(x) = -\nabla\, \mathrm{sd}(x, \partial\Omega)$$

for $x \in \overline{\Omega} \setminus \Omega_\sigma$ with $\sigma > 0$ small. By continuity of ν, it is then clear that

$$\langle \nu(x), \nabla\, \mathrm{sd}(x, \partial\Omega)\rangle \le -\frac{1}{2} \tag{7.3.6}$$

for all $x \in \overline{\Omega}$ near $\partial\Omega$. Fix $\varepsilon > 0$ small. We next set, for any $x \in \overline{\Omega}$,

$$\eta_\varepsilon(x) = \min\left\{1, \frac{\mathrm{sd}(x, \partial\Omega)}{\sqrt{2}\varepsilon}\right\} \tag{7.3.7}$$

and

$$\zeta_\varepsilon = 1 - \eta_\varepsilon^2 \qquad \text{in } \overline{\Omega}. \tag{7.3.8}$$

Game 3 Our game here is similar to Game 1-a in Sect. 7.1.1. Fix $1 < \theta < 2$ arbitrarily. For any $(x, t) \in \overline{\Omega} \times [0, \infty)$, the game starts from a fixed state $x \in \overline{\Omega}$ with the total of steps is again $\ell = [t/\varepsilon^2]$. Let y_0 denote the initial game position x. The following game rules determine a sequence of game positions $y_0(= x), y_1, y_2, \cdots, y_\ell \in \overline{\Omega}$. At the k-th step,

- Player I chooses a unit vector $v_k \in \mathbf{R}^2$;
- Player II then picks $b_k = \pm 1$;
- Once the choices of both players are determined, the game position is moved from y_{k-1} to $y_k = y_{k-1} + \sqrt{2}\varepsilon\eta_\varepsilon(y_{k-1})b_k v_k - \varepsilon^\theta \zeta_\varepsilon(y_{k-1})\nu(y_{k-1}) \in \overline{\Omega}$, $0 \le \eta_\varepsilon \le 1$ is the function given as in (7.3.7).

  Owing to the inclusion of the function η_ε, we can easily verify that

$$x + \sqrt{2}\varepsilon\eta_\varepsilon(x)bv - \varepsilon^\theta\zeta_\varepsilon(x)\nu(x) \in \overline{\Omega}$$

for all $x \in \overline{\Omega}$ and for $\varepsilon > 0$ sufficiently small. Indeed,

$$\begin{aligned}\mathrm{sd}\left(x + \sqrt{2}\varepsilon\eta_\varepsilon(x)bv - \varepsilon^\theta\zeta_\varepsilon(x)\nu(x), \partial\Omega\right) &\ge \mathrm{sd}\left(x + \sqrt{2}\varepsilon\eta_\varepsilon(x)bv, \partial\Omega\right)\\ &\ge \mathrm{sd}(x, \partial\Omega) - \sqrt{2}\varepsilon\eta_\varepsilon(x) \ge 0.\end{aligned}$$

We thus can repeat the game rules till the final step. Suppose that Player I aims to minimize the cost $u_0(y_\ell)$ while Player II is to maximize it. Then we can define the game value as in (7.1.3), that is,

$$u^\varepsilon(x,t) = \min_{|v_1|=1}\max_{b_1=\pm1}\dots\min_{|v_\ell|=1}\max_{b_\ell=\pm1} u_0(y_\ell).$$

Then one can obtain the convergence $u^\varepsilon \to u$, where u is the unique solution of (NP).

Theorem 7.3.1 (Game Convergence for Neumann Problem) *Let $\Omega \subset \mathbf{R}^2$ be a bounded C^2 domain. Assume that u_0 is Lipschitz continuous in $\overline{\Omega}$. Let u^ε be the value function associated to Game 3. Then $u^\varepsilon \to u$ locally uniformly in $\overline{\Omega}\times[0,\infty)$ as $\varepsilon \to 0$, where u is the unique viscosity solution of* (NP).

Remark 7.3.2 It is possible to consider the inverse game for of Game 3, replacing $\min_v \max_b$ by $\max_v \min_b$. One can show that its game value converges to the solution of the same problem.

The DPP of this game is

$$u^\varepsilon(x,t) = \min_{|v|=1}\max_{b=\pm1} u^\varepsilon\left(x+\sqrt{2}\varepsilon\eta_\varepsilon(x)bv - \varepsilon^\theta\zeta_\varepsilon(x)\nu(x), t-\varepsilon^2\right) \tag{7.3.9}$$

for all $x \in \overline{\Omega}$ and $t \geq \varepsilon^2$. We next formally verify the convergence near $\partial\Omega$. Assume that u^ε are smooth with $\nabla u^\varepsilon \neq 0$.

Our formal proof is based on a stability argument near $\partial\Omega$. Assume that u^ε is smooth and converges in a strong sense to a smooth function u with $|\nabla u| \neq 0$. Suppose that $x \in \partial\Omega$ and there is an approximating sequence $x_\varepsilon \in \overline{\Omega}$ with $x_\varepsilon \to x$ as $\varepsilon \to 0$. By Taylor expansion on (7.3.9), we have

$$\begin{aligned}0 = \min_{|v|=1}\Big\{ &\max_{b=\pm1}\sqrt{2}\varepsilon\left\langle \nabla u^\varepsilon(x_\varepsilon,t),\ \eta_\varepsilon(x_\varepsilon)bv\right\rangle + \varepsilon^2\eta_\varepsilon^2(x_\varepsilon)\left\langle \nabla^2 u^\varepsilon(x_\varepsilon,t)v, v\right\rangle\Big\}\\ &- \varepsilon^\theta\zeta_\varepsilon(x_\varepsilon)\left\langle\nabla u^\varepsilon(x_\varepsilon,t),\nu(x_\varepsilon)\right\rangle - \varepsilon^2 u_t^\varepsilon(x_\varepsilon,t) + o(\varepsilon^2).\end{aligned} \tag{7.3.10}$$

It follows that the minimizer v is nearly perpendicular to $\nabla u^\varepsilon(x_\varepsilon,t)$. Thus we have at (x_ε,t)

$$0 = \varepsilon^2\eta_\varepsilon^2|\nabla u^\varepsilon|\operatorname{div}\left(\frac{\nabla u^\varepsilon}{|\nabla u^\varepsilon|}\right) - \varepsilon^\theta\zeta_\varepsilon\left\langle\nabla u^\varepsilon,\nu\right\rangle - \varepsilon^2 u_t^\varepsilon + o(\varepsilon^2), \tag{7.3.11}$$

which by (7.3.8) yields that

$$\varepsilon^2\eta_\varepsilon^2\left(u_t^\varepsilon - |\nabla u^\varepsilon|\operatorname{div}\left(\frac{\nabla u^\varepsilon}{|\nabla u^\varepsilon|}\right)\right) + \varepsilon^2\zeta_\varepsilon u_t^\varepsilon + \varepsilon^\theta\zeta_\varepsilon\left\langle\nabla u^\varepsilon,\nu\right\rangle = 0. \tag{7.3.12}$$

Note that $\eta_\varepsilon(x_\varepsilon)$ and $\zeta(x_\varepsilon)$ are nonnegative and cannot be both equal to zero.

We discuss three different cases below. If

$$\liminf_{\varepsilon\to0}\frac{\zeta_\varepsilon(x_\varepsilon)}{\varepsilon^{2-\theta}\eta_\varepsilon^2(x_\varepsilon)} = \infty,$$

then dividing (7.3.12) by $\varepsilon^{\theta}\zeta(x_\varepsilon)$ and passing to the limit as $\varepsilon \to 0$, we get the Neumann condition

$$\langle \nabla u(x,t), \nu(x)\rangle = 0.$$

If, on the other hand,

$$\limsup_{\varepsilon\to 0} \frac{\zeta_\varepsilon(x_\varepsilon)}{\varepsilon^{2-\theta}\eta_\varepsilon^2(x_\varepsilon)} = 0,$$

then this time we can divide (7.3.12) by $\varepsilon^2\eta_\varepsilon(x_\varepsilon)$ and send the limit. We deduce that the curvature flow equation holds at (x,t).

If there exists $c > 0$ such that

$$\frac{\zeta_\varepsilon(x_\varepsilon)}{\varepsilon^{2-\theta}\eta_\varepsilon^2(x_\varepsilon)} \to c \quad \text{as } \varepsilon \to 0$$

along a subsequence, then sending the limit in (7.3.12) via this subsequence, we have at $(x,t) \in \partial\Omega \times (0,\infty)$

$$u_t - |\nabla u| \operatorname{div}\left(\frac{\nabla u}{|\nabla u|}\right) + c\,\langle \nabla u, \nu\rangle = 0,$$

which indicates that u still formally satisfies the Neumann condition in the viscosity sense, that is, at each $(x,t) \in \partial\Omega \times (0,\infty)$ we have

$$\min\left\{u_t - |\nabla u| \operatorname{div}\left(\frac{\nabla u}{|\nabla u|}\right),\ \langle \nabla u, \nu\rangle\right\} \le 0$$

and

$$\max\left\{u_t - |\nabla u| \operatorname{div}\left(\frac{\nabla u}{|\nabla u|}\right),\ \langle \nabla u, \nu\rangle\right\} \ge 0.$$

7.3.2 Convergence of Game Values

Our rigorous proof of convergence of game values is similar to that in the case of Cauchy problems. We first establish a consistency result analogous to Lemma 7.1.5. Fix $1 < \theta < 2$. We define a monotone operator $H_N^\varepsilon : C(\overline{\Omega}) \to C(\overline{\Omega})$ to be

$$H_N^\varepsilon[\psi](x) = \min_{|v|=1}\max_{b=\pm 1} \psi\left(x + \sqrt{2}\varepsilon\eta_\varepsilon(x)bv - \varepsilon^{\theta}\zeta_\varepsilon(x)\nu(x)\right),$$
$$x \in \overline{\Omega},\ \psi \in C^2(\mathbf{R}^2). \tag{7.3.13}$$

Lemma 7.3.3 (Consistency) *Suppose that $\Omega \subset \mathbf{R}^2$ is a bounded domain of C^2 class. Let η_ε and ζ_ε be given by* (7.3.7) *and* (7.3.8). *Fix $1 < \theta < 2$. Let $\psi \in C^2(\mathbf{R}^2)$ and H_N^ε be the operator defined in* (7.3.13). *Then for all $x \in \overline{\Omega}$*

$$H_N^\varepsilon[\psi](x) - \psi(x) \leq -\varepsilon^2\eta_\varepsilon^2(x)F_*(\nabla\psi(x), \nabla^2\psi(x)) - \varepsilon^\theta\zeta_\varepsilon(x)\langle\nabla\psi(x), \nu\rangle + o(\varepsilon^2), \tag{7.3.14}$$

$$H_N^\varepsilon[\psi](x) - \psi(x) \geq -\varepsilon^2\eta_\varepsilon^2(x)F^*(\nabla\psi(x), \nabla^2\psi(x)) - \varepsilon^\theta\zeta_\varepsilon(x)\langle\nabla\psi(x), \nu\rangle + o(\varepsilon^2), \tag{7.3.15}$$

where $F = F_f$ in (2.6.6) *and the constants implicit in the error terms above are uniform in $\overline{\Omega}$.*

Proof Fix arbitrarily $x \in \overline{\Omega}$. By Taylor expansion on (7.3.13), we have

$$H_N^\varepsilon[\psi](x) = \psi(x) + \min_{|v|=1}\left\{\sqrt{2}\varepsilon\eta_\varepsilon(x)|\langle\nabla\psi(x), v\rangle| + \varepsilon^2\eta_\varepsilon^2(x)\left\langle\nabla^2\psi(x)v, v\right\rangle\right\} - \varepsilon^\theta\zeta_\varepsilon(x)\langle\nabla\psi(x), \nu(x)\rangle + o(\varepsilon^2).$$

We may handle the terms in the minimum by the same arguments in the proof of Lemma 7.1.5, since the only difference here lies at the extra coefficient η_ε. In fact, we can derive

$$\min_{|v|=1}\left\{\sqrt{2}\varepsilon\eta_\varepsilon(x)|\langle\nabla\psi(x), v\rangle| + \varepsilon^2\eta_\varepsilon^2(x)\left\langle\nabla^2\psi(x)v, v\right\rangle\right\} \leq -\varepsilon^2\eta_\varepsilon^2(x)F_*(\nabla\psi(x), \nabla^2\psi(x)) - \varepsilon^\theta\zeta_\varepsilon(x)\langle\nabla\psi(x), \nu(x)\rangle + o(\varepsilon^2)$$

and

$$\min_{|v|=1}\left\{\sqrt{2}\varepsilon\eta_\varepsilon(x)|\langle\nabla\psi(x), v\rangle| + \varepsilon^2\eta_\varepsilon^2(x)\left\langle\nabla^2\psi(x)v, v\right\rangle\right\} \geq -\varepsilon^2\eta_\varepsilon^2(x)F^*(\nabla\psi(x), \nabla^2\psi(x)) - \varepsilon^\theta\zeta_\varepsilon(x)\langle\nabla\psi(x), \nu(x)\rangle + o(\varepsilon^2).$$

Thus (7.3.14) and (7.3.15) follow easily. □

Since Ω is bounded and u_0 is continuous, by definition it is easily seen that u^ε is bounded in $\overline{\Omega} \times [0, \infty)$ uniformly in ε. We thus take the relaxed limits $\overline{u}$ and $\underline{u}$ as in (7.1.10) for the value function u^ε associated to Game 3.

Proposition 7.3.4 (Sub- and Supersolution Properties) *Let $\Omega \subset \mathbf{R}^2$ be a bounded C^2 domain. Assume that u_0 is Lipschitz continuous in $\overline{\Omega}$. Let u^ε be the value function associated to Game 3. Let $\overline{u}$ and $\underline{u}$ be defined as in* (7.1.10). *Then $\overline{u}$ and $\underline{u}$ are respectively a subsolution and a supersolution of* (7.3.1)–(7.3.2).

Proof Let us first show that $\overline{u}$ is a subsolution. Suppose that there exist $(x_0, t_0) \in \overline{\Omega} \times (0, \infty)$ and $\phi \in C^2(\overline{\Omega} \times (0, \infty))$ such that $u - \phi$ attains a strict maximum on $\overline{\Omega} \times (0, \infty)$ at (x_0, t_0). Then there exists $r > 0$ small such that

$$\overline{u}(x_0, t_0) - \phi(x_0, t_0) > \overline{u}(x, t) - \phi(x, t)$$

for all $(x, t) \in Q_r$ with $(x, t) \neq (x_0, t_0)$, where $Q_r = B_r(x_0, t_0) \cap \left(\overline{\Omega} \times (0, \infty)\right)$. It follows that there exists $(x_\varepsilon, t_\varepsilon) \in \overline{\Omega} \times (0, \infty)$ such that $(x_\varepsilon, t_\varepsilon) \to (x_0, t_0)$ as $\varepsilon \to 0$ and

$$u^\varepsilon(x_\varepsilon, t_\varepsilon) - \phi(x_\varepsilon, t_\varepsilon) \geq \sup_{Q_r}(u^\varepsilon - \phi) - \varepsilon^3.$$

By (7.3.9), we have

$$\phi(x_\varepsilon, t_\varepsilon) \leq \min_{|v|=1} \max_{b=\pm 1} \phi\left(x_\varepsilon + \sqrt{2}\varepsilon\eta_\varepsilon(x_\varepsilon) - \varepsilon^\theta \zeta_\varepsilon(x)\nu(x_\varepsilon), t_\varepsilon - \varepsilon^2\right).$$

Adopting Lemma 7.3.3 with $\psi = \phi\left(\cdot, t_\varepsilon - \varepsilon^2\right)$ and $x = x_\varepsilon$, we get

$$\begin{aligned}\varepsilon^2\phi_t(x_\varepsilon, t_\varepsilon - \varepsilon^2) \leq -\varepsilon^2\eta_\varepsilon^2(x)F_*(\nabla\phi(x_\varepsilon, t_\varepsilon - \varepsilon^2), \nabla^2\phi(x_\varepsilon, t_\varepsilon - \varepsilon^2))\\ - \varepsilon^\theta\zeta_\varepsilon(x)\left\langle\nabla\phi(x_\varepsilon, t_\varepsilon - \varepsilon^2), \nu(x_\varepsilon)\right\rangle + o(\varepsilon^2).\end{aligned}$$

Dividing the inequality above by ε^2, we are led to

$$\phi_t + \eta_\varepsilon^2 F_*(\nabla\phi, \nabla^2\phi) + \varepsilon^{\theta-2}\zeta_\varepsilon \langle\nabla\phi, \nu\rangle \leq o(1) \quad \text{at } (x_\varepsilon, t_\varepsilon - \varepsilon^2), \tag{7.3.16}$$

which, by (7.3.8), can be rewritten as

$$\zeta_\varepsilon\phi_t + \eta_\varepsilon^2\left(\phi_t + F_*(\nabla\phi, \nabla^2\phi)\right) + \varepsilon^{\theta-2}\zeta_\varepsilon \langle\nabla\phi, \nu\rangle \leq o(1) \quad \text{at } (x_\varepsilon, t_\varepsilon - \varepsilon^2). \tag{7.3.17}$$

We can easily deduce the viscosity inequality (7.1.20) by sending $\varepsilon \to 0$ in (7.3.17) when $x_0 \in \Omega$. We next discuss two different cases in the case $x_0 \in \partial\Omega$. If

$$\frac{\varepsilon^{\theta-2}\zeta_\varepsilon(x_\varepsilon)}{\eta_\varepsilon^2(x_\varepsilon)} \to \infty \quad \text{as } \varepsilon \to 0,$$

we divide (7.3.17) by $\varepsilon^{\theta-2}\zeta_\varepsilon(x_\varepsilon)$ and send $\varepsilon \to 0$ so that we can get $\langle\nabla\phi(x_0, t_0), \nu(x_0)\rangle \leq 0$. Suppose that there exists a subsequence, still indexed by ε for our convenience, such that

$$\frac{\varepsilon^{\theta-2}\zeta_\varepsilon(x_\varepsilon)}{\eta_\varepsilon^2(x_\varepsilon)} \to c \quad \text{as } \varepsilon \to 0$$

for some $c \geq 0$. Dividing (7.3.17) by $\eta_\varepsilon^2(x_\varepsilon)$ and then letting $\varepsilon \to 0$, we have

$$\phi_t(x_0, t_0) + F_*(\nabla\phi(x_0, t_0), \nabla^2\phi(x_0, t_0)) + c\,\langle\nabla\phi(x_0, t_0), \nu(x_0)\rangle \leq 0,$$

which implies that

$$\min\left\{\phi_t(x_0, t_0) + F_*(\nabla\phi(x_0, t_0), \nabla^2\phi(x_0, t_0)), \langle\nabla\phi(x_0, t_0), \nu(x_0)\rangle\right\} \leq 0.$$

We complete the verification that $\overline{u}$ is a subsolution. The proof for $\underline{u}$ is symmetric and therefore omitted here. □

Proposition 7.3.5 (Initial Value) *Let $\Omega \subset \mathbf{R}^2$ be a bounded C^2 domain. Assume that u_0 is Lipschitz continuous in $\overline{\Omega}$. Let u^ε be the value function associated to Game 3. Let $\overline{u}$ and $\underline{u}$ be defined as in* (7.1.10)*. Then*

$$\overline{u}(x, 0) \leq u_0(x) \leq \underline{u}(x, 0) \quad \textit{for all } x \in \overline{\Omega}.$$

Proof Fix $x_0 \in \overline{\Omega}$. Since u_0 is Lipschitz continuous in $\overline{\Omega}$, there exists $L > 0$ such that

$$u_0(x) \leq u_0(x_0) + L|x - x_0| \leq u_0(x_0) + \frac{L^2}{\delta}|x - x_0|^2 + \frac{\delta}{4} \tag{7.3.18}$$

for any $\delta > 0$ small. Set, for any $x \in \overline{\Omega}$,

$$\psi(x) = u_0(x_0) + \frac{L^2}{\delta}|x - x_0|^2 + \frac{\delta}{4}, \quad \psi_K(x) = \psi(x) - 2K\,\mathrm{sd}(x, \partial\Omega) + c_{K,r},$$

where $r > 0$ is arbitrarily small and

$$K = \sup_{x \in \overline{\Omega}\setminus\Omega_{\sqrt{2}\varepsilon}} |\nabla\psi(x)||\nu(x)|, \quad c_{K,r} = 2K \sup_{B_r(x_0)\cap\overline{\Omega}} \mathrm{sd}(\cdot, \partial\Omega).$$

One can show that $u_0 \leq \psi_K$ in $\overline{\Omega}$ for $\delta > 0$ small (depending on K and r). Indeed, by (7.3.18) and the choice of $c_{K,r}$ above, we have $u_0 \leq \psi_K$ in $B_r(x_0) \cap \overline{\Omega}$. We may also let $\delta > 0$ sufficiently small to obtain $u_0 \leq \psi_K$ in $\overline{\Omega} \setminus B_r(x_0)$.

Recall that $\mathrm{sd}(x, \partial\Omega) \in C^2(\overline{\Omega})$ due to the C^2 regularity of Ω. By Lemma 7.3.3 and the fact that $0 \leq \eta_\varepsilon \leq 1$, we get

$$\begin{aligned} H_N^\varepsilon[\psi_K](x) &\leq \psi_K(x) - \varepsilon^2\eta_\varepsilon^2(x)F_*(\nabla\psi_K(x), \nabla^2\psi_K(x)) \\ &\quad - \varepsilon^\theta\zeta_\varepsilon(x)\,\langle\nabla\psi_K(x), \nu(x)\rangle + o(\varepsilon^2) \\ &\leq \psi(x) + C\varepsilon^2 - \varepsilon^\theta\zeta_\varepsilon(x)\,\langle\nabla\psi(x) - 2K\nabla\,\mathrm{sd}(x, \partial\Omega), \nu(x)\rangle + o(\varepsilon^2) \end{aligned}$$

for any $x \in \overline{\Omega}$, where $C > 0$ depends on δ, L, K and the size of the domain. Since (7.3.6) holds for all x near $\partial\Omega$, the inequality above implies that

$$\begin{aligned} H_N^\varepsilon[\psi_K](x) &\le \psi_K(x) + C\varepsilon^2 - \varepsilon^\theta \zeta_\varepsilon(x)\left(\langle \nabla\psi(x), \nu(x)\rangle + K\right) + o(\varepsilon^2) \\ &\le \psi_K(x) + C\varepsilon^2 + o(\varepsilon^2). \end{aligned}$$

Since H_N^ε is monotone, by DPP we have

$$u^\varepsilon(\cdot, \varepsilon^2) = H_N^\varepsilon[u_0] \le H_N^\varepsilon[\psi_K] \le \psi_K + C\varepsilon^2 + o(\varepsilon^2).$$

Using DPP to repeat the estimate, we have

$$\begin{aligned} u^\varepsilon(\cdot, 2\varepsilon^2) = H_N^\varepsilon[u^\varepsilon(\cdot, \varepsilon^2)] &= H_N^\varepsilon\left[\psi_K + C\varepsilon^2 + o(\varepsilon^2)\right] \\ &\le H_N^\varepsilon[\psi_K] + C\varepsilon^2 + o(\varepsilon^2) \le \psi_K + 2C\varepsilon^2 + o(2\varepsilon^2). \end{aligned}$$

Iterating this argument, we are led to

$$u^\varepsilon(x, t) \le \psi_K(x) + Ct + o(1)$$

for all $(x, t) \in \overline{\Omega} \times [0, \infty)$, which implies

$$\overline{u}(x_0, 0) \le \psi_K(x_0) = u_0(x_0) + \frac{\delta}{4} - 2K \operatorname{dist}(x_0, \partial\Omega) + 2K \sup_{B_r(x_0)\cap\overline{\Omega}} \operatorname{dist}(\cdot, \partial\Omega).$$

Letting $\delta \to 0$ and then $r \to 0$, we end up with $\overline{u}(x_0, 0) \le u_0(x_0)$. We omit the proof for the part with $\underline{u}(x, 0)$, since it is symmetric. □

We now present the proof of Theorem 7.3.1.

Proof of Theorem 7.3.1 By Proposition 7.3.4, $\overline{u}$ and $\underline{u}$ are respectively a subsolution and a supersolution of (7.3.1)–(7.3.2). Since $\overline{u}(\cdot, 0) \le \underline{u}(\cdot, 0)$ in $\overline{\Omega}$ due to Proposition 7.1.10, we may apply comparison principle Theorem 4.7.4 to obtain $\overline{u} \le \underline{u}$ in $\overline{\Omega} \times [0, \infty)$. This in particular implies the locally uniform convergence of u^ε to the unique viscosity solution in $\overline{\Omega} \times [0, \infty)$. □

Remark 7.3.6 In our game interpretation, the C^2 regularity of ν is not essential; indeed, one can take ν to be any continuous function in $\overline{\Omega}$ provided

$$\max_{x\in\partial\Omega} \langle \nu(x), \nabla \operatorname{sd}(x, \partial\Omega)\rangle < 0$$

In this case, as a substitute of (7.3.6), we deduce that for $\sigma > 0$ small,

$$\max_{x\in\overline{\Omega}\setminus\Omega_\sigma} \langle \nu(x), \nabla \operatorname{sd}(x, \partial\Omega)\rangle < 0.$$

7.3.3 *Other Boundary Conditions*

The game-theoretic interpretation for the Neumann boundary problem can be easily generalized for other boundary value problems. We give several generalizations below. In this section we still assume that Ω is a bounded domain of class C^2 and let ν be the unit outward normal to $\partial\Omega$.

7.3.3.1 Prescribed Contact Angle Boundary Condition

Besides the Neumann problem, we may also consider the curvature flow equation (7.3.1) with the prescribed contact angle boundary condition, which can be expressed in the level-set formulation by

$$\langle \nabla u(x,t), \nu(x)\rangle + z(x)|\nabla u(x,t)| = 0 \quad \text{for } (x,t) \in \partial\Omega \times (0,\infty), \tag{7.3.19}$$

where z is a given continuous function on $\partial\Omega$ satisfying

$$\max_{x\in\partial\Omega} |z(x)| < 1; \tag{7.3.20}$$

recall the geometric meaning of this boundary condition in Sect. 2.5.4. Given a continuous initial value (7.3.3), uniqueness of solutions of the oblique boundary problem is established in the work of H. Ishii and M.-H. Sato [283].

Before studying the games, we first reformulate the boundary condition (7.3.19) as

$$\langle \nabla u(x,t), \gamma(x)\rangle = 0 \quad \text{for } (x,t) \in \partial\Omega \times (0,\infty), \tag{7.3.21}$$

where γ denotes a unit outward oblique normal to $\partial\Omega$. More precisely, if u satisfies (7.3.21) for $x \in \partial\Omega$ and $t > 0$, then we can take a unit vector $\nu^{\perp}(x)$ perpendicular to $\nu(x)$ such that setting

$$\gamma(x) = \sqrt{1 - z^2(x)}\,\nu(x) + z(x)\nu^{\perp}(x)$$

yields

$$\begin{aligned}\langle \nabla u(x,t), \gamma(x)\rangle &= \left\langle \nabla u(x,t), \sqrt{1 - z^2(x)}\,\nu(x) + z(x)\nu^{\perp}(x)\right\rangle \\ &= -\sqrt{1 - z^2(x)}\,z(x)|\nabla u(x,t)| + \sqrt{1 - z^2(x)}\,z(x)|\nabla u(x,t)| = 0.\end{aligned}$$

Moreover, let us assume that γ can be extended to a continuous function in $\overline{\Omega} \setminus \Omega_\sigma$ for $\sigma > 0$ small, where Ω_σ is given by (7.3.5). In view of Remark 7.3.6, we emphasize that the C^2 smoothness of γ is not necessary. But a variant of (7.3.6)

$$\min_{x \in \partial\Omega} \langle \gamma(x), \nu(x) \rangle > 0 \tag{7.3.22}$$

holds thanks to the assumption (7.3.20).

Based on the game interpretation for the Neumann problem, the games in the current case can be constructed easily. In fact, we only need to replace the normal ν by γ in the game rules in Sect. 7.3. We fix $1 < \theta < 2$ and define η_ε and ζ_ε in the same way as in (7.3.7) and (7.3.8). We can show that the game positions $y_1, y_2, \ldots, y_N$ stay in $\overline{\Omega}$ thanks to the definition of ζ_ε and the condition (7.3.22).

One can similarly define the game value as in Game 3. In this case, the DPP reads

$$u^\varepsilon(x,t) = \min_{|v|=1} \max_{b=\pm 1} u^\varepsilon\left(x + \sqrt{2}\varepsilon\eta_\varepsilon(x)bv - \varepsilon^\theta\zeta_\varepsilon(x)\gamma(x), t - \varepsilon^2\right) \tag{7.3.23}$$

Theorem 7.3.7 *Let $\Omega \subset \mathbf{R}^2$ be a bounded C^2 domain. Assume that u_0 is Lipschitz continuous in $\overline{\Omega}$. Consider Game 3 with ν replaced by the oblique normal γ. Let u^ε be the value function in this case. Then $u^\varepsilon \to u$ locally uniformly in $\overline{\Omega} \times [0, \infty)$ as $\varepsilon \to 0$, where u is the unique viscosity solution of* (7.3.1) *and* (7.3.21) *with the initial value* (7.3.3).

We omit the detailed proof, since it is essentially the same as that of Theorem 7.3.1. Using (7.3.23), one can show that the relaxed limits are respectively a sub- and a supersolution. We finally use the comparison principle given in Theorem 2.1 of H. Ishii and M.-H. Sato [283] to conclude the convergence of u^ε.

7.3.3.2 Dynamic Boundary Condition

The game setting also applies to the *dynamic boundary condition*

$$u_t(x,t) + \langle \nabla u(x,t), \nu(x) \rangle = 0 \quad \text{on } \partial\Omega \times (0, \infty) \tag{7.3.24}$$

interpreted again in the viscosity sense. Note that the definition of viscosity solutions to the dynamic boundary value problem is not covered by Definition 3.3.1 or Remark 3.3.2, since the boundary operator $B(x, t, u, \nabla u, \nabla^2 u)$ there is assumed to be independent of the time derivative u_t. However, one can define viscosity subsolutions and supersolutions in a similar manner, requiring either the equation or the dynamic boundary condition holds on the boundary in the viscosity sense. For example, in the case of mean curvature flow equation (7.3.1), we ask locally bounded upper semicontinuous subsolutions to satisfy either

$$\varphi_t - |\nabla\varphi| \operatorname{div}\left(\frac{\nabla\varphi}{|\nabla\varphi|}\right) \leq 0$$

or

$$\varphi_t + \langle \nabla\varphi, \nu\rangle \leq 0$$

at any $\hat{z} \in \partial\Omega \times (0, \infty)$ whenever there exists a function $\varphi \in C^2(\mathcal{O})$ with a neighborhood $\mathcal{O}$ of $\hat{z}$ in $\mathbf{R}^N \times (0, \infty)$ such that $u - \varphi$ attains a maximum in $\mathcal{O} \cap (\overline{\Omega} \times (0, \infty))$. One can define analogously supersolutions and thus solutions as well.

For the curvature flow equation (7.3.1) with the boundary condition (7.3.24), we can also slightly modify the rules of Game 3 to establish a game interpretation. In this case, we keep the whole setting of Game 3 but set $\theta = 2$; in other words, at the k-th step, the game position moves from $y_{k-1} \in \overline{\Omega}$ to y_k given by

$$y_k = y_{k-1} + \sqrt{2}\varepsilon\eta_\varepsilon(y_{k-1})b_k v_k - \varepsilon^2\zeta_\varepsilon(y_{k-1})\nu(y_{k-1}).$$

It is not difficult to show that $y_k \in \overline{\Omega}$ so that the game position stays in $\overline{\Omega}$ after each move. Define the value function as usual. It satisfies the following DPP:

$$u^\varepsilon(x, t) = \min_{|v|=1} \max_{b=\pm 1} u^\varepsilon\left(x + \sqrt{2}\varepsilon\eta_\varepsilon(x)bv - \varepsilon^2\zeta_\varepsilon(x)\gamma(x), t - \varepsilon^2\right). \tag{7.3.25}$$

Thus we have the following result.

Theorem 7.3.8 *Let $\Omega \subset \mathbf{R}^2$ be a bounded C^2 domain. Assume that u_0 is Lipschitz continuous in $\overline{\Omega}$. Consider Game 3 with $\theta = 2$. Let u^ε be the value function in this case. Let $\overline{u}$ and $\underline{u}$ be the relaxed limits of u^ε as in* (7.1.10). *Then $\overline{u}$ and $\underline{u}$ are respectively a subsolution and a supersolution of* (7.3.1) *and* (7.3.24) *with*

$$\overline{u}(\cdot, 0) \leq u_0 \leq \underline{u}(\cdot, 0) \quad \text{in } \overline{\Omega}.$$

If the comparison principle holds, then $u^\varepsilon \to u$ locally uniformly in $\overline{\Omega} \times [0, \infty)$ as $\varepsilon \to 0$, where u is the unique viscosity solution of (7.3.1) *and* (7.3.24) *with the initial value* (7.3.3).

We omit the rigorous proof but briefly describe the heuristics of this result based on the formal proof in Sect. 7.3.1. We again assume that u^ε is smooth and converges to u with $|\nabla u| \neq 0$.

Let $x \in \partial\Omega$. Suppose that there exists $x_\varepsilon \in \overline{\Omega}$ such that $x_\varepsilon \to x$ as $\varepsilon \to 0$. Applying Taylor expansion on (7.3.25) and following the estimates (7.3.10) and (7.3.11), we get (7.3.12) with $\theta = 2$, i.e.,

$$\varepsilon^2\eta_\varepsilon^2\left(u_t^\varepsilon - |\nabla u^\varepsilon| \operatorname{div}\left(\frac{\nabla u^\varepsilon}{|\nabla u^\varepsilon|}\right)\right) + \varepsilon^2\zeta_\varepsilon\left(u_t^\varepsilon + \langle\nabla u^\varepsilon, \nu\rangle\right) = 0 \quad \text{at } (x_\varepsilon, t). \tag{7.3.26}$$

In view of (7.3.8), by taking a subsequence we have $\eta_\varepsilon^2 \to c$ and $\zeta_\varepsilon \to 1 - c$ as $\varepsilon \to 0$ for some $0 \leq c \leq 1$. Dividing (7.3.26) by ε^2 and letting $\varepsilon \to 0$, we deduce that at (x, t)

$$c\left(u_t - |\nabla u| \operatorname{div}\left(\frac{\nabla u}{|\nabla u|}\right)\right) + (1-c)\,(u_t + \langle \nabla u, \nu\rangle) = 0.$$

This implies that u satisfies the dynamic boundary condition at $x \in \partial\Omega$.

The comparison principle for (7.3.1)(7.3.24) is an open question for general domains. In the case of a half space, the comparison principle is established in Theorem 3.1 of Y. Giga and N. Hamamuki [212]. As an immediate result of this comparison theorem, we obtain the following convergence of game values in a half plane. Denote by $K_\alpha(\overline{\Omega})$ the space of all continuous functions that equal α outside some compact set in $\overline{\Omega}$.

Corollary 7.3.9 *Let* $\Omega = \{(x_1, x_2) \in \mathbf{R}^2; x_2 > 0\}$. *Assume that* $u_0 \in K_\alpha(\overline{\Omega})$ *for some* $\alpha \in \mathbf{R}$ *and is Lipschitz continuous in* $\overline{\Omega}$. *Consider Game 3 with* $\theta = 2$. *Let* u^ε *be the value function in this case. Then,* $u^\varepsilon \to u$ *locally uniformly in* $\overline{\Omega} \times [0, \infty)$ *as* $\varepsilon \to 0$, *where* u *is the unique viscosity solution of* (7.3.1) *and* (7.3.24) *with the initial value* (7.3.3).

7.4 Notes and Comments

The deterministic game interpretation for motion by mean curvature was proposed by R. V. Kohn and S. Serfaty [312]. Their approach applies to both the initial value problem for the parabolic equation and the stationary problem in a strictly convex domain with homogeneous Dirichlet boundary condition. It was a remarkably surprising result, establishing a purely deterministic representation formula for the solution of a second order elliptic or parabolic PDE, which is typically interpreted through certain stochastic processes. Related ideas about numerical approximation for motion by curvature appeared in F. Catté et al. [93]. A related continuous-time stochastic representation for mean curvature flow was provided by H. M. Soner and N. Touzi [403, 404]. In this chapter, we focus on our main results of the game convergence in two space dimensions. The original paper by R. V. Kohn and S. Serfaty [312] explained how to extend this game-theoretic approach to higher dimensional case and more general curvature flow equations, including the curvature flow of power type. See also Y. Giga and Q. Liu [216] for related results on the generalization. More recently, K. Misu [351] extended the game-theoretic approach to an obstacle problem for the mean curvature flow equation.

Compared to the parabolic case, the stationary Dirichlet boundary value problem (SDP) demonstrates more challenging difficulty in its analysis via discrete games. While it is relatively easy to show that the upper and lower relaxed limits of the game values are respectively a subsolution and a supersolution in the domain, certain convexity of the domain is needed for us to verify the homogeneous boundary condition.

The convergence rate in Theorem 7.2.7 is again due to R. V. Kohn and S. Serfaty [312]. The C^3 regularity of the solution U to (SDP) in two dimensions plays a key role in the proof. In general, the solution $U(x)$ can be regarded as the time of first arrival to x of mean curvature flow starting from the hypersurface $\partial\Omega$. In the work G. Huisken [260, 261], it is shown that the arrival time is C^2 when the initial hypersurface is convex. N. Sesum [394] further showed that U may not be C^3 for a convex initial hypersurface in dimensions higher than 2. For mean convex initial hypersurfaces, T. Ilmanen [266] gave an example in $\mathbf{R}^3$, showing that the arrival time may not be C^2. More recently, the twice differentiability of the arrival time for mean convex initial hypersurfaces is established by T. H. Colding and W. P. Minicozzi, II [122]. A sufficient and necessary condition on the structure of singularity is provided for the C^2 regularity of the arrival time in T. H. Colding and W. P. Minicozzi, II [123].

The case with a nonconvex domain for the stationary problem (SDP) is more challenging to treat. The main difficult lies at the possible loss of boundary data, i.e, the first exit time may be positive on the boundary of the domain. As shown in Example 7.2.10, the loss of boundary data may also cause discontinuity of the exit time. Example 7.2.10 is an adaptation of an example constructed by G. Barles and F. Da Lio, presented in Appendix C.3 of the paper by R. V. Kohn and S. Serfaty [312]. We therefore cannot expect the standard comparison principle to hold in general, as shown in Example 7.2.16. This example, constructed in Q. Liu [327], is closely related to the fattening phenomenon for the motion by curvature of a figure-eight shaped initial curve as described in Sect. 5.5. If the domain is assumed to be star-shaped, we obtain the game convergence to a unique possibly discontinuous viscosity solution by using the weak comparison principle, Theorem 7.2.14. These results are also due to Appendix C.3 of R. V. Kohn and S. Serfaty [312].

Further study is needed for a domain that is neither convex nor star-shaped. In Sect. 7.2.4, we examine the situation when Ω is a domain in $\mathbf{R}^2$ and $\partial\Omega$ is given as the graph of a smooth periodic function in $\mathbf{R}$. In this case, one can redefine the boundary values of the stationary problem (SDP) by utilizing the arrival time of the positive curvature flow. The new boundary data are compatible with the discrete games and enable us to justify the convergence of value functions. Our presentation of this part is a simplified version of Q. Liu [328].

Regarding the Neumann boundary value problem for the curvature flow equation, a game-theoretic approach was established in Y. Giga and Q. Liu [217], incorporating billiard dynamics within the games to generate the Neumann boundary condition. In Sect. 7.3, we employ a different game-theoretic interpretation of the Neumann boundary condition. This approach, introduced by N. Hamamuki and Q. Liu [251], accommodates more general boundary conditions for the curvature flow equation, including the oblique boundary conditions and dynamic boundary conditions.

It is worth mentioning that such a deterministic game-theoretic approach is not limited to curvature flow equations and other geometric evolution equations. Under the comparison principle, R. V. Kohn and S. Serfaty [313] developed deterministic discrete games who value functions converge uniformly to the unique viscosity solutions of general fully nonlinear elliptic and parabolic equations. The key again is to choose different scaling for the space and time step sizes. Such game-theoretic approximation for the fully nonlinear equations was extended to the Neumann boundary value problem by J.-P. Daniel [134] and to dynamic boundary problems by N. Hamamuki and Q. Liu [250]. We refer to P. Cheridito et al. [110] for a stochastic representation for the viscosity solutions of fully nonlinear parabolic equations.

The game-theoretic interpretation for PDEs has recently found exciting applications in the machine learning theory as well. Motivated by a model problem from the machine learning literature known as "prediction with expert advice", a repeated two-person game involving a predictor and his adversarial opponent was studied in N. Drenska and R. V. Kohn [149, 150]. In this setting, the predictor has access to guidance from experts, whose outcomes are chosen by the adversary. At each step, the predictor must combine the experts' guidance using a mixed strategy and aim to minimize his worst-case final-time shortfall, called regret, relative to the best-performing expert. The optimal policies and the predictor's worst-case regret are determined by a dynamic programming principle. To understand the asymptotic behavior of the game over many time steps, one can derive a scaled version of the dynamic programming principle, which leads to a characterization of the predictor's worst-case regret as the unique viscosity solution of a nonlinear parabolic PDE. See V. A. Kobzar et al. [311], J. Calder and N. Drenska [79], N. Drenska and J. Calder [148] for related discussions on this topic.

Alongside with the aforementioned development of game theory for nonlinear partial differential equations, a class of stochastic discrete games, known as tug-of-war, also attracted significant attention. Introduced by Y. Peres et al. [379], these stochastic two-player games approximate the viscosity solution of the normalized infinity Laplace equation:

$$-\Delta_\infty^N u := -\frac{1}{|\nabla u|^2}\left\langle \nabla^2 u \nabla u, \nabla u \right\rangle = 0.$$

The game rules are simple. In a bounded domain Ω, Player I seeks to minimize, while Player II aims to maximize, the expected boundary value of the final game state that is determined by repeated fair coin tosses. The winner of coin toss at each round chooses a direction and moves the state by a distance $\varepsilon > 0$. The game concludes when the state exits Ω, with the value function U^ε defined as the expected value of all possible boundary values at the boundary hit. At least intuitively, it is not difficult to see that U^ε satisfies the dynamic programming principle:

$$U^\varepsilon(x) = \frac{1}{2} \sup_{y \in \mathrm{int}\, B_\varepsilon(x)} U^\varepsilon(y) + \frac{1}{2} \inf_{y \in \mathrm{int}\, B_\varepsilon(x)} U^\varepsilon(y) \tag{7.4.1}$$

for any $x \in \Omega$ and $\varepsilon > 0$ sufficiently small. A formal Taylor expansion with $\varepsilon \to 0$ then yields the infinity Laplace equation easily. The game convergence with the Dirichlet boundary condition can be rigorously proved in the framework of viscosity solutions as with the games for curvature flows discussed in this chapter. We refer to F. Charro et al. [100] and T. Antunović et al. [32] for similar game-theoretic approximations involving the Neumann boundary condition and to Y. Peres et al. [378] for a general type of tug-of-war games.

By incorporating random walk to each step, Y. Peres and S. Sheffield [380] constructed a variant of the tug-of-war games that give rise to the homogeneous or inhomogeneous equations associated to the normalized p-Laplacian with $1 < p < \infty$:

$$- \Delta_p^N u := -|\nabla u|^{2-p} \operatorname{div}(|\nabla u|^{p-2} \nabla u).$$

The associated dynamic programming principle was further interpreted by J. J. Manfredi et al. [344] as an asymptotic mean value property for p-harmonic functions. See also J. J. Manfredi et al. [345] for results on the parabolic counterpart of the normalized p-Laplace equation. Note that the normalized p-Laplacian reduces to the mean curvature operator in the special case where $p = 1$. The asymptotic mean value property for the case $p = 1$ was discussed by M. B. Rudd and H. A. Van Dyke [385] using the notion of median instead of the standard mean in the formula. A similar approach utilizing the function median was adopted to study p-harmonic functions in the work D. Hartenstine and M. Rudd [255].

The stochastic game-theoretic interpretation and asymptotic mean value property for elliptic and parabolic equations related to the p-Laplacian $(1 < p \leq \infty)$ have been extensively studied. In addition to the previously mentioned work, see for example the papers Y. Yu [433], S. N. Armstrong and C. K. Smart [35, 36], A. M. Oberman [367] and H. Luiro et al. [340] for further insights into related games and schemes. Another asymptotic mean value property for p-Laplacian with a different notion of means was introduced in M. Ishiwata et al. [290]. These topics are beyond the scope of this chapter, but readers interested in this novel perspective may consult the surveys by L. C. Evans [163] and by J. D. Rossi [382], as well as the monographs by P. Blanc and J. D. Rossi [66] and by M. Lewicka [318], for detailed discussions.

We conclude this chapter by noting that discrete game-theoretic approximations can be extended to nonlinear elliptic and parabolic equations on sub-Riemannian manifolds and more general metric measure spaces. Related results can be found, for example, in F. Ferrari et al. [171, 172], F. Ferrari and A. Pinamonti [173], M. Lewicka et al. [319], and T. Adamowicz et al. [2]. The paper by F. Ferrari et al. [172] provides a deterministic game interpretation for the horizontal mean curvature flow in the Heisenberg group, a direction initiated by L. Capogna and G. Citti [83] using the level set method. See also N. Dirr et al. [145], along with N. Dirr et al. [144] for an approach to sub-Riemannian curvature flow equations based on stochastic optimal control.

Chapter 8
Applications of Game-Theoretic Approach

We provide two applications of the game-theoretic interpretation. We first use the games to discuss the fattening phenomena for the curvature flow equation. Another application of games is for the convexity preserving properties. We provide an alternative proof for the properties by adopting the game strategies and the dynamic programming principle.

8.1 A Game Interpretation of Fattening

In this section, we present several examples illustrating the game-theoretic approach to the *fattening phenomenon* for motion by curvature. Recall that fattening refers to the situation in which a level set of a solution to the level set equation develops nonempty interior, even though the corresponding level set of the initial data has no interior points. A rigorous proof for the occurrence of fattening is given in Sect. 6.2.1 based on the parabolic theory. We present in this section an alternative elementary proof based only on games.

8.1.1 Crossing Lines

Let us first revisit Example 6.2.1 in Sect. 6.2.1 and give a different proof based on the game-theoretic interpretation. Let

$$D_0 = \{(x_1, x_2) \in \mathbf{R}^2;\ |x_1| > |x_2|)\}. \tag{8.1.1}$$

We consider the initial value problem (MCF) for the level set curvature flow equation with u_0 being the signed distance to ∂D_0, i.e.,

Y. Giga, Q. Liu, *Surface Evolution Equations*, Monographs in Mathematics 99,
https://doi.org/10.1007/978-3-032-20303-8_8

$$u_0(x) = \mathrm{sd}(x, \partial D_0) = \mathrm{dist}\,(x, \mathbf{R}^2 \setminus D_0) - \mathrm{dist}\,(x, D_0), \qquad \mathrm{x} \in \mathbf{R}^2. \tag{8.1.2}$$

The zero level curve

$$\Gamma_0 = \{x \in \mathbf{R}^2;\, u_0(x) = 0\}$$

consists of two straight lines intersecting orthogonally. Such an initial set is known to develop interior instantly, and our result is thus as follows.

Theorem 8.1.1 (Fattening from Crossing Lines) *Let D_0 and u_0 be given by* (8.1.1) *and* (8.1.2). *Let u be the solution of* (MCF). *Then the zero level set*

$$\Gamma_t = \{x \in \mathbf{R}^2;\, u(x, t) = 0\} \tag{8.1.3}$$

has nonempty interior for every $t > 0$.

Proof We consider Game 1-a and use the approximation of the solution $u(x, t)$ by the game value $u^\varepsilon(x, t)$ for any given $(x, t) \in \mathbf{R}^2 \times [0, \infty)$.

(1) Suppose that $x \notin D_0$. Since Player I attempts to minimize the value of u_0 after $\ell = [t/\varepsilon^2]$ steps, one strategy he can choose is to take $v_k = x/|x|$ at the k-th step. Then this guarantees that $u_0(y_\ell) \le 0$, which in turn implies that $u^\varepsilon(x, t) \le 0$ for all $x \in D_0^c$ and $t \ge 0$.
(2) Suppose that $x = (x_1, x_2) \in D_0$. Let us assume that $x_1 > 0$. Define

$$D_0^+ = \{(x_1, x_2) \in D_0;\, x_1 > 0\},$$
$$D_0^- = \{(x_1, x_2) \in D_0;\, x_1 < 0\}.$$

Clearly, $x \in D_0^+$ under our assumption. In this case, Player I can utilize the concentric-circle strategy described in Example 7.1.3 to obtain an upper bound of the game value $u^\varepsilon(x, t)$. Consider circles in D_0^+ centered at the point $z = (\sqrt{2}r_0, 0)$, where we take

$$r_0 = \sqrt{2}x_1 + \sqrt{x_1^2 - x_2^2}.$$

It is not difficult to see that $x \in \partial B_{r_0}(z)$, since

$$|z - x|^2 = \left(\sqrt{2}r_0 - x_1\right)^2 + x_2^2 = 3x_1^2 - x_2^2 + 2\sqrt{2}x_1\sqrt{x_1^2 - x_2^2} = r_0^2.$$

Moreover, it is easily seen that $\mathrm{int}\, B_{r_0}(z)$ is inscribed in D_0^+, tangent to the boundary of D_0^+ at points $(r_0, \pm r_0)$. Denote by S_x the open set enclosed by the boundary of D_0^+ and the minor arc of $B_{r_0}(z)$ between $(r_0, \pm r_0)$.

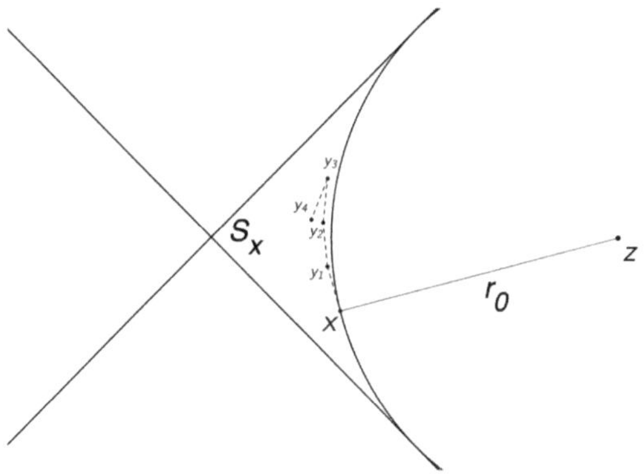

Fig. 8.1 The game trajectory in S_x

By choosing a direction v_1 tangential to the circle $\partial B_{r_0}(z)$ at x, regardless of the response of Player II, Player I ensures a next game position y_1 satisfying

$$|y_1 - z|^2 = r_0^2 + \varepsilon^2.$$

No matter how Player II plays, Player I can repeat this strategy, selecting tangential directions to concentric circles $B_{r_{j-1}}(z)$, to generate a sequence of positions y_j such that

$$r_j^2 = |y_j - z|^2 = r_0^2 + j\varepsilon^2.$$

See Fig. 8.1. Since the game position cannot enter $B_{r_0}(z)$ before exiting S_x, and

$$|y - z|^2 < 2r_0^2$$

for all $y \in S_x$, it follows that $y_j \notin D_0^+$ when $r_0^2 + j\varepsilon^2 > 2r_0^2$. In particular, under this strategy, the game position exits D_0^+ by the k-round with

$$k = \frac{r_0^2}{2\varepsilon^2} + 1.$$

Noting that at each $y_j \in S_x$, the tangent to the concentric circle does not intersect D_0^-, we see that the game position enters D_0^c immediately after leaving D_0^+. Upon exiting D_0^+, Player I can then switch to the strategy in (1) to keep the game position outside D_0. Consequently,

$$u^\varepsilon(x, t) \leq 0 \quad \text{for all } t \geq \frac{r_0^2}{2} + o(\varepsilon).$$

The case with $x_1 < 0$ is symmetric, leading to the same estimate above for u^ε.

Combining the above two estimates and applying Theorem 7.1.1, we obtain $u(x,t) \le 0$ for all $x \in \mathbf{R}^2$ and $t \ge 0$ with

$$t \ge \frac{1}{2}\left(\sqrt{2}|x_1| + \sqrt{x_1^2 - x_2^2}\right)^2. \tag{8.1.4}$$

Using the inverse game, Game 1-a', we can show that the game value $\tilde{u}^\varepsilon$ defined by (7.1.24) satisfies $\tilde{u}^\varepsilon(x,t) \ge 0$ for $x \in \mathbf{R}^2$ and $t \ge 0$ where (8.1.4) holds. In this case, Player II can adopt the same strategy as (1) above to guarantee this estimate. Since $\tilde{u}^\varepsilon$ converges to the same solution u, as explained in Remark 7.1.11, it follows that $u(x,t) \ge 0$ for all (x,t) satisfying (8.1.4). Consequently, $u(x,t) = 0$ if whenever (8.1.4) holds.

By exchanging the roles of x_1 and x_2, one can extend the argument above to show that $u(x,t) = 0$ for $x = (x_1, x_2)$ and $t \ge 0$ satisfying $S(x_1,x_2)^2 \le 2t$, where

$$S(x_1,x_2) := \sqrt{2}\max\{|x_1|, |x_2|\} + \sqrt{\max\left\{x_1^2, x_2^2\right\} - \min\left\{x_1^2, x_2^2\right\}};$$

in other words, we have

$$\left\{(x_1,x_2) \in \mathbf{R}^2;\ S(x_1,x_2) \le \sqrt{2t}\right\} \subset \Gamma_t$$

for all $t \ge 0$, as indicated in Fig. 6.3. □

The game-theoretic proof above is elementary and direct, requiring no parabolic PDE theory.

8.1.2 Figure-Eight Shape

We continue to investigate a slightly more complicated situation. The initial surface is in the shape of *figure eight*. Our analysis is similar to Example 7.2.16.

Fix a constant $R > 0$. Let $P_1 = (-R, 0)$, $P_2 = (R, 0)$ and

$$D_0 = \operatorname{int} B_R(P_1) \cup \operatorname{int} B_R(P_2). \tag{8.1.5}$$

We take the initial value u_0 in the same way as in (8.1.2).

Theorem 8.1.2 (Fattening from a Figure Eight Curve) *Let D_0 be as in* (8.1.5) *and u_0 be defined by* (8.1.2). *Let u be the solution of* (MCF). *Then the zero level set Γ_t given by* (8.1.3) *has nonempty interior for every* $0 < t < R^2/2$.

To prove this theorem, we again estimate the values of Game 1-a and its inverse Game 1-a' at a fixed time $t > 0$, as is done in Sect. 8.1.1. Let us first use Game 1-a to get the following estimate.

Proposition 8.1.3 *Let u_0 given in* (8.1.2) *with D_0 in* (8.1.5). *Let u be the unique solution of* (MCF). *Then $u(x,t) \leq 0$ for all $(x,t) \in \mathbf{R}^2 \times [0,\infty)$ satisfying*

$$2t \geq 2R|x_1| - x_1^2 - x_2^2. \tag{8.1.6}$$

Proof Denote the strip region

$$S = \{x = (x_1, x_2);\ |x_1| \leq \sqrt{2}\varepsilon\}.$$

We recall in Game 1-a Player I tries to minimize the value $u_0(y_\ell)$ for $\ell = [t/\varepsilon^2]$. The condition (8.1.6) amounts to saying that $x \notin B_r(P_1) \cup B_r(P_2)$, where $r = \sqrt{R^2 - 2t}$. Using the concentric strategy introduced in Example 7.1.3, Player I can make the game position exit either $\operatorname{int} B_R(P_1)$ or $\operatorname{int} B_R(P_2)$ in ℓ steps. It follows that $y_K \in S \cap (\mathbf{R}^2 \setminus D_0)$ for some $K \leq \ell$.

We claim that y_k stays in $S \cap (\mathbf{R}^2 \setminus D_0)$ for all $k \geq K$. In fact, Player I can continue to apply the concentric strategy so that either $y_\ell \in \mathbf{R}^2 \setminus (D_0 \cup S)$ or $y_k \in S$ for some $k \geq K$. If $y_k \in S$, then Player I switch to take the vector $v = (0, 1)$ for the rest of the game. This yields $y_\ell \in S$ with no regard to the responses of Player II. In either case, we have

$$u_0(y_\ell) \leq -\sqrt{2}\varepsilon.$$

Considering Player I may have better strategies, we have $u^\varepsilon(x,t) \leq -\sqrt{2}\varepsilon$ for (x,t) satisfying (8.1.6). Sending $\varepsilon \to 0$, we get $u(x,t) \leq 0$ for all $(x,t) \in \mathbf{R}^2 \times [0,\infty)$ that satisfies (8.1.6) by Theorem 7.1.1. □

We next use Game 1-a' to obtain the following result.

Proposition 8.1.4 *Let u_0 given in* (8.1.2) *with D_0 in* (8.1.5). *Let u be the unique solution of* (MCF). *Then $u(x,t) \geq 0$ for all $x = (x_1, x_2) \in \mathbf{R}^2$ and $t \geq 0$ satisfying*

$$x_1^2 + x_2^2 - R|x_1| \leq 0, \quad 2t(x_1^2 + x_2^2) \leq x_1^2 R^2. \tag{8.1.7}$$

Proof Let $x = (x_1, x_2)$ satisfy (8.1.7). Consider Game 1-a' and consider its game value $\tilde{u}^\varepsilon$ as in (7.1.24). In what follows we denote by u^ε instead of $\tilde{u}^\varepsilon$ for convenience. We first assume that $x \neq 0$. Suppose that Player I keeps choosing $v_k = x/|x|$ starting from x so that the game position stays on the straight line. Take the points

$$P_3 = \left(-\frac{Rx_1^2}{x_1^2 + x_2^2}, -\frac{Rx_1^2 x_2}{x_1(x_1^2 + x_2^2)}\right), \quad P_4 = \left(\frac{Rx_1^2}{x_1^2 + x_2^2}, \frac{Rx_1^2 x_2}{x_1(x_1^2 + x_2^2)}\right)$$

on the same line. Moreover, by (8.1.7) we see that x is on the line segment between P_3 and P_4. Now Player II has the following three options to run the game.

(1) Player II chooses b_k to let the position stay on the line segment joining P_3 and P_4. Then it is clear that $u_0(y_N) \geq 0$ under this strategy.
(2) Suppose that Player II moves the position towards P_3. Then after k steps with k finite, we have $y_k \in \operatorname{int} B_{\sqrt{2}\varepsilon}(P_3)$. From that point, Player I switches to the strategy of concentric circles in Example 7.1.3 around the center P_1. Then, regardless of the choices of Player II, it costs at least $K - 1$ steps to leave $B_R(P_1)$, where

$$K = \frac{x_1^2 R^2}{2\varepsilon^2(x_1^2 + x_2^2)}.$$

It follows that $u_0(y_\ell) \geq 0$ for all $t < K\varepsilon^2$. Note that $2K\varepsilon^2 = x_1^2 R^2/(x_1^2 + x_2^2)$ for all $\varepsilon > 0$ small.
(3) If Player II moves the position to P_4, then Player I can apply a strategy similar to the case above but with center at P_2. In this case, we still have $u_0(y_\ell) \geq 0$ for $t < K\varepsilon^2$.

Altogether, we deduce that $u^\varepsilon(x, t) \geq 0$ provided that $2t(x_1^2 + x_2^2) < x_1^2 R^2$. In view of Theorem 7.1.1 and the continuity of u, passing to the limit as $\varepsilon \to 0$, we are led to $u(x, t) \geq 0$ for all (x, t) satisfying (8.1.7). □

Remark 8.1.5 In our proof above, for $x = (x_1, x_2) \neq 0$ satisfying (8.1.7), we have

$$K\varepsilon^2 = \frac{x_1^2 R^2}{2(x_1^2 + x_2^2)} \geq \frac{x_1^2 + x_2^2}{2}.$$

This shows that, starting the game from such x, Player II needs at least $|x|^2/(2\varepsilon^2) - 1$ steps to expel the game position outside $\overline{D_0}$.

Proof of Theorem 8.1.2 Combining Proposition 8.1.3 and Proposition 8.1.4, we have $u(x, t) = 0$ for all $0 \leq t < R^2/2$ and $x \in E_t$, where

$$E_t := \left\{(x_1, x_2) \in \mathbf{R}^2;\ R|x_1| - 2t \leq x_1^2 + x_2^2 - R|x_1| \leq 0,\ 2t(x_1^2 + x_2^2) \leq x_1^2 R^2\right\}. \tag{8.1.8}$$

We complete the proof by noticing that E_t has nonempty interior for all $0 < t < R^2/2$. □

Besides the set E_t, the zero level set Γ_t in fact also contains

$$E_t' := \left\{(x_1, x_2) \in \mathbf{R}^2;\ \frac{2R|x_2|\sqrt{2t}}{\sqrt{R^2 - 2t}} - 2t < x_1^2 + x_2^2 < 2t,\ x_1^2 R^2 < 2t(x_1^2 + x_2^2)\right\}. \tag{8.1.9}$$

See Fig. 8.2.

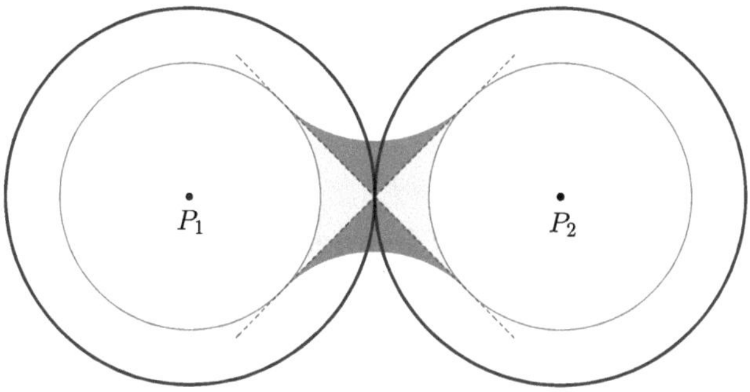

Fig. 8.2 Light gray region E_t and dark gray region E'_t

Proposition 8.1.6 *Let D_0 be as in* (8.1.5) *and u_0 be defined by* (8.1.2). *Let u be the solution of* (MCF). *Then $u(x,t) \geq 0$ for all $0 < t < R^2/2$ and $x \in E'_t$, where E'_t is given by* (8.1.9).

Proof Fix $x = (x_1, x_2) \in E'_t$ and assume $x_2 \geq 0$. We again let u^ε be the game value of Game 1-a'. Take $r = 2t/\sqrt{R^2 - 2t}$ and $y = (0,\ R\sqrt{2t}/\sqrt{R^2 - 2t})$. Then by definition of E'_t, we have $x \notin B_r(y)$. If Player I adopts the strategy in Example 7.1.3 with center at y, then the game position remains inside E'_t unless it gets into the set E_t, which is given by (8.1.8). Suppose that Player II takes a strategy to let $y_k \in E_t$ for some $k = 1, 2, \ldots, \ell$, then we have

$$|y_k - y|^2 - r^2 = 2k\varepsilon^2.$$

On the other hand, in view of Remark 8.1.5, Player I has a strategy to guarantee $y_\ell \in \overline{D_0}$ if

$$\ell - k \leq \frac{|y_k|^2}{2\varepsilon^2} - 1.$$

In other words, under the combination of the strategies of Player I above, we get $u_0(y_\ell) \geq -\sqrt{2}\varepsilon$ provided that

$$\ell\varepsilon^2 \leq \frac{|y_k|^2 + |y_k - y|^2 - r^2}{2}.$$

However, the last inequality always holds due to the choices of y and r; indeed, it is easily seen that

$$|y_k|^2 + |y_k - y|^2 - r^2 \geq |y|^2 - r^2 = 2t.$$

Hence, we conclude that $u^\varepsilon(x,t) \geq -\sqrt{2}\varepsilon$; passing to the limit as $\varepsilon \to 0$, we obtain and therefore $u(x,t) \geq 0$ for all $x \in E'_t$. □

In addition to the examples above, we can take advantage of this point of view to understand fattening for more geometric evolutions by the game interpretations introduced in Sect. 7.1.3.

8.1.3 Fattening for Dynamic Boundary Problems

We can further use the game interpretation studied in Sect. 7.3 to construct examples of fattening for various boundary value problems. We here give a simple example demonstrating the use of games to study dynamic boundary problems. We choose an initial value u_0 whose zero level set is tangent to the domain boundary at a point. This configuration can be viewed as an adaptation of a figure-eight type initial curve with a *dynamic boundary condition.*

Let us consider

$$\Omega = \{x = (x_1, x_2) \in \mathbf{R}^2;\, x_2 > 0\}. \tag{8.1.10}$$

For convenience, we denote $e_1 = (1, 0)$ and $e_2 = (0, 1)$. It is clear that the outward unit normal vector to $\partial\Omega$ in this case is $\nu = -e_2$.

Fix $R > 0$. Let $z = (0, R)$ and $E = B_R(z)$. We take a triangular region Q contained in E symmetric about the x_2-axis, i.e.,

$$Q := \{(x_1, x_2) \in \Omega;\, L|x_1| \le x_2 \le L\mu\} \subset E \tag{8.1.11}$$

for some $L > 1$ and $\mu > 0$. For $M > 0$ large, we take

$$u_0(x) = \max\{\min\{\mathrm{sd}(x, \partial E),\ M\},\ -M\} \quad \text{for } x \in \overline{\Omega}, \tag{8.1.12}$$

where $\mathrm{sd}(x, \partial E)$ denotes the signed distance to ∂E as defined in (7.3.4). Our truncation on the signed distance function of E by $\pm M$ is not essential to the example but only for the boundedness of u_0. Under these conditions, we can show that the zero level set of the solution u to (7.3.1) and (7.3.24) with the initial value (7.3.3). A precise statement is as follows.

Theorem 8.1.7 (Fattening for Curvature Flow with Dynamic Boundary Condition) *Suppose that Ω is the half plane given by* (8.1.10). *Let $z = (0, R)$ and $E = B_R(z)$ for $R > 0$. Let Q be given by* (8.1.11) *with $\mu > 0$ and $L > 1$. Let u_0 be defined as in* (8.1.12) *with $M > 0$ large and let u be the unique solution of* (7.3.1) *and* (7.3.24) *with the initial value* (7.3.3). *Then for any $r > 0$ small, there exist $\tau_2 > \tau_1 > 0$ and an open subset $\mathcal{O}$ of $\mathrm{int}\, B_r(0) \cap Q$ such that $u(x, t) = 0$ for all $x \in \mathcal{O}$ and $\tau_1 \le t \le \tau_2$.*

Proof We consider Game 3 in Sect. 7.3.1 with $\theta = 2$ for the dynamic boundary problem. Below we use B_r to denote $B_r(0)$ for simplicity. Suppose that the game starts at $x \in \mathrm{int}\, B_r$ for some $0 < r < R$. Let

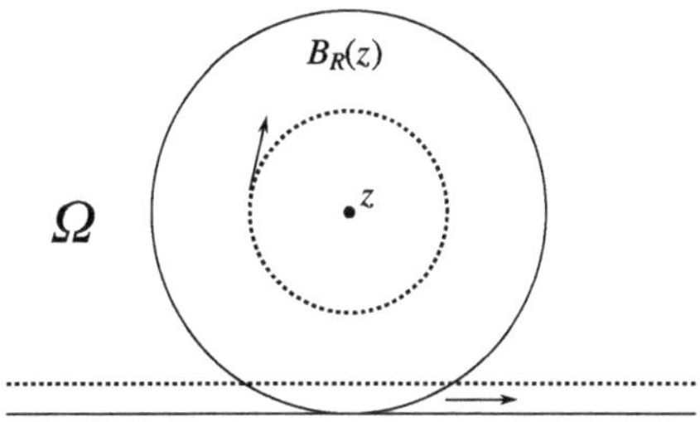

Fig. 8.3 A strategy for Player I. Reprinted without changes under the Creative Commons Attribution 4.0 International License (https://creativecommons.org/licenses/by/4.0/) from N. Hamamuki and Q. Liu, A Game-theoretic Approach to Dynamic Boundary Problems for Level-set Curvature Flow Equations and Applications, Partial Differ. Equ. Appl., 2 (2021), Article 30

$$S_\varepsilon = \mathbf{R} \times [0, \sqrt{2}\varepsilon]. \tag{8.1.13}$$

We divide the proof into several steps.

1. A Strategy for Player I

Since Player I aims to minimize the final value $u_0(y_\ell)$ with $\ell = [t/\varepsilon^2]$, he may take the following strategy at the k-th step for all $k = 1, 2, \ldots, \ell$ (Fig. 8.3):

1. if $y_{k-1} \in S_\varepsilon$, then Player I chooses $v_k = e_1$;
2. if $y_{k-1} \in \overline{\Omega} \setminus S_\varepsilon$, then he takes a unit vector v_k tangent to the circle concentric to $B_R(z)$, i.e., v_k satisfies

$$\langle v_k, y_{k-1} - z \rangle = 0.$$

Regardless of the choices of Player II, the strategy above enables us to make the next two assertions regarding y_k.

- Once $y_k \in S_\varepsilon$ for some $k = 1, 2, \ldots, \ell$, then despite the effect due to the normal vector ν in the game, the game trajectory will never leave the region S_ε and consequently $y_\ell \in S_\varepsilon$.
- If $y_k \in \overline{\Omega} \setminus S_\varepsilon$ for all $k = 1, 2, \ldots, \ell$, then

$$|y_\ell - z|^2 = |x - z|^2 + 2\varepsilon^2\ell \geq (R - r)^2 + 2\varepsilon^2\ell.$$

By definition of u^ε, it follows that either

$$u^\varepsilon(x, t) \leq \sqrt{2}\varepsilon$$

or

$$u^\varepsilon(x, t) \leq R - \sqrt{(R - r)^2 + 2t}.$$

In particular, we have

$$u^{\varepsilon}(x,t) \le \sqrt{2}\varepsilon \quad \text{for } x \in \operatorname{int} B_r \text{ and } t \ge \frac{1}{2}R^2 - \frac{1}{2}(R-r)^2. \tag{8.1.14}$$

2. A Strategy for Player II

On the other hand, we may also consider a special strategy in favor of Player II. However, in this case we need to further restrict the game starting point x. We are interested in the first exit time

$$\tau_{\varepsilon}(x) := \min\left\{k\varepsilon^2;\ y_k \in \overline{\Omega} \setminus E_{\varepsilon}\right\},$$

where E_ε is given by

$$E_{\varepsilon} = \{x \in \overline{\Omega};\ \operatorname{dist}(x, E) \le \sqrt{2}\varepsilon\}.$$

Let $L' = 2L$ and

$$Q' = \left\{(x_1, x_2) \in \Omega;\ L'|x_1| - L'\sqrt{2}\varepsilon \le x_2 \le \frac{L\mu}{2}\right\}.$$

We next take an open ball $\operatorname{int} B_{r_0}(z_0) \subset \operatorname{int} B_r \cap Q' \cap Q$, where z_0 is on the x_2-axis and $r_0 < \min\{r, \mu/4\}$. The game starts at an arbitrary point $x \in \operatorname{int} B_{r_0}(z_0)$.

Suppose that Player II adopts the following strategy at the k-th step for all $k = 1, 2, \ldots, \ell$ (Fig. 8.4). Let v_k be an arbitrary unit vector chosen by Player I. If $y_{k-1} \in Q'$, then Player II chooses $b_k = \pm 1$ such that

$$b_k \langle v_k, \xi \rangle \le 0,$$

where the vector $\xi \in \mathbf{R}^2$ is given by

$$\xi = \begin{cases} (L', -1) & \text{if } \langle y_{k-1}, e_1 \rangle \ge 0, \\ (-L', -1) & \text{if } \langle y_{k-1}, e_1 \rangle \le 0. \end{cases}$$

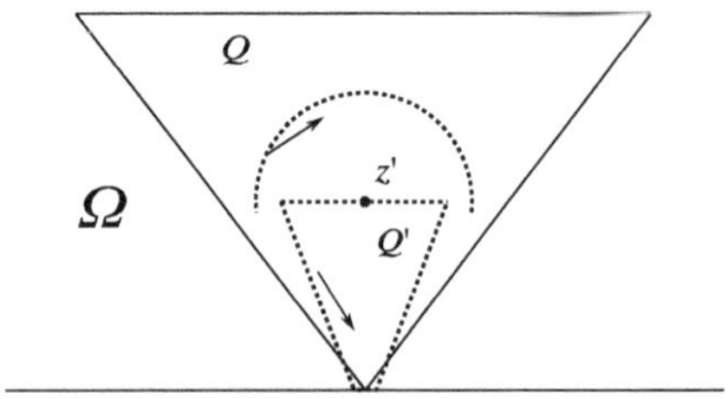

Fig. 8.4 A strategy for Player II. Reprinted without changes under the Creative Commons Attribution 4.0 International License (https://creativecommons.org/licenses/by/4.0/) from N. Hamamuki and Q. Liu, A Game-theoretic Approach to Dynamic Boundary Problems for Level-set Curvature Flow Equations and Applications, Partial Differ. Equ. Appl., 2 (2021), Article 30

If $y_{k-1} \in Q \setminus Q'$, then Player II takes $b_k = \pm 1$ such that

$$b_k \left\langle v_k, y_{k-1} - z' \right\rangle \le 0,$$

where $z' = (0, L'\mu/4) \in \partial Q'$. Then

$$y_k = y_{k-1} + \sqrt{2}\varepsilon \eta_\varepsilon(y_{k-1}) b_k v_k + \varepsilon^2 \zeta_\varepsilon(y_{k-1}) e_2$$

satisfies

$$\begin{aligned} |y_k - z'|^2 &\le |y_{k-1} - z'|^2 + 2\varepsilon^2 \eta_\varepsilon^2(y_{k-1}) + 2\sqrt{2}\varepsilon \eta_\varepsilon(y_{k-1}) b_k \left\langle v_k, y_{k-1} - z' \right\rangle \\ &\quad + 2\varepsilon^2 \zeta_\varepsilon(y_{k-1}) \left\langle e_2, y_{k-1} - z' \right\rangle + \zeta_\varepsilon(y_{k-1}) o(\varepsilon^2) \\ &\le |y_{k-1} - z'|^2 + 2\varepsilon^2 \end{aligned} \tag{8.1.15}$$

when $\varepsilon > 0$ is small. Adopting this strategy in the game, we may observe the following consequences:

- The game position $y_k \in Q'$ provided that $y_{k-1} \in Q'$ unless such a move crosses the top side of Q' and y_k enters the set

$$\mathcal{N} = \left\{ y \in \Omega \setminus Q';\, |y - x| < \sqrt{2}\varepsilon \text{ for some } x \in [-\mu/4, \mu/4] \times \{L'\mu/4\} \right\}.$$

- If y_k exits Q' for some k and never comes back into Q' again afterwards, then the first exit time $\tau_\varepsilon(x)$ satisfies

$$\begin{aligned} \tau_\varepsilon(x) &\ge \frac{1}{2} \operatorname{dist}^2(z', \partial Q) - \frac{1}{2} \left(\frac{\mu}{4} + \sqrt{2}\varepsilon \right)^2 \\ &= \frac{L^2 \mu^2}{8(1 + L^2)} - \frac{\mu^2}{32} + O(\varepsilon) \ge \frac{L^2 \mu^2}{16(1 + L^2)} \end{aligned}$$

  for any $\varepsilon > 0$ small. In fact, starting the game from any point in $\mathcal{N}$, Player II can use the concentric strategy (2) above to ensure a lower bound of the exit time τ_ε from Q as the right hand side above, thanks to the estimate (8.1.15).

It follows that

$$u^\varepsilon(x, t) \ge -\sqrt{2}\varepsilon \quad \text{for } x \in \operatorname{int} B_{r_0}(z_0) \text{ and } t \le \frac{L^2 \mu^2}{16(1 + L^2)}. \tag{8.1.16}$$

3. Application of the Game Approximation

Hence, by using the estimates (8.1.14), (8.1.16) and Corollary (7.3.9), we can let $r > 0$ sufficiently small and $\varepsilon \to 0$ to deduce that $u(x, t) = 0$ for any $x \in \operatorname{int} B_{r_0}(z_0) \subset \operatorname{int} B_r$ and any $\tau_1 \le t \le \tau_2$, where

$$\tau_1 = \frac{1}{2}R^2 - \frac{1}{2}(R-r)^2, \quad \tau_2 = \frac{L^2\mu^2}{16(1+L^2)}.$$

We thus can take $\mathcal{O} = \operatorname{int} B_{r_0}(z_0)$ to complete the proof. □

8.2 Convexity Preserving Properties

In this section, we give another application of the game-theoretic approach. We use the discrete games to study the convexity preserving property as discussed in Sect. 4.5.

8.2.1 Convexity of Level Sets

Let us begin with a quick proof of the convexity of level sets for the stationary problem (SDP) in two dimensions.

Theorem 8.2.1 *Assume that $\Omega \subset \mathbf{R}^2$ is bounded and strictly convex. Let U be the unique viscosity solution of* (SDP). *Then the level set $\{x \in \overline{\Omega} : U(x) \geq c\}$ is convex for all $c \geq 0$.*

Proof We recall that the value function $\tilde{U}^\varepsilon$ of Game 2-a', the inverse of Game 2-a (in Remark 7.2.5), converges to U uniformly in $\overline{\Omega}$ as $\varepsilon \to 0$. We will show that all level sets of $\tilde{U}^\varepsilon$ are convex up to an error for all $\varepsilon > 0$ small.

Fix $c \geq 0$ arbitrarily. Suppose that $x, x' \in \overline{\Omega}$ (with $x \neq x'$) satisfy

$$\min\{U(x), U(x')\} \geq c.$$

Then by the continuity of U, for $\delta > 0$ small, we have

$$\min\{U(y), U(y')\} \geq c - \frac{\delta}{2} \quad \text{for all } y \in B_r(x),\ y' \in B_r(x')$$

when $r > 0$ is sufficiently small. Then, by the uniform convergence of $\tilde{U}^\varepsilon$ to U, it follows that

$$\tilde{U}^\varepsilon(y), \tilde{U}^\varepsilon(y') \geq c - \delta$$

for all $y \in B_r(x)$, $y' \in B_r(x')$ and $\varepsilon > 0$ small. By the definition of $\tilde{U}^\varepsilon$, there exists a particular strategy of Player I, denoted by S^I, such that it costs at least time $c - \delta$ for the game position to leave Ω from any $y \in B_r(x)$ or from $y' \in B_r(x')$.

We next consider the game starting from $(x + x')/2 \in \Omega$. Player I can keep choosing a direction parallel to $x' - x$ before the game position is moved into

$B_{\sqrt{2\varepsilon}}(x)$ or $B_{\sqrt{2\varepsilon}}(x')$. Suppose Player II let it come into $B_{\sqrt{2\varepsilon}}(x)$ before the end of the game, then Player I may switch to use the strategy S^I from x afterwards. Our estimate for the strategy S^I can guarantee a total exit time from $(x+x')/2$ at least $c-\delta$. We can use a similar argument to show the same lower bound of the total exit time when Player II allows the game position to enter $B_{\sqrt{2\varepsilon}}(x')$. This yields

$$\tilde{U}^{\varepsilon}\left(\frac{x+x'}{2}\right) \geq c-\delta.$$

Sending $\varepsilon \to 0$ and then $\delta \to 0$, we are led to

$$U\left(\frac{x+x'}{2}\right) \geq c,$$

which completes the proof. □

It is not difficult to verify that $u(x,t) = U(x) - t$ is a solution of (7.1.1) for $(x,t) \in \Omega \times (0,\infty)$ provided that U is a solution of (SDP) in $\overline{\Omega}$. The result above amounts to saying that the superlevel set $\{x \in \mathbf{R}^2 : u(x,t) \geq 0\}$ is convex for all $t \geq 0$ if it is strictly convex when $t = 0$. In order to obtain a convexity preserving property with only convexity assumption (rather than strict convexity) of the initial level set, we may consider the parabolic problem (MCF) directly.

Theorem 8.2.2 (Convexity of Level Sets) *Suppose that u_0 is Lipschitz continuous in $\mathbf{R}^2$ with Lipschitz constant $L > 0$. Assume that the superlevel set*

$$D_0^c = \{x \in \mathbf{R}^2; u_0(x) \geq c\}. \tag{8.2.1}$$

is convex for all $c \in \mathbf{R}$. Let u^ε be the value function associated to Game 1-a'. Then every superlevel set of $u^\varepsilon(\cdot,t)$ is almost convex for any $t \geq 0$ and $\varepsilon > 0$ in the sense that for any $x, y \in \mathbf{R}^2$ and $t \geq 0$,

$$u^{\varepsilon}\left(\frac{x+x'}{2}, t\right) \geq c - \sqrt{2}L\varepsilon \tag{8.2.2}$$

provided that $u^\varepsilon(x,t) \geq c$ and $u^\varepsilon(x',t) \geq c$ for any $c \in \mathbf{R}$. In particular, all superlevel sets of the solution u of (MCF) *are convex in space for every $t \geq 0$.*

We remark that it is not restrictive to assume the convexity of all superlevel sets even if one intends to show the convexity preserving property for the level set at a particular level. The level set solutions are independent of the choice of the auxiliary function u, as elaborated in Sect. 5.2.

We use Game 1-a', the inverse of Game 1-a, to prove Theorem 8.2.2. Let us first establish a monotonicity result.

Lemma 8.2.3 (Monotonicity) *Suppose that u_0 is Lipschitz continuous in $\mathbf{R}^2$. Assume that the superlevel set D_0^c given by* (8.2.1) *for all $c \in \mathbf{R}$. Let u^ε be the value function associated to Game 1-a'. Then*

$$u^\varepsilon(x, s) \le u^\varepsilon(x, t) \text{ for all } x \in \mathbf{R}^2,\ s \ge t \ge 0 \text{ and } \varepsilon > 0. \tag{8.2.3}$$

In particular, $u(x, s) \le u(x, t)$ for all $x \in \mathbf{R}^2$, $s \ge t \ge 0$.

Proof Let us fix $\varepsilon > 0$. We begin our proof by claiming that for any $x \in \mathbf{R}^2$ and $v \in \mathbf{R}^2$ with $|v| = 1$, there exists $b = \pm 1$ such that

$$u_0\left(x + \sqrt{2}\varepsilon b v\right) \le u_0(x). \tag{8.2.4}$$

Indeed, assume by contradiction that for some $v \in \mathbf{R}^2$ with $|v| = 1$ and some $c > u_0(x)$,

$$u_0\left(x - \sqrt{2}\varepsilon v\right),\ u_0\left(x + \sqrt{2}\varepsilon v\right) \ge c.$$

By convexity of the superlevel set D_0^c,

$$x = \frac{1}{2}(x - \sqrt{2}\varepsilon v) + \frac{1}{2}(x + \sqrt{2}\varepsilon v) \in D_0^c,$$

which contradicts the assumption that $u_0(x) < c$.

Now let us incorporate the claim in our games. Suppose that $\ell = [t/\varepsilon^2]$ and $\ell' = [s/\varepsilon^2]$. Notice that at the k-th step in Game 1-a' ($k \le \ell'$), Player II may choose b_k as above to ensure

$$\max_{|v|=1} \min_{b=\pm 1} u_0\left(y_{k-1} + \sqrt{2}\varepsilon b_k v_k\right) \le u_0(y_{k-1}).$$

By applying this strategy from the step ℓ to $\ell' = [t/\varepsilon^2]$, we obtain

$$\max_{v_{\ell+1}} \min_{b_{\ell+1}} \ldots \max_{v_{\ell'}} \min_{b_{\ell'}} u_0 \left(y_\ell + \sqrt{2}\varepsilon \sum_{i=\ell+1}^{\ell'} b_i v_i \right) \le u_0(y_\ell).$$

Taking the extrema on both sides of the inequality above over b_ℓ, v_ℓ, $b_{\ell-1}$, $v_{\ell-1}$, …, b_1 and v_1 in order, we get (8.2.3) by definition. It follows immediately that $u(x, t) \le u(x, s)$ for all $t \ge s$, by passing to the limit as $\varepsilon \to 0$ in (8.2.3) with application of the result on game convergence stated in Remark 7.1.11. □

We next prove Theorem 8.2.2.

Proof of Theorem 8.2.2 We assume $x \neq x'$, since otherwise the statements are trivial. Since $u^\varepsilon(x,t) \geq c$ and $u^\varepsilon(x',t) \geq c$, we have $u^\varepsilon(x,s) \geq c$ and $u^\varepsilon(x',s) \geq c$ for $s \leq t$ as well in virtue of the monotonicity shown in Lemma 8.2.3. In particular, $u_0(x) \geq c$ and $u_0(y) \geq c$.

For any $s \leq t$, there must exist maximizing strategies S_x^I and $S_{x'}^I$ of Player I such that regardless of the choices of Player II, we have $u_0(y_j) \geq c$ and $u_0(y_j') \geq c$ if S_x^I and $S_{x'}^I$ are applied respectively in the games starting from x and x'. Here y_j and y_j' respectively denote the game positions from x and x' after j steps with $j = [s/\varepsilon^2]$. For any

We next consider the game started from $\hat{x} := (x + x')/2$ with duration t. In this case, Player I has the following possible of move: he keeps choosing $v = (x - x')/|x - x'|$ until the game position enters the closed ball $B_{\sqrt{2}\varepsilon}(x)$ or $B_{\sqrt{2}\varepsilon}(x')$. Without loss of generality, suppose that Player II allows the game state $\hat{y}_k \in B_{\sqrt{2}\varepsilon}(x)$ after k steps from $\hat{x}$ with $k \leq \ell$. Player I then turns to use S_x^I for the remaining $j = \ell - k$ steps, moving the game position to a point $z \in \mathbf{R}^2$, which depends also on Player II's responses to S_x^I. However, the same choices of Player II in response to S_x^I would send x to $y_j \in B_{\sqrt{2}\varepsilon}(z)$. Since $u_0(y_j) \geq c$, we thus get the following estimate for the possible game outcome starting from $\hat{x}$ with duration t:

$$u_0(z) \geq u_0(y_j) - \sqrt{2}L\varepsilon \geq c - \sqrt{2}L\varepsilon.$$

The other possibility is that Player II chooses to let the game position move away from the neighborhoods of x and x'. In this situation, the final position must remain on the line segment joining x and x'. Consequently, the outcome of the game is at least c due to the conditions that $u_0(x) \geq c$, $u_0(x') \geq c$ and the superlevel sets of u_0 are convex.

Since the possible outcomes above correspond to one particular strategy of Player I, and Player I's optimal strategy can only improve upon it, we conclude that (8.2.2) holds.

Now in order to prove the statement regarding convexity of superlevel sets of u, we only need to passing to the limit, using the convergence result for Game 1-a' given in Remark 7.1.11. More precisely, we have shown that, for any $c \in \mathbf{R}$ and $\delta > 0$, there exists $\varepsilon > 0$ such that

$$u^\varepsilon(x,t) \geq c - \delta, \quad u^\varepsilon(x',t) \geq c - \delta.$$

Our argument above yields

$$u^\varepsilon\left(\frac{x + x'}{2}, t\right) \geq c - \delta - \sqrt{2}L\varepsilon.$$

We conclude the proof by letting $\varepsilon \to 0$ and then $\delta \to 0$. □

8.2.2 Convexity Preserving for Curve Shortening Equation

We next use the game-theoretic approach to show the convexity preserving property for the solution itself. We give an alternative proof for Theorem 4.5.2 in the special case of (MCF) in two space dimensions. Our main result in this section is as below.

Theorem 8.2.4 (Convexity Preserving) *Suppose that u_0 is Lipschitz continuous in $\mathbf{R}^2$. Let u be the unique viscosity solution of* (MCF). *If u_0 is convex in $\mathbf{R}^2$. Then $x \mapsto u(x,t)$ is convex for any $t \geq 0$.*

In order to prove Theorem 8.2.4, we introduce another variant of Game 1-a by modifying its game rules as below. As usual, let the step size be $\varepsilon > 0$ and the total steps of the game be $\ell = [t/\varepsilon^2]$. We obtain a sequence $y_k \in \mathbf{R}^2$ $(k = 0, 1, 2, \ldots, \ell)$ with $y_0 = x$, following the rules below.

The game rules are as follows. At the k-th step, we now ask that

- Player I chooses $v_k, v_k' \in \mathbf{R}^2$ with $|v_k| = |v_k'| = 1$;
- Player II determines $b_k = \pm 1$ and $b_k' = \pm 1$;
- The game position is moved from y_{k-1} to

$$y_k = y_{k-1} + \frac{\sqrt{2}\varepsilon}{2}(b_k v_k + b_k' v_k').$$

Suppose that Player I and Player II aim to minimize and maximize $u_0(y_\ell)$ respectively.

Under the new rules above, we may define the value function to be

$$w^\varepsilon(x,t) := \min_{v_1, v_1'} \max_{b_1, b_1'} \ldots \min_{v_\ell, v_\ell'} \max_{b_\ell, b_\ell'} u_0(y_\ell), \tag{8.2.5}$$

It is clear that the new dynamic programming principle is

$$w^\varepsilon(x,t) = \min_{|v|=|v'|=1} \max_{b,b'=\pm 1} w^\varepsilon\left(x + \frac{\sqrt{2}\varepsilon}{2}(bv + b'v'), t - \varepsilon^2\right) \tag{8.2.6}$$

for any $x \in \mathbf{R}^2$ and $t \geq \varepsilon^2$ while $u^\varepsilon(\cdot, t) = u_0$ in $\mathbf{R}^2$ for all $0 \leq t < \varepsilon^2$.

Proposition 8.2.5 *Suppose that u_0 is Lipschitz continuous in $\mathbf{R}^2$. Assume that u_0 is convex in $\mathbf{R}^2$. Let u^ε and w^ε be respectively the value functions of Game 1-a and the modified game. Then $u^\varepsilon(x,t)$ satisfies*

$$u^\varepsilon(x+h,t) + u^\varepsilon(x-h,t) \geq 2w^\varepsilon(x,t), \tag{8.2.7}$$

for any $x, h \in \mathbf{R}^2$, $t \geq 0$ and $\varepsilon > 0$.

Proof Fix $\varepsilon > 0$ arbitrarily. In virtue of the definition of u^ε, there exist $v_\pm \in \mathbf{R}^2$ with $|v_\pm| = 1$ such that for any $x, h \in \mathbf{R}^2$

$$u^\varepsilon(x \pm h, \varepsilon^2) \geq u_0(x \pm h + \sqrt{2}\varepsilon b v_\pm)$$

for any $b = \pm 1$. By the convexity of u_0, we have

$$u^\varepsilon(x + h, \varepsilon^2) + u^\varepsilon(x - h, \varepsilon^2) \geq 2u_0\left(x + \frac{\sqrt{2}\varepsilon}{2}(bv_+ + b'v_-)\right)$$

for any $b, b' = \pm 1$. By (8.2.5), it follows that for any $x, h \in \mathbf{R}^2$

$$u^\varepsilon(x + h, \varepsilon^2) + u^\varepsilon(x - h, \varepsilon^2) \geq 2w^\varepsilon(x, \varepsilon^2). \tag{8.2.8}$$

Now we apply DPP as in (7.1.4) to get

$$u^\varepsilon(x \pm h, 2\varepsilon^2) \geq \min_{|v|=1} u^\varepsilon(x \pm h + \sqrt{2}\varepsilon b v)$$

for all $b = \pm 1$, which by (8.2.8) yields

$$u^\varepsilon(x + h, 2\varepsilon^2) + u^\varepsilon(x - h, 2\varepsilon^2) \geq \min_{|v_\pm|=1} 2w^\varepsilon\left(x + \frac{\sqrt{2}\varepsilon}{2}(bv_+ + b'v_-)\right)$$

for any $b, b' = \pm 1$. By (8.2.6), it implies that

$$u^\varepsilon(x + h, 2\varepsilon^2) + u^\varepsilon(x - h, 2\varepsilon^2) \geq 2w^\varepsilon(x, 2\varepsilon^2).$$

Continuing iteration of this estimate, we are led to (8.2.7) for all $x, h \in \mathbf{R}^2$ and $t \geq 0$. □

We next show that the relaxed limit $\underline{w}$ of w^ε given by

$$\underline{w} = \liminf_{\varepsilon \to 0}{}_* w^\varepsilon \tag{8.2.9}$$

is a supersolution of (7.1.1).

Proposition 8.2.6 *Suppose that u_0 is Lipschitz continuous in $\mathbf{R}^2$. Let w^ε be the value function of the modified game as defined in* (8.2.5). *Then the limit $\underline{w}$ given by* (8.2.9) *is a supersolution of* (7.1.1).

Proof Let us assume that there exist $(x_0, t_0) \in \mathbf{R}^2 \times (0, \infty)$ and $\phi \in C^2(\mathbf{R}^2 \times (0, \infty))$ such that

$$\overline{w}(x, t) - \phi(x, t) > \overline{w}(x_0, t_0) - \phi(x_0, t_0) \text{ for all } (x, t) \neq (x_0, t_0).$$

Then by definition of $\overline{w}$, there exists $(x_\varepsilon, t_\varepsilon) \in \mathbf{R}^2 \times (0, \infty)$ such that $(x_\varepsilon, t_\varepsilon) \to (x_0, t_0)$, $w^\varepsilon(x_\varepsilon, t_\varepsilon) \to \overline{w}(x_0, t_0)$, and

$$(w^\varepsilon - \phi)(x_\varepsilon, t_\varepsilon) \le (w^\varepsilon - \phi)(x, t) - \varepsilon^3$$

for any (x, t) in a neighborhood of (x_0, t_0) with its size independent of ε. By (8.2.6), we then have

$$\phi(x_\varepsilon, t_\varepsilon) \ge \min_{|v|=|v'|=1} \max_{b,b'=\pm 1} \phi\left(x_\varepsilon + \frac{\sqrt{2}\varepsilon}{2}(bv + b'v'), t_\varepsilon - \varepsilon^2\right) + \varepsilon^3.$$

Applying Taylor expansion, we are led to

$$\begin{aligned}\varepsilon^2 \phi_t \ge \min_{|v|=|v'|=1} \max_{b,b'=\pm 1} &\left\{\frac{\sqrt{2}\varepsilon}{2}\langle \nabla\phi, bv + b'v'\rangle\right. \\ &\left. + \frac{\varepsilon^2}{4}\langle \nabla^2\phi(bv + b'v'), (bv + b'v')\rangle\right\} + o(\varepsilon^2) \quad \text{at } (x_\varepsilon, t_\varepsilon).\end{aligned} \tag{8.2.10}$$

One may first compare $\langle \nabla\phi(x_\varepsilon, t_\varepsilon), v\rangle$ and $\langle \nabla\phi(x_\varepsilon, t_\varepsilon), v'\rangle$ to determine b or b'. For example, if

$$|\langle \nabla\phi(x_\varepsilon, t_\varepsilon), v\rangle| \le |\langle \nabla\phi(x_\varepsilon, t_\varepsilon), v'\rangle|, \tag{8.2.11}$$

then we may choose b_0' such that

$$|\langle \nabla\phi(x_\varepsilon, t_\varepsilon), v'\rangle| = \langle \nabla\phi(x_\varepsilon, t_\varepsilon), b_0'v'\rangle$$

and thus $\langle \nabla\phi, bv + b_0'v'\rangle \ge 0$ for either $b = \pm 1$ by (8.2.11). Therefore we have at $(x_\varepsilon, t_\varepsilon)$

$$\begin{aligned}\varepsilon^2\phi_t \ge \min_{v,v'} \max_{b} &\left\{\frac{\sqrt{2}\varepsilon}{2}|\langle \nabla\phi, bv + b_0'v'\rangle| + \frac{\varepsilon^2}{4}\langle \nabla^2\phi(bv + b_0'v'), (bv + b_0'v')\rangle\right\} \\ &+ o(\varepsilon^2).\end{aligned} \tag{8.2.12}$$

In what follows, we always assume (8.2.11) and keep the choice $b' = b_0'$.

Case A Suppose that $\nabla\phi(x_0, t_0) \ne 0$, which in turn implies that $\nabla\phi(x_\varepsilon, t_\varepsilon) \ne 0$ for all $\varepsilon > 0$ small. We apply Lemma 7.1.6 and Remark 7.1.7 with

$$p = |bv + b_0'v'|\frac{\nabla\phi(x_\varepsilon, t_\varepsilon)}{|\nabla\phi(x_\varepsilon, t_\varepsilon)|}, \quad \xi = bv + b_0'v', \quad X = \nabla^2\phi(x_\varepsilon, t_\varepsilon),$$

and get

$$\begin{aligned}&\left|\langle \nabla^2\phi(bv + b_0'v'), (bv + b_0'v')\rangle - |bv + b_0'v'|^2\left\langle \nabla^2\phi\frac{\nabla^\perp\phi}{|\nabla\phi|}, \frac{\nabla^\perp\phi}{|\nabla\phi|}\right\rangle\right| \\ &\le \frac{M|bv + b_0'v'|}{|\nabla\phi|}|\langle \nabla\phi, bv + b_0'v'\rangle|\end{aligned}$$

for any $v, v' \in \mathbf{R}^2$ with $|v| = |v'| = 1$. Hence, we have from (8.2.12)

$$\varepsilon^2\phi_t \geq \varepsilon^2|\nabla\phi|\,\mathrm{div}\left(\frac{\nabla\phi}{|\nabla\phi|}\right) + \varepsilon\Psi + o(\varepsilon^2) \text{ at } (x_\varepsilon, t_\varepsilon), \tag{8.2.13}$$

where the function Ψ is given by

$$\Psi = \min_{v,v'}\max_{b}\left\{\frac{\sqrt{2}}{2}|\langle\nabla\phi, bv + b_0'v'\rangle| + \varepsilon\left(\frac{|bv + b_0'v'|^2}{4} - 1\right)|\nabla\phi|\,\mathrm{div}\left(\frac{\nabla\phi}{|\nabla\phi|}\right)\right.$$
$$\left. + \frac{\varepsilon M}{4|\nabla\phi|}|bv + b_0'v'||\langle\nabla\phi, bv + b_0'v'\rangle|\right\}. \tag{8.2.14}$$

We next estimate $\Psi(x_\varepsilon, t_\varepsilon)$ for $\varepsilon > 0$ sufficiently small. Indeed, we discuss two different cases. If there exists a subsequence ε_k such that the maximum attains at v and v' (depending on ε_k) and as $k \to \infty$, either $|\langle\nabla\phi(x_{\varepsilon_k}, t_{\varepsilon_k}), v\rangle| \to c$ or $|\langle\nabla\phi(x_{\varepsilon_k}, t_{\varepsilon_k}), v'\rangle| \to c$ holds for some $c > 0$, then it is clear that there exists $b = \pm 1$ also depending on ε_k satisfying

$$|\langle\nabla\phi(x_{\varepsilon_k}, t_{\varepsilon_k}), bv + b_0'v'\rangle| > c,$$

since

$$\max_b |\langle\nabla\phi(x,t), bv{+}b_0'v'\rangle| \geq \frac{1}{2}(|\langle\nabla\phi(x,t), v{+}b_0'v'\rangle| {+} |\langle\nabla\phi(x,t), -v{+}b_0'v'\rangle|)$$
$$= \max\{|\langle\nabla\phi(x,t), v\rangle|, |\langle\nabla\phi(x,t), b_0'v'\rangle|\}.$$

In view of (8.2.14), this yields immediately that

$$\liminf_{\varepsilon_k\to 0}\Psi(x_{\varepsilon_k}, t_{\varepsilon_k}) \leq -\frac{\sqrt{2}c}{2}.$$

The remaining case is that, for any $\varepsilon > 0$ small, the minimum in Ψ is attained at v and v' such that

$$|\langle\nabla\phi(x_\varepsilon, t_\varepsilon), v\rangle| \to 0 \text{ and } |\langle\nabla\phi(x_\varepsilon, t_\varepsilon), v'\rangle| \to 0 \text{ as } \varepsilon \to 0.$$

We denote $\eta_\varepsilon = \nabla\phi(x_\varepsilon, t_\varepsilon)/|\nabla\phi(x_\varepsilon, t_\varepsilon)|$ and $\eta_\varepsilon^\perp = \nabla^\perp\phi(x_\varepsilon, t_\varepsilon)/|\nabla\phi(x_\varepsilon, t_\varepsilon)|$. Let us take a modulus of continuity ω such that

$$\max\{|\langle\eta_\varepsilon, v\rangle|, |\langle\eta_\varepsilon, v'\rangle|\} \leq \omega(\varepsilon).$$

Without loss of generality, we may assume

$$\left|b_0'v' - \eta_\varepsilon^\perp\right| \leq \omega(\varepsilon).$$

(Otherwise, we have $|b_0'v' + \eta_\varepsilon^\perp| \le \omega(\varepsilon)$ and apply a similar argument below.) Then we may pick b such that

$$\left|bv - \eta_\varepsilon^\perp\right| \le \omega(\varepsilon).$$

It follows that under this choice,

$$\left|\frac{1}{4}|bv + b_0'v'|^2 - 1\right| = \frac{1}{4}|bv - b_0'v'|^2 \le \frac{1}{2}\left(|bv - \eta_\varepsilon^\perp|^2 + |b_0'v' - \eta_\varepsilon^\perp|^2\right) \le \omega^2(\varepsilon).$$

It follows from (8.2.14) that

$$\Psi(x_\varepsilon, t_\varepsilon) \ge -C\varepsilon\omega^2(\varepsilon) - C\varepsilon\omega(\varepsilon)$$

for some $C > 0$ independent of ε. Hence, in either case, by (8.2.13), we have

$$\varepsilon^2\phi_t \ge \varepsilon^2|\nabla\phi|\operatorname{div}\left(\frac{\nabla\phi}{|\nabla\phi|}\right) + o(\varepsilon^2) \quad \text{at } (x_\varepsilon, t_\varepsilon). \tag{8.2.15}$$

Dividing the inequality by ε^2 and sending $\varepsilon \to 0$, we obtain

$$\phi_t - |\nabla\phi|\operatorname{div}\left(\frac{\nabla\phi}{|\nabla\phi|}\right) \ge 0 \quad \text{at } (x_0, t_0).$$

Case B It remains to show the viscosity inequality

$$\phi_t(x_0, t_0) \ge 0 \tag{8.2.16}$$

under the conditions that $\nabla\phi(x_0, t_0) = 0$ and $\nabla^2\phi(x_0, t_0) = O$.
Case B-1. Suppose that there exists a subsequence $\nabla\phi(x_{\varepsilon_k}, t_{\varepsilon_k}) \ne 0$ for k arbitrarily large. Then we may repeat the argument as in Case A and reach (8.2.15) again. We then obtain (8.2.16) immediately, since in the present case $\nabla^2\phi(x_\varepsilon, t_\varepsilon) \to \nabla^2\phi(x_0, t_0) = O$ as $\varepsilon \to 0$.
Case B-2. If $\nabla\phi(x_\varepsilon, t_\varepsilon) = 0$ for all $\varepsilon > 0$, then it follows immediately from (8.2.10) that

$$\varepsilon^2\phi_t(x_\varepsilon, t_\varepsilon) \ge -C\varepsilon^2\|\nabla^2\phi(x_\varepsilon, t_\varepsilon)\| + o(\varepsilon^2),$$

for some $C > 0$ independent of ε, which yields (8.2.16).

□

Proposition 8.2.7 *Suppose that u_0 is Lipschitz continuous in $\mathbf{R}^2$. Let w^ε be the value function of the modified game as defined in* (8.2.5). *Then the limit $\underline{w}$ given by* (8.2.9) *satisfies $\underline{w}(x, 0) \ge u_0(x)$ for all $x \in \mathbf{R}^2$.*

Proof We construct a game barrier at every fixed $x_0 \in \mathbf{R}^2$. It is easily seen that for any $\delta > 0$, there exits $K_\delta > 0$ such that

$$u_0(x) \geq w_0(x) = u_0(x_0) - \delta - K_\delta |x - x_0|^2 \text{ for all } x \in \mathbf{R}^2.$$

We now play the games with objective function w_0 starting from any x in a neighborhood of x_0. In the first step, we have

$$\left| x - x_0 + \frac{\sqrt{2}\varepsilon}{2}(bv + b'v') \right|^2 = |x - x_0|^2 + \sqrt{2}\varepsilon\langle x - x_0, bv + b'v'\rangle + \frac{\varepsilon^2}{2}|bv + b'v'|^2.$$

Note that for any v, v', Player II can pick b' (or b) so that

$$\langle x - x_0, bv + b'v'\rangle \leq 0$$

for any b (or any b'). This can be done by fixing b' such that

$$\langle b'v', x - x_0\rangle = -|\langle v', x - x_0\rangle|$$

when

$$|\langle v', x - x_0\rangle| \geq |\langle v, x - x_0\rangle|.$$

Such a strategy yields

$$\left| x - x_0 + \frac{\sqrt{2}\varepsilon}{2}(bv + b'v') \right|^2 \leq 2\varepsilon^2 + |x - x_0|^2$$

for any v, v'. Repeating this strategy of Player II regardless of choices by Player I, we get

$$w^\varepsilon(x, t) \geq w_0(x) - 2K_\delta t.$$

This implies $\underline{w}(x_0, 0) \geq w_0(x_0) \geq u_0(x_0) - \delta$. We conclude the proof by letting $\delta \to 0$. □

Let us complete the proof of Theorem 8.2.4.

Proof of Theorem 8.2.4 Since (8.2.7) holds for all $x, h \in \mathbf{R}^2$ and $\varepsilon > 0$ due to Proposition 8.2.5, by passing to the limits as $\varepsilon \to 0$ and applying Theorem 7.1.1, we get

$$u(x + h, t) + u(x - h, t) \geq 2\underline{w}(x, t) \tag{8.2.17}$$

for all $x, h \in \mathbf{R}^2$ and $t \geq 0$, where $\underline{w}$ is given by (8.2.9).

On the other hand, by Proposition 8.2.6, Proposition 8.2.7 and the comparison principle (Theorem 4.1.4), we obtain $\underline{w} \geq u$ in $\mathbf{R}^2 \times [0, \infty)$, which, together with (8.2.17), implies the convexity of $x \mapsto u(x, t)$ for all $t \geq 0$. □

We finally remark that the convexity preserving property holds only in the continuum limit as $\varepsilon \to 0$. It does not precisely hold on the discrete level in general. In other words, one cannot expect in general that

$$u^\varepsilon(x - h, t) + u^\varepsilon(x + h, t) \geq 2u^\varepsilon(x, t)$$

for all $x, h \in \mathbf{R}^2$, $t > 0$ and $\varepsilon > 0$. We give an example below to show this.

Example 8.2.8 Let u_0 be Lipschitz and concave in $\mathbf{R}^2$. Suppose that u_0 is the distance function to the first quadrant. Let

$$x_0 = \left(\sqrt{2}\varepsilon/2, \sqrt{2}\varepsilon/2\right), \quad h = \left(\sqrt{2}\varepsilon/2, -\sqrt{2}\varepsilon/2\right).$$

When $\varepsilon > 0$ is taken small, it is easily seen that $u^\varepsilon(x_0 \pm h, \varepsilon^2) \leq 0$, since the minimizing choices of v for Player I to start the game at $x_0 + h$ or $x_0 - h$ can be the ones parallel to the axes at which they are located. On the other hand, at the point x_0, the outcome after one step is always positive no matter what choice Player I makes, i.e., $u^\varepsilon(x_0, \varepsilon^2) > 0$. We therefore have

$$u^\varepsilon(x_0 + h, \varepsilon^2) + u^\varepsilon(x_0 - h, \varepsilon^2) \leq 2u^\varepsilon(x_0, \varepsilon^2),$$

although u_0 is convex.

8.3 Notes and Comments

In this chapter, we present two representative applications of the game-theoretic approach to understanding geometric properties of the level-set mean curvature flow equation. The fattening phenomenon in the examples of Sects. 8.1.1 and 8.1.2 was first observed by L. C. Evans and J. Spruck [166], and its rigorous justification, originally due to H. M. Soner [399], typically requires tools from parabolic theory, as demonstrated in Example 6.2.1. In contrast, our proof of fattening is based on appropriate choices of strategies and estimates of the associated value functions for both the original and inverse games. The convergence of value functions in the game-theoretic framework then yields a more elementary proof of fattening, avoiding reliance on parabolic theory altogether. This interpretation was first proposed in Q. Liu [327]. A similar method was used in the proof of fattening for dynamic boundary problems in Sect. 8.1.3, taken from N. Hamamuki and Q. Liu [251].

The preservation of convexity is another fundamental property of parabolic equations, particularly in the context of geometric evolution equations. For the mean curvature flow, this result was first established by G. Huisken [259], and later extended to viscosity solutions for the level set formulation by Y. Giga et al. [210]. We discussed the case of curve shortening equation in Theorem 8.2.4. See Theorem 4.5.2 for a result of convexity preservation for general fully nonlinear parabolic equations with a detailed PDE proof. The convexity of level sets of solutions to the stationary problem (SDP) presented in Theorem 8.2.1 is also related to the convexity preserving property of mean curvature flow. It was first proved by L. C. Evans and J. Spruck [166]. An alternative proof of Theorem 8.2.1 was later provided by E. N. Barron et al. [57]. We revisited the above convexity results for level set curvature flow equation in two space dimensions from a game-theoretic perspective. This approach was introduced by Q. Liu et al. [333]. The proof of Theorem 8.2.1 is based on a particular choice of game strategies that enables us to compare the game values at points on a line segment. Generalization of this argument to higher dimensions is however unclear.

Our game-theoretic proof of Theorem 8.2.4 is more involved. For any $x, h \in \mathbf{R}^2$ and $t > 0$, we use the dynamic programming principle to connect the game values at $(x \pm h, t)$ to a modified game that yields an estimate of the solution value at (x, t) by the comparison principle. It is possible to generalize this method for more general parabolic equations. In fact, the game-theoretic interpretation for the convexity preserving property of the parabolic normalized p-Laplace equation $(1 < p \leq \infty)$ is even simpler. For example, in the case of $p = \infty$, the equation reads

$$u_t - \operatorname{trace}\left(\frac{\nabla u \otimes \nabla u}{|\nabla u|^2}\nabla^2 u\right) = 0 \quad \text{in } \mathbf{R}^N \times (0, \infty),$$

and the dynamic programming principle for the value function u^ε associated to the tug-of-war game is given by

$$u^\varepsilon(x, t) = \frac{1}{2} \sup_{y \in \operatorname{int} B_\varepsilon(x)} u^\varepsilon\left(y, t - \frac{\varepsilon^2}{2}\right) + \frac{1}{2} \inf_{y \in \operatorname{int} B_\varepsilon(x)} u^\varepsilon\left(y, t - \frac{\varepsilon^2}{2}\right) \tag{8.3.1}$$

for $x \in \mathbf{R}^N$, $t \geq \varepsilon^2/2$, which can be viewed a parabolic version of (7.4.1). Here, we let $u^\varepsilon(\cdot, 0) = u_0$ and assume that the initial value u_0 is convex in $\mathbf{R}^N$. For any $x \in \mathbf{R}^N$ and $\delta > 0$, we can take $\hat{v} \in \operatorname{int} B_\varepsilon(0)$ such that

$$\sup_{\operatorname{int} B_\varepsilon(x)} u_0 \leq u_0(x + \hat{v}) + \delta. \tag{8.3.2}$$

For any $h \in \mathbf{R}^N$, applying the dynamic programming principle, we have

$$u^\varepsilon(x + h, \varepsilon^2/2) \geq \frac{1}{2} u_0(x + h + \hat{v}) + \frac{1}{2} u_0(x + v_+) - \delta,$$

$$u^\varepsilon(x - h, \varepsilon^2/2) \geq \frac{1}{2} u_0(x - h + \hat{v}) + \frac{1}{2} u_0(x + v_-) - \delta$$

for some $v_\pm \in \operatorname{int} B_\varepsilon(0)$. Using the convexity of u_0, we thus get

$$u^\varepsilon(x+h, \varepsilon^2/2) + u^\varepsilon(x-h, \varepsilon^2/2) \geq \frac{1}{2}u_0(x+\hat{v}) + \frac{1}{2}u_0\left(x + \frac{1}{2}(v_+ + v_-)\right) - 2\delta.$$

It follows from (8.3.1) and (8.3.2) that

$$\begin{aligned} &u^\varepsilon(x+h, \varepsilon^2/2) + u^\varepsilon(x-h, \varepsilon^2/2) \\ &\qquad \geq \frac{1}{2} \sup_{y \in \operatorname{int} B_\varepsilon(x)} u_0(y) + \frac{1}{2} \inf_{y \in \operatorname{int} B_\varepsilon(x)} u_0(y) - 3\delta = u^\varepsilon(x, \varepsilon^2) - 3\delta, \end{aligned}$$

which yields the convexity of $u^\varepsilon(\cdot, \varepsilon^2/2)$ is convex in $\mathbf{R}^N$ by sending $\delta \to 0$. One can repeat the argument above to show that $u^\varepsilon(\cdot, t)$ is convex in $\mathbf{R}^N$ for all $t > 0$. The convexity preserving property for the solution u of the parabolic normalized infinity Laplace equation is an immediate result of the convergence of u^ε to u. In this proof, we do not need to construct any modified game, since the dynamic programming principle exhibits a finer convexity preserving property than that for the curvature flow equation. A similar property holds for the dynamic programming principles corresponding to the parabolic normalized p-Laplace equations with $1 < p < \infty$. We refer the reader to Q. Liu et al. [333] for detailed discussions.

We remark that a similar idea was employed in Q. Liu [331] to establish the semiconvexity of viscosity solutions to general fully nonlinear parabolic equations (without vanishing gradient singularities), using the game-theoretic approximation framework developed by R. V. Kohn and S. Serfaty [313]. Moreover, the game-theoretic approach to convexity and concavity of viscosity solutions is not restricted to the Euclidean setting. A recent work by A. Kijowski et al. [301] applied the game interpretation to study the convexity preserving property of the horizontal mean curvature flow in the Heisenberg group.

Beyond the applications discussed in this chapter, several other works also adopted the game-theoretic interpretation to understand properties of curvature flow equations. A notable example is the study by H. Gao et al. [186], which investigates quantitative homogenization of mean curvature-type equations arising from a model of turbulent combustion. The level set equation involves the positive part of a curvature term, which poses difficulties for standard PDE methods. In a main step in proving convergence rates for homogenization, the authors utilized the game-theoretic interpretation of the stationary problem and establish partial reachability of game trajectories to the boundary. Furthermore, K. Misu [351] generalized the Kohn–Serfaty framework to obstacle problems for curvature flow equations, including cases with an additional driving force. His results are closely related to the work of Y. Giga et al. [221, 222] on the asymptotic behavior of forced mean curvature flow. More recently, T. Sato [391] extended the game setting to a free boundary problem associated to the anisotropic forced mean curvature flow and used the game to study the large time behavior of the solution.

As a stochastic counterpart to the game-theoretic interpretation, the tug-of-war game, with or without random noise, has played an important role in studying various properties of the p-Laplace operator for $1 < p \leq \infty$. The first major application of such stochastic discrete games appeared in the work of S. N. Armstrong and C. K. Smart [35], where the authors established the uniqueness of infinity-harmonic functions solely through the tug-of-war framework introduced by Y. Peres et al. [379]. Later, H. Luiro et al. [339] provided a novel proof of Harnack's inequality for the p-Laplacian via stochastic games. Since then, this regularity approach has been further developed to show Hölder continuity of solutions and game values in various PDE contexts associated with the p-Laplacian. We refer to E. Ruosteenoja [383], M. Parviainen and E. Ruosteenoja [377], Á. Arroyo et al. [37, 38], among many others, for further progress in this direction.

Exercises

Chapter 2

Problem 1 Write the mean curvature flow equation $V = H$ for $u = u(x_1, t)$, when $\Gamma_t \subset \mathbf{R}^N$ is a hypersurface of rotation of the form

$$\Gamma_t = \left\{ (x_1, \ldots, x_N) \in \mathbf{R}^N; r = u(x_1, t),\ r = \left(\sum_{j=2}^{N} x_j^2 \right)^{1/2} \right\}.$$

Problem 2 For $p \in \mathbf{R}^N \setminus \{0\}$ and $X \in \mathbf{S}^N$, set

$$F(p, X) = -\operatorname{trace}\left(I - \frac{p \otimes p}{|p|^2} \right) X.$$

Let $\lambda_1 \leq \lambda_2 \leq \cdots \leq \lambda_N$ denote the eigenvalues of X. Show that

$$F_*(0, O) = -\sum_{i=2}^{N} \lambda_i, \quad F^*(0, O) = -\sum_{i=1}^{N-1} \lambda_i.$$

Chapter 3

Problem 3 Let φ be a C^1 function in a neighborhood of $0 \in \mathbf{R}^N$. Show that

$$\varphi(x) - \varphi(0) = \langle \nabla\varphi(0), x \rangle + \omega(|x|) \quad \text{near} \quad x = 0,$$

Y. Giga, Q. Liu, *Surface Evolution Equations*, Monographs in Mathematics 99,
https://doi.org/10.1007/978-3-032-20303-8

where $\omega(\sigma)/\sigma \to 0$ as $\sigma \to 0$. Let φ be a $C^{2,1}$ function in a neighborhood of $(0, 0) \in \mathbf{R}^N \times \mathbf{R}$. Show that

$$\psi(x,t) = \psi(0,0) + \langle \nabla\psi(0,0), x\rangle + \frac{1}{2}\left\langle \nabla^2\psi(0,0)x, x\right\rangle + \partial_t\psi(0,0)t + \omega\left(|x|^2 + t\right)$$

near $(x, t) = (0, 0)$.

Problem 4 Let Ω be an open set in $\mathbf{R}^N$ and $T > 0$. Assume that for each $\varepsilon > 0$, u^ε is upper semicontinuous in $\overline{\Omega} \times (0, T)$ and is a viscosity subsolution of

$$\begin{cases} u_t - |\nabla u| \operatorname{div}(\nabla u/|\nabla u|) = 0 & \text{in } \Omega \times (0, T), \\ \langle \nabla u, \nu\rangle = 0 & \text{on } \partial\Omega \times (0, T), \end{cases}$$

where ν denotes the outward unit normal to $\partial\Omega$. Show that $\overline{u} = \operatorname{limsup}^*_{\varepsilon\to 0} u^\varepsilon$ in $\overline{\Omega} \times (0, T)$ is a viscosity subsolution of the same Neumann boundary value problem provided that $\overline{u} < \infty$ on $\overline{\Omega} \times (0, T)$.

Chapter 4

Problem 5 Assume that $\Omega \subset \mathbf{R}^N$ is a bounded domain and $H : \Omega \times \mathbf{R}^N \to \mathbf{R}$ is a continuous function. Assume further that there exists $C > 0$ such that

$$|H(x, p) - H(y, p)| \le C(1 + |p|)|x - y|,$$

for all $x, y \in \Omega$ and $p \in \mathbf{R}^N$. Let u and v be respectively a subsolution and a supersolution of

$$u_t + H(x, \nabla u) = 0 \quad \text{in } Q = \Omega \times (0, T)$$

with $T > 0$ given. If $u^* \le v_*$ on $\partial_p Q$, then $u^* \le v_*$.

Problem 6 Let Ω be a bounded open set in $\mathbf{R}^N$ and $Q = \Omega \times (0, T)$ for $T > 0$. Let u and $-v$ be upper semicontinuous with values in $\mathbf{R} \cup \{-\infty\}$. For $z = (x, t)$ and $z' = (y, s) \in Q$ we set

$$w(z, z') = u^*(z) - v_*(z'), \quad z, z' \in \overline{Q}.$$

Let M be the maximum value of w over $\overline{Q} \times \overline{Q}$, that is,

$$M = \max\{w(z, z'); z \in \overline{Q},\ z' \in \overline{Q}\}.$$

Consider barrier functions

$$\Phi_\zeta(z, z', \varepsilon, \sigma, \gamma, \gamma') = \frac{|x - y - \zeta|^2}{\varepsilon} + \frac{|t - s|^2}{\sigma} + \frac{\gamma}{T - t} + \frac{\gamma'}{T - s}$$

for parameters $\varepsilon, \sigma, \gamma, \gamma' > 0$ and $\zeta \in \mathbf{R}^N$. We set

$$\Phi_\zeta(z, z') = w(z, z') - \Phi_\zeta(z, z').$$

Let $(z_\zeta . z'_\zeta) = (x_\zeta, t_\zeta, y_\zeta, s_\zeta)$ be a maximizer of Φ_ζ over $\overline{Q} \times \overline{Q}$. Assume that

$$m_0 = \sup\{u(z) - v(z); z \in Q\} > 0.$$

(1) Prove that for each $m'_0 \in (0, m_0)$ there are $\gamma_0, \gamma'_0 > 0$ such that

$$\sup \Phi_\zeta > m'_0 \quad \text{for all } \varepsilon, \sigma > 0, \gamma_0 > \gamma > 0, \gamma'_0 > \gamma' > 0$$

and $|\zeta| \le \kappa_0(\varepsilon) = (\varepsilon(m_0 - m'_0))^{1/2}/2$.

(2) Prove that, for $\gamma_0, \gamma'_0, \kappa_0$ given in (1),

$$|t_\zeta - s_\zeta| \le (M\sigma)^{1/2}, \quad |x_\zeta - y_\zeta - \zeta| \le (M\varepsilon)^{1/2}$$

for all $\varepsilon, \sigma > 0$, $\gamma_0 > \gamma > 0$ and $\zeta \in \mathbf{R}^N$ with $|\zeta| \le \kappa_0(\varepsilon)$.

Chapter 5

Problem 7 Let u be a viscosity solution of

$$-\Delta u - f = 0 \quad \text{in} \quad \Omega,$$

where $f \in C(\Omega)$ and Ω is a domain in $\mathbf{R}^N$. Let θ be a real-valued convex function defined on $\mathbf{R}$. Show that $v = \theta \circ u$ is a viscosity subsolution of

$$-\Delta v - \theta'(u) f = 0 \quad \text{in} \quad \Omega.$$

This assertion yields

$$-\Delta |u| - (\operatorname{sgn} u) f \le 0 \quad \text{in} \quad \Omega$$

in viscosity sense by setting $\theta(\sigma) = |\sigma|$. Here the signature function sgn is defined as

$$\operatorname{sgn}\sigma = \begin{cases} 1, & \sigma > 0, \\ 0, & \sigma = 0, \\ -1, & \sigma < 0. \end{cases}$$

(This inequality is known as Kato's inequality when it is interpreted in the distributional sense.)

Problem 8 For $\alpha \in \mathbf{R}$, let K_α be the space of all real-valued continuous functions u satisfying $u = \alpha$ outside a ball. Let $E_0 \subset \mathbf{R}^N$ be a closed bounded set. Suppose that $u_0 \in K_\alpha$ with some $\alpha < 0$ such that

$$E_0 = \{x \in \mathbf{R}^N; u_0(x) \geq 0\}.$$

Consider the Cauchy problem

$$\begin{cases} u_t + |\nabla u| = 0 & \text{in } \mathbf{R}^N \times (0, \infty), \\ u(\cdot, 0) = u_0 & \text{in } \mathbf{R}^N. \end{cases}$$

Prove that, for any such u_0, there exists a unique viscosity solution $u \in C(\mathbf{R}^N \times [0, \infty))$ to this Cauchy problem, and the associated closed evolution at $t > 0$, given by

$$E(t) = \{x \in \mathbf{R}^N; u(x, t) > 0\},$$

is the t-parallel set of E_0, that is,

$$E(t) = \{x \in \mathbf{R}^N; \operatorname{dist}(x, E_0) \leq t\}.$$

Chapter 6

Problem 9 For the surface evolution equation $V = 1$ on Γ_t, find an example of a set-theoretic subsolution $G \subset \mathbf{R}^N \times [0, T)$ with $T > 0$ such that $\overline{G}(0) = \overline{G(0)}$ holds, but $\overline{G}(t) = \overline{G(t)}$ fails at some $t \in (0, T)$.

Problem 10 Consider the surface evolution equation

$$V = f(z, \mathbf{n}, \nabla \mathbf{n}) \quad \text{on } \Gamma_t,$$

where f satisfies the assumptions (f1) and (f2). Suppose that in $\mathbf{R}^N \times (0, T)$, G_1, G_2 are both set-theoretic solutions of this evolution equation. Show that $G_1 \cup G_2$ is a set-theoretic subsolution and $G_1 \cap G_2$ is a set-theoretic supersolution of the same equation.

Chapter 7

Problem 11 Let A be a compact set in $\mathbf{R}^d$. Assume that $f : \mathbf{R}^N \times A \to \mathbf{R}^N$ is bounded and continuous, and there exists $L > 0$ such that

$$|f(x,a) - f(y,a)| \leq L|x - y|$$

for $x, y \in \mathbf{R}^N$ and $a \in A$. Fix $x \in \mathbf{R}^N$ and $t > 0$. Let $\varepsilon > 0$ and set $\Lambda = [t/\varepsilon]$. For given $a_1, a_2, \ldots a_\Lambda \in A$, define $y_j \in \mathbf{R}^N$ inductively by

$$y_0 = x, \quad y_j = y_{j-1} + \varepsilon f(y_{j-1}, a_j), \quad j = 1, 2, \ldots, \Lambda.$$

Assume that $u_0 \in K_\alpha$ for some $\alpha \in \mathbf{R}$, that is, u_0 is continuous and satisfies $u = \alpha$ outside a ball. Define

$$u^\varepsilon(x,t) = \min_{a_1, \ldots a_\Lambda \in A} u_0(y_\Lambda).$$

Show that $u^\varepsilon \to u$ locally uniformly in $\mathbf{R}^N \times [0,\infty)$ as $\varepsilon \to 0$, where u is a unique viscosity solution of

$$u_t - \min_{a \in A} \langle f(x,a), \nabla u \rangle = 0 \quad \text{in } \mathbf{R}^N \times (0,\infty)$$

with initial value $u(\cdot, 0) = u_0$ in $\mathbf{R}^N$.

Problem 12 Suppose that for each $\varepsilon > 0$ small, $u_\varepsilon : \mathbf{R}^N \times [0,\infty) \to \mathbf{R}$ satisfies

$$u_\varepsilon(x,t) = \frac{1}{2} \sup_{y \in \operatorname{int} B_\varepsilon(x)} u_\varepsilon\left(y, t - \frac{\varepsilon^2}{2}\right) + \frac{1}{2} \inf_{y \in \operatorname{int} B_\varepsilon(x)} u_\varepsilon\left(y, t - \frac{\varepsilon^2}{2}\right)$$

for all $x \in \mathbf{R}^N$ and $t \geq \varepsilon^2$. Assume that there exists $M > 0$ such that $|u_\varepsilon| \leq M$ in $\mathbf{R}^N \times (0,\infty)$. Let

$$\overline{u} = \operatorname*{limsup^*}_{\varepsilon \to 0} u_\varepsilon, \quad \underline{u} = \operatorname*{liminf_*}_{\varepsilon \to 0} u_\varepsilon \quad \text{in } \mathbf{R}^N \times (0,\infty).$$

Show that $\overline{u}$ and $\underline{u}$ are respectively a subsolution and a supersolution of the parabolic (normalized) infinity-Laplace equation

$$u_t - \frac{1}{|\nabla u|^2} \left\langle \nabla^2 u \nabla u, \nabla u \right\rangle = 0 \quad \text{in } \mathbf{R}^N \times (0,\infty).$$

Chapter 8

Problem 13 Let $p \geq 2$ and

$$\alpha = \frac{p-2}{p+N}, \quad \beta = \frac{N+2}{p+N}, \quad \sigma = \frac{1}{2(p+N)}.$$

Suppose that there exists a solution u_ε of the dynamic programming equation

$$u_\varepsilon(x,t) = \frac{\alpha}{2} \sup_{y \in \text{int}\, B_\varepsilon(x)} u_\varepsilon(y, t - \sigma\varepsilon^2) + \frac{\alpha}{2} \inf_{y \in \text{int}\, B_\varepsilon(x)} u_\varepsilon(y, t - \sigma\varepsilon^2) + \frac{\beta}{|B_\varepsilon(x)|} \int_{B_\varepsilon(x)} u_\varepsilon(y, t - \sigma\varepsilon^2)\, dy \tag{7.3.3}$$

for $x \in \mathbf{R}^N$ and $t \geq \sigma\varepsilon^2$. Here, $|B_\varepsilon(x)|$ denotes the Lebesgue measure of the ball $B_\varepsilon(x)$. Assume that u_ε converges to $u \in C(\mathbf{R}^N \times [0, \infty))$ locally uniform in $\mathbf{R}^N \times [0, \infty)$ as $\varepsilon \to 0$.

(1) Show that u is a solution of the parabolic normalized p-Laplace equation

$$u_t - |\nabla u|^{2-p} \text{div}(|\nabla u|^{p-2} \nabla u) = 0 \quad \text{in } \mathbf{R}^N \times (0, \infty). \tag{7.3.4}$$

(2) Use (7.3.3) to show that $u(\cdot, t)$ is convex in $\mathbf{R}^N$ for each $t > 0$ provided that $u(\cdot, 0)$ is convex in $\mathbf{R}^N$.

Problem 14 Let Ω be a bounded domain in $\mathbf{R}^N$. For $\varepsilon > 0$, let

$$\Omega_\varepsilon = \left\{ x \in \mathbf{R}^N;\, \text{dist}(x, \overline{\Omega}) < \varepsilon \right\}.$$

Suppose that $u_\varepsilon \in C(\Omega_\varepsilon)$ satisfies the dynamic programming principle

$$u_\varepsilon(x) = \frac{1}{2} \sup_{\text{int}\, B_\varepsilon(x)} u_\varepsilon + \frac{1}{2} \inf_{\text{int}\, B_\varepsilon(x)} u_\varepsilon, \quad x \in \Omega.$$

Prove that if u_ε attains a maximum at some point in Ω, then u_ε is constant in Ω.

Appendix A
Proof of Theorem 4.3.3

We give an outline of a proof of Theorem 4.3.3 for completeness. The most part of the following presentation is taken from a note written by N. Hamamuki when he was an undergraduate student at the University of Tokyo. The authors are grateful to him for allowing them to include his note and providing them valuable comments.

A.1 Reduction to a Simpler Problem

To simplify the presentation, we only give a proof when $k = 2$, $\ell = 2$ and $N_1 = N_2 =: n$ for Theorem 4.3.3. We write $x = z_1$, $y = z_2$. By translation, we may assume $\hat{z} = (0, 0) \in \mathbf{R}^n \times \mathbf{R}^n$. By taking

$$\begin{aligned} u(x) &:= u_1(x) - u_1(0) - \langle D_x\varphi(0,0), x\rangle, \\ v(y) &:= u_2(y) - u_2(0) - \left\langle D_y\varphi(0,0), y\right\rangle, \\ \Phi(x,y) &:= \varphi(x,y) - \varphi(0,0) - \langle D_x\varphi(0,0), x\rangle - \left\langle D_y\varphi(0,0), y\right\rangle, \end{aligned}$$

we may assume that $\Phi(0,0) = u(0) = v(0) = 0$, $D\Phi(0,0) = (0,0)$. Since φ is C^2 near $(0,0)$,

$$\left|\Phi(x,y) - \frac{1}{2}\left\langle A\begin{pmatrix} x \\ y \end{pmatrix}, \begin{pmatrix} x \\ y \end{pmatrix}\right\rangle\right| \le o\left(|x|^2 + |y|^2\right)$$

as $x, y \to 0$, where $A = D^2\varphi(0,0) \in \mathbf{S}^{2n}$. Thus, the assumption

$$\max_Z (w - \Phi) = (w - \Phi)(0,0), \quad w(x,y) = u(x) + v(y)$$

Y. Giga, Q. Liu, *Surface Evolution Equations*, Monographs in Mathematics 99,
https://doi.org/10.1007/978-3-032-20303-8

implies that

$$(x, y) \mapsto u(x) + v(y) - \frac{1}{2}\left\langle (A + \eta I)\begin{pmatrix} x \\ y \end{pmatrix}, \begin{pmatrix} x \\ y \end{pmatrix} \right\rangle$$

takes locally strict maximum at $(0, 0)$ for any $\eta > 0$.

By a rotation of coordinates of $\mathbf{R}^n$, we may assume that for m, $0 \le m \le n$, the projections P_1 and P_2 in Theorem 4.3.3 are

$$P_1 = \left(\begin{array}{cc|cc} I_m & O & \multicolumn{2}{c}{O} \\ O & O & & \\ \hline \multicolumn{2}{c|}{O} & I_m & O \\ & & O & O \end{array}\right), \quad P_2 = \left(\begin{array}{cc|cc} O & O & \multicolumn{2}{c}{O} \\ O & I_{n-m} & & \\ \hline \multicolumn{2}{c|}{O} & O & O \\ & & O & I_{n-m} \end{array}\right) \in \mathbf{S}^{2n}, \tag{A.1.1}$$

where I_m denotes the $m \times m$ identity matrix. Since $W_j = P_j \mathbf{S}^{2n}$ is invariant space of A, the matrix A is of the form

$$A = \left(\begin{array}{cc|cc} A_{11} & O & A_{21} & O \\ O & A_{12} & O & A_{22} \\ \hline {}^tA_{21} & O & A_{41} & O \\ O & {}^tA_{22} & O & A_{42} \end{array}\right), \quad A_{11}, A_{41} \in \mathbf{S}^m, \quad A_{12}, A_{42} \in \mathbf{S}^{n-m}, \tag{A.1.2}$$

where A_{21} is $m \times m$ matrix and A_{22} is $(n - m) \times (n - m)$ matrix.

Here is a simplified version of Theorem 4.3.3 for $k = 2$, $\ell = 2$ and $N_1 = N_2 = n$.

Theorem A.1.1 *Let u and v be upper semicontinuous functions defined in a neighborhood $\mathcal{O}$ of 0 with values $\mathbf{R} \cup \{-\infty\}$. Assume that P_i $(i = 1, 2)$ and A are of the form* (A.1.1), (A.1.2). *Assume that*

$$\begin{cases} u(0) = v(0) = 0, & \text{(A.1.3)} \\ u(x) + u(y) \le \dfrac{1}{2}\left\langle A\begin{pmatrix} x \\ y \end{pmatrix}, \begin{pmatrix} x \\ y \end{pmatrix} \right\rangle \quad \textit{for all} \quad x, y \in \mathcal{O} \times \mathcal{O}. & \text{(A.1.4)} \end{cases}$$

Then for any $\varepsilon_1, \varepsilon_2 > 0$, there exist $X, Y \in \mathbf{S}^n$ satisfying

$$\begin{cases} (0, X) \in \bar{J}^{2,+}_{\mathcal{O}} u(0), & \text{(A.1.5)} \\ (0, Y) \in \bar{J}^{2,+}_{\mathcal{O}} v(0), & \text{(A.1.6)} \\ -\displaystyle\sum_{j=1}^{2}\left(\frac{1}{\varepsilon_j} + |AP_j|\right) P_j \le \begin{pmatrix} X & O \\ O & Y \end{pmatrix} \le A + \sum_{j=1}^{2} \varepsilon_j A^2 P_j. & \text{(A.1.7)} \end{cases}$$

Proof of Theorem 4.3.3 for $k = \ell = 2$, $N_1 = N_2 = n$ ***Admitting Theorem A.1.1***
As we discuss in the beginning of this section, we define $\hat{z}$, u, v and Φ so that $\hat{z} = (0, 0)$, $\Phi(0, 0) = 0$, $D_x\Phi(0, 0) = (0, 0)$ and $u(0) = v(0) = 0$ and that

$$(x, y) \mapsto u(x) + v(y) - \frac{1}{2}\left\langle (A + \eta I)\begin{pmatrix} x \\ y \end{pmatrix}, \begin{pmatrix} x \\ y \end{pmatrix} \right\rangle$$

takes local strict maximum at $(0, 0)$ for any $\eta > 0$, where $A = D^2\varphi(0, 0)$. This enables us to apply Theorem A.1.1 to have (A.1.5)–(A.1.7) with A replaced by $A + \eta I$. Note that X, Y in (A.1.5)–(A.1.7) may depend on η so we write $X = X_\eta$, $Y = Y_\eta$. Fortunately, the estimate (A.1.7) gives a bound

$$-CI \leq \begin{pmatrix} X_\eta & O \\ O & Y_\eta \end{pmatrix} \leq CI$$

with positive constant C independent of η. By compactness (see Proposition A.1.2 below), $\{X_\eta\}$ and $\{Y_\eta\}$ have convergent subsequences. Their limits denoted by X, Y satisfy (A.1.5)–(A.1.7) for $A = D^2\varphi(0, 0)$. The proof is now complete. □

Proposition A.1.2 *For a constant* $C > 0$, *the set*

$$K = \{X \in \mathbf{S}^n; -CI \leq X \leq CI\}$$

is compact in $\mathbf{S}^n$.

Proof Since X is symmetric, there exists a rotation $R\ (\in SO(n))$ and a diagonal matrix

$$D = \begin{pmatrix} \lambda_1 & & O \\ & \ddots & \\ O & & \lambda_n \end{pmatrix}$$

such that

$$X = R^{-1}DR.$$

The condition $-CI \leq X \leq CI$ implies $|\lambda_i| \leq C$ for $i = 1, \ldots, n$. Since the totality of rotations $SO(n)$ is compact, K is compact. □

A.2 Jets of Semi-convex Functions

For later purpose, we recall properties of semijets.

Proposition A.2.1 *Let $\mathcal{O}$ be a neighborhood of $\hat{x}$ in $\mathbf{R}^n$. Assume that $x_k \to \hat{x}$ in $\mathbf{R}^n$, $(p_k, X_k) \in \bar{J}^{2,+}_{\mathcal{O}} u(x_k)$ and $(u(x_k), p_k, X_k) \to \big(u(\hat{x}), p, X\big)$ as $k \to \infty$. Then $(p, X) \in \bar{J}^{2,+}_{\mathcal{O}} u(\hat{x})$.*

This easily follows from the definition. In fact, the mapping

$$x \mapsto \left\{(u(x), p, X)\,;\, (p, X) \in \bar{J}^{2,+}_{\mathcal{O}} u(x)\right\}$$

is closed as a set-valued function as remarked in one paragraph above Remark 4.2.5.

Proposition A.2.2 *Let $\mathcal{O}$ be a neighborhood of $\hat{x}$. Assume that $f : \mathcal{O} \to \mathbf{R}\cup\{-\infty\}$ takes its maximum over $\mathcal{O}$ at $\hat{x}$. Then, $(p, X) \in \bar{J}^{2,-}_{\mathcal{O}} f(\hat{x})$ implies $p = 0$, $X \le 0$.*

This is easy to prove since $(p, X) \in J^{2,-}_{\mathcal{O}} f(\hat{x})$ is equivalent to saying that

$$f(x) \ge f(\hat{x}) + \langle p, x - \hat{x}\rangle + \frac{1}{2}\big\langle X(x - \hat{x}), x - \hat{x}\big\rangle + o\left(|x - \hat{x}|^2\right)$$

as $x \to \hat{x}$.

Next, let us state a deep lemma for semi-convex functions. A function u in $\mathbf{R}^n$ is *semi-convex* if

$$u(x) + C|x|^2$$

is convex for some $C \ge 0$.

Lemma A.2.3 *Let f be a (real-valued) semi-convex function defined in $\mathbf{R}^n$. Assume that $f(x) < f(0)$ for all $x \in \mathbf{R}^n\backslash\{0\}$. Then, there are sequences $\{x_k\}$ and $\{p_k\}$ converging to 0 as $k \to \infty$ such that*

$$\begin{cases} J^2_{\mathbf{R}^n} f(x_k) := J^{2,+}_{\mathbf{R}^n} f(x_k) \cap J^{2,-}_{\mathbf{R}^n} f(x_k) \ne \emptyset, \\ x \mapsto f(x) - \langle p_k, x\rangle \text{ attains its local maximum at } x_k. \end{cases}$$

The assertion on jets implies that f has the Taylor's expansion at x_k up to the second order. This is a combination of Aleksandrov's almost everywhere differentiability result [5] for convex functions and Jensen's estimate [294] for the size of the set p such that $f(x) - \langle p, x\rangle$ takes a maximum near 0 for a semi-convex function f which takes its strict maximum at 0.

A.3 Sup-Convolution

To prove Theorem A.1.1, we regularize u by taking a sup-convolution. We begin with its definition.

Definition A.3.1 Let u be a function defined on $\mathbf{R}^n$ with values in $\mathbf{R} \cup \{-\infty\}$. Assume that u is bounded from above. For given positive numbers λ and μ, we set

$$u^{\lambda,\mu}(x',t) := \sup\left\{u(y',s) - \frac{1}{2\lambda}|x'-y'|^2 - \frac{1}{2\mu}|t-s|^2;\right.$$
$$\left.(y',s) \in \mathbf{R}^m \times \mathbf{R}^{n-m}\right\} \quad \text{for} \quad (x',t) \in \mathbf{R}^m \times \mathbf{R}^{n-m}.$$

We call $u^{\lambda,\mu}$ is a two-parameter *convolution* of u. (If $\mu = \lambda$, $u^{\lambda,\lambda}$ is called a standard convolution of u.)

We recall its basic property.

Proposition A.3.2

(i) *For each* $(x',t) \in \mathbf{R}^n$,

$$u^{\lambda,\mu}(x',t) \geq u(x',t).$$

(ii) *The function*

$$x \mapsto u^{\lambda,\mu}(x) + \frac{1}{2\lambda}|x'|^2 + \frac{1}{2\mu}|t|^2, \quad x = (x',t) \in \mathbf{R}^n$$

is a convex function on $\mathbf{R}^n$. *In particular* $u^{\lambda,\mu}$ *is semi-convex.*

Proof

(i) This follows immediately by setting $(y',s) = (x',t)$ in the definition of $u^{\lambda,\mu}$.
(ii) We first note that

$$|x'|^2 - |x'-y'|^2 = 2\langle x', y'\rangle - |y'|^2, \quad |t|^2 - |t-s|^2 = 2\langle t, s\rangle - |s|^2.$$

Using these identities in the definition of $u^{\lambda,\mu}$, we observe that

$$u^{\lambda,\mu}(x) + \frac{1}{2\lambda}|x'|^2 + \frac{1}{2\mu}|t|^2 = \sup\left\{u(y',s) + \frac{1}{\lambda}\langle x', y'\rangle + \frac{1}{\mu}\langle t, s\rangle\right.$$
$$\left. - \frac{1}{2\lambda}|y'|^2 - \frac{1}{2\mu}|s|^2;\ (y',s) \in \mathbf{R}^m \times \mathbf{R}^{n-m}\right\}.$$

Since the right-hand side is the supremum of a family of linear (affine) functions of (x',t'), it is convex. Thus, $u^{\lambda,\mu}$ must be semi-convex.

□

We next derive an estimate for semijets of a two-parameter convolution $u^{\lambda,\mu}$ of u. We often suppress the subscript $\mathbf{R}^n$ of $J^{2,\pm}_{\mathbf{R}^n}$. Let $Q_1, Q_2 \in \mathbf{S}^n$ be projections of $\mathbf{R}^n$ of the form

$$Q_1 = \begin{pmatrix} I_m & O \\ O & O \end{pmatrix}, \quad Q_2 = \begin{pmatrix} O & O \\ O & I_{n-m} \end{pmatrix}.$$

Proposition A.3.3 *Let u be a bounded above in $\mathbf{R}^n$ and $u^{\lambda,\mu}$ be its two-parameter convolution with $\lambda, \mu > 0$. Then $(p, X) \in \bar{J}^{2,+}u^{\lambda,\mu}(\hat{x})$ implies $X \geq -Q_1/\lambda - Q_2/\mu$, where $\hat{x} = (\hat{x}', \hat{t}) \in \mathbf{R}^m \times \mathbf{R}^{n-m}$.*

Proof By Proposition A.3.2 (ii), the function

$$f(x) := u^{\lambda,\mu}(x) + \frac{1}{2\lambda}|x'|^2 + \frac{1}{2\mu}|t|^2, \quad x = (x', t)$$

is convex in $\mathbf{R}^n$. By convexity, for any $x \in \mathbf{R}^n$, there exists $q \in \mathbf{R}^n$ such that

$$f(y) - f(x) \geq \langle g, y - x\rangle \quad \text{for all} \quad y \in \mathbf{R}^n.$$

For $x = \hat{x}$, we take such q and denote it by $\hat{q}$.

We begin with the case $(p, X) \in J^{2,+}u^{\lambda,\mu}(\hat{x})$. By definition of $J^{2,+}$, we observe

$$u^{\lambda,\mu}(y) \leq u^{\lambda,\mu}(\hat{x}) + \langle p, y - \hat{x}\rangle + \frac{1}{2}\langle X(y - \hat{x}), y - \hat{x}\rangle + o\left(|y - \hat{x}|^2\right)$$

as $y \to \hat{x}$. Rewriting this inequality for f, we have

$$\begin{aligned} f(y) - f(\hat{x}) \leq{}& \frac{1}{2\lambda}\left(|y'|^2 - |x'|^2\right) + \frac{1}{2\mu}\left(|s|^2 - |t|^2\right) + \langle p, y - \hat{x}\rangle \\ &+ \frac{1}{2}\langle X(y - \hat{x}), y - \hat{x}\rangle + o\left(|y - \hat{x}|^2\right), \quad y = (y', s). \end{aligned} \tag{A.3.1}$$

By the choice of $\hat{q}$, the left-hand side is dominated from below by $\langle \hat{q}, y - \hat{x}\rangle$. Since

$$\begin{aligned} |y'|^2 - |x'|^2 &= \langle y' - \hat{x}', y' - \hat{x}'\rangle + 2\langle \hat{x}', y' - \hat{x}'\rangle, \\ |s|^2 - |t|^2 &= \langle s - \hat{t}, s - \hat{t}\rangle + 2\langle \hat{t}, s - \hat{t}\rangle, \end{aligned}$$

we now observe that

$$\langle \hat{q}, y - \hat{x}\rangle \leq f(y) - f(\hat{x}) \leq \frac{1}{2\lambda}\left(\langle y' - \hat{x}', y - \hat{x}'\rangle + 2\langle \hat{x}', y' - \hat{x}'\rangle\right)$$

$$
\begin{aligned}
&+\frac{1}{2\mu}\left(\langle s-\hat{t}, s-\hat{t}\rangle+2\langle\hat{t}, s-\hat{t}\rangle\right)\\
&+\langle p, y-\hat{x}\rangle+\frac{1}{2}\langle X(y-\hat{x}), y-\hat{x}\rangle+o\left(|y-\hat{x}|^2\right)\\
&=\left\langle p+\left(\frac{1}{\lambda}Q_1+\frac{1}{\mu}Q_2\right)\hat{x}, y-\hat{x}\right\rangle\\
&+\frac{1}{2}\left\langle\left(X+\frac{1}{\lambda}Q_1+\frac{1}{\mu}Q_2\right)(y-\hat{x}), y-\hat{x}\right\rangle+o\left(|y-\hat{x}|^2\right).
\end{aligned}
$$

Thus,

$$
\hat{q}=p+\left(\frac{1}{\lambda}Q_1+\frac{1}{\mu}Q_2\right)\hat{x}, \quad X+\frac{1}{\lambda}Q_1+\frac{1}{\mu}Q_2\geq O.
$$

Here, we invoke the property that

$$
\langle p, x\rangle+\langle Xx, x\rangle=o\left(|x|^2\right) \quad \text{as} \quad x\to 0
$$

implies $p=0$ and $X\leq 0$. We have proved

$$
X\geq-\frac{Q_1}{\lambda}-\frac{Q_2}{\mu}
$$

when $(p, X)\in J^{2,+}u^{\lambda,\mu}(\hat{x})$.

For $(p, X)\in \bar{J}^{2,+}u^{\lambda,\mu}(\hat{x})$, there are sequences $\{x_k\}\subset\mathbf{R}^n$ and $\{(p_k, X_k)\}\subset\mathbf{R}^n\times\mathbf{S}^n$ such that

$$
\begin{aligned}
&x_k\to\hat{x}, \quad (p_k, X_k)\in J^{2,+}u^{\lambda,\mu}(\hat{x}) \quad \text{and}\\
&(u(x_k), p_k, X_k)\to\left(u(\hat{x}), p, X\right) \quad \text{as} \quad k\to\infty.
\end{aligned}
$$

Since $X_k+Q_1/\lambda+Q_2/\mu\geq O$, sending $k\to\infty$ yields $X+Q_1/\lambda+Q_2/\mu\geq O$. □

We finally discuss relations of semijets of u and $u^{\lambda,\mu}$.

Lemma A.3.4 *Let u be a upper semicontinuous function on $\mathbf{R}^n$ with values in $\mathbf{R}\cup\{-\infty\}$. Assume that u is bounded from above. Let $u^{\lambda,\mu}$ be a two-parameter convolution of u with $\lambda, \mu>0$. Let $(p, X)\in J^{2,+}u^{\lambda,\mu}(\hat{x})$. Then the following statement holds.*

(i) $(p, X)\in J^{2,+}u(\hat{x}+\lambda Q_1 p+\mu Q_2 p)$.
(ii) $u^{\lambda,\mu}(\hat{x})=u(\hat{x}+\lambda Q_1 p+\mu Q_2 p)-\frac{\lambda}{2}|Q_1 p|^2-\frac{\mu}{2}|Q_2 p|^2$.

Proof We first prove (i). Since $(p, X) \in J^{2,+}u^{\lambda,\mu}(\hat{x})$, by Lemma 4.2.7 there is $\phi \in C^2(\mathbf{R}^n)$ such that

$$u^{\lambda,\mu} - \phi \text{ takes its maximum at } \hat{x}, \tag{A.3.2}$$

$$\left(D\phi(\hat{x}), D^2\phi(\hat{x})\right) = (p, X). \tag{A.3.3}$$

Since $|x' - y'|^2$, $|t - s|^2$ tends to $+\infty$ as $|y'| \to \infty$, $|s| \to \infty$, for fixed x', t,

$$u(y', s) - \frac{1}{2\lambda}|x' - y'|^2 - \frac{1}{2\mu}|t - s|^2$$

attains its maximum in $\mathbf{R}^n$. Let $\hat{y} = (\hat{y}', \hat{s}) \in \mathbf{R}^n$ be a maximizer so that

$$u^{\lambda,\mu}(\hat{x}) = u(\hat{y}) - \frac{1}{2\lambda}|\hat{x}' - \hat{y}'|^2 - \frac{1}{2\mu}|\hat{t} - \hat{s}|^2. \tag{A.3.4}$$

We set

$$\Psi(x, y) := u(y) - \frac{1}{2\lambda}|x' - y'|^2 - \frac{1}{2\mu}|t - s|^2 - \phi(x).$$

Then Ψ is maximized at $(\hat{x}, \hat{y})$. Indeed, for any $(x, y) \in \mathbf{R}^n \times \mathbf{R}^n$,

$$\begin{aligned}
\Psi(x, y) &= \left\{u(y) - \frac{1}{2\lambda}|x' - y'|^2 - \frac{1}{2\mu}(t - s)^2\right\} - \phi(x) \\
&\le u^{\lambda,\mu}(x) - \phi(x) \\
&\le u^{\lambda,\mu}(\hat{x}) - \phi(\hat{x}) \quad \text{by (A.3.2)} \\
&= u(\hat{y}) - \frac{1}{2\lambda}|\hat{x}' - \hat{y}'|^2 - \frac{1}{2\mu}|\hat{t} - \hat{s}|^2 - \phi(\hat{x}) \quad \text{by (A.3.4)} \\
&= \Psi(\hat{x}, \hat{y}).
\end{aligned}$$

We set for $y \in \mathbf{R}^n$,

$$\begin{aligned}
\Phi_1(y) &:= \Psi(y + \hat{x} - \hat{y}, y) \\
&= u(y) - \psi(y), \\
\psi(y) &:= \frac{1}{2\lambda}|\hat{x}' - \hat{y}'|^2 + \frac{1}{2\mu}|\hat{t} - \hat{s}|^2 + \phi(y + \hat{x} - \hat{y}).
\end{aligned}$$

Since Φ_1 takes its maximum at $\hat{x}$, by Lemma 4.2.7,

$$\left(D\psi(\hat{y}),\, D^2\psi(\hat{y})\right) \in J^{2,+}u(\hat{y})$$

and

$$\begin{aligned}\left(D\psi(\hat{y}),\, D^2\psi(\hat{y})\right) &= \left(D\phi(\hat{y}+\hat{x}-\hat{y}),\, D^2\phi(\hat{y}+\hat{x}-\hat{y})\right)\\ &= \left(D\phi(\hat{x}),\, D^2\phi(\hat{x})\right)\\ &= (p, X) \quad \text{by (A.3.3).}\end{aligned}$$

We thus conclude that

$$(p, X) \in J^{2,+}u(\hat{y}). \tag{A.3.5}$$

We next set

$$\Phi_2(x) := \Psi(x, \hat{y}) = u(\hat{y}) - \frac{1}{2\lambda}|x' - \hat{y}'|^2 - \frac{1}{2\mu}|t - \hat{s}|^2 - \phi(x)$$

for $x \in \mathbf{R}^n$. By definition, Φ_2 is in $C^2(\mathbf{R}^n)$. Since Φ_2 takes its maximum at $x = \hat{x}$,

$$0 = D\Phi_2(\hat{x}) = -\frac{1}{\lambda}Q_1(\hat{x} - \hat{y}) - \frac{1}{\mu}Q_2(\hat{x} - \hat{y}) - D\phi(\hat{x}).$$

Since $p = D\phi(\hat{x})$, this is equivalent to

$$\hat{y} = \hat{x} + \lambda Q_1 p + \mu Q_2 p. \tag{A.3.6}$$

Combining (A.3.5) and (A.3.6), we conclude that

$$(p, X) \in J^{2,+}u(\hat{x} + \lambda Q_1 p + \mu Q_2 p).$$

It remains to prove (ii). Recall (A.3.4), i.e.,

$$u^{\lambda,\mu}(\hat{x}) = u(\hat{y}) - \frac{1}{2\lambda}|\hat{x}' - \hat{y}'|^2 - \frac{1}{2\mu}|\hat{t} - \hat{s}|^2.$$

By (A.3.6), the right-hand side equals

$$\begin{aligned}&u(\hat{x} + \lambda Q_1 p + \mu Q_2 p) - \frac{1}{2\lambda}| - \lambda Q_1 p|^2 - \frac{1}{2\mu}| - \mu Q_2 p|^2\\ &= u(\hat{x} + \lambda Q_1 p + \mu Q_2 p) - \frac{\lambda}{2}|Q_1 p|^2 - \frac{\mu}{2}|Q_2 p|^2.\end{aligned}$$

We thus proved (ii). □

A.4 Semijets with Useful Estimates

We shall prove Theorem A.1.1. We begin with a few basic estimates for a quadratic form.

Proposition A.4.1 *The estimate*

$$\langle Qw, w\rangle \le \left(|Q| + \varepsilon^{-1}\right)|w-\rho|^2 + \left\langle (Q+\varepsilon Q^2)\rho, \rho\right\rangle \tag{A.4.1}$$

holds for all $\varepsilon > 0$, $Q \in \mathbf{S}^n$, $w, \rho \in \mathbf{R}^n$.

Proof We proceed

$$\begin{aligned}\langle Qw, w\rangle &= \langle Q(w-\rho) + Q\rho,\ w-\rho+\rho\rangle \\ &= \langle Q(w-\rho), w-\rho\rangle + 2\langle Q\rho, w-\rho\rangle + \langle Q\rho, \rho\rangle.\end{aligned}$$

By the Schwarz inequality, we observe that

$$\begin{aligned}\langle Q(w-\rho), w-\rho\rangle &\le |Q||w-\rho|^2 \\ 2\langle Q\rho, w-\rho\rangle &\le 2|Q\rho||w-\rho|.\end{aligned}$$

By the Cauchy-Schwarz inequality ($2ab \le \varepsilon a^2 + \varepsilon^{-1}b^2$, $a, b \in \mathbf{R}$),

$$2\langle Q\rho, w-\rho\rangle \le \varepsilon|Q\rho|^2 + \varepsilon^{-1}|w-\rho|^2.$$

Combining these estimates, we obtain

$$\langle Qw, w\rangle \le |Q||w-\rho|^2 + \varepsilon^{-1}|w-\rho|^2 + \varepsilon|Q\rho|^2 + \langle Q\rho, \rho\rangle$$

which is the same as (A.4.1). □

Proposition A.4.2 *Let* P_i $(i = 1, 2)$ *be the projection defined by* (A.1.1). *Let* $A \in \mathbf{S}^{2n}$ *be as in* (A.1.2). *Then for any* $\varepsilon_1, \varepsilon_2 > 0$, *the estimate*

$$\begin{aligned}\left\langle A\begin{pmatrix}x\\y\end{pmatrix}, \begin{pmatrix}x\\y\end{pmatrix}\right\rangle &\le \frac{1}{\lambda}\left(|x'-\xi'|^2 + |y'-\eta'|^2\right) + \frac{1}{\mu}\left(|t-\tau|^2 + |s-\sigma|^2\right) \\ &\quad + \left\langle (A + \varepsilon_1 A^2 P_1 + \varepsilon_2 A^2 P_2)\begin{pmatrix}\xi\\\eta\end{pmatrix}, \begin{pmatrix}\xi\\\eta\end{pmatrix}\right\rangle\end{aligned}$$

holds for all $x = (x', t)$, $y = (y', s)$, $\xi = (\xi', \tau)$, $\eta = (\eta', \sigma) \in \mathbf{R}^m \times \mathbf{R}^{n-m}$ *provided that* λ, μ *are taken as*

$$\lambda = \left(|AP_1| + \varepsilon_1^{-1}\right)^{-1}, \quad \mu = \left(|AP_2| + \varepsilon_2^{-1}\right)^{-1}.$$

Proof We set

$$B_1 = \begin{pmatrix} A_{11} & A_{21} \\ {}^tA_{21} & A_{41} \end{pmatrix} \in \mathbf{S}^{2m}, \quad B_2 = \begin{pmatrix} A_{12} & A_{22} \\ {}^tA_{22} & A_{42} \end{pmatrix} \in \mathbf{S}^{2(n-m)}.$$

By definition,

$$\begin{aligned} |B_1| = |AP_1|, \quad & \left\langle B_1^k \begin{pmatrix} \xi' \\ \eta' \end{pmatrix}, \begin{pmatrix} \xi' \\ \eta' \end{pmatrix} \right\rangle = \left\langle A^k P_1 \begin{pmatrix} \xi \\ \eta \end{pmatrix}, \begin{pmatrix} \xi \\ \eta \end{pmatrix} \right\rangle, \\ |B_2| = |AP_2|, \quad & \left\langle B_2^k \begin{pmatrix} \tau \\ \sigma \end{pmatrix}, \begin{pmatrix} \tau \\ \sigma \end{pmatrix} \right\rangle = \left\langle A^k P_2 \begin{pmatrix} \xi \\ \eta \end{pmatrix}, \begin{pmatrix} \xi \\ \eta \end{pmatrix} \right\rangle \end{aligned} \tag{A.4.2}$$

for $k = 1, 2, \ldots$. We decompose

$$\left\langle A \begin{pmatrix} x \\ y \end{pmatrix}, \begin{pmatrix} x \\ y \end{pmatrix} \right\rangle = \left\langle B_1 \begin{pmatrix} x' \\ y' \end{pmatrix}, \begin{pmatrix} x' \\ y' \end{pmatrix} \right\rangle + \left\langle B_2 \begin{pmatrix} t \\ s \end{pmatrix}, \begin{pmatrix} t \\ s \end{pmatrix} \right\rangle$$

and apply (A.4.1) for each term, i.e.,

$$\begin{aligned} \left\langle B_1 \begin{pmatrix} x' \\ y' \end{pmatrix}, \begin{pmatrix} x' \\ y' \end{pmatrix} \right\rangle &\le \left(|B_1| + \varepsilon_1^{-1} \right) \left| \begin{pmatrix} x' \\ y' \end{pmatrix} - \begin{pmatrix} \xi' \\ \eta' \end{pmatrix} \right|^2 \\ &\quad + \left\langle (B_1 + \varepsilon_1 B_1^2) \begin{pmatrix} \xi' \\ \eta' \end{pmatrix}, \begin{pmatrix} \xi' \\ \eta' \end{pmatrix} \right\rangle, \\ \left\langle B_2 \begin{pmatrix} t \\ s \end{pmatrix}, \begin{pmatrix} t \\ s \end{pmatrix} \right\rangle &\le \left(|B_2| + \varepsilon_2^{-1} \right) \left| \begin{pmatrix} t \\ s \end{pmatrix} - \begin{pmatrix} \tau \\ \sigma \end{pmatrix} \right|^2 \\ &\quad + \left\langle \left(B_2 + \varepsilon_2 B_2^2 \right) \begin{pmatrix} \tau \\ \sigma \end{pmatrix}, \begin{pmatrix} \tau \\ \sigma \end{pmatrix} \right\rangle \end{aligned}$$

to get the desired estimates if we notice (A.4.2) for $k = 1, 2$. □

Proof of Theorem A.1.1 We may assume that u and v are defined in $\mathbf{R}^n$ by setting the value $-\infty$ outside $\mathcal{O}$; since we may assume that $\mathcal{O}$ is closed, these extended functions are also upper semicontinuous. Since we may assume that $\mathcal{O}$ is bounded, u and v are bounded from above in $\mathbf{R}^n$.

By Proposition A.4.2, this inequality (A.1.4) yields

$$\begin{aligned} &\left(u(x) - \frac{1}{2\lambda}|x' - \xi'|^2 - \frac{1}{2\mu}|t - \tau|^2 \right) + \left(v(y) - \frac{1}{2\lambda}|y' - \eta'|^2 - \frac{1}{2\mu}|s - \sigma|^2 \right) \\ &\quad \le \frac{1}{2} \left\langle \left(A + \sum_{j=1}^{2} \varepsilon_j A^2 P_j \right) \begin{pmatrix} \xi \\ \eta \end{pmatrix}, \begin{pmatrix} \xi \\ \eta \end{pmatrix} \right\rangle, \end{aligned}$$

where λ and μ are constant defined in Proposition A.4.2. Since this inequality holds for all $x, y \in \mathbf{R}^n$, taking supremum with respect to x and y, we get

$$u^{\lambda,\mu}(\xi) + v^{\lambda,\mu}(\xi) \le \frac{1}{2}\left\langle \left(A + \sum_{j=1}^{2} \varepsilon_j A^2 P_j\right)\begin{pmatrix}\xi\\ \eta\end{pmatrix}, \begin{pmatrix}\xi\\ \eta\end{pmatrix}\right\rangle. \tag{A.4.3}$$

In particular

$$u^{\lambda,\mu}(0) + v^{\lambda,\mu}(0) \le 0.$$

By Proposition A.3.2 (i), we see

$$u^{\lambda,\mu}(0) \ge u(0) = 0, \quad v^{\lambda,\mu}(0) \ge v(0) = 0.$$

We thus conclude that

$$u^{\lambda,\mu}(0) = v^{\lambda,\mu}(0) = 0. \tag{A.4.4}$$

For $\delta > 0$, we define

$$f(x, y) := u^{\lambda,\mu}(x) + v^{\lambda,\mu}(y) - \frac{1}{2}\left\langle \left(A + \sum_{j=1}^{2} \varepsilon_j A^2 P_j\right)\begin{pmatrix}x\\ y\end{pmatrix}, \begin{pmatrix}x\\ y\end{pmatrix}\right\rangle - \delta\left(|x|^2 + |y|^2\right).$$

The reason we introduce δ-term is that we would like to make sure that f takes the maximum over $\mathbf{R}^n \times \mathbf{R}^n$ only at $(x, y) = (0, 0)$.

Application of Lemma A.2.3 The function f is semi-convex in $\mathbf{R}^{2n}$. Indeed, $u^{\lambda,\mu}$ and $v^{\lambda,\mu}$ are semi-convex by Proposition A.3.2 (ii). Moreover, a function on $\mathbf{R}^{2n}$ defined by $z \mapsto \langle Qz, z\rangle$ for $Q \in \mathbf{S}^{2n}$ is always semi-convex. Thus f is semi-convex as a sum of semi-convex functions.

We next observe that $f(0, 0) = 0$ by (A.4.4) and $f(x, y) < 0 = f(0, 0)$ for $(x, y) \neq (0, 0)$ by (A.4.3) and δ-term. We now apply Lemma A.2.3 to conclude that there are sequences $\{x_k\}, \{y_k\}, \{p_k\}, \{q_k\} \subset \mathbf{R}^n$ (depending on δ) such that $x_k, y_k, p_k, q_k \to 0$ (as $k \to \infty$) and satisfying:

$$J^2 f(x_k, y_k) \neq \emptyset;$$

$$(x, y) \mapsto f(x, y) - \langle p_k, x\rangle - \langle q_k, y\rangle \text{ takes its local maximum at } x_k. \tag{A.4.5}$$

Application of Proposition A.2.2 Since $J^2 f(x_k, y_k) \neq \emptyset$, we see, by definition of f, that

$$J^2 u^{\lambda,\eta}(x_k) \neq \emptyset, \quad J^2 v^{\lambda,\eta}(y_k) \neq \emptyset.$$

We take

$$(\xi_k, X_k) \in J^2 u^{\lambda,\eta}(x_k), \quad (\eta_k, Y_k) \in J^2 v^{\lambda,\eta}(y_k),$$
$$\left(\begin{pmatrix} \bar{p}_k \\ \bar{q}_k \end{pmatrix}, \begin{pmatrix} \bar{X}_k & \bar{Y}_k \\ {}^t\bar{Y}_k & \bar{Z}_k \end{pmatrix} \right) \in J^2 f(x_k, y_k).$$

Since f has a Taylor's expansion up to order 2, we obtain that

$$\begin{pmatrix} \bar{p}_k \\ \bar{q}_k \end{pmatrix} = \begin{pmatrix} \xi_k \\ \eta_k \end{pmatrix} - \left(A + \sum_{j=1}^{2} \varepsilon_j A^2 P_j \right) \begin{pmatrix} x_k \\ y_k \end{pmatrix} - 2\delta \begin{pmatrix} x_k \\ y_k \end{pmatrix},$$

$$\begin{pmatrix} \bar{X}_k & \bar{Y}_k \\ {}^t\bar{Y}_k & \bar{Z}_k \end{pmatrix} = \begin{pmatrix} X_k & O \\ O & Y_k \end{pmatrix} - \left(A + \sum_{j=1}^{2} \varepsilon_j A^2 P_j \right) - 2\delta I_{2n}.$$

By (A.4.5), applying Proposition A.2.2 yields

$$\begin{pmatrix} \bar{p}_k \\ \bar{q}_k \end{pmatrix} - \begin{pmatrix} p_k \\ q_k \end{pmatrix} = 0 \quad \text{and} \quad \begin{pmatrix} \bar{X}_k & \bar{Y}_k \\ {}^t\bar{Y}_k & \bar{Z}_k \end{pmatrix} \leq O.$$

Thus, we obtain

$$\begin{pmatrix} \xi_k \\ \eta_k \end{pmatrix} = \left(A + \sum_{j=1}^{2} \varepsilon_j A^2 P_j \right) \begin{pmatrix} x_k \\ y_k \end{pmatrix} + 2\delta \begin{pmatrix} x_k \\ y_k \end{pmatrix} + \begin{pmatrix} p_k \\ q_k \end{pmatrix}, \tag{A.4.6}$$

$$\begin{pmatrix} X_k & O \\ O & Y_k \end{pmatrix} \leq A + \sum_{j=1}^{2} \varepsilon_j A^2 P_j + 2\delta I_{2n}. \tag{A.4.7}$$

Since $x_k, y_k, p_k, q_k \to 0$ as $k \to \infty$, the identity (A.4.6) implies that $\xi_k, \eta_k \to 0$ as $k \to \infty$.

Application of Lemma A.3.4 Since u and v are upper semicontinuous on $\mathbf{R}^n$ and

$$(\xi_k, X_k) \in J^{2,+} u^{\lambda,\mu}(x_k), \quad (\eta_k, Y_k) \in J^{2,+} v^{\lambda,\mu}(y_k),$$

applying Lemma A.3.4 yields

$$(\xi_k, X_k) \in J^{2,+} u(x_k + \lambda Q_1 \xi_k + \mu Q_2 \xi_k), \tag{A.4.8}$$

$$u^{\lambda,\mu}(x_k) = u(x_k + \lambda Q_1 \xi_k + \mu Q_2 \xi_k) - \frac{\lambda}{2} |Q_1 \xi_k|^2 - \frac{\mu}{2} |Q_2 \xi_k|^2. \tag{A.4.9}$$

and

$$(\eta_k, Y_k) \in J^{2,+}v(y_k + \lambda Q_1\eta_k + \mu Q_2\eta_k),$$

$$v^{\lambda,\mu}(y_k) = v(y_k + \lambda Q_1\eta_k + \mu Q_2\eta_k) - \frac{\lambda}{2}|Q_1\eta_k|^2 - \frac{\mu}{2}|Q_2\eta_k|^2.$$

Application of Proposition A.3.3 Since $(\xi_k, X_k) \in J^{2,+}u^{\lambda,\mu}(x_k)$ and $(\eta_k, Y_k) \in J^{2,+}v^{\lambda,\mu}(y_k)$, Proposition A.3.3 yields

$$X_k \geq -\frac{Q_1}{\lambda} - \frac{Q_2}{\mu}, \quad Y_k \geq -\frac{Q_1}{\lambda} - \frac{Q_2}{\mu}. \tag{A.4.10}$$

Completion of the Proof By (A.4.7) and (A.4.10), we observe that

$$-\frac{1}{\lambda}P_1 - \frac{1}{\mu}P_2 \leq \begin{pmatrix} X_k & 0 \\ 0 & Y_k \end{pmatrix} \leq A + \sum_{j=1}^{2} \varepsilon_j A^2 P_j + 2\delta I_{2n}. \tag{A.4.11}$$

In particular, by Proposition A.1.2, we may assume $X_k \to X_\delta$, $Y_k \to Y_\delta$ as $k \to \infty$ by taking a subsequence. Moreover, applying Proposition A.1.2 again, there is a convergent subsequence of X_δ and also Y_δ as $\delta \downarrow 0$. By taking a subsequence of δ, we may assume $X_\delta \to X$, $Y_\delta \to Y$ as $\delta \downarrow 0$. Sending $k \to \infty$ and $\delta \downarrow 0$ in the matrix inequality (A.4.11), we conclude

$$-\frac{1}{\lambda}P_1 - \frac{1}{\mu}P_2 \leq \begin{pmatrix} X & O \\ O & Y \end{pmatrix} \leq A + \sum_{j=1}^{2} \varepsilon_j A^2 P_j$$

which is exactly (A.1.7).

It remains to prove (A.1.5) and (A.1.6). By (A.4.9),

$$u(x_k + \lambda Q_1\xi_k + \mu Q_2\xi_k) = u^{\lambda,\mu}(x_k) + \frac{\lambda}{2}|Q_1\xi_k|^2 + \frac{\mu}{2}|Q_2\xi_k|^2.$$

Since $u^{\lambda,\mu}$ is continuous, $\lim_{k\to\infty} u^{\lambda,\mu}(x_k) = u^{\lambda,\mu}(0)$. By (A.4.4), $u^{\lambda,\mu}(0) = 0 = u(0)$, Thus

$$\lim_{k\to\infty} u(x_k + \lambda Q_1\xi_k + \mu Q_2\xi_k) = u(0).$$

Since $\xi_k \to 0$, $X_k \to X_\delta$ $(k \to \infty)$, the definition of $\bar{J}^{2,+}$ together with (A.4.8) implies that $(0, X_\delta) \in \bar{J}^{2,+}u(0)$. Sending $\delta \downarrow 0$ so that $X_\delta \to X$, we conclude that $(0, X) \in \bar{J}^{2,+}u(0)$ by Proposition A.2.1. Similarly, we obtain $(0, Y) \in \bar{J}^{2,+}v(0)$. We thus complete the proof of (A.1.5), (A.1.6). □

Bibliography

1. D. Adalsteinsson, L. C. Evans and H. Ishii, The level set method for etching and deposition, *Math. Models Methods Appl. Sci.,* **7** (1997), 1153–1186. https://doi.org/10.1142/s0218202597000578
2. T. Adamowicz, A. Kijowski and E. Soultanis, Asymptotically mean value harmonic functions in doubling metric measure spaces, *Anal. Geom. Metr. Spaces,* **10** (2022), 344–372. https://doi.org/10.1515/agms-2022-0143
3. K. Ahara and N. Ishimura, On the mean curvature flow of "thin" doughnuts, In: *Nonlinear PDE-Japan Symposium 2, 1991 (Kyoto 1991), Lecture Notes Numer. Appl. Anal,* **12** (eds. K. Masuda, M. Mimura and T. Nishida), 1–33, Kinokuniya Company Ltd., Tokyo, 1993.
4. E. S. Al-Aidarous, E. O. Alzahrani, H. Ishii and A. M. M. Younas, Asymptotic analysis for the eikonal equation with the dynamical boundary conditions, *Math. Nachr.,* **287** (2014), 1563–1588. https://doi.org/10.1002/mana.201300224
5. A. D. Aleksandrov, Almost everywhere existence of the second differential of a convex function and some properties of convex surfaces connected with it (Russian), Leningrad State Univ. Annals [Uchenye Zapiski], *Math. Ser.* **6** (1939), 3–35.
6. A. D. Aleksandrov, Uniqueness theorems for surfaces in the large. I [translation of *Vestnik Leningrad Univ. Math.,* **11** (1956), 5–17], *Amer. Math. Soc. Transl. (2),* **21** (1962), 341–354. https://doi.org/10.1090/trans2/021/09
7. A. D. Aleksandrov, Uniqueness theorems for surfaces in the large. V [translation of *Vestnik Leningrad Univ. Math.,* **13** (1958), 5–8], *Amer. Math. Soc. Transl. (2),* **21** (1962), 412–416. https://doi.org/10.1090/trans2/021
8. S. M. Allen and J. W. Cahn, A microscopic theory for antiphase boundary motion and its application to antiphase domain coarsening, *Acta Metall.,* **27** (1979), 1085–1095. https://doi.org/10.1016/0001-6160(79)90196-2
9. F. Almgren, J. E. Taylor and L. Wang, Curvature-driven flows: a variational approach, *SIAM J. Control Optim.,* **31** (1993), 387–438. https://doi.org/10.1137/0331020
10. S. Altschuler, S. B. Angenent and Y. Giga, Mean curvature flow through singularities for surfaces of rotation, *J. Geom. Anal.,* **5** (1995), 293–358. https://doi.org/10.1007/bf02921800
11. L. Alvarez, F. Guichard, P.-L. Lions and J.-M. Morel, Axioms and fundamental equations of image processing, *Arch. Rational Mech. Anal.,* **123** (1993), 199–257. https://doi.org/10.1007/bf00375127
12. L. Ambrosio and N. Dancer, Calculus of variations and partial differential equations, Topics on geometrical evolution problems and degree theory, *Papers from the Summer School held in Pisa, September 1996* (eds. G. Buttazzo, A. Marino and M. K. V. Murthy), Springer-Verlag, Berlin, 2000.

Y. Giga, Q. Liu, *Surface Evolution Equations*, Monographs in Mathematics 99,
https://doi.org/10.1007/978-3-032-20303-8

13. L. Ambrosio and J. Feng, On a class of first order Hamilton-Jacobi equations in metric spaces, *J. Differential Equations,* **256** (2014), 2194–2245. https://doi.org/10.1016/j.jde.2013.12.018
14. L. Ambrosio and H. M. Soner, Level set approach to mean curvature flow in arbitrary codimension, *J. Differential Geom.,* **43** (1996), 693–737. https://doi.org/10.4310/jdg/1214458529
15. K. Anada, Contraction of surfaces by harmonic mean curvature flows and nonuniqueness of their self similar solutions, *Calc. Var. Partial Differential Equations,* **12** (2001), 109–116. https://doi.org/10.1007/pl00009908
16. F. Andreu-Vaillo, V. Caselles and J. M. Mazón, Parabolic quasilinear equations minimizing linear growth functionals, *Progr. Math.,* **223**, Birkhäuser Basel, 2004. https://doi.org/10.1007/978-3-0348-7928-6
17. B. Andrews, Contraction of convex hypersurfaces in Euclidean space, *Calc. Var. Partial Differential Equations,* **2** (1994), 151–171. https://doi.org/10.1007/bf01191340
18. B. Andrews, Contraction of convex hypersurfaces by their affine normal, *J. Differential Geom.,* **43** (1996), 207–230. https://doi.org/10.4310/jdg/1214458106
19. B. Andrews, Evolving convex curves, *Calc. Var. Partial Differential Equations,* **7** (1998), 315–371. https://doi.org/10.1007/s005260050111
20. B. Andrews, Motion of hypersurfaces by Gauss curvature, *Pacific J. Math.,* **195** (2000), 1–34. https://doi.org/10.2140/pjm.2000.195.1
21. B. Andrews, Non-convergence and instability in the asymptotic behaviour of curves evolving by curvature, *Comm. Anal. Geom.,* **10** (2002), 409–449. https://doi.org/10.4310/cag.2002.v10.n2.a8
22. B. Andrews, Singularities in crystalline curvature flows, *Asian J. Math.,* **6** (2002), 101–122. https://doi.org/10.4310/AJM.2002.v6.n1.a6
23. B. Andrews and P. Bryan, Curvature bound for curve shortening flow via distance comparison and a direct proof of Grayson's theorem, *J. Reine Angew. Math.,* **653** (2011), 179–187. https://doi.org/10.1515/CRELLE.2011.026
24. B. Andrews, B. Chow, C. Guenther and M. Langford, Extrinsic geometric flows, *Grad. Stud. Math.,* **206**, American Mathematical Society, Providence, RI, 2020.
25. B. Andrews and M. Feldman, Nonlocal geometric expansion of convex planar curves, *J. Differential Equations,* **182** (2002), 298–343. https://doi.org/10.1006/jdeq.2001.4107
26. S. B. Angenent, Shrinking doughnuts, In: *Nonlinear diffusion equations and their equilibrium states, 3 (Gregynog, 1989)* (eds. N. G. Lloyd, W. M. Ni, L. A. Peletier and J. Serrin), 21–38, Birkhäuser Boston, Inc., Boston, MA, 1992.
27. S. Angenent and M. E. Gurtin, Multiphase thermomechanics with interfacial structure. 2. Evolution of an isothermal interface, *Arch. Rational Mech. Anal.,* **108** (1989), 323–391. https://doi.org/10.1007/bf01041068
28. S. Angenent and M. E. Gurtin, General contact-angle conditions with and without kinetics, *Quart. Appl. Math.,* **54** (1996), 557–569. https://doi.org/10.1090/qam/1402410
29. S. Angenent, T. Ilmanen and D. L. Chopp, A computed example of nonuniqueness of mean curvature flow in $\mathbf{R}^3$, *Comm. Partial Differential Equations,* **20** (1995), 1937–1958. https://doi.org/10.1080/03605309508821158
30. S. B. Angenent, T. Ilmanen and J. J. L. Velázquez, Fattening from smooth initial data in mean curvature flow, *unpublished note* (2002), *preprint* (2017). University of Wisconsin-Madison. https://people.math.wisc.edu/~angenent/preprints/unfinished/fattening.pdf
31. S. B. Angenent and J. J. L. Velázquez, Degenerate neckpinches in mean curvature flow, *J. Reine Angew. Math.,* **482** (1997), 15–66. https://doi.org/10.1515/crll.1997.482.15
32. T. Antunović, Y. Peres, S. Sheffield and S. Somersille, Tug-of-war and infinity Laplace equation with vanishing Neumann boundary condition, *Comm. Partial Differential Equations,* **37** (2012), 1839–1869. https://doi.org/10.1080/03605302.2011.642450
33. M. Arisawa and Y. Giga, Anisotropic curvature flow in a very thin domain, *Indiana Univ. Math. J.,* **52** (2003), 257–281. https://doi.org/10.1512/iumj.2003.52.2099
34. S. Armstrong and P. Cardaliaguet, Stochastic homogenization of quasilinear Hamilton-Jacobi equations and geometric motions, *J. Eur. Math. Soc. (JEMS),* **20** (2018), 797–864. https://doi.org/10.4171/JEMS/777

35. S. N. Armstrong and C. K. Smart, An easy proof of Jensen's theorem on the uniqueness of infinity harmonic functions, *Calc. Var. Partial Differential Equations,* **37** (2010), 381–384. https://doi.org/10.1007/s00526-009-0267-9
36. S. N. Armstrong and C. K. Smart, A finite difference approach to the infinity Laplace equation and tug-of-war games, *Trans. Amer. Math. Soc.,* **364** (2012), 595–636. https://doi.org/10.1090/s0002-9947-2011-05289-x
37. Á. Arroyo, P. Blanc and M. Parviainen, Local regularity estimates for general discrete dynamic programming equations, *J. Math. Pures Appl. (9),* **167** (2022), 225–256 https://doi.org/10.1016/j.matpur.2022.09.006
38. Á. Arroyo, P. Blanc and M. Parviainen, Hölder regularity for stochastic processes with bounded and measurable increments, *Ann. Inst. H. Poincaré C Anal. Non Linéaire,* **40** (2023), 215–258 https://doi.org/10.4171/aihpc/41
39. D. Azagra, M. Jiménez-Sevilla and F. Macià, Generalized motion of level sets by functions of their curvatures on Riemannian manifolds. *Calc. Var. Partial Differential Equations,* **33** (2008), 133–167. https://doi.org/10.1007/s00526-008-0160-y
40. M. Bardi and I. Capuzzo-Dolcetta, Optimal control and viscosity solutions of Hamilton-Jacobi-Bellman equations, *Systems Control Found. Appl.,* Birkhäuser, Boston, Inc., Boston, MA, 1997.
41. M. Bardi, M. G. Crandall, L. C. Evans, H. M. Soner and P. E. Souganidis, Viscosity solutions and applications, In: *Lecture Notes in Math.,* **1660**, *Lectures given at the 2nd C.I.M.E. Session (Montecatini Terme, 1995)* (eds. I. Capuzzo Dolcetta and P. L. Lions), Springer-Verlag, Berlin, 1997. https://doi.org/10.1007/bfb0094293
42. M. Bardi and F. Da Lio, On the strong maximum principle for fully nonlinear degenerate elliptic equations, *Arch. Math. (Basel),* **73** (1999), 276–285. https://doi.org/10.1007/s000130050399
43. M. Bardi and Y. Giga, Right accessibility of semicontinuous initial data for Hamilton-Jacobi equations, *Commun. Pure Appl. Anal.,* **2** (2003), 447–459. https://doi.org/10.3934/cpaa.2003.2.447
44. G. Barles, Remarks on a flame propagation model, *[Research Report] RR-0451, INRIA* (1985), inria-00076104. https://inria.hal.science/inria-00076104/
45. G. Barles, Fully nonlinear Neumann type boundary conditions for second-order elliptic and parabolic equations, *J. Differential Equations,* **106** (1993), 90–106. https://doi.org/10.1006/jdeq.1993.1100
46. G. Barles, Solutions de viscosité des équations de Hamilton-Jacobi, *Math. Appl. (Berlin),* **17**, Springer-Verlag, Paris, 1994.
47. G. Barles, Nonlinear Neumann boundary conditions for quasilinear degenerate elliptic equations and applications, *J. Differential Equations,* **154** (1999), 191–224. https://doi.org/10.1006/jdeq.1998.3568
48. G. Barles, A. Cesaroni and M. Novaga, Homogenization of fronts in highly heterogeneous media, *SIAM J. Math. Anal.,* **43** (2011), 212–227. https://doi.org/10.1137/100800014
49. G. Barles and F. Da Lio, A geometrical approach to front propagation problems in bounded domains with Neumann type boundary conditions, *Interfaces Free Bound.,* **5** (2003), 239–274. https://doi.org/10.4171/ifb/79
50. G. Barles and C. Georgelin, A simple proof of convergence for an approximation scheme for computing motions by mean curvature, *SIAM J. Numer. Anal.,* **32** (1995), 484–500. https://doi.org/10.1137/0732020
51. G. Barles, H. Ishii and H. Mitake, On the large time behavior of solutions of Hamilton-Jacobi equations associated with nonlinear boundary conditions, *Arch. Ration. Mech. Anal.,* **204** (2012), 515–558. https://doi.org/10.1007/s00205-011-0484-1
52. G. Barles and B. Perthame, Discontinuous solutions of deterministic optimal stopping time problems, *RAIRO Modél. Math. Anal. Numér.,* **21** (1987), 557–579. https://doi.org/10.1051/m2an/1987210405571
53. G. Barles and B. Perthame, Exit time problems in optimal control and vanishing viscosity method, *SIAM J. Control Optim.,* **26** (1988), 1133–1148. https://doi.org/10.1137/0326063

54. G. Barles and J.-M. Roquejoffre, Large time behaviour of fronts governed by eikonal equations, *Interfaces Free Bound.*, **5** (2003), 83–102. https://doi.org/10.4171/ifb/73
55. G. Barles, H. M. Soner and P. E. Souganidis, Front propagation and phase field theory, *SIAM J. Control Optim.*, **31** (1993), 439–469. https://doi.org/10.1137/0331021
56. G. Barles and P. E. Souganidis, A new approach to front propagation problems: theory and applications, *Arch. Rational Mech. Anal.*, **141** (1998), 237–296. https://doi.org/10.1007/s002050050077
57. E. N. Barron, R. K. Goebel and R. R. Jensen, Quasiconvex functions and nonlinear PDEs, *Trans. Amer. Math. Soc.*, **365** (2013), 4229–4255. https://doi.org/10.1090/S0002-9947-2013-05760-1
58. E. Baspinar and G. Citti, Uniqueness of viscosity mean curvature flow solution in two sub-Riemannian structures, *SIAM J. Math. Anal.*, **51** (2019), 2633–2659. https://doi.org/10.1137/17M1150797
59. G. Bellettini, Lecture notes on mean curvature flow: barriers and singular perturbations, *Appunti. Sc. Norm. Super. Pisa (N.S.)*, **12**, Scuola Normale Superiore di Pisa, Edizioni della Normale, Pisa, 2013. https://doi.org/10.1007/978-88-7642-429-8
60. G. Bellettini and S. Y. Kholmatov, Minimizing movements for mean curvature flow of droplets with prescribed contact angle, *J. Math. Pures Appl. (9)*, **117** (2018), 1–58. https://doi.org/10.1016/j.matpur.2018.06.003
61. G. Bellettini and M. Novaga, Comparison results between minimal barriers and viscosity solutions for geometric evolutions, *Ann. Scuola Norm. Sup. Pisa Cl. Sci. (4)*, **26** (1998), 97–131. http://www.numdam.org/item/ASNSP_1998_4_26_1_97_0/
62. G. Bellettini and M. Paolini, Two examples of fattening for the curvature flow with a driving force, *Atti Accad. Naz. Lincei Cl. Sci. Fis. Mat. Natur. Rend. Lincei (9) Mat. Appl.*, **5** (1994), 229–236.
63. G. Bellettini and M. Paolini, Some results on minimal barriers in the sense of De Giorgi applied to driven motion by mean curvature, *Rend. Accad. Naz. Sci. XL Mem. Mat. Appl. (5)*, **19** (1995), 43–67.
64. X. Bian, Y. Giga and H. Mitake, A level-set method for a mean curvature flow with a prescribed boundary, *Adv. Differential Equations*, **30** (2025), 1–34.
65. X. Bian, Y. Giga, and H. Mitake, Distance comparison for curve shortening flow with a prescribed boundary, *Adv. Math. Sci. Appl.*, **34**, (2025), 101–113.
66. P. Blanc and J. D. Rossi, Game theory and partial differential equations, *De Gruyter Ser. Nonlinear Anal. Appl.*, **31**, De Gruyter, Berlin, 2019.
67. K. A. Brakke, The motion of a surface by its mean curvature, *Math. Notes*, **20**, Princeton University Press, Princeton, NJ, 1978.
68. S. Brendle, Two-point functions and their applications in geometry, *Bull. Amer. Math. Soc. (N.S.)*, **51** (2014), 581–596. https://doi.org/10.1090/s0273-0979-2014-01461-2
69. L. Bronsard and R. V. Kohn, Motion by mean curvature as the singular limit of Ginzburg-Landau dynamics, *J. Differential Equations*, **90** (1991), 211–237. https://doi.org/10.1016/0022-0396(91)90147-2
70. R. Buckdahn, P. Cardaliaguet and M. Quincampoix, A representation formula for the mean curvature motion, *SIAM J. Math. Anal.*, **33** (2001), 827–846. https://doi.org/10.1137/s0036141000380334
71. L. A. Caffarelli, Interior a priori estimates for solutions of fully non-linear equations, *Ann. of Math. (2)*, **130** (1989), 189–213. https://doi.org/10.2307/1971480
72. L. A. Caffarelli and X. Cabré, Fully nonlinear elliptic equations, *Amer. Math. Soc. Colloq. Publ.*, **43** (1995), American Mathematical Society, Providence, RI, 1995. https://doi.org/10.1090/coll/043
73. L. Caffarelli, M. G. Crandall, M. Kocan and A. Święch, On viscosity solutions of fully nonlinear equations with measurable ingredients, *Comm. Pure Appl. Math.*, **49** (1996), 365–398. https://doi.org/10.1002/(SICI)1097-0312(199604)49:4\T1\textless365::AID-CPA3\T1\textgreater3.0.CO;2-A

74. L. A. Caffarelli and R. Monneau, Counter-example in three dimension and homogenization of geometric motions in two dimension, *Arch. Ration. Mech. Anal.,* **212** (2014), 503–574. https://doi.org/10.1007/s00205-013-0712-y
75. L. Caffarelli, L. Nirenberg and J. Spruck, The Dirichlet problem for nonlinear second-order elliptic equations. III. Functions of the eigenvalues of the Hessian, *Acta Math.,* **155** (1985), 261–301. https://doi.org/10.1007/bf02392544
76. L. Caffarelli, L. Nirenberg and J. Spruck, Nonlinear second-order elliptic equations. V. The Dirichlet problem for Weingarten hypersurfaces, *Comm. Pure Appl. Math.,* **41** (1988), 47–70. https://doi.org/10.1002/cpa.3160410105
77. J. W. Cahn and D. W. Hoffman, A vector thermodlnamics for anisotropic surfaces–II. Curved and faceted surfaces, *Acta Metallurgica,* **22** (1974), 1205–1214. https://doi.org/10.1016/0001-6160(74)90134-5
78. J. W. Cahn, J. E. Taylor and C. A. Handwerker, Evolving crystal forms: Frank's characteristics revisited, *Sir Charles Frank, OBE, FRS: An eightieth birthday tribute* (eds. R. G. Chambers, J. E. Enderby, A. Keller, A. R. Lang and J. W. Steeds), 88–118, Adam Hilger, Bristol, Philadelphia and New York, 1991.
79. J. Calder and N. Drenska, Asymptotically optimal strategies for online prediction with history-dependent experts, *J. Fourier Anal. Appl.,* **27** (2021), Paper No. 20. https://doi.org/10.1007/s00041-021-09815-4
80. F. Camilli, A stability property for the generalized mean curvature flow equation, *Adv. Differential Equations,* **3** (1998), 815–846. https://doi.org/10.57262/ade/1366292550
81. F. Cao, Partial differential equations and mathematical morphology, *J. Math. Pures Appl.,* **77** (1998), 909–941. https://doi.org/10.1016/s0021-7824(01)80003-9
82. F. Cao, Geometric curve evolution and image processing, *Lecture Notes in Math,* Springer Berlin, Heidelberg, 2003. https://doi.org/10.1007/b10404
83. L. Capogna and G. Citti, Generalized mean curvature flow in Carnot groups, *Comm. Partial Differential Equations,* **34** (2009), 937–956. https://doi.org/10.1080/03605300903050257
84. L. Capogna, G. Citti and C. Senni Guidotti Magnani, Sub-Riemannian heat kernels and mean curvature flow of graphs, *J. Funct. Anal.,* **264** (2013), 1899–1928. https://doi.org/10.1016/j.jfa.2013.01.020
85. L. Capogna, D. Danielli, S. D. Pauls and J. T. Tyson, An introduction to the Heisenberg group and the sub-Riemannian isoperimetric problem, *Progr. Math.,* **259**, Birkhäuser Verlag, Basel, 2007. https://link.springer.com/book/10.1007/978-3-7643-8133-2
86. I. Capuzzo-Dolcetta, F. Leoni and A. Vitolo, The Alexandrov-Bakelman-Pucci weak maximum principle for fully nonlinear equations in unbounded domains, *Comm. Partial Differential Equations,* **30** (2005), 1863–1881. https://doi.org/10.1080/03605300500300030
87. P. Cardaliaguet, On front propagation problems with nonlocal terms, *Adv. Differential Equations,* **5** (2000), 213–268. https://doi.org/10.57262/ade/1356651384
88. P. Cardaliaguet, Front propagation problems with nonlocal terms. II, *J. Math. Anal. Appl.,* **260** (2001), 572–601. https://doi.org/10.1006/jmaa.2001.7483
89. P. Cardaliaguet and O. Ley, Some flows in shape optimization, *Arch. Ration. Mech. Anal.,* **183** (2007), 21–58. https://doi.org/10.1007/s00205-006-0002-z
90. P. Cardaliaguet, P.-L. Lions and P. E. Souganidis, A discussion about the homogenization of moving interfaces, *J. Math. Pures Appl. (9),* **91** (2009), 339–363. https://doi.org/10.1016/j.matpur.2009.01.014
91. P. Cardaliaguet and E. Rouy, Viscosity solutions of increasing flows of sets. Application of the Hele-Shaw problem for power-law fluids, *SIAM J. Math. Anal.,* **38** (2006), 143–165. https://doi.org/10.1137/040613366
92. V. Caselles and A. Chambolle, Anisotropic curvature-driven flow of convex sets, *Nonlinear Anal.,* **65** (2006), 1547–1577. https://doi.org/10.1016/j.na.2005.10.029
93. F. Catté, F. Dibos and G. Koepfler, A morphological scheme for mean curvature motion and applications to anisotropic diffusion and motion of level sets, *SIAM J. Numer. Anal.,* **32** (1995), 1895–1909. https://doi.org/10.1137/0732085

94. A. Chambolle, An algorithm for mean curvature motion, *Interfaces Free Bound.*, **6** (2004), 195–218. https://doi.org/10.4171/IFB/97
95. A. Chambolle, D. De Gennaro and M. Morini, Minimizing movements for anisotropic and inhomogeneous mean curvature flows, *Adv. Calc. Var.*, **17** (2024), 1095–1129. https://doi.org/10.1515/acv-2022-0102
96. A. Chambolle, M. Morini, M. Novaga and M. Ponsiglione, Existence and uniqueness for anisotropic and crystalline mean curvature flows, *J. Amer. Math. Soc.*, **32** (2019), 779–824. https://doi.org/10.1090/jams/919
97. A. Chambolle, M. Morini and M. Ponsiglione, Nonlocal curvature flows, *Arch. Rational Mech. Anal.*, **218** (2015), 1263–1329. https://doi.org/10.1007/s00205-015-0880-z
98. A. Chambolle, M. Morini and M. Ponsiglione, Existence and uniqueness for a crystalline mean curvature flow, *Comm. Pure Appl. Math.*, **70** (2017), 1084–1114. https://doi.org/10.1002/cpa.21668
99. A. Chambolle and M. Novaga, Approximation of the anisotropic mean curvature flow, *Math. Models Methods Appl. Sci.*, **17** (2007), 833–844. https://doi.org/10.1142/S0218202507002121
100. F. Charro, J. García Azorero and J. D. Rossi, A mixed problem for the infinity Laplacian via tug-of-war games, *Calc. Var. Partial Differential Equations*, **34** (2009), 307–320. https://doi.org/10.1007/s00526-008-0185-2
101. R. M. Chen and Q. Liu, A nonlinear parabolic equation with discontinuity in the highest order and applications, *J. Differential Equations*, **260** (2016), 1200–1227. https://doi.org/10.1016/j.jde.2015.09.022
102. Xinfu Chen, Generation and propagation of interfaces for reaction-diffusion equations, *J. Differential Equations*, **96** (1992), 116–141. https://doi.org/10.1016/0022-0396(92)90146-e
103. Xinfu Chen, Global asymptotic limit of solutions of the Cahn-Hilliard equation, *J. Differential Geom.*, **44** (1996), 262–311. https://doi.org/10.4310/jdg/1214458973
104. Xu-Yan Chen, Dynamics of interfaces in reaction diffusion systems, *Hiroshima Math. J.*, **21** (1991), 47–83. https://doi.org/10.32917/hmj/1206128923
105. Y.-G. Chen, Y. Giga and S. Goto, Uniqueness and existence of viscosity solutions of generalized mean curvature flow equations, *Proc. Japan Acad. Ser. A Math. Sci.*, **65** (1989), 207–210. https://doi.org/10.3792/pjaa.65.207
106. Y.-G. Chen, Y. Giga and S. Goto, Uniqueness and existence of viscosity solutions of generalized mean curvature flow equations, *J. Differential Geom.*, **33** (1991), 749–786. https://doi.org/10.4310/jdg/1214446564
107. Y.-G. Chen, Y. Giga and S. Goto, Remarks on viscosity solutions for evolution equations, *Proc. Japan Acad. Ser. A Math. Sci.*, **67** (1991), 323–328. https://doi.org/10.3792/pjaa.67.323
108. Y.-G. Chen, Y. Giga and S. Goto, Analysis toward snow crystal growth, In: *Proc. International Symposium on Functional Analysis & Related Topics (Sapporo, 1990)* (ed. S. Koshi), 43–57, World Scientific Publishing Co., Inc., River Edge, NJ, 1991.
109. Y.-G. Chen, Y. Giga , T. Hitaka and M. Honma, A stable difference scheme for computing motion of level surfaces by the mean curvature, In: *Proc. the third GARC symposium on pure and applied mathematics (Seoul, 1994)* (eds. D. Kim, J. S. Bae and S. H. Kye), 1–19.
110. P. Cheridito, H. M. Soner, N. Touzi and N. Victoir, Second-order backward stochastic differential equations and fully nonlinear parabolic PDEs, *Comm. Pure Appl. Math.*, **60** (2007), 1081–1110. https://doi.org/10.1002/cpa.20168
111. A. A. Chernov, Crystal growth, *Springer series in solid-state sciences* **36**, Modern crystallography III, Springer Berlin, Heidelberg, 1984. https://doi.org/10.1007/978-3-642-81835-6
112. B. Choi, K. Choi and P. Daskalopoulos, Convergence of curve shortening flow to translating soliton, *Amer. J. Math.*, **143** (2021), 1043–1077. https://doi.org/10.1353/ajm.2021.0027
113. B. Choi, K. Choi and P. Daskalopoulos, Convergence of Gauss curvature flows to translating solitons, *Adv. Math.*, **397** (2022), Paper No. 108207. https://doi.org/10.1016/j.aim.2022.108207
114. K. Choi, P. Daskalopoulos, L. Kim and K.-A. Lee, The evolution of complete non-compact graphs by powers of Gauss curvature, *J. Reine Angew. Math.*, **757** (2019), 131–158. https://doi.org/10.1515/crelle-2017-0032

115. D. L. Chopp, Computation of self-similar solutions for mean curvature flow, *Experiment. Math.,* **3** (1994), 1–15.
116. D. Chopp, L. C. Evans and H. Ishii, Waiting time effects for Gauss curvature flows, *Indiana Univ. Math. J.,* **48** (1999), 311–334. https://doi.org/10.1512/iumj.1999.48.1556
117. K.-S. Chou and X.-P. Zhu, Anisotropic flows for convex plane curves, *Duke Math. J.,* **97** (1999), 579–619. https://doi.org/10.1215/S0012-7094-99-09722-3
118. K.-S. Chou and X.-P. Zhu, A convexity theorem for a class of anisotropic flows of plane curves, *Indiana Univ. Math. J.,* **48** (1999), 139–154. https://doi.org/10.1512/iumj.1999.48.1273
119. K.-S. Chou and X.-P. Zhu, The curve shortening problem, Chapman & Hall/CRC, Boca Raton, FL, 2001. https://doi.org/10.1201/9781420035704
120. B. Chow, Deforming convex hypersurfaces by the nth root of the Gaussian curvature, *J. Differential Geom.,* **22** (1985), 117–138. https://doi.org/10.4310/jdg/1214439724
121. T. H. Colding and W. P. Minicozzi, II, Uniqueness of blowups and Łojasiewicz inequalities, *Ann. of Math.,* **182** (2015), 221–285. https://doi.org/10.4007/annals.2015.182.1.5
122. T. H. Colding and W. P. Minicozzi, II, Differentiability of the arrival time, *Comm. Pure Appl. Math.,* **69** (2016), 2349–2363. https://doi.org/10.1002/cpa.21635
123. T. H. Colding and W. P. Minicozzi, II, Regularity of the level set flow, *Comm. Pure Appl. Math.,* **71** (2018), 814–824. https://doi.org/10.1002/cpa.21703
124. T. H. Colding, W. P. Minicozzi, II and E. K. Pedersen, Mean curvature flow, *Bull. Amer. Math. Soc.,* **52** (2015), 297–333. https://doi.org/10.1090/s0273-0979-2015-01468-0
125. R. Courant and D. Hilbert, Methods of mathematical physics. Vol. I, Interscience Publishers, Inc., New York, 1953.
126. M. G. Crandall, L. C. Evans and P.-L. Lions, Some properties of viscosity solutions of Hamilton-Jacobi equations, *Trans. Amer. Math. Soc.,* **282** (1984), 487–502. https://doi.org/10.2307/1999247
127. M. G. Crandall and H. Ishii, The maximum principle for semicontinuous functions, *Differential Integral Equations,* **3** (1990), 1001–1014.
128. M. G. Crandall, H. Ishii and P.-L. Lions, User's guide to viscosity solutions of second order partial differential equations, *Bull. Amer. Math. Soc. (N.S.),* **27** (1992), 1–67. https://doi.org/10.1090/s0273-0979-1992-00266-5
129. M. G. Crandall and P.-L. Lions, Condition d'unicité pour les solutions généralisées des équations de Hamilton-Jacobi du premier ordre, *C. R. Acad. Sci. Paris Sér. I Math.,* **292** (1981), 183–186.
130. M. G. Crandall and P.-L. Lions, Viscosity solutions of Hamilton-Jacobi equations, *Trans. Amer. Math. Soc.,* **277** (1983), 1–42. https://doi.org/10.2307/1999343
131. M. G. Crandall and P.-L. Lions, Convergent difference schemes for nonlinear parabolic equations and mean curvature motion, *Numer. Math.,* **75** (1996), 17–41. https://doi.org/10.1007/s002110050228
132. B. Dacorogna and C.-E. Pfister, Wulff theorem and best constant in Sobolev inequality, *J. Math. Pures Appl. (9),* **71** (1992), 97–118.
133. F. Da Lio, C. I. Kim and D. Slepčev, Nonlocal front propagation problems in bounded domains with Neumann-type boundary conditions and applications, *Asymptot. Anal.,* **37** (2004), 257–292. https://doi.org/10.3233/asy-2004-609
134. J.-P. Daniel, A game interpretation of the Neumann problem for fully nonlinear parabolic and elliptic equations, *ESAIM Control Optim. Calc. Var.,* **19** (2013), 1109–1165. https://doi.org/10.1051/cocv/2013047
135. K. Deckelnick, Error bounds for a difference scheme approximating viscosity solutions of mean curvature flow, *Interfaces Free Bound.,* **2** (2000), 117–142. https://doi.org/10.4171/ifb/15
136. K. Deckelnick and G. Dziuk, A finite element level set method for anisotropic mean curvature flow with space dependent weight, In: *Geometric Analysis and Nonlinear Partial Differential Equations* (eds. S. Hildebrandt and H. Karcher), 249–264, Springer, Berlin, Heidelberg, 2003. https://doi.org/10.1007/978-3-642-55627-2_15

137. K. Deckelnick and G. Dziuk, Numerical approximation of mean curvature flow of graphs and level sets, In: *Mathematical aspects of evolving interfaces, Lecture Notes in Math.*, **1812** (2003), 53–87, Springer, Berlin, Heidelberg. https://doi.org/10.1007/978-3-540-39189-0_2
138. E. De Giorgi, New conjectures on flow by mean curvature, In: *Nonlinear variational problems and partial differential equations (Isola d'Elba, 1990),* 120–128. *Pitman Res. Notes Math. Ser.,* **320**, Longman Scientific & Technical, Harlow, 1995.
139. M. G. Delgadino and F. Maggi, Alexandrov's theorem revisited, *Anal. PDE,* **12** (2019), 1613–1642. https://doi.org/10.2140/apde.2019.12.1613
140. P. de Mottoni and M. Schatzman, Geometrical evolution of developed interfaces, *Trans. Amer. Math. Soc.,* **347** (1995), 1533–1589. https://doi.org/10.2307/2154960 announcement: Évolution géométrique d'interfaces, *C. R. Acad. Sci. Paris Sér. I Math.,* **309** (1989), 453–458.
141. G. De Philippis and T. Laux, Implicit time discretization for the mean curvature flow of outward minimizing sets, *Ann. Sc. Norm. Super. Pisa Cl. Sci.,* **21** (special, 1–2) (2020), 911–930. https://doi.org/10.2422/2036-2145.201810_003
142. A. De Rosa, S. Kolasiński and M. Santilli, Uniqueness of critical points of the anisotropic isoperimetric problem for finite perimeter sets, *Arch. Ration. Mech. Anal.,* **238** (2020), 1157–1198. https://doi.org/10.1007/s00205-020-01562-y
143. A. Dinghas, Über einen geometrischen Satz von Wulff für die Gleichgewichtsform von Kristallen, *Z. Kristallogr., Mineral. Petrogr. Abt. A,* **105** (1944), 304–314.
144. N. Dirr, F. Dragoni and R. Grande, Asymptotics for optimal controls for horizontal mean curvature flow, *Nonlinear Anal.,* **202** (2021), Paper No. 112076. https://doi.org/10.1016/j.na.2020.112076
145. N. Dirr, F. Dragoni and M.-K. von Renesse, Evolution by mean curvature flow in sub-Riemannian geometries: a stochastic approach, *Commun. Pure Appl. Anal.,* **9** (2010), 307–326. https://doi.org/10.3934/cpaa.2010.9.307
146. C. Dohmen and Y. Giga, Selfsimilar shrinking curves for anisotropic curvature flow equations, *Proc. Japan Acad. Ser. A Math. Sci.,* **70** (1994), 252–255. https://doi.org/10.3792/pjaa.70.252
147. C. Dohmen, Y. Giga and N. Mizoguchi, Existence of selfsimilar shrinking curves for anisotropic curvature flow equations, *Calc. Var. Partial Differential Equations,* **4** (1996), 103–119. https://doi.org/10.1007/bf01189949
148. N. Drenska and J. Calder, Online prediction with history-dependent experts: the general case, *Comm. Pure Appl. Math.,* **76** (2023), 1678–1727. https://doi.org/10.1002/cpa.22049
149. N. Drenska and R. V. Kohn, Prediction with expert advice: a PDE perspective, *J. Nonlinear Sci.,* **30** (2020), 137–173. https://doi.org/10.1007/s00332-019-09570-3
150. N. Drenska and R. V. Kohn, A PDE approach to the prediction of a binary sequence with advice from two history-dependent experts, *Comm. Pure Appl. Math.,* **76** (2023), 843–897. https://doi.org/10.1002/cpa.22071
151. K. Ecker, Regularity theory for mean curvature flow, *Progr. Nonlinear Differential Equations Appl.,* **57**, Birkhäuser Boston, Inc., Boston, MA, 2004.
152. K. Ecker and G. Huisken, Mean curvature evolution of entire graphs, *Ann. of Math. (2),* **130** (1989), 453–471. https://doi.org/10.2307/1971452
153. S.-I. Ei, M.-H. Sato and E. Yanagida, Stability of stationary interfaces with contact angle in a generalized mean curvature flow, *Amer. J. Math.,* **118** (1996), 653–687.
154. C. M. Elliott, Approximation of curvature dependent interface motion, *Inst. Math. Appl. Conf. Ser. New Ser.,* **63** (1997), 407–440.
155. C. M. Elliott and H. Garcke, Existence results for diffusive surface motion laws, *Adv. Math. Sci. Appl.,* **7** (1997), 467–490.
156. C. M. Elliott, Y. Giga and S. Goto, Dynamic boundary conditions for Hamilton-Jacobi equations, *SIAM J. Math. Anal.,* **34** (2003), 861–881. https://doi.org/10.1137/S003614100139957X
157. J. Escher, U. F. Mayer and G. Simonett, The surface diffusion flow for immersed hypersurfaces, *SIAM J. Math. Anal.,* **29** (1998), 1419–1433. https://doi.org/10.1137/s0036141097320675

158. T. Eto and Y. Giga, On a minimizing movement scheme for mean curvature flow with prescribed contact angle in a curved domain and its computation, *Ann. Mat. Pura Appl. (4),* **203** (2024), 1195–1221. https://doi.org/10.1007/s10231-023-01398-9
159. T. Eto and Y. Giga, A convergence result for a minimizing movement scheme for mean curvature flow with prescribed contact angle in a curved domain, *Adv. Math. Sci. Appl.,* **34** (2025), 115–147.
160. T. Eto, Y. Giga and K. Ishii, An area-minimizing scheme for anisotropic mean-curvature flow, *Adv. Differential Equations,* **17** (2012), 1031–1084. https://doi.org/10.57262/ade/1355702938
161. L. C. Evans, Periodic homogenisation of certain fully nonlinear partial differential equations, *Proc. Roy. Soc. Edinburgh Sect. A,* **120** (1992), 245–265. https://doi.org/10.1017/S0308210500032121
162. L. C. Evans, Convergence of an algorithm for mean curvature motion, *Indiana Univ. Math. J.,* **42** (1993), 533–557. https://doi.org/10.1512/iumj.1993.42.42024
163. L. C. Evans, The 1-Laplacian, the ∞-Laplacian and differential games, In: *Perspectives in nonlinear partial differential equations,* 245–254, *Contemp. Math.,* 446, American Mathematical Society, Providence, RI, 2007.
164. L. C. Evans, H. M. Soner and P. E. Souganidis, Phase transitions and generalized motion by mean curvature, *Comm. Pure Appl. Math.,* **45** (1992), 1097–1123. https://doi.org/10.1002/cpa.3160450903
165. L. C. Evans and P. E. Souganidis, Differential games and representation formulas for solutions of Hamilton-Jacobi-Isaacs equations, *Indiana Univ. Math. J.,* **33** (1984), 773–797. https://doi.org/10.1512/iumj.1984.33.33040
166. L. C. Evans and J. Spruck, Motion of level sets by mean curvature. I, *J. Differential Geom.,* **33** (1991), 635–681. https://doi.org/10.4310/jdg/1214446559
167. L. C. Evans and J. Spruck, Motion of level sets by mean curvature. II, *Trans. Amer. Math. Soc.,* **330** (1992), 321–332. https://doi.org/10.2307/2154167
168. L. C. Evans and J. Spruck, Motion of level sets by mean curvature. III, *J. Geom. Anal.,* **2** (1992), 121–150. https://doi.org/10.1007/bf02921385
169. L. C. Evans and J. Spruck, Motion of level sets by mean curvature. IV, *J. Geom. Anal.,* **5** (1995), 77–114. https://doi.org/10.1007/bf02926443
170. A. Fathi, Sur la convergence du semi-groupe de Lax-Oleinik, C. R. Acad. Sci. Paris Sér. I Math. **327** (1998), 267–270 https://doi.org/10.1016/S0764-4442(98)80144-4
171. F. Ferrari, Q. Liu and J. J. Manfredi, On the characterization of p-harmonic functions on the Heisenberg group by mean value properties, *Discrete Contin. Dyn. Syst.,* **34** (2014), 2779–2793. https://doi.org/10.3934/dcds.2014.34.2779
172. F. Ferrari, Q. Liu and J. J. Manfredi, On the horizontal mean curvature flow for axisymmetric surfaces in the Heisenberg group, *Commun. Contemp. Math.,* **16** (2014), 1350027. https://doi.org/10.1142/S0219199713500272
173. F. Ferrari and A. Pinamonti, Characterization by asymptotic mean formulas of q-harmonic functions in Carnot groups, *Potential Anal.,* **42** (2015), 203–227. https://doi.org/10.1007/s11118-014-9430-9
174. J. Fischer, S. Hensel, T. Laux and T. Simon, The local structure of the energy landscape in multiphase mean curvature flow: Weak-strong uniqueness and stability of evolutions. arXiv: 2003.05478 (First part: to appear in *J. Eur. Math. Soc.*)
175. W. J. Firey, Shapes of worn stones, *Mathematika,* **21** (1974), 1–11. https://doi.org/10.1112/S0025579300005714
176. W. H. Fleming and H. M. Soner, Controlled Markov processes and viscosity solutions, *Stochastic Modelling and Applied Probability,* Springer New York, NY, 1993, 2nd edition 2006.
177. I. Fonseca, The Wulff theorem revisited, *Proc. Roy. Soc. London Ser. A,* **432** (1991), 125–145. https://doi.org/10.1098/rspa.1991.0009
178. I. Fonseca and M. A. Katsoulakis, Γ-convergence, minimizing movements and generalized mean curvature evolution, *Differential Integral Equations,* **8** (1995), 1619–1656.

179. I. Fonseca and S. Müller, A uniqueness proof for the Wulff theorem, *Proc. Roy. Soc. Edinburgh Sect. A,* **119** (1991), 125–136. https://doi.org/10.1017/s0308210500028365
180. T. Fukui and Y. Giga, Motion of a graph by nonsmooth weighted curvature, In: *World Congress of Nonlinear Analysts '92, Vol. I–IV (Tampa, FL, 1992)* (ed. V. Lakshmikantham), 47–56, Walter de Gruyter & Co., Berlin, 1996. https://doi.org/10.1515/9783110883237.47
181. M. E. Gage, Evolving plane curves by curvature in relative geometries, *Duke Math. J.,* **72** (1993), 441–466. https://doi.org/10.1215/s0012-7094-93-07216-x
182. M. Gage and R. S. Hamilton, The heat equation shrinking convex plane curves, *J. Differential Geom.,* **23** (1986), 69–96. https://doi.org/10.4310/jdg/1214439902
183. M. E. Gage and Y. Li, Evolving plane curves by curvature in relative geometries. II, *Duke Math. J.,* **75** (1994), 79–98. https://doi.org/10.1215/s0012-7094-94-07503-0
184. W. Gangbo and A. Święch, Optimal transport and large number of particles, *Discrete Contin. Dyn. Syst.,* **34** (2014), 1397–1441. https://doi.org/10.3934/dcds.2014.34.1397
185. W. Gangbo and A. Święch, Metric viscosity solutions of Hamilton-Jacobi equations depending on local slopes, *Calc. Var. Partial Differential Equations,* **54** (2015), 1183–1218. https://doi.org/10.1007/s00526-015-0822-5
186. H. Gao, Z. Long, J. Xin and Y. Yu, Existence of an effective burning velocity in a cellular flow for the curvature G-equation proved using a game analysis, *J. Geom. Anal.,* **34** (2024), Paper No. 81. https://doi.org/10.1007/s12220-023-01523-3
187. M.-H. Giga and Y. Giga, Evolving graphs by singular weighted curvature, *Arch. Rational Mech. Anal.,* **141** (1998), 117–198. https://doi.org/10.1007/s002050050075
188. M.-H. Giga and Y. Giga, Motion driven by nonlocal curvature in the plane, In: *Proc. the International Conference on Asymptotics in Nonlinear Diffusive Systems (Sendai, 1997)* (eds. Y. Nishiura, I. Takagi and E. Yanagida), 75–83, *Tohoku Math. Publ.,* **8**, Tohoku University, Mathematical Institute, Sendai, 1998.
189. M.-H. Giga and Y. Giga, Stability for evolving graphs by nonlocal weighted curvature, *Comm. Partial Differential Equations,* **24** (1999), 109–184. https://doi.org/10.1080/03605309908821419
190. M.-H. Giga and Y. Giga, Crystalline and level set flow–convergence of a crystalline algorithm for a general anisotropic curvature flow in the plane, In: *Free boundary problems: theory and applications, I (Chiba, 1999)* (ed. N. Kenmochi), 64–79, *GAKUTO Internat. Ser. Math. Sci. Appl.,* **13**, Gakkōtosho Co., Ltd., Tokyo, 2000.
191. M.-H. Giga and Y. Giga, Generalized motion by nonlocal curvature in the plane, *Arch. Ration. Mech. Anal.,* **159** (2001), 295–333. https://doi.org/10.1007/s002050100154
192. M.-H. Giga and Y. Giga, Minimal vertical singular diffusion preventing overturning for the Burgers equation, In: *Recent advances in scientific computing and partial differential equations (Hong Kong, 2002),* 73–88, *Contemp. Math.,* **330**, American Mathematical Society, Providence, RI, 2003. https://doi.org/10.1090/conm/330/05885
193. M.-H. Giga and Y. Giga, Very singular diffusion equations: second and fourth order problems, *Jpn. J. Ind. Appl. Math.,* **27** (2010), 323–345. https://doi.org/10.1007/s13160-010-0020-y
194. M.-H. Giga and Y. Giga, Crystalline surface diffusion flow for graph-like curves, *Discrete Contin. Dyn. Syst.,* **43** (2023), 1436–1468. https://doi.org/10.3934/dcds.2022160
195. M.-H. Giga and Y. Giga, A basic guide to uniqueness problems for evolutionary differential equations, *Compact Textb. Math.,* Birkhäuser/Springer, Cham, 2023.
196. M.-H. Giga, Y. Giga and N. Požár, Anisotropic total variation flow of non-divergence type on a higher dimensional torus, *Adv. Math. Sci. Appl.,* **23** (2013), 235–266.
197. M.-H. Giga, Y. Giga and N. Požár, Periodic total variation flow of non-divergence type in $\mathbb{R}^n$, *J. Math. Pures Appl.,* **102** (2014), 203–233. https://doi.org/10.1016/j.matpur.2013.11.007
198. M.-H. Giga, Y. Giga and J. Saal, Nonlinear Partial Differential Equations. Asymptotic Behavior of Solutions and Self-Similar Solutions, *Progress in Nonlinear Differential Equations and their Applications,* **79**, Birkhäuser, Boston-Basel-Berlin, 2010. https://doi.org/10.1007/978-0-8176-4651-6
199. Y. Giga, Motion of a graph by convexified energy, *Hokkaido Math. J.,* **23** (1994), 185–212. https://doi.org/10.14492/hokmj/1381412492

200. Y. Giga, A level set method for surface evolution equations [translation of *Sūgaku,* **47** (1995), 321–340], *Sugaku Expositions,* **10** (1997), 217–241.
201. Y. Giga, Evolving curves with boundary conditions, In: *Curvature flows and related topics (Levico, 1994)* (eds. A. Damlamian, J. Spruck and A. Visintin), 99–109, *GAKUTO Internat. Ser. Math. Sci. Appl.,* **5**, Gakkōtosho Co., Ltd., Tokyo, 1995.
202. Y. Giga, Anisotropic curvature effects in interface dynamics [translation of *Sūgaku,* **52** (2000), 113–127], *Sugaku Expositions,* **16** (2003), 135–152.
203. Y. Giga, Viscosity solutions with shocks, *Comm. Pure Appl. Math.,* **55** (2002), 431–480. https://doi.org/10.1002/cpa.3015
204. Y. Giga, Surface evolution equations: a level set method, *Hokkaido Univ. Tech. Rep. Ser. Math.,* **71** (2002). https://eprints.lib.hokudai.ac.jp/dspace/handle/2115/705 (a preliminary version of the first edition of this book)
205. Y. Giga, Shocks and very strong vertical diffusion, In: *Lectures on partial differential equations: Proc. in honor of Louis Nirenberg's 75th Birthday* (eds. S.-Y. A. Chang, C.-S. Lin and H.-T. Yau), 95–102, *New Stud. Adv. Math.,* **2**, International Press, Somerville, MA, 2003.
206. Y. Giga, On large time behavior of growth by birth and spread, In: *Proc. International Congress of Mathematicians (Rio de Janeiro, 2018),* Vol. Ⅲ, 2287–2310. https://doi.org/10.1142/11060
207. Y. Giga and S. Goto, Motion of hypersurfaces and geometric equations, *J. Math. Soc. Japan,* **44** (1992), 99–111. https://doi.org/10.2969/jmsj/04410099
208. Y. Giga and S. Goto, Geometric evolution of phase-boundaries, In: *On the evolution of phase boundaries (Minneapolis, MN, 1990–91)* (eds. M. E. Gurtin and G. B. McFadden), 51–65, *IMA Vol. Math. Appl.,* **43**, Springer-Verlag, New York, 1992.
209. Y. Giga, S. Goto and H. Ishii, Global existence of weak solutions for interface equations coupled with diffusion equations, *SIAM J. Math. Anal.,* **23** (1992), 821–835. https://doi.org/10.1137/0523043
210. Y. Giga, S. Goto, H. Ishii and M.-H. Sato, Comparison principle and convexity preserving properties for singular degenerate parabolic equations on unbounded domains, *Indiana Univ. Math. J.,* **40** (1991), 443–470. https://doi.org/10.1512/iumj.1991.40.40023
211. Y. Giga and N. Hamamuki, Hamilton-Jacobi equations with discontinuous source terms, *Comm. Partial Differential Equations,* **38** (2013), 199–243. https://doi.org/10.1080/03605302.2012.739671
212. Y. Giga and N. Hamamuki, On a dynamic boundary condition for singular degenerate parabolic equations in a half space, *NoDEA Nonlinear Differential Equations Appl.,* **25** (2018), Paper No. 51. https://doi.org/10.1007/s00030-018-0542-6
213. Y. Giga, N. Hamamuki and A. Nakayasu, Eikonal equations in metric spaces, *Trans. Amer. Math. Soc.,* **367** (2015), 49–66. https://doi.org/10.1090/S0002-9947-2014-05893-5
214. Y. Giga and K. Ito, On pinching of curves moved by surface diffusion, *Commun. Appl. Anal.,* **2** (1998), 393–405.
215. Y. Giga and R. V. Kohn, Asymptotically self-similar blow-up of semilinear heat equations, *Comm. Pure Appl. Math.,* **38** (1985), 297–319. https://doi.org/10.1002/cpa.3160380304
216. Y. Giga and Q. Liu, A remark on the discrete deterministic game approach for curvature flow equations, In: *Proc. Nonlinear phenomena with energy dissipation: Mathematical analysis, modeling and simulation* (eds. A. Damlamian, N. Kenmochi, M. Mimura and J. Sprekels), *GAKUTO Internat. Ser. Math. Sci. Appl.,* **29** (2008), 103–115.
217. Y. Giga and Q. Liu, A billiard-based game interpretation of the Neumann problem for the curve shortening equation, *Adv. Differential Equations,* **14** (2009), 201–240. https://doi.org/10.57262/ade/1355867265
218. Y. Giga, Q. Liu and H. Mitake, Large-time asymptotics for one-dimensional Dirichlet problems for Hamilton-Jacobi equations with noncoercive Hamiltonians, *J. Differential Equations* **252** (2012), 1263–1282. https://doi.org/10.1016/j.jde.2011.10.010
219. Y. Giga, Q. Liu and H. Mitake, Singular Neumann problems and large-time behavior of solutions of noncoercive Hamilton-Jacobi equations, *Trans. Amer. Math. Soc.,* **366** (2014), 1905–1941. https://doi.org/10.1090/s0002-9947-2013-05905-3

220. Y. Giga, H. Mitake, T. Ohtsuka and H. V. Tran, Existence of asymptotic speed of solutions to birth-and-spread type nonlinear partial differential equations, *Indiana Univ. Math. J.*, **70** (2021), 121–156. https://doi.org/10.1512/iumj.2021.70.8305
221. Y. Giga, H. Mitake and H. V. Tran, On asymptotic speed of solutions to level-set mean curvature flow equations with driving and source terms, *SIAM J. Math. Anal.*, **48** (2016), 3515–3546 https://doi.org/10.1137/15M1052755
222. Y. Giga, H. Mitake and H. V. Tran, Remarks on large time behavior of level-set mean curvature flow equations with driving and source terms, *Discrete Contin. Dyn. Syst. Ser. B,* **25** (2020), 3983–3999. https://doi.org/10.3934/dcdsb.2019228
223. Y. Giga and T. Namba, Well-posedness of Hamilton-Jacobi equations with Caputo's time fractional derivative, *Comm. Partial Differential Equations,* **42** (2017), 1088–1120. https://doi.org/10.1080/03605302.2017.1324880
224. Y. Giga and M. Ohnuma, On strong comparison principle for semicontinuous viscosity solutions of some nonlinear elliptic equations, *Int. J. Pure Appl. Math.,* **22** (2005), 165–184.
225. Y. Giga, M. Ohnuma and M.-H. Sato, On the strong maximum principle and the large time behavior of generalized mean curvature flow with the Neumann boundary condition, *J. Differential Equations,* **154** (1999), 107–131. https://doi.org/10.1006/jdeq.1998.3569
226. Y. Giga and N. Požár, A level set crystalline mean curvature flow of surfaces, *Adv. Differential Equations,* **21** (2016), 631–698. https://doi.org/10.57262/ade/1462298654
227. Y. Giga and N. Požár, Approximation of general facets by regular facets with respect to anisotropic total variation energies and its application to the crystalline mean curvature flow, *Comm. Pure Appl. Math.,* **71** (2018), 1461–1491. https://doi.org/10.1002/cpa.21752
228. Y. Giga and N. Požár, Viscosity solutions for the crystalline mean curvature flow with a nonuniform driving force term, *SN Partial Differ. Equ. Appl.,* **1** (2020), Article number: 39. https://doi.org/10.1007/s42985-020-00040-0
229. Y. Giga and N. Požár, Motion by crystalline-like mean curvature: a survey, *Bull. Math. Sci.,* **12** (2022), 2230004-1-68. https://doi.org/10.1142/S1664360722300043
230. Y. Giga and M.-H. Sato, Neumann problem for singular degenerate parabolic equations, *Differential Integral Equations,* **6** (1993), 1217–1230.
231. Y. Giga and M.-H. Sato, A level set approach to semicontinuous viscosity solutions for Cauchy problems, *Comm. Partial Differential Equations,* **26** (2001), 813–839. https://doi.org/10.1081/PDE-100002379
232. Y. Giga, H. V. Tran and L. Zhang, On obstacle problem for mean curvature flow with driving force, *Geom. Flows* **4** (2019), 9–29. https://doi.org/10.1515/geofl-2019-0002
233. Y. Giga and K. Yama-uchi, On a lower bound for the extinction time of surfaces moved by mean curvature, *Calc. Var. Partial Differential Equations,* **1** (1993), 417–428. https://doi.org/10.1007/BF01206960
234. Y. Giga and J. Zhai, Uniqueness of constant weakly anisotropic mean curvature immersion of the sphere S^2 in $\mathbb{R}^3$, *Adv. Differential Equations,* **14** (2009), 601–619. https://doi.org/10.57262/ade/1355867227
235. D. Gilbarg and N. S. Trudinger, Elliptic partial differential equations of second order, Revised third printing, *Grundlehren der Mathematischen Wissenschaften,* 224, Springer, Berlin 1998; available as *Classics in Mathematics,* **224**, Springer Berlin, Heidelberg, 2001. https://doi.org/10.1007/978-3-642-61798-0
236. P. M. Girão, Convergence of a crystalline algorithm for the motion of a simple closed convex curve by weighted curvature, *SIAM J. Numer. Anal.,* **32** (1995), 886–899. https://doi.org/10.1137/0732041
237. P. M. Girão and R. V. Kohn, Convergence of a crystalline algorithm for the heat equation in one dimension and for the motion of a graph by weighted curvature, *Numer. Math.,* **67** (1994), 41–70. https://doi.org/10.1007/s002110050017
238. P. M. Girão and R. V. Kohn, The crystalline algorithm for computing motion by curvature, In: *Variational methods for discontinuous structures (Como, 1994)* (eds. R. Serapioni and F. Tomarelli), 7–18, *Progr. Nonlinear Differential Equations Appl.,* **25**, Birkhäuser Verlag, Basel, 1996.

239. S. Goto, Generalized motion of hypersurfaces whose growth speed depends superlinearly on the curvature tensor, *Differential Integral Equations,* **7** (1994), 323–343.
240. S. Goto, M. Nakagawa and T. Ohtsuka, Uniqueness and existence of generalized motion for spiral crystal growth, *Indiana Univ. Math. J.,* **57** (2008), 2571–2599. https://doi.org/10.1512/iumj.2008.57.3350
241. M. A. Grayson, The heat equation shrinks embedded plane curves to round points, *J. Differential Geom.,* **26** (1987), 285–314. https://doi.org/10.4310/jdg/1214441371
242. M. A. Grayson, A short note on the evolution of a surface by its mean curvature, *Duke. Math. J.,* **58** (1989), 555–558. https://doi.org/10.1215/s0012-7094-89-05825-0
243. F. Guichard and J.-M. Morel, Image analysis and P.D.E.'s, In: *IPAM Geometrically Based Motions Tutorials (2001),* University of California, Los Angeles, 2001.
244. M. E. Gurtin, Toward a nonequilibrium thermodynamics of two-phase materials, *Arch. Rational Mech. Anal.,* **100** (1988), 275–312. https://doi.org/10.1007/bf00251518
245. M. E. Gurtin, Multiphase thermomechanics with interfacial structure. I. Heat conduction and the capillary balance law, *Arch. Rational Mech. Anal.,* **104** (1988), 195–221. https://doi.org/10.1007/bf00281354
246. M. E. Gurtin, Thermomechanics of evolving phase boundaries in the plane, *Oxford Math. Monogr.,* The Clarendon Press, Oxford University Press, New York, 1993.
247. M. E. Gurtin, H. M. Soner and P. E. Souganidis, Anisotropic motion of an interface relaxed by the formation of infinitesimal wrinkles, *J. Differential Equations,* **119** (1995), 54–108. https://doi.org/10.1006/jdeq.1995.1084
248. N. Hamamuki, On large time behavior of Hamilton-Jacobi equations with discontinuous source terms, In: *Proc. Fifth Polish-Japanese Days on nonlinear analysis in interdisciplinary sciences: modellings, theory and simulations,* 83–112. *GAKUTO Internat. Ser. Math. Sci. Appl.,* **36** Gakkōtosho Co., Ltd., Tokyo, 2013.
249. N. Hamamuki, Uniqueness and existence of viscosity solutions under degenerate dynamic boundary condition, Hokkaido University Preprint Series in Mathematics #1131 (2020). https://doi.org/10.14943/92820
250. N. Hamamuki and Q. Liu, A deterministic game interpretation for fully nonlinear parabolic equations with dynamic boundary conditions, *ESAIM Control Optim. Calc. Var.,* **26** (2020), Paper No. 13. https://doi.org/10.1051/cocv/2019076
251. N. Hamamuki and Q. Liu, A game-theoretic approach to dynamic boundary problems for level-set curvature flow equations and applications, *Partial Differ. Equ. Appl.,* **2** (2021), Paper No. 30. https://doi.org/10.1007/s42985-021-00076-w
252. R. S. Hamilton, Isoperimetric estimates for the curve shrinking flow in the plane, In: *Modern methods in complex analysis (Princeton, NJ, 1992),* 201–222. *Ann. of Math. Stud.,* **137**, Princeton University Press, Princeton, NJ, 1995. https://doi.org/10.1016/1053-8127(94)00130-3
253. Q. Han and F. Lin, Elliptic partial differential equations, *Courant Lect. Notes in Math.,* **1**, New York University, Courant Institute of Mathematical Sciences, New York; American Mathematical Society, Providence, RI, 1997. https://doi.org/10.1090/cln/001
254. A. Handlovičová and K. Mikula, Analysis of a semidiscrete scheme for solving image smoothing equation of mean curvature flow type, *Acta Math. Univ. Comenian. (N.S.),* **71** (2002), 93–111.
255. D. Hartenstine and M. Rudd, Asymptotic statistical characterizations of p-harmonic functions of two variables, *Rocky Mountain J. Math.,* **41** (2011), 493–504. https://doi.org/10.1216/rmj-2011-41-2-493
256. Y. He, H. Li, H. Ma and J. Ge, Compact embedded hypersurfaces with constant higher order anisotropic mean curvatures, *Indiana Univ. Math. J.,* **58** (2009), 853–868. https://doi.org/10.1512/iumj.2009.58.3515
257. S. Hensel and T. Laux, Weak-strong uniqueness for the mean curvature flow of double bubbles, *Interfaces Free Bound.,* **25** (2023), 37–107. https://doi.org/10.4171/IFB/484
258. S. Hensel and T. Laux, A new varifold solution concept for mean curvature flow: convergence of the Allen-Cahn equation and weak-strong uniqueness, *J. Differential Geom.,* **130** (2025), 209–268. https://doi.org/10.4310/jdg/1747065796

259. G. Huisken, Flow by mean curvature of convex surfaces into spheres, *J. Differential Geom.,* **20** (1984), 237–266. https://doi.org/10.4310/jdg/1214438998
260. G. Huisken, Asymptotic behavior for singularities of the mean curvature flow, *J. Differential Geom.,* **31** (1990), 285–299. https://doi.org/10.4310/jdg/1214444099
261. G. Huisken, Local and global behaviour of hypersurfaces moving by mean curvature, In: *Differential geometry: partial differential equations on manifolds (Los Angeles, 1990),* 175–191, *Proc. Sympos. Pure Math.,* **54**, Part 1, American Mathematical Society, Providence, RI, 1993.
262. G. Huisken, A distance comparison principle for evolving curves, *Asian J. Math.,* **2** (1998), 127–134. https://doi.org/10.4310/AJM.1998.v2.n1.a2
263. G. Huisken and T. Ilmanen, The Riemannian Penrose inequality, *Internat. Math. Res. Notices,* **1997**, No. 20 (1997), 1045–1058. https://doi.org/10.1155/s1073792897000664
264. G. Huisken and A. Polden, Geometric evolution equations for hypersurfaces, In: *Calculus of variations and geometric evolution problems (Cetraro, 1996),* 45–84, *Lecture Notes in Math.,* **1713**, Fond. CIME/CIME Found. Subser., Springer-Verlag, Berlin, 1999.
265. G. Huisken and C. Sinestrari, Convexity estimates for mean curvature flow and singularities of mean convex surfaces, *Acta Math.,* **183** (1999), 45–70. https://doi.org/10.1007/bf02392946
266. T. Ilmanen, Generalized flow of sets by mean curvature on a manifold, *Indiana Univ. Math. J.,* **41** (1992), 671–705. https://doi.org/10.1512/iumj.1992.41.41036
267. T. Ilmanen, The level-set flow on a manifold, In: *Differential geometry: partial differential equations on manifolds (Los Angeles, 1990),* 193–204, *Proc. Sympos. Pure Math.,* **54**, Part I (eds. R. Greene and S.-T. Yau), American Mathematical Society, Providence, RI, 1993.
268. T. Ilmanen, Convergence of the Allen-Cahn equation to Brakke's motion by mean curvature, *J. Differential Geom.,* **38** (1993), 417–461. https://doi.org/10.4310/jdg/1214454300
269. H. Ishii, Hamilton-Jacobi equations with discontinuous Hamiltonians on arbitrary open sets, *Bull. Fac. Sci. Engrg. Chuo. Univ.,* **28** (1985), 33–77.
270. H. Ishii, Perron's method for Hamilton-Jacobi equations, *Duke Math. J.,* **55** (1987), 369–384. https://doi.org/10.1215/s0012-7094-87-05521-9
271. H. Ishii, On uniqueness and existence of viscosity solutions of fully nonlinear second-order elliptic PDEs, *Comm. Pure Appl. Math.,* **42** (1989), 15–45. https://doi.org/10.1002/cpa.3160420103
272. H. Ishii, A boundary value problem of the Dirichlet type for Hamilton-Jacobi equations, *Ann. Scuola Norm. Sup. Pisa Cl. Sci. (4),* **16** (1989), 105–135. http://www.numdam.org/item/ASNSP_1989_4_16_1_105_0/
273. H. Ishii, Fully nonlinear oblique derivative problems for nonlinear second-order elliptic PDEs, *Duke Math. J.,* **62** (1991), 633–661. https://doi.org/10.1215/s0012-7094-91-06228-9
274. H. Ishii, Viscosity solutions for nonlinear PDE [translation of *Sūgaku,* **46** (1994), 144–157], *Sugaku Expositions,* **9** (1996), 135–152.
275. H. Ishii, A generalization of the Bence, Merriman and Osher algorithm for motion by mean curvature, In: *Curvature flows and related topics (Levico, 1994)* (eds. A. Damlamian, J. Spruck and A. Visintin), 111–127, *GAKUTO Internat. Ser. Math. Sci. Appl.,* **5**, Gakkōtosho Co., Ltd., Tokyo, 1995.
276. H. Ishii, Degenerate parabolic PDEs with discontinuities and generalized evolutions of surfaces, *Adv. Differential Equations,* **1** (1996), 51–72. https://doi.org/10.57262/ade/1366896314
277. H. Ishii and P.-L. Lions, Viscosity solutions of fully nonlinear second-order elliptic partial differential equations, *J. Differential Equations,* **83** (1990), 26–78. https://doi.org/10.1016/0022-0396(90)90068-z
278. H. Ishii and T. Mikami, A mathematical model of the wearing process of a nonconvex stone, *SIAM J. Math. Anal.,* **33** (2001), 860–876. https://doi.org/10.1137/s0036141001384958
279. H. Ishii and T. Mikami, A level set approach to the wearing process of a nonconvex stone, *Calc. Var. Partial Differential Equations,* **19** (2004), 53–93. https://doi.org/10.1007/s00526-003-0196-y
280. H. Ishii and T. Mikami, Motion of a graph by R-curvature, *Arch. Ration. Mech. Anal.,* **171** (2004), 1–23. https://doi.org/10.1007/s00205-003-0294-1

281. H. Ishii and T. Mikami, Convexified Gauss curvature flow of sets: a stochastic approximation, *SIAM J. Math. Anal.,* **36** (2004), 552–579. https://doi.org/10.1137/s0036141002420509
282. H. Ishii, G. E. Pires and P. E. Souganidis, Threshold dynamics type approximation schemes for propagating fronts, *J. Math. Soc. Japan,* **51** (1999), 267–308. https://doi.org/10.2969/jmsj/05120267
283. H. Ishii and M.-H. Sato, Nonlinear oblique derivative problems for singular degenerate parabolic equations on a general domain, *Nonlinear Anal.,* **57** (2004), 1077–1098. https://doi.org/10.1016/j.na.2004.04.003
284. H. Ishii and P. Souganidis, Generalized motion of noncompact hypersurfaces with velocity having arbitrary growth on the curvature tensor, *Tohoku Math. J. (2),* **47** (1995), 227–250. https://doi.org/10.2748/tmj/1178225593
285. H. Ishii and Y. Yoshimura, Demi-eigenvalues for uniformly elliptic Isaccs operators, *preprint* (2005). published as Y. Yoshimura, A note on demi-eigenvalues for uniformly elliptic Isaacs operators, *Sūrikaisekikenkyūsho Kōkyūroku,* **1481** (2006), 106–114. http://hdl.handle.net/2433/58052
286. K. Ishii, An approximation scheme for the anisotropic and nonlocal mean curvature flow, *NoDEA Nonlinear Differential Equations Appl.,* **21** (2014), 219–252. https://doi.org/10.1007/s00030-013-0244-z
287. K. Ishii, Optimal rate of convergence of the Bence-Merriman-Osher algorithm for motion by mean curvature, *SIAM J. Math. Anal.,* **37** (2005), 841–866. https://doi.org/10.1137/04061862X
288. K. Ishii and H. M. Soner, Regularity and convergence of crystalline motion, *SIAM J. Math. Anal.,* **30** (1999), 19–37. https://doi.org/10.1137/s0036141097317347
289. N. Ishimura, Curvature evolution of plane curves with prescribed opening angle, *Bull. Austral. Math. Soc.,* **52** (1995), 287–296. https://doi.org/10.1017/s0004972700014714
290. M. Ishiwata, R. Magnanini and H. Wadade, A natural approach to the asymptotic mean value property for the p-Laplacian, *Calc. Var. Partial Differential Equations,* **56** (2017), Paper No. 97. https://doi.org/10.1007/s00526-017-1188-7
291. N. M. Ivochkina, S. I. Prokof'eva and G. V. Yakunina, On a class of equations of Monge-Ampère type, *J. Math. Sci.,* **73** (1995), 663–673. https://doi.org/10.1007/bf02364944
292. J. Jang, Capillary-type boundary value problems of mean curvature flows with force and transport terms on a bounded domain, *Calc. Var. Partial Differential Equations* **62** (2023), Paper No. 108. https://doi.org/10.1007/s00526-023-02450-5
293. J. Jang, D. Kwon, H. Mitake and H. V. Tran, Level-set forced mean curvature flow with the Neumann boundary condition, *J. Math. Pures Appl.,* (9) **168** (2022), 143–167. https://doi.org/10.1016/j.matpur.2022.11.002
294. R. Jensen, The maximum principle for viscosity solutions of fully nonlinear second order partial differential equations, *Arch. Rational Mech. Anal.,* **101** (1988), 1–27. https://doi.org/10.1007/bf00281780
295. T. Kagaya, Singular Dirichlet boundary problems for a class of fully nonlinear parabolic equations in one dimension, *Math. Ann.,* **394**, Paper No. 10 (2026). https://doi.org/10.1007/s00208-026-03334-7
296. T. Kagaya and Q. Liu, Singular Neumann boundary problems for a class of fully nonlinear parabolic equations in one dimension, *SIAM J. Math. Anal.,* **53** (2021), 4350–4385. https://doi.org/10.1137/20M1371646
297. T. Kato, A short introduction to perturbation theory for linear operators, Springer-Verlag, New York-Berlin, 1982.
298. M. Katsoulakis, G. T. Kossioris and F. Reitich, Generalized motion by mean curvature with Neumann conditions and the Allen-Cahn model for phase transitions, *J. Geom. Anal.,* **5** (1995), 255–279. https://doi.org/10.1007/bf02921677
299. M. A. Katsoulakis and P. E. Souganidis, Interacting particle systems and generalized evolution of fronts, *Arch. Rational Mech. Anal.,* **127** (1994), 133–157. https://doi.org/10.1007/bf00377658

300. B. Kawohl and N. Kutev, Comparison principle and Lipschitz regularity for viscosity solutions of some classes of nonlinear partial differential equations, *Funkcial. Ekvac.*, **43** (2000), 241–253.
301. A. Kijowski, Q. Liu, Y. Zhang and X. Zhou, A second-order operator for horizontal quasiconvexity in the Heisenberg group and application to convexity preserving for horizontal curvature flow, *Discrete Contin. Dyn. Syst.*, **45** (2025), 3483–3511. https://doi.org/10.3934/dcds.2025029
302. I. C. Kim, Uniqueness and existence results on the Hele-Shaw and the Stefan problems, *Arch. Ration. Mech. Anal.*, **168** (2003), 299–328. https://doi.org/10.1007/s00205-003-0251-z
303. I. C. Kim, A free boundary problem with curvature, *Comm. Partial Differential Equations*, **30** (2004), 121–138. https://doi.org/10.1081/pde-200044474
304. I. Kim and D. Kwon, Volume preserving mean curvature flow for star-shaped sets, *Calc. Var.*, **59** (2020), article number 81. https://doi.org/10.1007/s00526-020-01738-0
305. I. Kim, D. Kwon and N. Požár, On volume-preserving crystalline mean curvature flow, *Math. Ann.*, **384** (2022), 733–774. https://doi.org/10.1007/s00208-021-02286-4
306. L. Kim and Y. Tonegawa, On the mean curvature flow of grain boundaries, *Ann. Inst. Fourier (Grenoble)*, **67** (2017), 43–142. https://doi.org/10.5802/aif.3077
307. R. Kimmel, Numerical geometry of images. Theory, algorithms, and applications, Springer-Verlag, New York, 2004.
308. R. Kimmel, A. Amir and A. M. Bruckstein, Finding shortest paths on surfaces using level sets propagation, *IEEE Trans. Pattern Anal. Mach. Intell.*, **17** (1995), 635–640. https://doi.org/10.1109/34.387512Z
309. S. Kobayashi and K. Nomizu, Foundations of differential geometry. Vol. I, Reprint of the 1963 original, Wiley Classics Lib. Wiley-Intersci. Publ. John Wiley & Sons, Inc., New York, 1996.
310. S. Kobayashi and K. Nomizu, Foundations of differential geometry. Vol. II, Reprint of the 1969 original, Wiley Classics Lib. Wiley-Intersci. Publ. John Wiley & Sons, Inc., New York, 1996.
311. V. A. Kobzar, R. V. Kohn and Z. Wang, New potential-based bounds for prediction with expert advice. In: *Proc. Thirty Third Conference on Learning Theory*, 2370–2405, *Proc. Machine Learning Research*, **125**, 2020.
312. R. V. Kohn and S. Serfaty, A deterministic-control-based approach to motion by curvature, *Comm. Pure Appl. Math.*, **59** (2006), 344–407. https://doi.org/10.1002/cpa.20101
313. R. V. Kohn and S. Serfaty, A deterministic-control-based approach to fully nonlinear parabolic and elliptic equations, *Comm. Pure Appl. Math.*, **63** (2010), 1298–1350. https://doi.org/10.1002/cpa.20336
314. Y. Kohsaka, Free boundary problem for quasilinear parabolic equation with fixed angle of contact to a boundary, *Nonlinear Anal.*, **45** (2001), 865–894. https://doi.org/10.1016/s0362-546x(99)00422-8
315. S. Koike, A beginner's guide to the theory of viscosity solutions, *MSJ Mem.*, **13**, Mathematical Society of Japan, Tokyo, 2004.
316. M. Koiso and B. Palmer, Geometry and stability of surfaces with constant anisotropic mean curvature, *Indiana Univ. Math. J.*, **54** (2005), 1817–1852. https://doi.org/10.1512/iumj.2005.54.2613
317. O. A. Ladyženskaja, V. A. Solonnikov and N. N. Ural'ceva, Linear and quasilinear equations of parabolic type, Translated from the Russian by *S. Smith, Transl. Math. Monogr.*, **23**, American Mathematical Society, Providence, RI, 1968. https://doi.org/10.1090/mmono/023
318. M. Lewicka, A course on tug-of-war games with random noise, *Universitext*, Springer, Cham, 2020. https://doi.org/10.1007/978-3-030-46209-3
319. M. Lewicka, J. J. Manfredi and D. Ricciotti, Random walks and random tug of war in the Heisenberg group, *Math. Ann.*, **377** (2020), 797–84. https://doi.org/10.1007/s00208-019-01853-0
320. P.-L. Lions, Generalized solutions of Hamilton-Jacobi equations, *Res. Notes in Math.*, **69**, Pitman (Advanced Publishing Program), Boston, Mass.-London, 1982.

321. P.-L. Lions, Optimal control of diffusion processes and Hamilton-Jacobi-Bellman equations. I. The dynamic programming principle and applications, *Comm. Partial Differential Equations,* **8** (1983), 1101–1174. https://doi.org/10.1080/03605308308820297
322. P.-L. Lions, Optimal control of diffusion processes and Hamilton-Jacobi-Bellman equations. I I. Viscosity solutions and uniqueness, *Comm. Partial Differential Equations,* **8** (1983), 1229—1276. https://doi.org/10.1080/03605308308820301
323. P.-L. Lions, Some recent results in the optimal control of diffusion processes, In: *Stochastic analysis (Katata/Kyoto, 1982),* 333–367, *North-Holland Math. Library,* **32**, North-Holland Publishing Co., Amsterdam, 1984. https://doi.org/10.1016/s0924-6509(08)70399-4
324. P.-L. Lions, Neumann type boundary conditions for Hamilton-Jacobi equations, *Duke Math. J.,* **52** (1985), 793–820. https://doi.org/10.1215/s0012-7094-85-05242-1
325. P.-L. Lions, G. Papanicolaou and S. R. S. Varadhan, Homogenization of Hamilton-Jacobi equations, preprint (1987). https://www.researchgate.net/publication/246383838_Homogenization_of_Hamilton-Jacobi_equation
326. P.-L. Lions and P. E. Souganidis, Homogenization of degenerate second-order PDE in periodic and almost periodic environments and applications, *Ann. Inst. H. Poincaré C Anal. Non Linéaire,* **22** (2005), 667–677. https://doi.org/10.1016/j.anihpc.2004.10.009
327. Q. Liu, Fattening and comparison principle for level-set equations of mean curvature type, *SIAM J. Control Optim.,* **49** (2011), 2518–2541. https://doi.org/10.1137/100814330
328. Q. Liu, Waiting time effect for motion by positive second derivatives and applications, *NoDEA Nonlinear Differential Equations Appl.,* **21** (2014), 589–620. https://doi.org/10.1007/s00030-013-0259-5
329. Q. Liu, The vanishing exponent limit for motion by a power of mean curvature, *Interfaces Free Bound.,* **22** (2020), 51–84. https://doi.org/10.4171/IFB/432
330. Q. Liu, Large exponent behavior for power-type nonlinear evolution equations and applications, *J. Evol. Equ.,* **20** (2020), 777–810. https://doi.org/10.1007/s00028-019-00539-z
331. Q. Liu, Semiconvexity of viscosity solutions to fully nonlinear evolution equations via discrete games, In: *Geometric properties for parabolic and elliptic PDEs*, 205–231, *Springer INdAM Ser.,* **47**, Springer, Cham, 2021. https://doi.org/10.1007/978-3-030-73363-6_10
332. Q. Liu and A. Nakayasu, Convexity preserving properties for Hamilton-Jacobi equations in geodesic spaces, *Discrete Contin. Dyn. Syst.,* **39** (2019), 157–183. https://doi.org/10.3934/dcds.2019007
333. Q. Liu, A. Schikorra and X. Zhou, A game-theoretic proof of convexity-preserving properties for motion by curvature, *Indiana Univ. Math. J.,* **65** (2016), 171–197. https://doi.org/10.1512/iumj.2016.65.5740
334. Q. Liu, N. Shanmugalingam and X. Zhou, Equivalence of solutions of eikonal equation in metric spaces, *J. Differential Equations,* **272** (2021), 979–1014. https://doi.org/10.1016/j.jde.2020.10.018
335. Q. Liu, N. Shanmugalingam and X. Zhou, Discontinuous eikonal equations in metric measure spaces, *Trans. Amer. Math. Soc.,* **378** (2025), 695–729. https://doi.org/10.1090/tran/9294
336. Q. Liu and M. B. P. Wiranata, Monge solutions of time-dependent Hamilton-Jacobi equations in metric spaces, *ESAIM Control Optim. Calc. Var.,* **32**, Paper No. 2 (2026). https://doi.org/10.1051/cocv/2025087
337. Q. Liu and N. Yamada, An obstacle problem arising in large exponent limit of power mean curvature flow equation, *Trans. Amer. Math. Soc.,* **372** (2019), 2103–2141. https://doi.org/10.1090/tran/7717
338. S. Luckhaus and T. Sturzenhecker, Implicit time discretization for the mean curvature flow equation, *Calc. Var. Partial Differential Equations,* **3** (1995), 253–271. https://doi.org/10.1007/bf01205007
339. H. Luiro, M. Parviainen and E. Saksman, Harnack's inequality for p-harmonic functions via stochastic games, *Comm. Partial Differential Equations,* **38** (2013), 1985–2003. https://doi.org/10.1080/03605302.2013.814068
340. H. Luiro, M. Parviainen and E. Saksman, On the existence and uniqueness of p-harmonious functions, *Differential Integral Equations,* **27** (2014), 201–216. https://doi.org/10.57262/die/1391091363

341. A. Lunardi, Analytic semigroups and optimal regularity in parabolic problems, *Progr. Nonlinear Differential Equations Appl.,* **16**, Birkhäuser Verlag, Basel, 1995.
342. A. Magni and C. Mantegazza, A note on Grayson's theorem, *Rend. Semin. Mat. Univ. Padova,* **131** (2014), 263–279. https://doi.org/10.4171/RSMUP/131-16
343. S. Makida and A. Nakayasu, Stability of metric viscosity solutions under Hausdorff convergence, *Anal. Geom. Metr. Spaces* (to appear)
344. J. J. Manfredi, M. Parviainen and J. D. Rossi, An asymptotic mean value characterization for p-harmonic functions, *Proc. Amer. Math. Soc.,* **138** (2010), 881–889. https://doi.org/10.1090/s0002-9939-09-10183-1
345. J. J. Manfredi, M. Parviainen and J. D. Rossi, An asymptotic mean value characterization for a class of nonlinear parabolic equations related to tug-of-war games, *SIAM J. Math. Anal.,* **42** (2010), 2058–2081. https://doi.org/10.1137/100782073
346. C. Mantegazza, Lecture notes on mean curvature flow, *Progr. Math.,* **290**, Birkhäuser/Springer Basel AG, Basel, 2011. https://doi.org/10.1007/978-3-0348-0145-4
347. A. M. Meirmanov, The Stefan problem, *De Gruyter Exp. Math.,* **3**, Walter de Gruyter & Co., Berlin, 1992. https://doi.org/10.1515/9783110846720.245
348. G. Mercier, Mean curvature flow with obstacles: a viscosity approach, *preprint* (2014). arXiv: 1409.7657
349. G. Mercier and M. Novaga, Mean curvature flow with obstacles: existence, uniqueness and regularity of solutions, *Interfaces Free Bound.,* **17** (2015), 399–426. https://doi.org/10.4171/IFB/348
350. B. Merriman, J. Bence and S. Osher, Diffusion generated motion by mean curvature, *Selected Lectures in Math. The Comp. Crystal Grower Workshop* (ed. J. E. Taylor), 73–83, Amer. Math. Soc., Providence, RI, 1992.
351. K. Misu, A game-theoretic approach to the asymptotic behavior of solutions to an obstacle problem for the mean curvature flow equation, *preprint* (2024). arXiv: 2403.00358
352. H. Mitake, C. Mooney, H. V. Tran, J. Xin and Y. Yu, Bifurcation of homogenization and nonhomogenization of the curvature G-equation with shear flows, *Math. Ann.,* **391** (2025), 3077–3111.
353. H. Mitake and H. V. Tran, Asymptotic growth rate of solutions to level-set forced mean curvature flows with evolving spirals, *Bull. Lond. Math. Soc.,* **57** (2024), 748–770. https://doi.org/10.1112/blms.13227
354. D. S. Mitrinović, Analytic inequalities, In cooperation with P. M. Vasić, *Die Grundlehren der mathematischen Wissenschaften,* **165**, Springer-Verlag, New York, Berlin, 1970.
355. F. Morgan, Riemannian geometry. A beginner's guide. 2nd edition, A K Peters, Ltd., Wellesley, MA, 1998.
356. F. Morgan, Planar Wulff shape is unique equilibrium, *Proc. Amer. Math. Soc.,* **133** (2005), 809–813. https://doi.org/10.1090/S0002-9939-04-07661-0
357. P. B. Mucha and P. Rybka, A new look at equilibria in Stefan-type problems in the plane, *SIAM J. Math. Anal.,* **39** (2007), 1120–1134. https://doi.org/10.1137/060677124
358. L. Mugnai and M. Röger, The Allen-Cahn action functional in higher dimensions, *Interfaces Free Bound.,* **10** (2008), 45–78. https://doi.org/10.4171/IFB/179
359. H. Müller-Krumbhaar, T. W. Burkhardt and D. M. Kroll, A generalized kinetic equation for crystal growth, *J. Crystal Growth,* **38** (1977), 13–22. https://doi.org/10.1016/0022-0248(77)90367-0
360. W. W. Mullins, Two-dimensional motion of idealized grain boundaries, *J. Appl. Phys.,* **27** (1956), 900–904. https://doi.org/10.1063/1.1722511
361. A. Nakayasu, Metric viscosity solutions for Hamilton-Jacobi equations of evolution type, *Adv. Math. Sci. Appl.,* **24** (2014), 333–351.
362. A. Nakayasu and T. Namba, Stability properties and large time behavior of viscosity solutions of Hamilton-Jacobi equations on metric spaces, *Nonlinearity,* **31** (2018), 5147–5161. https://doi.org/10.1088/1361-6544/aadc02
363. G. Namah and J.-M.Roquejoffre, Remarks on the long time behaviour of the solutions of Hamilton-Jacobi equations. *Comm. Partial Differential Equations,* **24** (1999), 883—893. https://doi.org/10.1080/03605309908821451

364. T. Namba, On existence and uniqueness of viscosity solutions for second order fully nonlinear PDEs with Caputo time fractional derivatives, *NoDEA Nonlinear Differential Equations Appl.,* **25** (2018), Paper No. 23. https://doi.org/10.1007/s00030-018-0513-y
365. R. T. Newcomb, II and J. Su, Eikonal equations with discontinuities, *Differential Integral Equations,* **8** (1995), 1947–1960. https://doi.org/10.57262/die/1369056134
366. J. A. Oaks, Singularities and self-intersections of curves evolving on surfaces, *Indiana Univ. Math. J.,* **43** (1994), 959–981. https://doi.org/10.1512/iumj.1994.43.43042
367. A. M. Oberman, Finite difference methods for the infinity Laplace and p-Laplace equations, *J. Comput. Appl. Math.,* **254** (2013), 65–80. https://doi.org/10.1016/j.cam.2012.11.023
368. M. Ohnuma and K. Sato, Singular degenerate parabolic equations with applications to the p-Laplace diffusion equation, *Comm. Partial Differential Equations,* **22** (1997), 381–411. https://doi.org/10.1080/03605309708821268
369. M. Ohnuma and M.-H. Sato, Singular degenerate parabolic equations with applications to geometric evolutions, *Differential Integral Equations,* **6** (1993), 1265–1280. https://doi.org/10.57262/die/1370019755
370. T. Ohta, D. Jasnow and K. Kawasaki, Universal scaling in the motion of random interfaces, *Phys. Rev. Lett.,* **49** (1982), 1223–1226. https://doi.org/10.1103/PhysRevLett.49.1223
371. T. Ohtsuka, A level set method for spiral crystal growth, *Adv. Math. Sci. Appl.,* **13** (2003), 225–248.
372. T. Ohtsuka, Y.-H. R. Tsai and Y. Giga, A level set approach reflecting sheet structure with single auxiliary function for evolving spirals on crystal surfaces, *J. Sci. Comput.,* **62** (2015), 831–874. https://doi.org/10.1007/s10915-014-9877-2
373. T. Ohtsuka, Y.-H. R. Tsai and Y. Giga, Growth rate of crystal surfaces with several dislocation centers, *Cryst. Growth Des.,* **18** (2018), 1917–1929. https://doi.org/10.1021/acs.cgd.7b00833
374. S. Osher and R. Fedkiw, Level set methods and dynamic implicit surfaces, *Appl. Math. Sci.,* **153**, Springer-Verlag, New York, 2003.
375. S. Osher and B. Merriman, The Wulff shape as the asymptotic limit of a growing crystalline interface, *Asian J. Math.,* **1** (1997), 560–571. https://doi.org/10.4310/ajm.1997.v1.n3.a6
376. S. Osher and J. A. Sethian, Fronts propagating with curvature-dependent speed: algorithms based on Hamilton-Jacobi formulations, *J. Comput. Phys.,* **79** (1988), 12–49. https://doi.org/10.1016/0021-9991(88)90002-2
377. M. Parviainen and E. Ruosteenoja, Local regularity for time-dependent tug-of-war games with varying probabilities, *J. Differential Equations,* **261** (2016), 1357–1398. https://doi.org/10.1016/j.jde.2016.04.001
378. Y. Peres, G. Pete and S. Somersille, Biased tug-of-war, the biased infinity Laplacian, and comparison with exponential cones, *Calc. Var. Partial Differential Equations,* **38** (2010), 541–564. https://doi.org/10.1007/s00526-009-0298-2
379. Y. Peres, O. Schramm, S. Sheffield and D. B. Wilson, Tug-of-war and the infinity Laplacian, *J. Amer. Math. Soc.,* **22** (2009), 167–210. https://doi.org/10.1090/s0894-0347-08-00606-1
380. Y. Peres and S. Sheffield, Tug-of-war with noise: A game-theoretic view of the p-Laplacian, *Duke Math. J.,* **145** (2008), 91–120. https://doi.org/10.1215/00127094-2008-048
381. M. H. Protter and H. F. Weinberger, Maximum principles in differential equations, Prentice-Hall, Inc., Englewood Cliffs, NJ, 1967.
382. J. D. Rossi, Tug-of-war games and PDEs, *Proc. Roy. Soc. Edinburgh Sect. A,* **141** (2011), 319–369. https://doi.org/10.1017/s0308210510000041
383. E. Ruosteenoja, Local regularity results for value functions of tug-of-war with noise and running payoff, *Adv. Calc. Var.,* **9** (2016), 1–17. https://doi.org/10.1515/acv-2014-0021
384. L. I. Rubenšteĭn, The Stefan problem, *Transl. Math. Monogr.,* **27**, American Mathematical Society, Providence, RI, 1971.
385. M. B. Rudd and H. A. Van Dyke, Median values, 1-harmonic functions, and functions of least gradient, *Commun. Pure Appl. Anal.,* **12** (2013), 711–719. https://doi.org/10.3934/cpaa.2013.12.711
386. G. Sapiro, Geometric partial differential equations and image analysis, Reprint of the 2001 original, Cambridge University Press, Cambridge, 2006.

387. G. Sapiro and A. Tannenbaum, On affine plane curve evolution, *J. Funct. Anal.*, **119** (1994), 79–120. https://doi.org/10.1006/jfan.1994.1004
388. M.-H. Sato, Comparison principle for singular degenerate elliptic equations on unbounded domains, *Proc. Japan Acad. Ser. A Math. Sci.*, **66** (1990), 252–256. https://doi.org/10.3792/pjaa.66.252
389. M.-H. Sato, Interface evolution with Neumann boundary condition, *Adv. Math. Sci. Appl.*, **4** (1994), 249–264.
390. M.-H. Sato, Capillary problem for singular degenerate parabolic equations on a half space, *Differential Integral Equations*, **9** (1996), 1213–1224. https://doi.org/10.57262/die/1367846897
391. T. Sato, A game approach to free boundary problems of anisotropic forced mean curvature flow equations, *Calc. Var. Partial Differential Equations*, **65** (2026), 92. https://doi.org/10.1007/s00526-025-03236-7
392. R. Schneider, Convex bodies: the Brunn-Minkowski theory, *Encyclopedia Math. Appl.*, **44**, Cambridge University Press, Cambridge, 1993.
393. J. A. Sethian, Level set methods, Evolving interfaces in geometry, fluid mechanics, computer vision, and materials science, *Cambridge Monogr. Appl. Comput. Math.*, **3**, Cambridge University Press, Cambridge, 1996.
394. N. Sesum, Rate of convergence of the mean curvature flow, *Comm. Pure Appl. Math.*, **61** (2008), 464–485. https://doi.org/10.1002/cpa.20209
395. L. Simon, Lectures on geometric measure theory, *Proc. Centre Math. Anal. Austral. Nat. Univ.*, **3**, Australian National University, Centre for Mathematical Analysis, Canberra, 1983.
396. K. Smoczyk, The symmetric "doughnut" evolving by its mean curvature, *Hokkaido Math. J.*, **23** (1994), 523–547.
397. H. M. Soner, Optimal control with state-space constraint. I, *SIAM J. Control Optim.*, **24** (1986), 552–561. https://doi.org/10.1137/0324032
398. H. M. Soner, Optimal control with state-space constraint. II, *SIAM J. Control Optim.*, **24** (1986), 1110–1122. https://doi.org/10.1137/0324067
399. H. M. Soner, Motion of a set by the curvature of its boundary, *J. Differential Equations*, **101** (1993), 313–372. https://doi.org/10.1006/jdeq.1993.1015
400. H. M. Soner, Convergence of the phase-field equations to the Mullins-Sekerka problem with kinetic undercooling, *Arch. Rational Mech. Anal.*, **131** (1995), 139–197. yy10.1007/bf00386194
401. H. M. Soner, Ginzburg-Landau equation and motion by mean curvature. I. Convergence, *J. Geom. Anal.*, **7** (1997), 437–475. https://doi.org/10.1007/bf02921628
402. H. M. Soner and P. E. Souganidis, Singularities and uniqueness of cylindrically symmetric surfaces moving by mean curvature, *Comm. Partial Differential Equations*, **18** (1993), 859–894. https://doi.org/10.1080/03605309308820954
403. H. M. Soner and N. Touzi, Dynamic programming for stochastic target problems and geometric flows, *J. Eur. Math. Soc.*, **4** (2002), 201–236. https://doi.org/10.1007/s100970100039
404. H. M. Soner and N. Touzi, A stochastic representation for mean curvature type geometric flows, *Ann. Probab.*, **31** (2003), 1145–1165. https://doi.org/10.1214/aop/1055425773
405. P. Soravia, Generalized motion of a front propagating along its normal direction: a differential games approach, *Nonlinear Anal.*, **22** (1994), 1247–1262. https://doi.org/10.1016/0362-546x(94)90108-2
406. P. Sternberg and W. P. Ziemer, Generalized motion by curvature with a Dirichlet condition, *J. Differential Equations*, **114** (1994), 580–600. https://doi.org/10.1006/jdeq.1994.1162
407. M. Struwe, Geometric evolution problems, In: *Nonlinear partial differential equations in differential geometry (Park City, UT, 1992)*, 257–339, *IAS/Park City Math. Ser.*, **2**, American Mathematical Society, Providence, RI, 1996.
408. S. Stuvard and Y. Tonegawa, On the existence of canonical multi-phase Brakke flows, *Adv. Calc. Var.*, **17** (2024), 33–78. https://doi.org/10.1515/acv-2021-0093
409. S. Stuvard and Y. Tonegawa, Dynamical instability of minimal surfaces at flat singular points, *J. Differential Geom.*, **130** (2025), 477–516. https://doi.org/10.4310/jdg/1747158946

410. J. E. Taylor, Crystalline variational problems, *Bull. Amer. Math. Soc.,* **84** (1978), 568–588. https://doi.org/10.1090/s0002-9904-1978-14499-1
411. J. E. Taylor, Constructions and conjectures in crystalline nondifferential geometry, In: *Differential geometry (Rio de Janeiro, 1991)* (eds. B. Lawson and K. Tanenblat), 321–336, *Pitman Monogr. Surveys Pure Appl. Math.,* **52**, Longman Scientific & Technical, Harlow; copublished in the United States with, 1991.
412. J. E. Taylor, Mean curvature and weighted mean curvature, *Acta Metall.,* **40** (1992), 1475–1485. https://doi.org/10.1016/0956-7151(92)90091-R
413. J. E. Taylor, J. W. Cahn and C. A. Handwerker, Geometric models of crystal growth, *Acta Metall.,* **40** (1992), 1443–1474. https://doi.org/10.1016/0956-7151(92)90090-2
414. Y. Tonegawa, Some remarks on the level set flow by anisotropic curvature, *Calc. Var. Partial Differential Equations,* **10** (2000), 101–118. https://doi.org/10.1007/s005260050146
415. Y. Tonegawa, Brakke's mean curvature flow: An introduction, *SpringerBriefs Math.* Springer, Singapore, 2019. https://doi.org/10.1007/978-981-13-7075-5
416. Y. Tonegawa, An existence theorem for Brakke flow, *Adv. Stud. Pure Math.,* **85**, 429–439, Mathematical Society of Japan, Tokyo, 2020. https://doi.org/10.2969/aspm/08510429
417. E. Topp and M. Yangari, Existence and uniqueness for parabolic problems with Caputo time derivative, *J. Differential Equations,* **262** (2017), 6018–6046. https://doi.org/10.1016/j.jde.2017.02.024
418. N. S. Trudinger, The Dirichlet problem for the prescribed curvature equations, *Arch. Rational Mech. Anal.,* **111** (1990), 153–179. https://doi.org/10.1007/bf00375406
419. Y.-H. R. Tsai, Y. Giga and S. Osher, A level set approach for computing discontinuous solutions of Hamilton-Jacobi equations, *Math. Comp.,* **72** (2003), 159–181. https://doi.org/10.1090/s0025-5718-02-01438-2
420. K. Tso (K.-S. Chou), Deforming a hypersurface by its Gauss-Kronecker curvature, *Comm. Pure Appl. Math.,* **38** (1985), 867–882. https://doi.org/10.1002/cpa.3160380615
421. J. Urbas, Complete noncompact self-similar solutions of Gauss curvature flows. I. Positive powers, *Math. Ann.,* **311** (1998), 251–274. https://doi.org/10.1007/s002080050187
422. J. Urbas, Self-similar solutions of Gauss curvature flows, In: *Monge Ampère equation: applications to geometry and optimization (Deerfield Beach, FL, 1997),* 157–172, Contemp. Math., **226**, *American Mathematical Society, Providence, RI,* 1999.
423. J. J. L. Velázquez, Curvature blow-up in perturbations of minimal cones evolving by mean curvature flow, *Ann. Scuola Norm. Sup. Pisa Cl. Sci. (4),* **21** (1994), 595–628. http://www.numdam.org/item/ASNSP_1994_4_21_4_595_0/
424. A. Visintin, Models of phase transitions, Progr. Nonlinear Differential Equations Appl., **28**, Birkhäuser Boston, Inc., Boston, MA, 1996.
425. L. Wang, On the regularity theory of fully nonlinear parabolic equations, *Bull. Amer. Math. Soc. (N.S.),* **22** (1990), 107–114. https://doi.org/10.1090/s0273-0979-1990-15854-9
426. B. White, The size of the singular set in mean curvature flow of mean-convex sets, *J. Amer. Math. Soc.,* **13** (2000), 665–695. https://doi.org/10.1090/s0894-0347-00-00338-6
427. B. White, Evolution of curves and surfaces by mean curvature, In: *Proc. the International Congress of Mathematicians, Vol. I (Beijing, 2002),* 525–538, Higher Education Press, Beijing, 2002.
428. B. White, The nature of singularities in mean curvature flow of mean-convex sets, *J. Amer. Math. Soc.,* **16** (2003), 123–138. https://doi.org/10.1090/s0894-0347-02-00406-x
429. G. Wulff, Zur Frage der Geschwindigkeit des Wachsthums und der Auflösung der Krystallflächen, *Zeit. Krystall. Min.,* **34** (1901), 449–530. https://doi.org/10.1524/zkri.1901.34.1.449
430. H. Yagisita, Non-uniqueness of self-similar shrinking curves for an anisotropic curvature flow, *Calc. Var. Partial Differential Equations,* **26** (2006), 49–55. https://doi.org/10.1007/s00526-005-0357-2
431. H. M. Yahia, E. G. Huot, I. L. Herlin and I. Cohen, Geodesic distance evolution of surfaces: a new method for matching surfaces, In: *Proc. IEEE Conference on Computer Vision and Pattern Recognition (CVPR),* 2000. https://doi.org/10.1109/CVPR.2000.855883

432. N. Yamada, Viscosity solutions for a system of elliptic inequalities with bilateral obstacles, *Funkcial. Ekvac.*, **30** (1987), 417–425.
433. Y. Yu, Maximal and minimal solutions of an Aronsson equation: L^∞ variational problems versus the game theory, *Calc. Var. Partial Differential Equations*, **37** (2010), 63–74. https://doi.org/10.1007/s00526-009-0251-4
434. X.-P. Zhu, Asymptotic behavior of anisotropic curve flows, *J. Differential Geom.*, **48** (1998), 225–274. https://doi.org/10.4310/jdg/1214460796
435. X.-P. Zhu, Lectures on mean curvature flows, AMS/IP Stud. Adv. Math., **32**, *American Mathematical Society, Providence, RI; International Press, Somerville, MA*, 2002. https://doi.org/10.1090/amsip/032

Notation

Y. Giga, Q. Liu, *Surface Evolution Equations*, Monographs in Mathematics 99,
https://doi.org/10.1007/978-3-032-20303-8

Index

Y. Giga, Q. Liu, *Surface Evolution Equations*, Monographs in Mathematics 99,
https://doi.org/10.1007/978-3-032-20303-8

MIX
Papier aus verantwortungsvollen Quellen
Paper from responsible sources
FSC® C105338

If you have any concerns about our products,
you can contact us on
ProductSafety@springernature.com

In case Publisher is established outside the EU,
the EU authorized representative is:
Springer Nature Customer Service Center GmbH
Europaplatz 3, 69115 Heidelberg, Germany

Printed by Libri Plureos GmbH
in Hamburg, Germany